Infectious Blood Diseases of Man and Animals

Diseases Caused by Protista

VOLUME I
Special Topics and General Characteristics

Infectious Blood Diseases
of Man and Animals

Diseases Caused by Protista

Edited by

DAVID ⌊WEINMAN

Department of Microbiology
Yale University
New Haven, Connecticut

MIODRAG RISTIC

College of Veterinary Medicine
University of Illinois
Urbana, Illinois

VOLUME I

Special Topics and General Characteristics

1968

ACADEMIC PRESS New York and London

ACADEMIC PRESS INC.
111 Fifth Avenue, New York, New York 10003

United Kingdom Edition published by
ACADEMIC PRESS INC. (LONDON) LTD.
Berkeley Square House, London W-1

LIBRARY OF CONGRESS CATALOG CARD NUMBER: 68-18685

PRINTED IN THE UNITED STATES OF AMERICA

List of Contributors

Numbers in parentheses indicate the pages on which the authors' contributions begin.

EUGENE C. BOVEE (393), *Department of Zoology, University of California, Los Angeles, California*

R. S. BRAY (123), *London School of Hygiene and Tropical Medicine, London, England*

JOHN O. CORLISS (139), *Department of Biological Sciences, University of Illinois, Chicago, Illinois*

HUGUETTE FROMENTIN (175), *Service de Parasitologie, Institut Pasteur, Paris, France*

S. H. HUTNER (175), *Haskins Laboratories, New York, New York*

THEODORE L. JAHN (393), *Department of Zoology, University of California, Los Angeles, California*

NORMAN D. LEVINE (3), *College of Veterinary Medicine, University of Illinois, Urbana, Illinois*

H. T. MERYMAN (343), *Biophysics Division, U.S. Navy Medical Research Institute, National Naval Medical Center, Bethesda, Maryland*

R. BARCLAY McGHEE (307), *Department of Zoology, University of Georgia, Athens, Georgia*

KATHLEEN M. O'CONNELL (175), *Haskins Laboratories, New York, New York*

MIODRAG RISTIC (63, 79), *College of Veterinary Medicine, University of Illinois, Urbana, Illinois*

MARIA A. RUDZINSKA (217), *The Rockefeller University, New York, New York*

WILLIAM F. SCHROEDER (63), *Hemotropic Parasite Research Project, Center for Veterinary Investigations, Ministry of Agriculture and Animal Breeding, Maracay, Venezuela*

DIEGO SEGRE (37), *College of Veterinary Medicine, University of Illinois, Urbana, Illinois*

WILLIAM TRAGER (149), *The Rockefeller University, New York, New York*

KEITH VICKERMAN (217), *Department of Zoology, University College, London, England*

PETER J. WALKER (367), *Department of Zoology, The University of Exeter, The Hatherly Laboratories, Exeter, Devon, England*

DAVID WEINMAN (343), *Department of Microbiology, Yale University, New Haven, Connecticut*

AVIVAH ZUCKERMAN (23, 79), *Department of Parasitology, Hadassah Medical School, The Hebrew University, Jerusalem, Israel*

Foreword

Reader: "I wish to congratulate you on your monograph. It is invaluable."

Author: "Thank you. It is very pleasant of you to say so."

Reader: "But why did you write it?"

Author: "What do you mean?"

Reader: "Well, now that it is published, everyone will know as much about the subject as you do."

This book is intended to be useful. It will have accomplished its aims if: (1) it provides a critical review of the subject matter, (2) it provokes the solution of unsolved problems, and (3) it can assist in overcoming the amputation of microbiology into the various heads, arms, and legs which today characterize it.

The protistan blood diseases are here defined as that group caused by microorganisms which pass a major portion of their lifetime in the blood. Viruses are excluded as not being protista, which is the consensus of virologists today.

The group has not been reviewed for many years. It is hoped that this book will provide a source of pertinent material and literature with emphasis on new acquisitions, insights, and syntheses. Suggestions for productive lines of research have been solicited and, if these volumes are successful, the surest evidence will be that they become rapidly outdated.

Preface

This book is intended to be a modern critical source of material and literature on the protistan blood diseases. It is anticipated that the book will provide a unique source of information for professional microbiologists, graduate students in the medical and microbiological fields, and others interested in the biological and medical sciences.

The first volume consists of a summary and critical analysis of metabolic, immuno-pathologic, taxonomic, ultrastructural, locomotor, ecologic, zoonotic, and other aspects of the protistan blood diseases. The second volume includes pertinent disease aspects as they occur in man and animals infected with the pathogens belonging to this group.

There are interesting features associated with the protistan blood diseases. All are transmitted by arthropod vectors, some are zoonotic in nature. Certain of these diseases cause devastating losses to the livestock industry throughout the world, and others are of great importance to human health, e.g., malaria, still considered a major disease problem. Attempts to develop immunologic rather than chemotherapeutic control measures have been greatly handicapped by inadequate information about immune mechanisms as well as other aspects of the host-parasite relationships in these diseases.

The last decade has witnessed great expansion in the application of immunologic principles and techniques to the study of the agents of these diseases and in an understanding of their interactions with the host they invade. We are now beginning to comprehend that factors other than the parasite per se may cause the extensive erythrocyte destruction which is a striking feature of many of these diseases.

Specialists with varied scientific backgrounds are interested in studying these diseases, and their number has increased tremendously in recent years. The efforts of protozoologists and microbiologists are now being added to by the contributions of biophysicists, biochemists, immunologists, and molecular geneticists using modern research tools for the analysis of hitherto unsolved problems. It is with these new approaches and results in mind that we have undertaken the task of assembling these volumes.

ix

It is a pleasure to be able to express our appreciation to the authors whose contributions have made the two volumes possible. For their editorial assistance beyond any mere call of duty our thanks go to Elizabeth G. Weinman and Mrs. Alys von Lehmden-Maslin. The staff of Academic Press has participated helpfully at all stages, and their comprehension, their skill, and their standards of excellence are reflected throughout.

February, 1968

DAVID WEINMAN
MIODRAG RISTIC

Contents

Part I
SPECIAL TOPICS

1. Ecology and Host-Parasite Relationship
NORMAN D. LEVINE

2. Basis of Host Cell-Parasite Specificity
AVIVAH ZUCKERMAN

3. Abrogation of Immunological Tolerance as a Model for Autoimmunity
DIEGO SEGRE

4. Autoimmune Response and Pathogenesis of Blood Parasite Diseases

WILLIAM F. SCHROEDER AND MIODRAG RISTIC

5. Blood Parasite Antigens and Antibodies

AVIVAH ZUCKERMAN AND MIODRAG RISTIC

6. Zoonotic Potential of Blood Parasites

R. S. BRAY

Part II

GENERAL CHARACTERISTICS OF THE BLOOD PROTISTS

7. Definition and Classification

JOHN O. CORLISS

8. Cultivation and Nutritional Requirements

WILLIAM TRAGER

9. Some Biological Leads to Chemotherapy of Blood Protista, Especially Trypanosomatidae

S. H. HUTNER, HUGUETTE FROMENTIN, KATHLEEN M. O'CONNELL

10. The Fine Structure

MARIA A. RUDZINSKA AND KEITH VICKERMAN

11. Development and Reproduction (Vertebrate and Arthropod Host)

R. BARCLAY McGHEE

12. Preservation and Storage *in Vitro*

H. T. MERYMAN AND DAVID WEINMAN

13. Investigational Problems and the Mechanisms of Inheritance in Blood Protozoa

PETER J. WALKER

14. Locomotion of Blood Protists

THEODORE L. JAHN AND EUGENE C. BOVEE

Contents of Volume II

Infectious Blood Diseases
of Man and Animals

Diseases Caused by Protista

VOLUME I
Special Topics and General Characteristics

I

SPECIAL TOPICS
AND GENERAL CHARACTERISTICS

1

Ecology and Host-Parasite Relationship

NORMAN D. LEVINE

I. TYPES OF HOST-PARASITE RELATIONS

Living things are found in every possible habitat, and it is natural that one such habitat should be other living things. The relations between the organism and its habitat in this case are the concern of parasitology, which can be defined as that branch of ecology in which one organism is the environment for another. The first organism is the host, and the second the parasite.

The difference is one of size; parasites are smaller than their hosts. If the dependent (or attacking) member of the partnership were the same size as the attacked member, then the former would no longer be called a parasite, but a predator. A wolf eating a deer is a predator, a reduviid bug seizing and sucking the juices from another insect is a predator; but a triatomin bug or a mosquito sucking the juices from a mammal is a

3

parasite, and so is a nematode living in the enteric cavity of its host or a protozoon living in its blood.

There is an intermediate group between parasites and predators, the so-called parasitoids. These are mostly insects, but some are of great economic importance. The tiny wasps and flies whose larvae live within other insects, eating out their insides and then pupating in or on them, are examples. They are common in caterpillars and grasshoppers, and help to control their numbers. Here, too, belong the large wasps which capture a spider or caterpillar, paralyze it, place it in a special cell, and deposit an egg in it. The egg hatches, and the larva which emerges feeds within its helpless host until it has consumed all of it and is ready to pupate.

Parasites are defined as organisms which live on or within some other living organism, deriving sustenance from it; the organism on which it lives is the host. Parasites may be either animals or plants of many phyla. They may live in many different habitats on or within the host—on its outer surface, its outer or inner respiratory organs (gills or lungs), in the lumen of its gut, in various organs or tissues, inside different types of cells, even inside cell nuclei. In this book we are concerned primarily with two of these habitats—the blood plasma and blood cells—but this concern forces us to consider also an additional type of parasite—those which suck blood or plasma and thus transfer blood parasites from one host to another.

Parasites may either harm or help their hosts, or they may have no appreciable effect on them. Each of these types of parasitism has its own name. Symbiosis is defined as the permanent association between two specifically distinct organisms so dependent upon each other that life apart is impossible under natural conditions. The relation between the more primitive termites and their intestinal protozoa is symbiotic. Mutualism is somewhat similar, but it differs from symbiosis in that it is not obligatory for both partners, although it may be for one. The relation between cattle or sheep and their rumen protozoa is mutualistic; the ruminants can survive perfectly well without their enteric fauna even though they are benefited by them in several ways.

Commensalism is the neutral situation; it is an association between host and parasite in which one partner (the parasite) is benefited and the other (the host) is neither benefited nor harmed. Most intestinal bacteria are commensals, although some of them are mutualistic in that they produce vitamins for their hosts, and a few are pathogenic.

If the parasite is pathogenic and harms its host, the condition is known as parasitosis. If it is potentially pathogenic but does not harm its host, the condition is known as parasitiasis. A hookworm infection in which there are too few worms to cause significant damage is a case of parasitiasis; a hookworm infection in which the worms cause clinical damage is a case of parasitosis. Active disease is parasitosis; the carrier state is parasitiasis, and so is the incubation period.

In blood diseases, both parasitosis and parasitiasis are important. In malaria, for instance, a person suffering from clinical illness has a parasitosis, while one between relapses has a parasitiasis. And the type of host may be important also. An argasid tick infected with relapsing fever spirochetes is never harmed by its parasites; the parasite is a commensal in it. But in a vertebrate, the same parasite causes disease, i.e., parasitosis.

Parasites may be monoxenous or heteroxenous. The former have a single type of host; among them are the enteric bacteria and protozoa. Heteroxenous parasites have two types of host, types belonging to different phyla. Most blood parasites of vertebrates are heteroxenous, for they make use of an invertebrate vector to transmit them from one vertebrate to another. However, some blood parasites are monoxenous; *Salmonella typhosa* and *Bacillus anthracis*, which live in the plasma and cause septicemia, are monoxenous. However, the present book is not concerned with them but with the heteroxenous blood parasites.

Another set of terms has to do with a parasite's host range, i.e., with the number of host species of a particular phylum in which it may occur. A stenoxenous parasite has a narrow host range, an euryxenous parasite a broad one, and a mesoxenous parasite an intermediate one. One type grades into another, of course, for nature is a continuum. However, for purposes of definition I should like to define stenoxenous parasites as those which occur within hosts of a single genus (or in some cases closely related genera), mesoxenous parasites as those which occur within hosts of a single order, and euryxenous parasites as those which occur within hosts of more than one order or class. *Trypanosoma lewisi* of rats, *Babesia bigemina* of cattle, and *Plasmodium falciparum* of man are stenoxenous in the vertebrate host, but *T. avium* of birds, *T. rhodesiense* of mammals, and *P. relictum* of birds are euryxenous.

There may or may not be a difference in host range between the vertebrate and invertebrate hosts of heteroxenous parasites. *Leishmania tropica* may infect man, the dog, or various wild rodents, but its invertebrate vector is always a species of *Phlebotomus*; it is euryxenous in the vertebrate host and stenoxenous in the invertebrate. *Plasmodium gallinaceum* is the reverse. It occurs naturally only in birds of the genus *Gallus*, but can be transmitted by mosquitoes of six genera (Huff, 1954). *Trypanosoma cruzi* is euryxenous in the vertebrate host, having been found in man, the raccoon, skunk, opossum, dog, cat, pig, woodrat, armadillo, and many other mammals, but it is mesoxenous in the invertebrate host, being transmitted by members of the subfamily Triatominae; in addition, it may be transmitted through the placenta.

Zoonoses are all caused by euryxenous parasites. They are defined as diseases which are naturally transmitted between vertebrate animals and man, a definition which implies a wide vertebrate host range. Some of the diseases considered in this book are zoonoses.

Since we are primarily concerned with the diseases of man or domestic animals, we think of other animals in which the same diseases exist as reservoirs. However, this point of view is essentially anthropocentric, and is a matter of convenience. We define a reservoir host as a vertebrate host in which a parasite or disease occurs naturally and which is a source of infection for man or domestic animals, as the case may be. Wild animals are reservoirs of infection for man of psittacosis, yellow fever, tularemia, plague, and Rhodesian sleeping sickness. Domestic animals are reservoirs of infection for man of salmonellosis, visceral larva migrans, trichinosis, and infantile leishmaniosis. Wild animals are reservoirs of infection for domestic animals of *Trypanosoma brucei*, while man is a reservoir of infection for domestic animals of *Entamoeba histolytica*.

Another definition of reservoir, preferred by some workers, is the total population of the disease agent, wherever it may be—in the human or domestic animal host, the wild host, or the vector. I prefer to call this the infection pool, parasite plenum, or some other term, but not everyone will agree.

Parasites and diseases may continue to exist indefinitely in their wild, "reservoir" hosts, and man or domestic animals may become infected when they intrude into the locality where the parasites or diseases exist. Such a locality is known as a nidus (literally, nest). This term is used primarily in connection with vector-borne diseases, but it is not restricted to them.

Natural nidi may be either elementary or diffuse. An elementary nidus is confined within narrow limits. A prairie dog burrow containing prairie dogs, fleas, and plague bacteria is an elementary nidus, as is a cave containing bats and rabies virus. In a diffuse nidus, the donor, vectors, and recipients are distributed more widely over the landscape. A wooded region in which tsetse flies circulate trypanosomes among the wild game is a diffuse nidus, as is an area where leptospiras are circulated among rodents, reptiles, farm animals, and man, or an area where ticks and deerflies circulate tularemia in the wild rabbits. The nidality of a disease refers to the distribution and characteristics of its nidi.

II. RELATION OF VERTEBRATE HOST GROUP
TO PARASITE GROUP

Parasite genera or even types are not distributed generally through the animal kingdom. Each group of hosts has its own group of parasites, and this relationship bears out the fact that parasites evolved along with their hosts. As a result, some major groups of parasites are confined to certain groups of hosts. Monogenetic trematodes occur almost entirely on fish; most digenetic trematodes, which are considered higher in the evolutionary scale than monogenetic ones, occur in higher vertebrates. Biting lice occur primarily on birds, but a few are found on mammals. Sucking lice, on the other hand, are found only on mammals. Of the 14 orders of tapeworms, only 2 are found in mammals. The more primitive tapeworm orders are found in the more primitive vertebrates. The rule is not universal. Many fish-eating birds and mammals have the same trematodes, but in this case fish serve as their intermediate hosts, so that the trematodes reflect the food habits of the definitive hosts rather than their taxonomic relationship. However, one can speak of parallel evolution between host and parasite as the usual situation. This situation generally holds true for blood parasites.

A. *Plasmodium* AND RELATED PROTOZOA

Six genera, of which two are very poorly known, comprise the protozoan family Plasmodiidae. These are *Plasmodium*, *Hepatocystis*, *Haemoproteus*, *Leucocytozoon*, *Nycteria*, and *Polychromophilus*. All are found in vertebrates, and indeed most are in the higher vertebrates. In what follows, I seem to speak confidently of valid species and numbers of species. It must be remembered, however, that these figures are really no more than educated guesses; they represent merely the present state of our knowledge and are subject to change as more information becomes available.

Of the 76 recognized species of *Plasmodium* (Levine, 1967), 2 occur in amphibia, 20 in reptiles, 25 in birds, and 29 in mammals. However, their distribution is far from uniform, even within a host class. Only 2 of the reptilian species have been reported from snakes; the remainder are from lizards. The avian species tend to be euryxenous, most being able to infect birds of several orders. The mammalian species, however, tend to be stenoxenous or at best mesoxenous, and they are not found in all mammalian orders. In the mammalian species, 2 are recognized from bats, 5 from rodents, 1 from ruminants, and 21 from primates (including 2 from lemurs, 8 from lower primates, 7 from anthropoid apes, and 4 from man).

Haemoproteus is limited to amphibia, reptiles, and birds, but actually there are only 3 species in amphibia (toads) and 1 in reptiles. The remaining 84 occur in birds. Because the gamonts of so many of them look alike and because the cross-transmission studies needed to establish host ranges have not been carried out, it is impossible to make any positive statements about host-parasite relations within the class. We feel that the avian species of *Haemoproteus* are stenoxenous or mesoxenous; Baker's (1966) finding that the domestic pigeon *Columba livia* and the British wood pigeon *C. palumbus* have different species of *Haemoproteus* tends to bear this out. However, further research may reveal that this opinion is incorrect.

Hepatocystis is even more limited in distribution than *Plasmodium* or *Haemoproteus*. The 13 recognized species all occur in mammals—4 in lower primates, 3 in rodents, 3 in bats, 2 in ruminants, and 1 in the hippopotamus.

Leucocytozoon is also limited to a single host class; all 58 recognized species occur in birds.

So little is known of *Nycteria* and *Polychromophilus* that generalizations are probably unsafe. However, all of the presently accepted species (1 of the former and 3 of the latter) occur in bats.

B. *Babesia, Theileria,* AND RELATED PROTOZOA

Members of this group, the class Piroplasmasida, are now recognized as sarcodines (Honigberg *et al.*, 1964). The genera *Babesia* (and its various subgenera, which some authors prefer to call genera), *Aegyptianella*, *Theileria*, and *Cytauxzoon* belong in this group, but I am hesitant about assigning other genera to it, and indeed am not sure how many species to include in each genus.

Some broad generalizations can be made, however. *Babesia, Theileria,* and *Cytauxzoon* occur only in mammals, and *Aegyptianella* only in birds. (Everyone will not agree with this statement, however; Laird and Lari, 1957, for instance, believed that *A. moshkovskii* actually belongs to the genus *Babesia*.)

Babesia occurs primarily in ruminants, equids, carnivores, and rodents. *Theileria* occurs primarily in ruminants. These two genera take the place in ruminants that *Plasmodium* does in primates, and because of this were thought for many years to be closely related to it.

C. *Trypanosoma, Leishmania,* AND RELATED PROTOZOA

Members of the flagellate family Trypanosomatidae undoubtedly evolved from intestinal parasites of insects, and the great majority of the more primitive genera still live only in these hosts. The author accepts nine genera of parasitic Trypanosomatidae: *Leptomonas, Phytomonas, Herpetomonas, Crithidia, Blastocrithidia, Rhynchoidomonas, Trypanosoma, Endotrypanum,* and *Leishmania* (Levine, 1967). Of these, *Leptomonas, Herpetomonas, Crithidia, Blastocrithidia,* and *Rhynchoidomonas* are monoxenous, while the others are heteroxenous. *Leptomonas* occurs predominantly in insects (54 of the 58 accepted species) as does *Herpetomonas* (9 of the 15 accepted species). *Crithidia, Blastocrithidia,* and *Rhynchoidomonas* occur only in arthropods (15, 30, and 6 species, respectively). Of the heteroxenous genera, *Phytomonas* (9 species) occurs in plants and is transmitted by insects; *Endotrypanum* (1 species) occurs in a mammal and is presumably transmitted by an insect; *Leishmania* (7 species?) occurs in mammals (3 species), reptiles (3 species), and a protozoon (1 species); the mammalian species are transmitted by insects (*Phlebotomus*), but the vectors of the other species are unknown. All but 4 of the 190 accepted species of *Trypanosoma* occur in vertebrates; 79 of them are in mammals, 2 in birds, 20 in reptiles, 21 in amphibia, and 64 in fish.

The avian species are euryxenous (cf. Bennett, 1961), members of the *lewisi* group in mammals are stenoxenous, and most of the other mammalian species are euryxenous. Practically all are transmitted by insects, primarily biting flies. Too little is known about the species in lower vertebrates to justify any positive statements about their host ranges; presumably they are transmitted by leeches and other invertebrates. (Again, I may be thought to speak rather confidently about the numbers of accepted species in each genus, but I am actually not very confident about the matter. I am simply giving the best estimates available; cf. Levine, 1967.)

D. *Anaplasma, Haemobartonella,* AND OTHER RICKETTSIAE

The rickettsiae presumably arose in ticks and mites, and are still transmitted by them (and also by fleas and lice). Members of the genus *Rickettsia* are euryxenous in man, rodents, and lagomorphs (and one species at least in man and carnivores), and are transmitted by ticks, mites, lice, and fleas, each species having its own vectors. The single species of *Coxiella* is euryxenous in man, ruminants, rodents, lagomorphs, etc., and is transmitted by ticks and in the air. The single species of *Neorickettsia* is also essentially euryxenous, but primarily in carnivores. It is most pathogenic in the Canidae, and is the only species transmitted by a helminth, the fluke *Nanophyetus salmincola.*

The single species of *Cowdria* is mesoxenous in ruminants and is transmitted by ticks. The single species of *Colettsia* is also monoxenous in ruminants but apparently has no vector. *Anaplasma* was once thought to be stenoxenous, each species having a different ruminant host, but most species are now known to be mesoxenous, infecting more than one ruminant host.

All the other rickettsial parasites of vertebrates are stenoxenous. These include species of *Ehrlichia, Chlamydia* (except that *C. oculogenitalis* is mesoxenous), *Colesiota, Ricolesia, Miyagawanella* (except for *M. psittaci* and *M. ornithosis,* which are euryxenous), *Grahamella, Haemobartonella,* and *Eperythrozoon.* Among the stenoxenous genera, *Ehrlichia* occurs in ruminants and carnivores; *Chlamydia* in man and other primates; *Colesiota* in sheep; *Ricolesia* in birds, swine, and ruminants; *Miyagawanella* in man, birds, rodents, carnivores, opossums, and ruminants; *Grahamella* in insectivores and rodents; *Haemobartonella* in rodents, carnivores, ruminants, and insectivores; and *Eperythrozoon* in rodents, cattle, sheep, and swine.

E. *Bartonella* AND OTHER BACTERIA

A book on infectious blood diseases could theoretically include all bacterial septicemias. That is not the scope of the present volume, which is limited primarily to vector-borne diseases. I shall limit myself, therefore, to two genera, *Bartonella* and *Pasteurella*.

There is a single species of *Bartonella*, *B. bacilliformis*. This organism is an epidemiological enigma. It occurs in man and has not been found in other vetebrates. It is thus presumably stenoxenous; but it can be transmitted to monkeys experimentally, which would make it mesoxenous; and the nagging belief persists that it must have some as yet undiscovered wild reservoir, which would make it euryxenous. Sandflies (*Phlebotomus*) are presumably its vectors.

The genus *Pasteurella*, taken in the broad sense, includes both stenoxenous, mesoxenous and euryxenous species. They fall into two groups, those which have no arthropod vector, and those which may or may not have one. Among the former (i.e., monoxenous) species are *P. anatipestifer* and *P. pfaffi*, stenoxenous in ducks and canaries, respectively; *P. haemolytica*, mesoxenous in ruminants; and *P. multocida* and *P. pseudotuberculosis*, euryxenous in mammals and birds, but rarely found in man.

Among the species of *Pasteurella* which may or may not have arthropod vectors (i.e., may be either heteroxenous or monoxenous, depending upon the circumstances) are *P. pestis* and *P. tularensis*. Both occur primarily in rodents and lagomorphs. Both cause a highly fatal disease in man. Both also occur in a wide variety of mammals as well, but the host range of *P. tularensis* is even broader than that of *P. pestis*; it has been found not only in mammals but also in birds, including the owl, pheasant, and grouse, and the bull snake, snapping turtle, catfish, and in water. Both may be transmitted directly without a vector, but both generally (or often) have one. For *P. pestis* it is a flea; the bacterium has been found in many species. For *P. tularensis* it may be either a tick or a tabanid fly; certain mosquitoes may also be vectors, and some fleas, lice, and mites may possibly serve as vectors in lower animals. Pneumonic plague due to direct transmission from man to man is a well-known form of *P. pestis* infection. Tularemia may be transmitted in the water or even through the air.

F. *Leptospira, Borrelia,* AND OTHER SPIROCHETES

Parasitic spirochetes presumably arose from forms living free in the water. Nonpathogenic species are known in shellfish and also in the intestinal tract of various higher animals including the dog. The pathogenic species fall into three genera, *Leptospira*, *Borrelia*, and *Treponema*.

So many serotypes of *Leptospira* have been reported that any list would be incomplete by the time it appeared in print. Through 1965, 113 named plus 23 unnamed serotypes belonging to 17 groups had been isolated in the world (Communicable Disease Center, 1966).

Most are euryxenous, and others which are known only from a single host or a few closely related ones may turn out to be euryxenous as well. We are still exploring their host ranges.

The genus *Borrelia* contains a few species such as *B. vincentii* which are stenoxenous and also not vector-borne, but the majority are vector-borne. *Borrelia anserina* is mesoxenous or euryxenous, causing disease in chickens, geese, and ducks but not in mammals. The other species all occur in mammals and do not affect birds. These include *B. recurrentis*, which causes louse-borne relapsing fever in man and which is naturally stenoxenous (although it may be transmitted experimentally to monkeys, mice, rats, and rabbits), and an as yet undefined welter of tick-borne species which are euryxenous in rodents, lagomorphs, and man. Tick-borne relapsing fevers are zoonoses, while louse-borne relapsing fever is not.

So far as is known, all species of *Treponema* are stenoxenous.

III. PATTERNS OF HOST RESPONSE TO BLOOD PARASITES

Warm-blooded animals have a rather restricted series of responses to infection. An infection of a particular organ or type of cell elicits a rather stereotyped response, regardless of the nature of the infecting organism. In the present volume we are interested in blood diseases; they will be dealt with in this section.

A. ERYTHROCYTE PARASITES

Erythrocyte parasites destroy their host cells. They thus cause anemia. But there are more similarities than this. The anemia is caused by the parasites multiplying within their host cells and destroying them when they emerge. It may also be due in part to destruction of intact, uninvaded erythrocytes by a type of autoimmune reaction (see later chapters). If the destruction is rapid or massive enough, hemoglobin or hematin may be passed in the urine, causing so-called redwater or blackwater fever. If the hemoglobin is broken down to hemozoin, as occurs in malaria, the hemozoin (malarial pigment) granules are taken up by the liver cells and accumulate in them; this, however, occurs only in the malarial diseases. *Babesia*, for instance, carries the breakdown beyond the hemozoin stage, so that no pigment granules are formed.

When the erythrocytes are destroyed, their contents are released into the plasma. This does not affect the total blood constituents, but it does affect the plasma constituents. Potassium is present within the erythrocytes in much higher concentration than in normal plasma, so that when the erythrocytes are ruptured the plasma or serum potassium level rises.

Fever is characteristic of erythrocytic infections, but its exact cause is uncertain.

Another characteristic of erythrocytic infections is premunition. The causative organisms may remain hidden in the host body, presumably continuing to multiply at a very low level, for many years. There is no disease, but the presence of the organisms protects the host against reinfection.

Most blood diseases are transmitted by arthropod vectors. In the case of malaria and equine infectious anemia, the vectors are mosquitoes. The vectors of *Babesia* and *Theileria* are ticks, and those of *Anaplasma* are primarily ticks and tabanid flies.

Erythrocyte parasites may be sporozoan protozoa (*Plasmodium*), sarcodine protozoa (*Babesia, Theileria*), rickettsiae (*Anaplasma, Haemobartonella*), bacteria (*Bartonella*), or viruses (equine infectious anemia). Whatever the form, the reactions that they elicit and the diseases that they cause are basically similar. Investigators once thought that they could use the type of disease as a clue to the taxonomic position of the disease agent. This was the reason that *Plasmodium, Babesia, Theileria,* and *Anaplasma* were formerly placed together under the Sporozoa. We know now that this is not necessarily true, and modern work, especially with the electron microscope, has revealed the true affinities of the causative organisms.

B. Plasma Parasites

Plasma parasites include the trypanosomes and spirochetes. It was once thought that they were related because of a similarity in their wiggle; the fact that both *Trypanosoma gambiense* and *Treponema pallidum* eventually invaded the central nervous system and caused disease there probably reinforced this opinion. And the similarity induced Ehrlich to use trypanosome infections in searching for a cure for syphilis. That he succeeded was a fortunate coincidence and not proof of similarity. No one now disputes the fact that trypanosomes are animals and spirochetes plants.

Plasma parasites may cause both fever and anemia. Hemoglobinuria occurs in leptospirosis but not in trypanosomosis or borreliosis. We do not know how they destroy the erythrocytes. Potassium may be released into the plasma and the serum potassium level may rise considerably, especially in trypanosomosis. The observed levels, however, are not particularly harmful. If the anemia is sufficiently great and sufficiently long-lasting, secondary edema may follow as a result of an altered salt balance between the blood and other tissues. If the disease agents invade the central nervous system, reticuloendothelial system, or muscles (as may *Trypanosoma*), they may persist for many years. If not, and if they do not kill the host, they die in due time.

IV. RELATION OF PARASITE GROUP TO VECTOR GROUP

Even though all the parasites included in this volume occur in the blood, they are not transmitted by whatever arthropod happens to suck the blood. Each genus or group of parasites is generally transmitted by a specific vector genus or a limited group of genera. This is because they are adapted to these vector hosts and because part of their life cycle must take place in them.

All bloodsuckers transmit one disease or another. Horsfall (1962) has given a good discussion of the relations between arthropods and disease, and should be consulted for further details. The general picture is given in Table I, but no attempt has been made to be complete.

Ixodid ticks carry more types of parasites than any other group of vectors. Taken all in all, however, mosquitoes probably cause more human illness than any other vectors. They carry malaria, still the world's most important disease, and a choice variety of virus diseases including not only the encephalitides, dengue, yellow fever, and Rift Valley fever given in the table, but also such more recently discovered and exotically named viruses as the Chikungunya, O'Nyong-nyong, Semliki Forest, Sindbis, West Nile, and Bunyamwere viruses. An extended discussion of the role of arthropods as vectors of disease would go beyond the limitations of this chapter.

V. INTERACTIONS IN DISEASE TRANSMISSION

A. INTERACTION SYSTEMS

Diseases may be transmitted from one vertebrate to another either directly or through a vector. There are actually five possible components or factors: the host, parasite, reservoir, vector and environment. The relationships between these components are shown in Table II.

We can speak of 3-component, 4-component, and 5-component systems. These correspond to May's (1958) 2-factor, 3-factor, and 4-factor complexes, respectively; May omitted the environment in his classification, but the environment is an extremely important component.

Note that there are two types of 4-component systems; one with a vector

TABLE I

RELATIONSHIP OF PARASITE GROUP TO VECTOR GROUP

Vector group	Parasite group				
	Protozoa	Bacteria	Spirochetes	Rickettsiae	Viruses
Triatomin bugs	Trypanosoma cruzi, T. rangeli, etc.				
Lice			Borrelia recurrentis	Epidemic typhus Trench fever Murine haemobartonellosis	
Fleas	Trypanosoma lewisi, T. duttoni, etc.	Pasteurella pestis		Endemic (murine) typhus Canine haemobartonellosis	
Mosquitoes	Plasmodium spp.				Encephalitides Dengue Yellow fever Rift Valley fever
Midges (Culicoides)	Haemoproteus Leucocytozoon Hepatocystis				Blue tongue
Sandflies (Phlebotomus)	Leishmania	Bartonella bacilliformis			Pappataci fever (sandfly fever)

Vector	Protozoa	Bacteria	Spirochaetes	Rickettsiae	Viruses
Blackflies (*Simulium*)	*Leucocytozoon*				
Tabanid flies (*Tabanus, Chrysops*, etc.)	*Trypanosoma evansi*	*Pasteurella tularensis*		Anaplasmosis	
Tsetse flies (*Glossina*)	*Trypanosoma brucei, T. gambiense, T. rhodesiense, T. congolense, T. vivax,* etc.				
Hippoboscid flies	*Haemoproteus*				
Ixodid ticks	*Babesia Theileria*	*Pasteurella tularensis*		Q fever Rocky Mountain spotted fever Boutonneuse fever Siberian tick typhus Anaplasmosis	Colorado tick fever Omsk hemorrhagic fever Russian spring-summer encephalitis
Argasid ticks	*Aegyptianella*		*Borrelia* spp.		
Mites other than ticks	*Hepatozoon Lankesterella*			Rickettsialpox Scrub typhus	

TABLE II

RELATIONSHIPS BETWEEN DISEASE TRANSMISSION COMPONENTS[a]

Type of system	Components		Interactions	
	Nature	No.	Nature	No.
HEPa	H, E, P	3	H-E, H-P, E-P	3
HEPaV	H, E, P, V	4	H-E, H-P, H-V, E-P, E-V, P-V	6
HEPaR	H, E, P, R	4	H-E, H-P, H-R, E-P, E-R, P-R	6
HEPaRV	H, E, P, R, V	5	H-E, H-P, H-R, H-V, E-P, E-R, E-V, P-R, P-V, R-V	10

[a] H = host; E = environment; P = parasite; V = vector; R = reservoir.

and the other with a reservoir. The latter involves euryxenous diseases, while the former includes primarily stenoxenous ones. The latter diseases, too, may be zoonoses while the former are not. The 5-component systems have a reservoir and are vector-borne; they may be zoonoses also.

Because these terms are rather clumsy, the use of mnemonics is helpful. By substituting "Pa" for "P" one arrives at HEPa, HEPaV, HEPaR and HEPaRV systems.

Actually, it is interactions which are important in epidemiology. In the 3-component systems there are 3 interactions—between the host and the environment, the host and the parasite, and the environment and the parasite. In the 4-component systems there are 6 interactions, and in the 5-component systems there are 10.

The host-parasite (H-P) interaction produces disease (or its absence), while all the other interactions have to do with disease transmission. (Of course, the P-V interaction or even the H-V interaction might be involved in disease also—for the vector or host.)

B. INTERACTION ANALYSIS

Even the simplest of these systems is exceedingly complex. As an example, one might take the essentially stenoxenous virus disease of man, measles. Its mode of transmission is apparently not certain, but it appears to be transmitted directly from man to man via aerosols or droplets from the respiratory tract.

But the dynamics of measles transmission are not easy to put in mathematical terms. When an infected person releases the virus into the air, what are the chances that it will infect another person? They will depend on the presence or availability of susceptible people. Here are some of the factors that will affect rate and extent of transmission:

1. The number of people per unit area, i.e., the concentration of people.

The more people, the greater the number of contacts and the higher the transmission rate. The rate does not increase linearly, but with the square of the number of people. As the population increases, we can expect disease transmission to increase much more. If the population doubles, the potential transmission rate will quadruple.

2. The proportions or numbers of susceptible and immune people.

3. In the immune group, the proportion which can harbor or transmit the parasite without having any disease (i.e., the proportion of healthy carriers), and the proportion which cannot be infected at all.

4. The length of the period during which the host disseminates the parasite—whether long or short.

5. The virulence of the particular strain of parasite.

6. The parasite's mode of transmission—whether by aerosol, droplet, dust, fomites, food or water contamination, etc.

7. The length of time the parasite can survive in the external environment, and the climatic and other environmental factors which influence its survival. This is a whole complex of environmental factors and not just one—temperature, precipitation, moisture, relative humidity, soil type, evapotranspiration, wind, sunlight and protection therefrom, etc.

8. The ability of the parasite to multiply in the environment, and the effect of the above environmental factors thereon. (In the case of measles, there is no multiplication, but there may be in some other diseases such as listeriosis.)

9. The distribution of the host population. The most common mistake in the mathematical models which have been worked out to explain disease transmission patterns has been the assumption that the population is randomly distributed. It is not. People are distributed in groups of various sizes and kinds. The basic group is the family, and families come in different sizes and with different component parts—adults and children of two sexes and varying ages. Similarly, in domestic or wild animals, the basic group is the herd or flock, or the nidus.

10. The types and numbers of interactions between individual hosts or basic groups. Families interact with each other in various ways, and so do their components. Two families living side by side may be in each other's kitchens half the time—or they may scarcely know each other. And there are many other interactions between family components, either from nearby, from the other end of town, or from out-of-town—children in school (the school, the public swimming pool, and the theater are excellent places for disease transmission), shopping, place and conditions of work, play habits, vacation patterns, etc. Each family has its own interactions with other families and individuals, and these may have an important bearing on the probability of a disease being introduced into it.

The herd of domestic animals is sometimes a simpler group than the family, but nevertheless it is far from truly simple. And, among wild animals, we have the family, the extended family, the herd or flock, etc., depending on the species of animal and the conditions.

11. The general type of environment, whether urban, suburban, or rural, and the subtype thereof.

12. The interactions between different parasites in the same host. One parasite already present may prevent establishment of another, or it may enhance the possibility of its establishment.

These are probably not all the factors in a 3-component system, but they are enough to show its complexity. The system is so complex that we have yet to arrive at a satisfactory mathematical analysis of the transmission process.

The HEPaV (4-component) system, malaria, is even more complex. Here a mosquito vector is introduced into the system. When a vector is introduced an entirely new series of complicating factors comes along with it. We have not only the factors affecting infectibility of the host by the parasite, but we have an equal number of factors affecting infectibility of the vector by the parasite. Then there is a large group of factors having to do with the propagation and survival of the vector—in this particular case, with mosquito-breeding conditions. And, finally, there is a group of factors having to do with the relations between the host and the vector. For instance, a particular species of mosquito might be ideal for the parasite to multiply in, but if it refused to bite man it would not be a suitable vector. The principal vector of malaria in the United States is *Anopheles quadrimaculatus*, a species which prefers cattle to man. Hence, the relative numbers of cattle and people affect the malaria transmission rate.

Another type of host-vector interaction has to do with coordination between the vector's flight habits (if any) and man's sleeping habits. The principal malaria vector in the Philippines, *A. minimus flavirostris*, does not fly high, so that people sleeping on the second story of a house are not bitten by it.

This, then, is a highly complicated system. Some attempts have been made to work out mathematical theories of malaria transmission, but none is really satisfactory. Too many factors are involved, and especially too many factors of unknown importance.

This discussion could be continued to include HEPaR and HEPaRV systems, or systems in which two vectors are necessary, as in the life cycles of many trematodes and pseudophyllidean tapeworms, but it is unnecessary. Table I indicates that there are 3 to 10 interactions per system, but each interaction comprises a whole group of factors. And these groups of factors are not simply additive; they are exponential. They are not merely added, but must be multiplied together. Mathematical analysis of the systems becomes increasingly difficult as one increases the number of components and interactions, and breaking down each interaction into its factors (some of which are unknown) makes the analysis impossible.

VI. LANDSCAPE EPIDEMIOLOGY

As already stated, each disease has its own particular group of vectors. Each vector has its own habitat. This habitat occurs in a particular landscape. Mosquitoes live in swampy areas, slow-moving streams, ponds, puddles, etc., depending on the species. Midges live in moist places or shallow water with ample organic matter such as the weedy margins of ponds, puddles, tree holes, crab holes, moist vegetation and manure. Blackflies live in mountainous areas with fast-running streams. Argasid ticks live in burrows, caves, etc., in desert or semidesert regions. Ixodid ticks live in forest glades, ground covered with small bushes and shrubs, etc.

The reservoirs of zoonotic diseases, too, live in specific habitats. Rabbits live in fields where there are shrubs and bushes. Ground squirrels live in rather similar localities, but their requirements are different. Prairie dogs live on the relatively dry prairie. Gerbils live in relatively dry, sandy areas.

If one knows what vectors and what reservoirs occur in each type of landscape, one can predict from the landscape what diseases to expect there. This is the basis of landscape epidemiology, a field which has been studied and exploited especially by the Russians (cf. Pavlovsky, 1963, 1966).

It is beyond the scope of this chapter to do more than mention the subject and give a few illustrations, but the reader can make out his own list of landscape-disease relationships.

A. TYPES OF LANDSCAPE

There are many different types of general landscape, and many subdivisions within each general type. Each has its own denizens. Often the characteristics which make a particular habitat suitable for one species and not for another are not clear to us.

In general, we can speak of (1) tundra, in the Far North, where there are no trees; (2) forest, which may be divided into coniferous forest, broadleaved forest, and mixed forest; the Russian taiga, which is entirely or partly coniferous, forms a broad band south of the tundra; (3) steppe; (4) prairie; (5) semidesert; (6) desert; (7) mountainous areas.

Each of the above grades into its neighbor, and the types and subtypes often interdigitate. Ecotones, contact zones between different landscapes, are especially important in epidemiology.

Furthermore, one cannot speak of landscape epidemiology without discussing climate. Different diseases are limited to different climates. Yellow fever is a tropical disease. It cannot maintain itself in the temperate zone except during the warm season. The same is true of dengue. Texas fever is essentially subtropical. Trypanosomosis (except for dourine) is tropical or subtropical, as are babesiosis and theileriosis. Borreliosis is essentially a disease of warm, dry regions. One could go on indefinitely.

A third factor which must be taken into account is geographic location. Diseases are not distributed uniformly throughout the world, and the diseases of a particular type of landscape in a particular climate are not the same everywhere. The American tropics and the African are not the same, even though their climates and appearance are similar. Tsetse flies occur only in Africa, and tsetse-borne trypanosomosis can occur nowhere else. The southwestern United States, southwestern USSR, and the Middle East are similar in many respects, but Oriental sore occurs only in the last two. There are no gerbils in the United States, and other disease reservoirs take their place there. The common "malarial" parasite of African monkeys is *Hepatocystis kochi*, that of New World monkeys is *Plasmodium brasilianum*, while Asian monkeys have a whole series of *Plasmodium* species but not *P. brasilianum*. Even in Asia, however, the picture is not uniform. Primate malaria occurs in southeast Asia, southern India, and Ceylon. The rhesus monkey, which inhabits northern India and Pakistan and which is the laboratory animal in which monkey malaria has been studied most, is essentially free of this disease in its native habitat.

B. RELATION OF LANDSCAPE TO DISEASE OCCURRENCE

As already mentioned, each type of landscape has its own diseases, and the expert can tell what diseases to expect from a knowledge of the landscape, climate, and geographic region.

Tick-borne encephalitis occurs in the entire mixed or broad-leaved forest zone in the USSR, along river and stream valleys, in wood-cutting areas, and along paths, roads, etc. Leptospirosis is characteristic of moist meadows around large lakes and in river floodlands. Cutaneous leishmaniosis occurs in the USSR in desert areas, highlands, and areas with warm cilmates. Tick-borne relapsing fever in the USSR occurs in similar localities, but in the United States it is found more in forested areas of the West. The name Rocky Mountain spotted fever indicates its primary locus even though it may be found outside this region. During World War II, troops learned to avoid grassy glades in New Guinea because that was where scrub typhus lurked; it did not occur in the tropical rain forest. Gambian sleeping sickness occurs along streams near human habitations, but Rhodesian trypanosomosis occurs on the savannah of East Africa, far from human dwellings.

The endemicity or the transmission of a particular disease depends upon a combination of suitable factors. Chagas' disease of man is a good example. It requires a combination of the parasite, *Trypanosoma cruzi*, the vector, a species of triatomin bug, and a vertebrate host. A reservoir may be important in some areas, but it is not necessary. The particular host in which the parasite is found will depend upon the ecological characteristics of the parasite, vector, and host. Chagas' disease occurs in man only in the American tropics and subtropics. Here it is transmitted from man to man by infected kissing bugs which live in the cracks and crevices of the houses. In the forested areas of the southeastern United States, the vertebrate hosts are raccoons, and again transmission is by triatomin bugs which live in conjunction with them. In the southwestern United States, which is much drier, an important vertebrate host is the wood rat, and the parasite is transmitted by triatomins living in wood rat nests. In neither region does the disease occur normally in man. The situation is still different in Illinois. Here both vertebrate and a few vector hosts occur, and the triatomins live in houses and bite people. But *T. cruzi* is absent and there is no Chagas' disease. Presumably the climate is too cold for the protozoon.

Malaria was once common in the temperate zone, but it is now confined primarily to the tropics and subtropics. The host and the anopheline vector are present in the temperate zone, and the parasite has in the past survived there. It could in the future, too. But the infection chain was broken by measures practiced against the vector and against the parasite. As a consequence, although the vector came back, the parasite never did, and it will not so long as adequate preventive measures continue to be practiced.

REFERENCES

Baker, J. A. (1966). *J. Protozool.* 13, 515–519.
Bennett, G. F. (1961). *Can. J. Zool.* 39, 17–33.
Communicable Disease Center, U.S. Dept. of Health, Education and Welfare. (1966). "Leptospiral Serotype Distribution Lists According to Host and Geographic Area." Natl. Communicable Disease Center, U.S. Dept. of Health, Education and Welfare, Atlanta, Georgia.
Honigberg, B. M. *et al.* (1964). *J. Protozool.* 11, 7–20.
Horsfall, W. R. (1962). "Medical Entomology. Arthropods and Human Diseases." Ronald Press, New York.
Huff, C. G. (1954). *Res. Rept. Naval Med. Res. Inst.* 12, 619–644.
Laird, M., and Lari, F. A. (1957). *Can. J. Zool.* 35, 783–795.
Levine, N. D. (1967). *In* "Protozoa in Biological Research" (T. T. Chen, ed.), Vol. 4. Pergamon Press, Oxford (in press).
May, J. M. (1958). "The Ecology of Human Disease." MD Publ., New York.
Pavlovskii, E. N. (1963). *In* "Natural Foci of Human Infections" (O. Theodor, ed.) (transl. by H. Brachyahu), p. 201. Israel Progr. Sci. Trans., Jerusalem, Israel.
Pavlovsky, E. N. (1966). *In* "Natural Nidality of Transmissible Diseases with Special Reference to the Landscape Epidemiology of Zooanthroponoses" (N. D. Levine, ed.) (transl. by F. K. Plous, Jr.), p. 261. Univ. of Illinois Press, Urbana, Illinois.

2

Basis of Host Cell-Parasite Specificity[*]

AVIVAH ZUCKERMAN

I. INTRODUCTION

Innate immunity to protozoan organisms was defined by Taliaferro (1949) as the sum total of nonspecific reactions against a parasite and its products inherent in a host without prior contact between the two organisms. Such innate immune reactions "account for the essentially total refractiveness or nonsusceptibility of certain otherwise possible hosts, and much of the so-called host specificity of the parasites." It would be quixotic to set out systematically to define all the possible inherent factors delimiting the relatively narrow adaptive range within which each host and each parasite are capable of establishing the reciprocal relationship which we term parasitism. Few have therefore devoted themselves to the systematic study of innate immunity, in contrast to the relatively many who have studied acquired immunity to parasites. Where successful studies of innate immunity have been done, they have focussed on a single aspect of the problem rather than on the field as a whole. Several such aspects are thus relatively well understood, while contiguous areas of this enormous jigsaw puzzle still remain totally unexplored. This accounts for the fragmentary nature of the present discussion of recent contributions to the topic.

Much interesting material has been discussed by Manwell (1963), with special emphasis on evolution and host ranges. Manwell quotes a much earlier statement by Huff as still true, that "in no case of parasitism has

[*] This study was aided by Grant AI-02859 from the United States Public Health Service.

the mechanism of host specificity ever yet been completely and indubitably explained."

The present paper is therefore merely a tentative outline.

II. FACTORS AFFECTING HOST CELL-PARASITE SPECIFICITY IN THE VERTEBRATE HOST

A. PHYSIOLOGICAL AND NUTRITIONAL CONDITION OF THE HOST

1. The Effects of Host Temperature on Parasitic Development

The fact that protozoan blood parasites are generally transmitted by poikilothermic arthropod vectors bears witness to the wide range of adaptability to temperature inherent in such parasites. Their development in arthropods is discussed in Section III.

Modifying the body temperature of the vertebrate host can significantly affect the course of a protozoan infection. Basu et al. (1962) studied rats infected with *Plasmodium berghei*, whose temperatures had been experimentally lowered (by the injection of chlorpromazine or by refrigeration) or raised (by the injection of an agent inducing pyrexia). Both manipulations depressed parasitemia. Kretschmar (1964) and Ringwald and Kretschmar (1965), working with *P. berghei* in mice, found that low temperatures were associated with low parasitemias, and that the survival rate varied inversely with the temperature at which the mice were maintained (29 % survival at 12°C; 0 % survival at 32°C). Thus, by depressing parasitemia with the aid of low temperatures during the period of the acquisition of active immunity it was possible to save the lives of animals otherwise doomed to die. Similarly, modifying the environmental temperature of mice infected with *Trypanosoma cruzi* (Trejos et al., 1965) affected the course of the disease. Mice maintained at 18°C had high parasitemias and numerous myocardial lesions, while those kept at 37°C had low parasitemias and negligible myocardial lesions. Beckman (1964b) caused *Aedes aegypti* infected with *Plasmodium gallinaceum* to feed upon chick skin chilled to 32°C, and maintained at this temperature for half an hour following the feed. The skin was then excised and inoculated intraperitoneally into clean chicks, 14 out of 15 of which became infected. He concluded that the insusceptibility of the mouse to the avian *P. gallinaceum* is not temperature-dependent.

2. The Effects of Host Diet and Vitamin Balance on Parasitic Development

Geiman (1958) and Lincicome (1963) have emphasized the fact that parasitism is being defined increasingly in terms of biochemical and biophysical deficiencies of the parasite for which the host is capable of compensating. When these compensatory capabilities of the host are impaired—as by dietary deficiencies—the host becomes an altered milieu. Should the alteration involve factors essential for parasitic development, changes in the parasite population would ensue. In general, a good nutritional state of the host encourages parasitemia. Where they have been sufficiently studied, qualitative and quantitative biochemical differences have rather closely matched taxonomic relationships (Geiman, 1964).

Numerous corroborative studies have followed the observations that a milk diet suppresses *P. berghei* malaria (Maegraith *et al.*, 1952), and that this suppression is due to the absence of *p*-aminobenzoic acid (PABA) from milk (Hawking, 1953). The suppressive effect of a milk diet has been confirmed by most of those working with rodent, simian, and avian malarias (Jacobs, 1964).

Adler (1958) has extended the observations on the effect of a milk diet to include rodents on a meat diet, which is also deficient in PABA, and which also suppresses parasitemia even if the mice are splenectomized. Again, the administration of PABA restored normal parasitemias. Mice on a meat diet, which have recovered from infection with *Plasmodium vinckei*, are later immune to challenge with a virulent strain (Adler and Gunders, 1965; Kuperman, 1966). As in the case of the suppression of parasitemia by lowering the temperature (Section II, A, 1), depressing parasitemia in a malaria-infected mouse during the period of acquisition of immunity (in this case by means of a dietary deficiency) permits this highly susceptible host to achieve latency.

Bastianelli (1959), in an extensive study of the interrelations between plasmodia and erythrocytes, pointed out that monkey malaria parasitemias, depressed when ascorbic acid was withheld, became normal when the vitamin was added to the diet.

Tolbert and McGhee (1960) reopened the question of the effect of host blood sugar levels on parasitemia. They observed that *P. berghei* was partially suppressed in rats with alloxan diabetes, a result which did not confirm earlier studies on *Plasmodium cathemerium* in diabetic canaries.

The literature on the effect of the host's diet on susceptibility to malaria has been reviewed by Geiman (1958) and Jacobs (1964).

Trypanosomes, like plasmodia, are influenced by the host's dietary state. Yaeger and Miller undertook to investigate the effects of a series of dietary deficiencies on *T. cruzi* infections in rats. Deficiencies of thiamine (1960a), riboflavin (1960b), pantothenate (1960c), pyridoxine (1960d), vitamin A (1963a), and lysine (1963b) were studied, as well as the effect of the quality of the dietary protein (1963c). Whereas thiamine, pantothenate, pyridoxine and lysine deficiencies markedly exacerbated *T. cruzi* infections in rats, vitamin A deficiency had a less adverse effect, and riboflavin deficiency led to no significant increase in susceptibility. Thus, as pointed out by Geiman (1958), parasites respond selectively to the host's dietary deficiencies. *Trypanosoma cruzi* parasitemia and cardiac lesions were less severe in rats on a diet of casein than on a diet of wheat gluten.

Thiamine deficiency of the host also resulted in higher parasitemias in rats infected with *Trypanosoma rhodesiense* (Singer, 1961).

Trypanosoma lewisi can be induced to develop in the mouse by the injection of normal rat serum; but serum from starved rats loses its ability to support parasitic development in the unusual host (Lincicome and Hinnant, 1962; Lincicome, 1963), probably due to the depletion of an essential trypanosomal growth factor from the serum exhausted by starvation. Similarly, normal sheep serum supports the development of *Trypanosoma vivax* in an unusual host, the rat (Desowitz, 1963). This effect was also attributed either to the presence in the serum supplement of an essential growth factor, or, alternatively, to interference with antibody activity by the serum supplement.

Geiman (1958) reviewed studies on the effect of the host's nutritional state on trypanosomal infections.

3. The Effects of Host Metabolites on Parasites

The extent of a parasitic population is affected not only by the nature of the host as a source of nutrients (Sections II, A, 2 and II, B, 2) but also by the parasite's interactions with the metabolic products of the host. Types of host metabolite which tend to limit parasite populations include growth inhibitors, antagonists, and antibodies (Geiman, 1964). Thus, strictly speaking, the entire spectrum of acquired immune reactions, dealt with extensively in other chapters, comes under this heading. The extreme diversity of the metabolic patterns involved has impeded the setting up of experimental models to unravel these relationships (Geiman, 1964). Such studies are urgently needed.

One obvious experimental model is the study of sex-associated reactions to parasitic disease. *Plasmodium berghei* infections in male and female rats were studied by Zuckerman and Yoeli (1954). Females shortly after gestation were less susceptible than virgin females of the same age. Virgin females and males of the same age were equally susceptible, and the extirpation of their sex organs did not affect susceptibility. Goble *et al.* (1965), working with *P. berghei* in mice, found that females had more effective innate immunity than males. Moreover, male mice were more susceptible than females to *Trypanosoma gambiense*, *Trypanosoma congolense*, and to *T. cruzi*.

Huff *et al.* (1958) concluded that the gradual diminution in the infectiousness of gametocytes observed in the course of avian malaria was due to a humoral immune factor rather than to the depletion of an essential growth factor.

The unusual host may sometimes be induced to support the growth of a trypanosome to which it is normally refractory by intervention in its functioning, such as splenectomy, reticuloendothelial blockade, or the introduction of foreign substances (drugs, hormones) (Desowitz, 1963). Profound changes in the biology of trypanosomes may ensue after they have been maintained for some time in an unusual host. These may include (*a*) the loss of the ability to undergo cyclical transmission; (*b*) morphological and physiological changes; (*c*) changes in virulence; and (*d*) changes in reactivity to drugs (Desowitz, 1963).

B. PARASITIC METABOLISM

1. The Effects of Parasite Metabolites on the Host

In extreme cases the presence of a parasite can be so physically or chemically traumatic that the host dies. All gradations exist between this lethal effect and harmful effects so subtle as to be barely demonstrable. Even beneficial effects of parasites on their hosts are on record.

The gross morphological changes in a host cell often accompanying intracellular parasitism must be based on physiological changes at the molecular level (Trager, 1960). The metabolic pathways inherent in a cell may be altered by the presence of an intracellular parasite. Bastianelli (1959) noted that erythrocytes infected with simian or fowl plasmodia had augmented titers of phospholipids, adenosine triphosphate (ATP), fatty acids, and adenine, which he regarded as changes in the chemistry of the invaded cell. Similarly, duck erythrocytes infected with *Plasmodium lophurae* contain greatly augmented quantities of folic and folinic acid, over and above those present in the parasites (Trager, 1960, 1963b; Siddiqui and Trager, 1964, 1965). This, incidentally, is of benefit to the parasite, which requires folic and folinic acid. Trager (1963a) has suggested that *P. lophurae* may parasitize the energy transfer mechanisms of the host cell.

In contrast to such changes in the host cell induced by parasites, a stable situation was observed by Sherman and Hull (1960) in respect to the types of hemoglobin present in the erythrocytes of chicks during infection with *P. lophurae*. Two chick hemoglobins can be differentiated by electrophoresis, and the chick continues to produce cells with normal proportions of the two hemoglobins even during acute infection.

Maegraith and his co-workers have investigated changes in the metabolism of host liver mitochondria during plasmodial infection. The enzymatic functions of mitochondria from mice with *P. berghei* are disturbed. In particular, the oxidation rates of substrates are depressed (Riley and Maegraith, 1962). The liver mitochondria of monkeys with *Plasmodium knowlesi* also undergo similar if less extensive changes (Maegraith *et al.*, 1962). The serum of infected hosts contains a nonspecific factor which inhibits oxidative phosphorylation by liver mitochondria *in vitro* by damaging the mitochondrial membranes (Maegraith, 1964; Maegraith and Fletcher, 1965). The addition of cofactors is capable of reducing this inhibition. Thus, a parasitic product paralyzes host cell respiration *in vitro*. Furthermore, the authors consider it likely that the same occurs *in vivo* during malaria, and that this is an important link in the chain of events constituting the disease.

Lincicome *et al.* (1963) discussed a well-documented instance of a beneficial effect of a parasite on its host. Rats infected with *T. lewisi* and mice with *Trypanosoma duttoni* (Lincicome and Shepperson, 1963) grew at a consistently faster rate than did uninfected controls. Lincicome and Shepperson (1965) concluded that these results were probably due to the stimulation of host cell metabolic activity by a factor supplied by the parasite. On the basis of studies with rats deficient in thiamine, they further suggested that the factor contributed by the parasite may be thiamine.

2. Growth Requirements of Protozoan Blood Parasites

The growth requirements of parasitic blood protozoa are probably not radically different from those of self-sufficient cells. However, parasites, unlike self-sufficient cells, are unable to produce all the substances which they require, and depend for certain of their constituents on their host cells. An intracellular parasite is thought of as having evolved from a free-living form by a progressive dwindling of the number of essential metabolic activities it is capable of carrying out on its own (Trager, 1960). The more dependent the parasite, the more rigidly specialized is its adaptation to a given host cell. To the extent that more than one host can satisfy a parasite's deficiencies, its host specificity will be broad. On the other hand, the fewer the hosts which can satisfy the deficiencies, the narrower the host specificity. While the principle of parasitic adaptation is thus clear, the exact adaptive ranges of specific parasites are still largely unknown (Geiman, 1964). Much valuable information in this general area is reviewed by Geiman (1958, 1964), Trager (1960), Moulder (1962), and Desowitz (1963).

Different species of plasmodia, all of which have very similar growth requirements, need to be supplied with pantothenate, PABA, ascorbic acid, thiamine, riboflavin, biotin, and methionine (Trager, 1960). Both the host cell and the tissue juices bathing it—plasma, in the present instance—may serve as a source of these substances. Thus, for example, methionine is supplied to a plasmodium both by the host erythrocyte and by the ambient plasma (Moulder, 1962). Plasmodia, ingesting erythrocytic cytoplasm by phagotrophy (or pinocytosis) (Rudzinska *et al.*, 1965; Aikawa *et al.*, 1966), obtain from it protein (the globin of hemoglobin), carbohydrates, and enzymes (Trager, 1960, 1963a). The same types of enzyme system and the same metabolic pathways are probably involved in the synthesis of both host and plasmodial cytoplasm (Moulder, 1962).

The growth requirements of different stages in the life cycle of the same species of parasite must differ markedly to explain the rhythmical changes in predilection for given host cells. Sometimes demonstrable structural changes in the parasite accompany such changes in nutritional habit. Thus, the bloodstream forms of *Trypanosoma brucei* have a simple mitochondrion, whereas insect midgut forms have an elaborate one. This change is associated with a different oxygen tension and with a concomitant change in respiratory behavior (Vickerman, 1962). In other cases, although structural changes may be inapparent, rhythmic alternation of host cells nevertheless presupposes underlying changes in the biochemical requirements of the parasite.

It is possible to gain additional insight into the growth requirements of intracellular parasites by attempting to maintain them either intracellularly or extracellularly *in vitro*. Trager (1964, 1965) has carried out an extensive series of studies of this sort on *P. lophurae*. The addition to the culture medium of glutathione, nicotinamide, ATP, pyruvate, malate, coenzyme A, folinic acid, and lactalbumin hydrolyzate all favored survival *in vitro*. Essential substances like ATP or coenzyme A are probably supplied *in vivo* by the host cell. Antimetabolites which curtail the supply of coenzyme A to the red cell *in vivo* thereby depress parasitemia, and thus serve as antimalarials.

C. Types of Barriers Affecting the Invasion of a Host Cell by a Parasite

Geiman's comment (1958) that "in very few cases are we in a position to explain precisely how parasites invade host tissues and cells" still applies today. We can still do little more than record the affinities of a parasite for a given cell or tissue. For example, Gaillard (1965) has distinguished strains of *T. cruzi* which have affinities for either the myocardium or the skeletal musculature. Geiman (1958) reviewed the earlier work on the metabolic pathways which may play a role in explaining the affinities of certain blood parasites for certain cells.

How a parasite gains entry into its host has been discussed by Trager (1960). When a parasite is passively taken up (e.g., ingested), no specific relationship need exist between it and its host. In contrast, when a blood-inhabiting parasite is required to adapt to existence either in the plasma or in a blood corpuscle in order to survive, certain specific problems have to be overcome. If it is an intracellular parasite, it must, for example, make its way into the host cell either by physical or by enzymatic means. The ability or the inability of a parasite to pass such barriers plays a role in defining the susceptibility of a given host to a given parasite. For example, the viable sporozoites of *P. gallinaceum*, deposited by mosquito bite into a chick's skin, could be recovered an hour later from its blood, whereas if a mouse received the infective bite, no viable parasites could be recovered (Beckman, 1963). Mice were also unsuitable hosts for the exoerythrocytic forms of *P. gallinaceum* (Beckman, 1964a).

Plasmodial merozoites often exhibit a marked predilection for penetrating cells of a definite age on the scale between erythroblast and definitive normocyte (reviewed by Zuckerman, 1957). Thus, for example, *P. berghei* merozoites favor reticulocytes which are neither the most mature nor the most immature in the series (Zuckerman, 1957). Other merozoites, like those of *P. vinckei*, enter available erythrocytes at random (Zuckerman, 1958). Warren *et al.* (1965), discussing similar preferences on the part of merozoites of simian malaria, have reiterated the view that such predilections are so fixed that they can actually aid in differentiating species. Yet we know nothing of the mechanism whereby merozoites are capable of assessing the age of the red cell before they enter it; or, alternatively, of the barriers to penetration possibly inherent in the uninvaded cells.

In Chapter 6, Bray discusses the zoonotic potential of blood parasites, and deals with natural infections in unusual hosts. Parasites have also been induced to develop in hosts other than those in which they occur in nature. (The term "abnormal host," often used in this context, is incorrect, since it is the situation of the parasite and not the host which is abnormal.) In a body of work reviewed by himself (1953), McGhee has shown that under given conditions plasmodia are capable of penetrating the red cells of species widely divergent from those in which they ordinarily occur. Thus, merozoites of *P. lophurae* in a chick embryo invaded washed mouse erythrocytes inoculated into the embryo, and also erythrocytes from young but not from older rats. All invaded erythrocytes from unusual hosts had a high K content. Trager (1960), referring to McGhee's work, states that "if we knew the biochemical basis . . . we would have made significant progress toward understanding host-parasite specificity at the cellular level." This comment can still be made today.

Attempts to explore the adaptability of blood parasites to unusual hosts include the human plasmodia. Porter and Young (1966) review earlier experiments on the transmission of *Plasmodium vivax* to monkeys. All except the chimpanzee (*Pan satyrus*) had hitherto proved unsusceptible. They obtained low parasitemias in titi marmosets (*Saguinus geoffroyi*) and in night monkeys (*Aotus trivingatus*). *Plasmodium falciparum* has been transmitted to simian hosts on several occasions (reviewed by Hickman *et al.*, 1966). Recent successful transmissions of *P. falciparum* include that of Sadun *et al.* (1966) to chimpanzees; Hickman *et al.* (1966) to splenectomized chimpanzees; and of Ward and Cadigan (1966) to gibbons (*Hylobates lar*). The importance of transmitting the human plasmodia to nonhuman hosts capable of serving as a source of plasmodial antigen has been pointed out by Zuckerman and Ristic in Chapter 5. Contacos and Coatney (1963) and Tobie *et al.* (1963) reviewed experimental trials to date in the transmission of simian plasmodia to human beings and to unusual simian hosts. Their interest lay chiefly in the possible importance of the simian malarias as zoonoses. Further transmission of *Plasmodium cynomolgi* to man was reported by Tobie and Coatney (1964).

McGhee (1953) failed to alter the susceptibility of red cells to *P. lophurae* merozoites by removing surface receptors with cholera vibrio filtrate, or by treating with steapsin or with carbon monoxide before inoculating the cells into infected chick embryos. However, partial lysis brought about a sharp reduction in invasion. Sherman (1966) has described a similar series of *in vitro* manipulations of erythrocytes which failed to affect their penetration by merozoites of *P. lophurae*. These included pretreatment with trypsin and chymotrypsin.

Huff *et al.* (1960) explored the effect of heterologous host body fluids on plasmodia in tissue culture. They implanted culture chambers containing chick embryo tissue infected with *P. gallinaceum* or *Plasmodium fallax* intraperitoneally in a series of host species. Exoerythrocytic merozoites could pass freely out of the chamber into the peritoneum. The implanted *P. gallinaceum* chambers induced infection in homologous hosts (chickens and turkeys), but not in ducks, which ordinarily resist infection with this parasite. The contents of chambers containing *P. fallax* remained viable to homologous avian hosts for 21 days, but the mice into which they were implanted resisted infection. Exoerythrocytic forms of *P. fallax* directly implanted (without culture chambers) into the mouse peritoneum were promptly destroyed.

D. GENETIC FACTORS INFLUENCING HOST SUSCEPTIBILITY

The genetic constitution of a host plays a role in its susceptibility to parasites. This is true not only of obvious genetic differences between species, but also of more subtle differences within a species.

During the past decade the importance of hereditary red cell traits as factors in susceptibility to malaria has been and is still being widely debated (Allison, 1963; Motulsky, 1964). The mechanism of some of these traits has been worked out to a remarkable degree. Thus, for example, the cause of the sickle cell trait, said to confer partial protection against *P. falciparum*, has been defined as the substitution of a single valine residue for one of glutamic acid in the hemoglobin molecule (Ingram, 1959). This substitution entails a distinct increase in the viscosity of the hemoglobin, which may interfere with phagotrophy of the erythrocyte cytoplasm by the parasite (Motulsky, 1964). Other hereditary blood traits under suspicion of affecting susceptibility to malaria are glucose-6-phosphate dehydrogenase deficiency, thalassemia, and reduced content of adenosine triphosphate in the erythrocytes (Brewer and Powell, 1965). Each is suspected of conferring a biological advantage vis-à-vis malaria.

With the aid of serial inbreeding, it is possible to select strains of vertebrate hosts with consistently high or low susceptibility to malaria. Thus, Ramakrishnan *et al.* (1964) selected a strain of mice increasingly resistant to infection with *P. berghei*. In a similar attempt to select rat strains of high and low susceptibilities to *P. vinckei* (Zuckerman, 1966), brother-and-sister matings were carried out through 16 inbred generations; this is, however, still far short of homozygosity. At this stage the same inoculum induced negligible parasitemias (often less than 0.1 %) in the low-susceptibility lines; while it regularly killed rats of the high-susceptibility lines (with peak parasitemias of 60 to 70 %) unless they received chemotherapeutic treatment. Where homozygous vertebrate lines as, for example, mice, are available for research, degrees of susceptibility to rodent malaria are fixed genetic characters, which may be manipulated in the usual manner in hybridization experiments (reviewed by Bruce-Chwatt, 1964). A case in point is the NMRI line of mice, many of which actually recover from infection with *P. berghei* (Kretschmar and Jerusalem, 1963).

III. SUSCEPTIBILITY OF ARTHROPOD VECTORS TO PROTOZOAN PARASITES

The protozoan blood diseases of man and animals are all arthropod-borne. The availability of suitable vectors, susceptible to the parasite and infesting the vertebrate host, is therefore an essential link in the establishment of each parasite within a vertebrate population.

In a comprehensive review, Garnham (1964) has analyzed the factors affecting the development of protozoa in their arthropod hosts under two main headings: external and internal factors. Huff (1965) has also discussed such factors in the susceptibility of mosquitoes to avian malarias.

Since arthropods are poikilotherms, the external temperature is of major importance in their biology. Insects can generally survive at higher temperatures than are borne by their parasites, each stage of which may have its own intrinsic temperature range. Within the limits of viability, the higher the temperature the more rapid the development of the parasite in its vector. Freezing techniques, which kill the vector, can preserve its parasites. A particularly narrow temperature range (18° to 22°C) delimits the sporogonic development of a strain of *P. berghei* (Yoeli *et al.*, 1966). Temperature is also an important factor in the *in vitro* culture of the arthropod phases of the protozoan parasites under discussion (Zuckerman, 1966b).

Each vector also has its own optimum range of humidity. These two factors alone, temperature and humidity, defining as they do the climatic range within which a vector is capable of existing in nature, yield a rough idea of the ecological niche within which their parasites are likely to occur.

Internal factors affecting the susceptibility of arthropods to protozoa include:

1. The infectivity of the parasite at the time of the infective feed. This may vary with the stage of the infection and with the parasite rate in the vertebrate host.

2. The innate susceptibility of the arthropod, which, in culicine mosquitoes, is a part of their genetic constitution (reviewed by Huff, 1965; Ward, 1963, 1966). With the exception of *Trypanosoma rangeli*, a vector-borne protozoan is generally less pathogenic in the arthropod than in the vertebrate host.

3. The age of the arthropod. Advancing age has generally been associated with diminishing susceptibility.

4. The stage of nutrition of the vector. For example, *A. aegypti* females were fed on sugar and a series of test substances before their infective *P. gallinaceum* meals. Mosquito susceptibility was enhanced by feeding on bases and acids, but was depressed by feeding on a series of salts (Terzian and Stahler, 1960).

The gonadotropic hormone cycle of *A. aegypti* does not affect susceptibility to *P. gallinaceum*, since decapitation, which interrupts the cycle and prevents ovarian development, does not significantly interfere with parasitic development if the decapitated mosquitoes are fed by capillary tube (Rozeboom, 1961).

Weathersby (1963) summarizes a series of studies to determine whether the susceptibility of mosquitoes to malaria is systemic or resides in the stomach wall. Gametocytes introduced into the hemocoel of susceptible mosquitoes, thus bypassing the stomach wall, developed into viable sporozoites. An antagonistic factor, depressing susceptibility to *P. gallinaceum*, was transferred by parabiosis from the hemolymph of *Culex pipiens*, a non-susceptible host, to that of *A. aegypti*, a susceptible one. Thus the stomach wall is not essential in determining susceptibility, and a systemic factor affecting susceptibility has been demonstrated.

The ability of a series of trypanosomatids to develop in an arthropod host other than their natural one was demonstrated by Hanson and McGhee (1963). Crithidias from a series of Diptera and Hemiptera flourished in the gut of the hemipteran *Oncopeltus fasciatus*, but failed to be transmitted, although flagellates and leishmaniform bodies were present in the feces.

Garnham (1964) concluded that the susceptibility of arthropods to protozoa will probably ultimately be defined in physical and chemical terms, but that we have barely reached the stage of defining the problem.

IV. CONCLUSION

Whether a parasite can continue to maintain itself in a given host, once it has established itself in the face of the inimical effects of innate immunity, further depends on its ability to withstand the specific factors associated with acquired immunity. Such factors, considered outside the scope of the present discussion, nevertheless contribute to host specificity. Thus a parasite which has invaded each of two hosts with equivalent innate immune mechanisms might persist in the face of weak acquired immunity in the first host, whereas it might be promptly eliminated by the more vigorous acquired immunity of the second.

REFERENCES

Adler, S. (1958). *Bull. Res. Council Israel* E7, 9–14.
Adler, S., and Gunders, A. E. (1965). *Israel J. Med. Sci.* 1, 441.
Aikawa, M., Huff, C. G., and Sprinz, H. (1966). *In* "Research in Malaria" (E. H. Sadun and H. S. Osborne, eds.), pp. 969–983. Walter Reed Army Inst. Res., Washington, D.C. (*Military Med.* 131, Suppl.).
Allison, A. C. (1963). *In* "Immunity to Protozoa" (P. C. C. Garnham, A. E. Pierce, and I. Roitt, eds.), pp. 109–122. Blackwell, Oxford.
Bastianelli, G. (1959). *Riv. Malariol.* 38, 9–26.
Basu, P. C., Mondal, M. M., and Chowdhury, D. S. (1962). *Indian J. Malariol.* 16, 33–40.
Beckman, H. (1963). *Am. J. Trop. Med. Hyg.* 12, 519–523.
Beckman, H. (1964a). *Proc. Soc. Exptl. Biol. Med.* 115, 559–560.
Beckman, H. (1964b). *Proc. Soc. Exptl. Biol. Med.* 117, 135–137.
Brewer, G. J., and Powell, R. D. (1965). *Proc. Natl. Acad. Sci. U. S.* 54, 741–745.
Bruce-Chwatt, L. J. (1964). *In* "Colloque Internat. sur le *Plasmodium berghei*," pp. 1–12. Inst. Med. Trop. Prince Leopold.
Contacos, P. G., and Coatney, G. R. (1963). *J. Parasitol.* 49, 912–918.
Desowitz, R. S. (1963). *Ann. N. Y. Acad. Sci.* 113, 74–87.
Galliard, H. (1965). *Proc. 2nd Intern. Conf. Protozool., London, 1965* Intern. Congr. Ser. No. 91, pp. 143–144. Excerpta Med. Found., Amsterdam.
Garnham, P. C. C. (1964). *Tech. Parasitol.* 2, 33–50.
Geiman, Q. M. (1958). *Vitamins Hormones* 16, 1–33.
Geiman, Q. M. (1964). *Ann. Rev. Physiol.* 26, 75–108.
Goble, F. C., Konopka, E. A., and Boyd, J. L. (1965). *Proc. 2nd Inter. Conf. Protozool., London, 1965* Intern. Congr. Ser. No. 91, pp. 54–55. Excerpta Med. Found., Amsterdam.
Hanson, W. L., and McGhee, R. B. (1963). *J. Protozool.* 10, 233–238.
Hawking, F. (1953). *Brit. Med. J.* I, 1201–1202.
Hickman, R. L., Gochenour, W. S., Marshall, J. D., and Guilloud, N. B. (1966). *In* "Research in Malaria" (E. H. Sadun and H. S. Osborne, eds.), pp. 935–943. Walter Reed Army Inst. Res., Washington, D.C. (*Military Med.* 131, Suppl.).
Huff, C. G. (1965). *Exptl. Parasitol.* 16, 107–132.
Huff, C. G., Marchbank, D. F., and Shiroishi, T. (1958). *Exptl. Parasitol.* 7, 399–417.
Huff, C. G., Weathersby, A. B., Pipkin, A. C., and Algire, G. H. (1960). *Exptl. Parasitol.* 9, 98–104.

Ingram, V. M. (1959). *Brit. Med. Bull.* **15**, 27–32.

Jacobs, R. L. (1964). *Exptl. Parasitol.* **15**, 213–225.

Kretschmar, W. (1964). *In* "Colloque Internationale sur le *Plasmodium berghei*," pp. 1–14. Inst. Med. Trop. Prince Leopold.

Kretschmar, W., and Jerusalem, C. (1963). *Z. Tropenmed. Parasitol.* **14**, 279–310.

Kuperman, O. (1966). M.S. Thesis, Hebrew University, Jerusalem, Israel.

Lincicome, D. R. (1963). *Ann. N. Y. Acad. Sci.* **113**, 360–380.

Lincicome, D. R., and Hinnant, J. A. (1962). *Exptl. Parasitol.* **12**, 128–133.

Lincicome, D. R., and Shepperson, J. R. (1963). *J. Parasitol.* **49**, 31–34.

Lincicome, D. R., and Shepperson, J. R. (1965). *Exptl. Parasitol.* **17**, 148–167.

Lincicome, D. R., Rossan, R. N., and Jones, W. C. (1963). *Exptl. Parasitol.* **14**, 54–65.

Maegraith, B. G. (1964). *In* "Colloque International sur le *Plasmodium berghei*," pp. 1–11. Inst. Med. Trop. Prince Leopold.

Maegraith, B. G., and Fletcher, K. A. (1965). *Proc. 2nd Intern. Conf. Protozool., London, 1965* Intern. Congr. Ser. No. 91, p. 167. Excerpta Med. Found., Amsterdam.

Maegraith, B. G., Deegan, T., and Jones, E. S. (1952). *Brit. Med. J.* **II**, 1382–1384.

Maegraith, B. G., Riley, M. V., and Deegan, T. (1962). *Ann. Trop. Med. Parasitol.* **56**, 483–491.

Manwell, R. D. (1963). *Ann. N. Y. Acad. Sci.* **113**, 332–342.

McGhee, R. B. (1953). *Ann. N. Y. Acad. Sci.* **56**, 1070–1073.

Motulsky, A. G. (1964). *Am. J. Trop. Med. Hyg.* **13**, Suppl., 147–158.

Moulder, J. W. (1962). "The Biochemistry of Intracellular Parasitism," pp. 1–42. Univ. of Chicago Press, Chicago, Illinois.

Porter, J. A., and Young, M. D. (1966) *In* "Research in Malaria" (E. H. Sadun and H. S. Osborne, eds.), pp. 952–958. Walter Reed Army Inst. Res., Washington D. C. (*Military Med.* **131**, Suppl.).

Ramakrishnan, S. P., Bruce-Chwatt, L. J., Prakash, S., Chowdhury, D. S., Pattanayak, S., Singh, D., and Dhar, S. K. (1964). *Bull. World Health Organ.* **31**, 393–397.

Riley, M. V., and Maegraith, B. G. (1962). *Ann. Trop. Med. Parasitol.* **56**, 473–482.

Ringwald, U., and Kretschmar, W. (1965). *Naturwissenschaften* **12**, 353–354.

Rozeboom, L. E. (1961). *J. Parasitol.* **47**, 597–599.

Rudzinska, M. A., Trager, W., and Bray, R. S. (1965). *J. Protozool.* **12**, 563–576.

Sadun, E. H., Hickman, R. L., Wellde, B. T., Moon, A. P., and Udeozo, I. O. K. (1966). *In* "Research in Malaria" (E. H. Sadun and H. S. Osborne, eds.), pp. 1250–1262. Walter Reed Army Inst. Res., Washington, D. C. (*Military Med.* **131**, Suppl.).

Sherman, I. W. (1966). *J. Parasitol.* **52**, 17–22.

Sherman, I. W., and Hull, R. W. (1960). *J. Parasitol.* **46**, 765–767.

Siddiqui, W. A., and Trager, W. (1964). *J. Parasitol.* **50**, 753–756.

Siddiqui, W. A., and Trager, W. (1965). *Proc. 2nd Intern. Conf. Protozool., London, 1965* Intern. Congr. Ser. No. 91, pp. 171–172. Excerpta Med. Found., Amsterdam.

Singer, I. (1961). *Exptl. Parasitol.* **11**, 391–401.

Taliaferro, W. H. (1949). *In* "Malariology" (M. F. Boyd, ed.), Vol. 2, pp. 935–965. Saunders, Philadelphia, Pennsylvania.

Terzian, L. A., and Stahler, N. (1960). *J. Infect. Diseases* **106**, 45–52.

Tobie, J. E., and Coatney, G. R. (1964). *Am. J. Trop. Med. Hyg.* **13**, 786–789.

Tobie, J. E., Kuvin, S. F., Contacos, P. G., Coatney, G. R., and Evans, C. B. (1963). *J. Am. Med. Assoc.* **184**, 945–947.

Tolbert, M. G., and McGhee, R. B. (1960). *J. Parasitol.* **46**, 552–558.

Trager, W. (1960). *In* "The Cell" (J. Brachet and A. E. Mirsky, eds.), Vol. 4, pp. 151–213. Academic Press, New York.

Trager, W. (1963a). *Proc. 7th Intern. Congr. Trop. Med. Malaria, 1963* Vol. 5, pp. 103–104. Congress Secretariat.

Trager, W. (1963b). *Proc. 1st Intern. Congr. Protozool., Prague, 1961* p. 174. Academic Press, New York.

Trager, W. (1964). *Am. J. Trop. Med. Hyg.* 13, Suppl., 162–166.

Trager, W. (1965). *Proc. 2nd Intern. Conf. Protozool., London, 1965* Intern. Congr. Ser. No. 91, pp. 97–98. Excerpta Med. Found., Amsterdam.

Trejos, A., de Urquilla, M. A., and Paredes, B. R. (1965). *Proc. 2nd Intern. Conf. Protozool., London, 1965* Intern. Congr. Ser. No. 91, p. 144. Excerpta Med. Found., Amsterdam.

Vickerman, K. (1962). *Trans. Roy. Soc. Trop. Med. Hyg.* 56, 487–495.

Ward, R. A. (1963). *Exptl. Parasitol.* 13, 328–341.

Ward, R. A. (1966). *In* "Research in Malaria" (E. H. Sadun and H. S. Osborne, eds.), pp. 923–928. Walter Reed Army Inst. Res., Washington, D. C. (*Military Med.* 131, Suppl.).

Ward, R. A., and Cadigan, F. C. (1966). *In* "Research in Malaria" (E. H. Sadun and H. S. Osborne, eds.), pp. 944–951. Walter Reed Army Inst. Res., Washington, D. C. (*Military Med.* 131, Suppl.).

Warren, M., Skinner, J. G., and Guinn, E. (1965). *J. Parasitol.* 51, Sect. 2, 17.

Weathersby, A. B. (1963). *Proc. 7th Intern. Congr. Trop. Med. Malaria, 1963* Vol. 5, pp. 30–31. Congress Secretariat.

Yaeger, R. G., and Miller, O. N. (1960a). *Exptl. Parasitol.* 9, 215–222.

Yaeger, R. G., and Miller, O. N. (1960b). *Exptl. Parasitol.* 10, 227–231.

Yaeger, R. G., and Miller, O. N. (1960c). *Exptl. Parasitol.* 10, 232–237.

Yaeger, R. G., and Miller, O. N. (1960d). *Exptl. Parasitol.* 10, 238–244.

Yaeger, R. G., and Miller, O. N. (1963a). *Exptl. Parasitol.* 14, 9–14.

Yaeger, R. G., and Miller, O. N. (1963b). *J. Nutr.* 81, 169–174.

Yaeger, R. G., and Miller, O. N. (1963c). *Proc. 7th Intern. Congr. Trop. Med. Malaria, 1963* Vol. 2, p. 235. Congress Secretariat.

Yoeli, M., Upmanis, R. S., Vanderberg, J., and Most, H. (1966). *In* "Research in Malaria" (E. H. Sadun, and H. S. Osborne, eds.), pp. 900–914. Walter Reed Army Inst. Res., Washington, D. C. (*Military Med.* 131, Suppl.).

Zuckerman, A. (1957). *J. Infect. Diseases* 100, 172–206.

Zuckerman, A. (1958). *J. Infect. Diseases* 103, 205–224.

Zuckerman, A. (1966a). Unpublished data.

Zuckerman, A. (1966b). *Ann. N. Y. Acad. Sci.* 139, 24–38.

Zuckerman, A., and Yoeli, M. (1954). *J. Infect. Diseases* 94, 225–236.

3

Abrogation of Immunological Tolerance as a Model for Autoimmunity

DIEGO SEGRE

I. INTRODUCTION

Very early in the history of immunology it became apparent that individual animals do not, as a rule, become immune to substances present in their own body. Many of these substances, however, are capable of inducing an immune response when they are administered to an animal of a different species, or even to a different individual of the same species as that of the animal from which they were derived. The inability of an individual to react immunologically against its own components was recognized by Ehrlich and Morgenroth (1900), who termed this phenomenon "horror autotoxicus." This concept was later reemphasized by Burnet and Fenner (1949) and Burnet (1959), who restated it in what is sometimes referred to as the axiom of Burnet: Any animal must be capable of discriminating between self and not-self.

Acquired immunity is a highly specialized form of an individual's adaptation to the environment. Burnet and Fenner (1949) recognized that the lack of immunity against self must also be a response to environmental stimuli, exerted by the self components under appropriate circumstances.

They specifically predicted that animals exposed to an antigen during embryonic life would fail to respond immunologically to the same antigen during adult life. This prediction was experimentally confirmed by Billingham *et al.* (1953), who found that mice injected *in utero* with living cells from mice of a different inbred strain were later unable to reject skin grafts from the cell-donor strain. This phenomenon was termed "acquired tolerance." A number of other phenomena, discovered both prior and after the description of acquired tolerance, and involving specific suppression or depression of the immune response by exposure to antigen during embryonic (Traub, 1936, 1938, 1939; Owen, 1945; Hašek, 1953), perinatal (Wells and Osborne, 1911; Hraba and Hašek, 1956; Hanan and Oyama, 1954; Dixon and Maurer, 1955a,b) or adult life (Sulzberger, 1929; Felton and Ottinger, 1942; Chase, 1946) were soon related to acquired tolerance. These phenomena covered a wide range of antigens, including living cells (Owen, 1945; Hašek, 1953; Hraba and Hašek, 1956), the virus of lymphocytic choriomeningitis of mice (Traub, 1936, 1938, 1939), defined haptens (Sulzberger, 1929; Chase, 1946), a bacterial polysaccharide (Felton and Ottinger, 1942), and proteins (Wells and Osborne, 1911; Hanan and Oyama, 1954; Dixon and Maurer, 1955a,b). They involved depression of various types of immune responses, including homograft rejection (Owen, 1945; Hašek, 1953; Hraba and Hašek, 1956), delayed (Sulzberger, 1929; Chase, 1946) and immediate (Wells and Osborne, 1911) hypersensitivity, and circulating antibodies (Felton and Ottinger, 1942; Hanan and Oyama, 1954; Dixon and Maurer, 1955a,b). The phenomena were variously designated as immunological paralysis (Felton and Ottinger, 1942), immunological unresponsiveness (Chase, 1946), and protein overloading (Dixon and Maurer, 1955a,b). Today, they are generally regarded as manifestations of the same basic mechanism, and the term "immunological tolerance" is most often used. A more complete coverage of the subject can be found in recent review articles (Chase, 1959; Hašek *et al.*, 1961; Smith, 1961) and proceedings of symposia (Hašek *et al.*, 1963; Bussard, 1963).

Immunological tolerance is often regarded as the experimental equivalent of the naturally occurring immunological unresponsiveness to self-components. Acceptance of this view is important to the present discussion, in which abrogation of immunological tolerance is considered as a model of autoimmunity. Experimental evidence in support of the equivalence of acquired immunological tolerance and natural tolerance to self-components was presented by Nachtigal and Feldman (1964), who found that the antibody response to sulfanilazo conjugates of human serum albumin (HSA) in rabbits tolerant to HSA was qualitatively similar to the antibody response of normal rabbits to sulfanilazo conjugates of rabbit serum albumin.

If the basic premise of the equivalence of acquired and natural tolerance is accepted, it can be hoped that experimental manipulations leading to breakdown of acquired tolerance may shed light on the events resulting in spontaneous autoimmunity. Before examining the experimental procedures which have been used to terminate tolerance, however, it may be well to review some of the theoretical models of immunity. Experimental abrogation of tolerance and its implications will then be related to the theoretical models.

II. THE IMMUNOLOGICAL RECOGNITION SYSTEM

A. RECOGNITION OF THE NOT-SELF

The existence of natural and acquired immunological tolerance implies that a mechanism must operate which enables an animal to differentiate between self and not-self. There are two possible ways in which such a differentiation can be accomplished: Either the animal can recognize its own macromolecules, or else it can recognize all foreign configurations. The first alternative was suggested by Burnet and Fenner (1949), who proposed that all self-components carry a marker identifying them as such. This possibility is unlikely, however, since tolerance, like immunity, is specific for individual antigenic determinants and not for whole molecules (Smith, 1961). As pointed out by Talmage (1959a), self-recognition would require a blocking at the site of antibody production of every part of the autologous molecule, which could not be rejected on the basis of a single marker representing a small portion of the molecule. Injection of autologous or isologous proteins coupled to haptens results in the production of antibodies specific for the hapten, but not of antibodies specific for the protein carrier. In this case, the isologous protein is still recognized as self, while a hapten attached to it induces antibody formation.

The second alternative, that foreign antigens are recognized as not-self, is generally preferred. Recognition of not-self requires the existence of a complete set of receptors capable of recognizing any configuration not present among the self-components. Since it is difficult to conceive of recognition of foreign antigenic determinants by means other than sterospecific combination with such determinants, a property which defines antibody, the set of not-self-recognizing receptors may be equated to a set of antibodies with specificities corresponding to all conceivable antigenic determinants. The number of potential antigens is unknown, but it must be very large, since it must include most naturally occurring macromolecules as well as many synthetic chemicals. The assumption that antibodies specific for all potential antigens preexist antigenic exposure may seem, at first sight, difficult to accept. It was this difficulty, in fact, that led to the abandonment of Ehrlich's early side-chain theory of antibody formation (Ehrlich, 1900) and to the proposal that the antigen, acting as a template, directs the formation of antibody (Breinl and Haurowitz, 1930; Alexander, 1931; Mudd, 1932; Pauling, 1940). However, Talmage (1959b) pointed out that the ability of antisera to distinguish between different antigens need not imply the existence of as many specific antibodies as there are antigens. Talmage (1959b) suggested that 5000 different types of specific antibodies may be sufficient to account for the distinguishing ability of immune sera.

It was suggested above that the immunologically competent animal must possess antibodies capable of combining with all conceivable foreign antigenic determinants, and lack antibodies specific for self-components and for those determinants to which the animal has acquired immunological tolerance. In this statement antibodies are regarded as molecules possessing combining sites specific for the antigenic determinants. No other descriptive characteristic is implied. Such antibodies may, but need not, be immunoglobulins. The combining site may be part of a circulating immunoglobulin molecule, similar or identical to antibodies found in immune sera, or it may be part of a fixed cellular structure, such as the cell membrane, or it may be part of a viruslike structure made up of nucleic acid carrying the genetic information relevant to antibody specificity and coated with protein possessing the combining site (Smithies, 1965). All of these alternatives have been suggested. Some of the models which have been proposed will be considered next. They will be classified according to the proposed location of the set of antibodies, or antibodylike receptors, which make up the recognition mechanism. Some consideration will be given to how the information embodied in the vast range of possible antibody specificities might be acquired. However, a complete account of the literature dealing with the biosynthetic mechanisms which lead to the acquisition of immunological specificity is outside the scope of this discussion.

B. Immunological Recognition at the Cellular Level

Burnet (1957, 1958, 1959) and Lederberg (1959) proposed that the specificity of antibodies is dependent on a unique sequence of amino acids in a portion of the immunoglobulin molecule. The specific primary structure of the antibody molecule would, in turn, be controlled by the characteristic nucleotide sequence of a structural gene, the "gene for globulin." The great diversity of antibody specificities would be generated through a high frequency of mutation of the gene for globulin. As a result, each antibody-forming cell, or its precursor cell, would carry genetic information for manufacturing one specific antibody (or at the most two, since most cells in noninbred animals may be assumed to be heterozygous). The antibody-combining site would be displayed at the surface of the cell carrying the corresponding gene, where it would combine specifically with the relevant antigen. Interaction of the precursor of the antibody-forming cell with the antigen would result in one of two opposite effects: If the cell has reached a sufficient stage of maturation, it will be stimulated by the antigen to undergo a number of divisions, followed by maturation to plasma cell, which produces antibodies; if the cell is immature, it will be killed or suppressed by the antigen. No mature, antibody-forming cells would be formed under these circumstances, and the animal will exhibit immunological tolerance.

In this model, which Burnet called the clonal selection theory, the recognition system is made up by the population of cells genetically capable of responding to foreign antigens, and by the absence or suppression of cells potentially able to respond to self-components and to tolerated antigens. The diversity of genetic information required to account for the variety of antibody specificities is generated by somatic mutation of the precursor cells, followed by elimination of those mutants which happen to carry information relevant to antiself antibodies. Somatic mutation as a means of diversification of genetic information has the advantage of being economical: Each cell devotes only one structural gene or two to each type of immunoglobulin. If each cell were to carry information for all possible specificities, even though only the information relevant to one specificity were expressed in the cell actively synthesizing antibody, each cell would have to devote a large fraction of its deoxyribonucleic acid (DNA) to the immunoglobulins.

The question of whether the antibody-producing cell carries a single "gene for globulin" or many such genes is difficult to approach experimentally because of technical difficulties in cloning antibody-forming cells or their precursors. However, it has been possible to take advantage of a malignant condition of plasma cells (multiple myeloma) occurring in men and mice. This disease apparently results from the clonal proliferation of a single cell. The resulting cell line reproduces itself indefinitely while secreting a specific homogeneous protein. The secreted protein has no known antibody function, but is structurally related to one or another of the various types of immunoglobulins with antibody function. Like antibodies, myeloma proteins are made of subunit polypeptide chains of two types, called H (or A) and L (or B) (Cohen and Porter, 1964). The C-terminal halves of myeloma L-chains appear to be identical within a given class of proteins, whereas the N-terminal halves vary in amino acid sequence from protein to protein (Titani et al., 1965; Potter et al., 1964; Bennet et al., 1965). Hood et al. (1966) found that the N-terminal sequences of several myeloma L-chains of human and mouse origin had remarkable similarities. The differences at any one position in the sequence were restricted to a small number of amino acids. The differences found were of the same type as those existing between homologous macromolecules in different species, such as cytochrome c or hemoglobin. This suggested that the observed changes have arisen by the accumulation of random mutations and selection of mutants over the evolutionary time scale, in a manner similar to that which has led to the species variations in other proteins, rather than through somatic mutation during differentiation of the immune system in each individual (Gray, *et al.*, 1967).

The clonal selection theory requires that the genetic information of a cell be restricted to that necessary to synthesize one specific antibody or at most two. If each cell were multipotent, loss or suppression of a clone potentially able to produce antiself antibodies would entail the loss or suppression of potential to synthesize other antibodies as well. Burnet (1958) recognized that, if it were possible to demonstrate that a single cell, or the progeny of a single cell, can produce antibodies of more than one specificity, the clonal selection theory would have to be abandoned or substantially modified. Evidence that a single cell can produce antibodies of two different specificities was presented by Attardi *et al.* (1959, 1964a,b,c). Single cells from lymph nodes of rabbits immunized with the unrelated bacteriophages T2, T5, and C were isolated in microdrops or in micropipettes. The fluids in which the single cells had been nurtured were then examined for their ability to neutralize the bacteriophages. Single cells neutralizing two bacteriophages were found with both the microdrop and the micropipette methods. The frequency of double producers was relatively low when two different bacteriophages were used as test antigens. However, in an experiment utilizing two antigens of the same bacteriophage which could be assayed independently, Attardi *et al.* (1964c) found that 65 % of the antibody-producing cells appeared to produce two types of antibody. A similar conclusion was reached by Hiramoto and Hamlin (1965), who examined cells from guinea pigs immunized with human immunoglobulin G. Intracellular antibodies were detected by the immunofluorescence technique. Antibodies directed against different portions of the antigen molecule could be distinguished by the use of monospecific antisera conjugated to fluorescein or rhodamine. About 45 % of the antibody-containing cells appeared to contain antibodies directed against the two portions of the antigen molecule. Much lower frequencies of double antibody-producing cells were found by Nossal and Mäkelä (1962) in microdrop experiments in which they tested single rat cells for antibodies against the H and O antigens of *Salmonella*. More recently, evidence for double antibody producers in the spleen of chickens immunized with erythrocytic isoantigens was obtained with the use of the sensitive localized hemolysis in gel technique (McBride and Schierman, 1966).

The results cited above do not seem compatible with the requirement of the clonal selection theory that each cell be capable to produce one, or at most two, antibody specificities. Even though triple producer cells were not found, the probability of testing for just the two antibodies for which a cell possesses genetic information should be negligible on the reasonable assumption that the two allelic genes for globulin are distributed at random among the immunologically competent cells. Additional evidence for the pluripotentiality of clones of immunologically competent cells is available. Trentin and Fahlberg (1963) injected lethally irradiated mice with a small number (4 × 10⁴) of mouse bone marrow cells. They then isolated a discrete colony of proliferating cells from the spleen of one of the irradiated mice 11 days after irradiation. It had been shown previously by Till and McCulloch (1961) that each splenic colony is derived from the clonal proliferation of a single donor cell. The cells of the splenic colony were then passed twice in lethally irradiated mice. The second passage mice were stimulated with four antigenically different *salmonellae* and produced antibodies to all four microorganisms. Since the immunologically competent cells of the latter mice were presumably all derived from a single cell of the original donor, this cell must have had, and transmitted to its progeny, the genetic information needed to produce all four types of antibodies. The possibility of such information having been acquired by somatic mutation during the repopulation of the recipient mice, however unlikely, cannot be ruled out. Similar results were obtained more recently by Feldman and Mekori (1966).

In view of the evidence cited above the original simple formulation of the clonal selection theory must be abandoned, as Burnet (1963) recognized. This does not mean that clonal proliferation is not involved in antibody production, for there is a great deal of evidence to this effect (e.g., Nakano and Braun, 1966; Kennedy *et al.*, 1966; Celada and Wigzell, 1966). But the immunological recognition mechanism may have to be moved from the population of the immunologically competent cells to another location. The immunologically competent cell may still be restricted in its potentialities to the production of less than the total number of antibody specificities. But the decision as to whether a macromolecule is going to induce an immune response or immunological tolerance may be made outside the cell.

C. IMMUNOLOGICAL RECOGNITION AT THE EXTRACELLULAR LEVEL

It was previously stated that the immunological recognition system must consist of a complete set of antibodylike receptors capable of binding all conceivable foreign antigens. If such receptors are not located at the surface of the immunologically competent cells, they may be located either in other cells or in circulation. Evidence has been presented that the antigen is first processed by macrophages, which then pass on the genetic information needed for making antibodies to cells of the lymphoid series (Fishman, 1961; Fishman *et al.*, 1965; Gallily and Feldman, 1967). Even if this were the case, however, cellular recognition of not-self would simply be moved from the immunologically competent cells to the macrophages. The objections to cellular recognition described in the previous section would remain valid.

The other alternative is that recognition takes place in the circulation. In this case, the antibodylike receptors that make up the recognition system may simply be circulating natural antibodies. Jerne (1955, 1960), Talmage and Pearlman (1963), and Eisen and Karush (1964) have all proposed models in which natural antibody plays a role in the regulation of tolerance and immunity. Although the three models differ considerably from one another, they will be considered in this section because they all locate the immunological recognition system at the level of circulating natural antibodies.

The natural selection theory of Jerne (1955, 1960) proposed that there is some degree of freedom in the assembly of a short amino acid sequence of the globulin molecule, corresponding to the combining sites of antibody. Globulin molecules of all possible specificities are spontaneously produced by random assembly of the critical amino acid sequence. Globulin molecules capable of binding self-components are then eliminated after complexing with the self-components, presumably during embryonic life. Thus, only antibodies specific for foreign antigens are left in circulation. Such antibodies constitute the immunological recognition mechanism. When antigen is introduced, it complexes with the corresponding natural antibody. The complex stimulates the immunologically competent cell to divide and to make copies of the antibody in the complex.

Certain portions of the natural selection model are not in line with current biological theory. For example, it is difficult to imagine that a particular amino acid sequence of the antibody molecule could direct its own reproduction by a cell, because this would imply flow of information from protein to nucleic acid to protein. A way out of this difficulty is to imagine that the immunologically competent cell already possesses the genetic information necessary to produce the amino acid sequence of the antibody involved, along with information for other antibody specificities. The function of the antibody would then be to selectively activate or derepress the structural gene corresponding to itself.

The central idea of the natural selection theory is that the antigen by itself is not immunogenic, but the antigen-antibody complex is. Tolerance is explained by the absence, from the pool of circulating natural antibodies, of antibody specific for the tolerated antigen. This last statement is so obviously in agreement with the facts that the idea seems worth retaining, whatever the inadequacies of other portions of the natural selection theory might be.

Eisen and Karush (1964) also assigned to natural antibodies the function of recognizing foreign antigens and forming immunogenic complexes with them. Their model has several attractive features, the principal one being that a specific meaning is given to the property of immunoglobulin G antibodies to bind two identical antigenic determinants (bivalence of antibody).

Eisen and Karush (1964) assumed that uptake of antigen by the immunologically competent cell is a prerequisite of antibody production. The relative concentrations of antigen and of specific antibody in circulation would determine whether antigen uptake will occur. Antigen and bivalent antibody may be free in circulation or may form the following types of complexes: bimolecular complexes consisting of one antigen molecule and one antibody molecule (AgAb); trimolecular complexes, made of two molecules of antigen linked to one molecule of antibody (Ag_2Ab); and multimolecular complexes consisting of one molecule of antigen linked to two or more molecules of antibody ($AgAb_{n+1}$). All these complexes and free antigen and antibody are in reversible equilibrium. The predominant type of complex will depend, other factors being equal, on the ratio of the concentration of antigen to the concentration of antibody. Thus, in antigen excess, the Ag_2Ab complexes will be favored. This type of complex is assumed to be incapable of penetrating the immunologically competent cells and to be degraded and eliminated from the organism. The Ag_2Ab complex, therefore, is not immunogenic, but will induce immunological tolerance, a condition known to result from the administration of large doses of antigen. Likewise, complexes formed in excess antibody ($AgAb_{n+1}$) and free antigen are assumed not to be immunogenic. However, bimolecular complexes (AgAb) are assumed to enter the appropriate cells and initiate the process of antibody formation. The model does not specify what are these processes or which cells take up the AgAb complexes. It is only concerned with the regulatory function of bivalent antibody in the induction of tolerance and immunity. However, the model is equally compatible with instructive (Karush, 1961, 1963) and selective (Burnet, 1959) theories of antibody formation, insofar as both types of theory provide for natural antibodies covering the range of possible antigens and for some kind of interaction between antigen and cell. There are two important corollaries to the model of Eisen and Karush: (1) The tolerant animal does not have in circulation antibodies specific for the tolerated antigen, since such antibodies will immediately form Ag_2Ab complexes with the tolerated antigen and be eliminated; and (2) the tolerant animal must contain cells capable of responding to the tolerated antigen. Such cells have not been eliminated, as in the clonal selection theory (Burnet, 1959; Lederberg, 1959); they have simply not been stimulated by the appropriate AgAb complex.

A comprehensive model of cellular differentiation (Talmage and Claman, 1964) and of antibody response (Talmage and Pearlman, 1963) was proposed, in which the concentration of natural antibody also plays an important role in determining whether immunity or tolerance will follow antigenic exposure. The main feature of Talmage's model is the postulate that the antigen exerts two separate and antagonistic actions, one specific and the other nonspecific. The specificity of the antibody response is thought to depend on a specific stimulus which is the reaction of antigen with preformed antibody at a strategic location within the cell. The cell is assumed to possess genetic information (potential) for several antibody specificities. The specific antigenic stimulus would activate the genetic information for the appropriate antibody. The cell so stimulated would differentiate into a plasma cell, synthesize antibody, and lose the ability to divide. The second, nonspecific action of the antigen is thought to depend on the ability of the antigen to aggregate antibody molecules on its surface. Aggregated antibody would then remove a complementlike intracellular repressor of cell division and cause the cell so stimulated to divide. While this stimulus may be mediated by the specific combination of antigen with antibody, it may also result from aggregation of globulin effected by nonspecific agents, such as bacterial endotoxin or other adjuvants of immunity. When both nonspecific and specific stimuli are applied by the antigen, cells with potential to form antibody specific for that antigen divide repeatedly. Only after the pool of such cells has expanded through clonal division, a portion of the cells matures and synthesizes antibody. The magnitude of the antibody response is a function of the number of cell divisions that precede maturation to plasma cells. If only the specific stimulus is applied, and the nonspecific stimulus is absent, all the cells with potential to produce the antibody corresponding to the stimulating antigen will mature without previous division. A small quantity of antibody will be produced, but the potential for that particular antibody will be lost with the death of the mature cells, which are incapable of further division. Tolerance will have been induced.

Contrary to the models of Jerne (1955, 1960) and of Eisen and Karush (1964), circulating natural antibodies alone do not constitute the immunological recognition system in Talmage's model. A completely tolerant animal should have no cells with antibody potential for the tolerated antigen. However, natural antibodies aggregated on the surface of the antigen provide the nonspecific stimulus necessary for a complete antibody response. Moreover, the specificity of natural antibody does not have to correspond to the antigenic determinant which applies the specific stimulus. Most foreign macromolecules have several different antigenic determinants on their surface. When the macromolecule enters the immature cell and applies the specific stimulus by binding intracellular antibody to one of its determinants, an adjacent but different determinant may have already combined with natural antibody while the macromolecule was in circulation. The two antibodies of different specificities will thus be aggregated on the surface of the macromolecule and the nonspecific stimulus will be applied. In contrast, if the determinant adjacent to the one which applies the specific stimulus happens to be one for which the animal is tolerant, it will find no corresponding antibody in circulation. There will be no aggregation of globulin at the surface of the macromolecule. The cells will be stimulated specifically, but not nonspecifically, and tolerance will result. Factors such as the number and arrangement of the determinants of the antigen and the coincidental occurrence of a nonspecific stimulus, such as bacterial endotoxin, play a role in determining whether tolerance or immunity will follow antigenic exposure. However, in this model, as in the model of Eisen and Karush (1964) discussed above, the relative concentrations of antigen and antibody in circulation, other factors being equal, provide a steering mechanism between the alternate pathways of tolerance and immunity.

The common feature of the three models discussed in this section is the hypothesis that only antigen-antibody complexes are immunogenic. Experimental evidence for this view was provided by the finding of Terres and Wolins (1959, 1961) that antigen-antibody precipitates made in antigen excess were more effective than antigen alone in immunizing mice. Experiments performed in the writer's laboratory also support a role of natural antibody in immunological competence. It was found that baby pigs deprived of colostrum were less able to form antibodies than colostrum-fed pigs of the same age. Baby pigs are known to acquire maternal gamma globulin through the colostrum. Colostrum-deprived baby pigs are hypogammaglobulinemic and presumably deficient in natural antibodies. Immunological competence was restored to colostrum-deprived baby pigs by administration of small quantities of antiserum specific for the antigen used (Segre and Kaeberle, 1962a,b; Segre, 1963). Colostrum-deprived baby pigs born of immune sows were capable of forming antibodies to the antigen against which the sow had been immunized (Myers and Segre, 1963). This phenomenon was ascribed to transplacental transfer of a small quantity of antibody from mother to fetus, with consequent increase of antibody concentration in the baby pigs' serum. Administration of large quantities of normal swine gamma globulin also increased the immunological competence of colostrum-deprived baby pigs (Segre and Myers, 1964). The activity of normal gamma globulin was ascribed to its content of natural antibody. Immunoglobulin G antibodies of sheep origin enhanced antibody formation in colostrum-deprived baby pigs, while immunoglobulin M antibodies of the same origin did not (Locke et al., 1964). Since immunoglobulin G antibodies are bivalent and immunoglobulin M antibodies are pentavalent (Onoue et al., 1965), this finding is in agreement with the model of Eisen and Karush (1964), which assigns special significance to the bivalence of antibody. All these experiments suggest that the relative immunological incompetence of colostrum-deprived baby pigs was caused by the lower than normal concentration of natural antibodies. Dawe et al. (1965) reported experiments suggesting that the enhanced antibody response which follows administration of antigen emulsified in Freund's adjuvant is mediated by increased concentration of natural antibodies. Donor rabbits were injected with Freund's adjuvant, but received no antigen. Serum obtained from the donor rabbits was injected to recipient rabbits. The latter also received antigen, but no adjuvant. The antibody response of the recipients of serum from adjuvant-treated rabbits was significantly greater than that of rabbits which received serum from untreated donors. A correlation was found between the magnitude of the antibody response of the recipients and the increase in concentration of gamma globulin in the serum of the corresponding donor rabbits. In recent work (Kerman et al., 1967) it was found that the dose of pneumococcal polysaccharide type III required to induce immunological paralysis in newborn offspring of immunologically paralyzed mice was one tenth of the corresponding paralyzing dose for newborn offspring of normal mice. Similarly, immunization of the offspring of paralyzed mice was accomplished with one tenth the dose of polysaccharide necessary to immunize normal newborn mice. Since the immunologically paralyzed mice would not have natural antibodies specific for the polysaccharide, they would not transfer such antibodies to their offspring. In contrast, normal newborn mice would acquire maternal natural antibodies specific for the polysaccharide transplacentally and through the colostrum (Brambell, 1958). Thus, the concentration of natural antipolysaccharide antibodies would be expected to be lower in offspring of paralyzed mice than in offspring of normal mice. Correspondingly lower doses of polysaccharide should be required to establish the antigen-antibody ratios leading to paralysis or immunity in the offspring of paralyzed mice. Transplacental transfer of polysaccharide was ruled out as a factor influencing the susceptibility of offspring of paralyzed mice to the induction of paralysis and immunity. The results of this experiment are in agreement with, and were specifically predicted from, the model of Eisen and Karush (1964).

Immunological recognition by natural antibodies at the extracellular level appears to be well founded in experimentation. Furthermore, it does not prejudge the issue of whether the antigen acts as a direct template for antibody (Pauling, 1940; Karush, 1961, 1963; Haurowitz, 1965b), or selects the cell possessing the relevant genetic information from among a population of genetically heterogeneous cells (Burnet, 1959; Lederberg, 1959), or selects the relevant genetic information from among many genes available in the cell (Dreyer and Bennet, 1965). In the latter case, mechanisms can be imagined by which the antigen or the antibody in the complex taken up by the cell may activate the appropriate gene by removing a specific gene repressor.

On the basis of the clonal selection view, however, extracellular antigen recognition by antibody might seem redundant, since cells genetically able to respond to self-components and tolerated antigens are presumed to be absent. However, self-tolerance is so obviously important to the welfare of the individual that separate and independent control mechanisms may well have evolved to preserve it.

III. EXPERIMENTAL ABROGATION OF TOLERANCE

Acquired immunological tolerance can be terminated by a number of experimental procedures. All experimental procedures leading to abrogation of tolerance are of great interest because they may shed some light on the nature of tolerance and on the processes underlying the spontaneous loss of self-tolerance presumably involved in autoimmunity. In many cases, however, the experimental conditions are such that a decision between immunological recognition at the cellular and extracellular levels cannot be made. Moreover, some of the means leading to breakdown of tolerance are highly artificial and have no relevance to the spontaneous conditions which may induce autoimmunity. Some of the means by which experimental abrogation of tolerance has been obtained will be considered next.

A. DEPLETION OF ANTIGEN

Much evidence indicates that persistence of the antigen is necessary to sustain tolerance (Smith, 1961). This is readily apparent in the case of tolerance to nonliving antigens, which is of finite duration unless antigen is periodically reinjected. An apparent exception to this rule is the long-lasting immunological paralysis induced by pneumococcal polysaccharides in mice (Felton and Ottinger, 1942). However, pneumococcal polysaccharides are not readily metabolized and can be demonstrated in the tissues of paralyzed animals as long as 1 year after administration (Kaplan *et al.*, 1950; Felton *et al.*, 1955; Stark, 1955).

In the case of tolerance to homografts it is generally assumed that the graft provides a continuous source of transplantation antigens to the host (Smith, 1961). The need for persisting antigen in tolerance to living cells is perhaps best demonstrated by the elegant experiment of Triplett (1962). The experiment consisted of removing the buccal component of the pituitary gland of embryos of the tree frog, *Hyla regilla*, at a stage of development preceding that in which the hypophysis begins to produce adult-type proteins. The extirpated hypophyses were then implanted in larvae of the same species, where they continued to develop. Each of the glands was then reimplanted in its original donor, which, by this time, had reached the metamorphic stage. In a significant percentage of cases the animals rejected their own hypophyses. Control experiments excluded the possibility that the antigenicity of the graft had changed or that the animals had lost the ability to accept autografts.

If the tolerated antigen is allowed to decay below a critical quantity, tolerance terminates. Smith (1960) and Humphrey (1960) calculated that from 10^{10} to 10^{12} molecules of bovine serum albumin remain at the end of the tolerant state in rabbits rendered tolerant to that antigen at birth. It is interesting that spontaneous immunity often follows the termination of the tolerant state, even if additional antigen is not administered (Coons, 1963; Thorbecke *et al.*, 1961; Terres and Hughes, 1959; Kaeberle and Segre, 1964).

Spontaneous immunity following decay of tolerance appears to be consistent with those theories which locate the immunological recognition mechanism at the level of circulating natural antibodies. Indeed, the model of Eisen and Karush (1964) requires that, when the antigen concentration becomes low enough, the formation of inductive, bimolecular antigen-antibody complexes should be favored. The model of Talmage and Pearlman (1963) also explains spontaneous immunity after decay of tolerance, since aggregation of antibody at the surface of the antigen would not be expected to occur in antigen excess, but may take place after enough antigen has been eliminated to substantially reduce the ratio of the concentration of antigen to that of antibody.

The clonal selection theory (Burnet, 1959; Lederberg, 1959) also provides for the need of persisting antigen in the maintenance of tolerance. After the antigen has been eliminated, new mutant cells carrying the relevant genetic information would be free to mature beyond the point where they would be killed by the antigen. The animal should then be capable of becoming immune upon injection of the antigen. However, it is difficult to see why such an animal should become immune *without* further administration of antigen. This difficulty is removed if one assumes that the antigen only imposes a temporary block on the maturation of a precursor cell in the tolerant animal, and that the block is released and maturation proceeds after the intracellular concentration of antigen declines below a critical level (Humphrey, 1964a,b). The further assumption must be made that the cell released from the block responds to the residual intracellular antigen or to antigen still present in circulation.

The relevance of abrogation of tolerance by depletion of antigen to autoimmune states is difficult to assess. The concentration of self-components is probably maintained at a fairly constant level by homeostatic controls. Such controls, however, might fail in pathological states. It is conceivable that a decrease in the normal rate of synthesis may cause the concentration of a tolerated self-component to decline below a critical level (Eisen and Karush, 1964).

B. ADJUVANTS OF IMMUNITY AND IONIZING RADIATION

Most of the experimentally induced states of autoimmunity, such as experimental allergic encephalomyelitis, uveitis, neuritis, thyroiditis, etc., require the use of Freund's adjuvant (Waksman, 1959). It was logical, therefore, to attempt to induce a breakdown of acquired tolerance by injections of antigens emulsified in Freund's adjuvant. When tolerance is solidly established, this is not generally accomplished. However, Neeper and Seastone (1963b) reported that immunological paralysis to pneumococcal polysaccharide type I in mice was abrogated by intraperitoneal injections of the polysaccharide emulsified in Freund's adjuvant, or even by Freund's adjuvant alone. Brooke (1966), however, was unable to confirm these results in mice tolerant to pneumococcal polysaccharide type III. Brooke pointed out that immunological paralysis with type I polysaccharide is not as solid as with type III polysaccharide. Paraf *et al.* (1963a) reported failure to induce tolerance to human serum albumin in rabbits injected with Freund's adjuvant at birth and with the antigen 5 days later. Paraf *et al.* (1963b) shortened the duration of tolerance to human serum albumin in adult mice by injecting Freund's adjuvant 10 days after induction of tolerance.

The mechanism by which Freund's adjuvant terminates or shortens the duration of tolerance, in the few cases when this phenomenon has been reported, is not well understood. It has been suggested that the adjuvant may induce the maturation of immunologically competent cells, which would then be stimulated by the antigen to produce antibodies (Paraf *et al.*, 1963a). The passive transfer of Freund's adjuvant action reported by Dawe *et al.* (1965) indicates that the adjuvant may act, in part, by increasing the concentration of natural antibodies. If this view is accepted, termination of tolerance by Freund's adjuvant may be explained by a decrease in the ratio of the concentration of antigen to that of antibody, with formation of inductive complexes (Eisen and Karush, 1964).

Termination of partial tolerance to heterologous erythrocytes and to bovine serum albumin in rats has been accomplished by whole-body X irradiation (Mäkelä and Nossal, 1962). It was suggested that this effect was due to the appearance of mutant cells genetically capable of responding to the tolerated antigen during the rapid compensatory proliferation following radiation-induced cell death. In addition, intracellular antigen needed to maintain tolerance may have been diluted by cell death and cell proliferation (Mäkelä and Nossal, 1962). A third possible explanation, not suggested by Mäkelä and Nossal, may be that radiation-induced cell destruction released intracellular natural antibodies, thus decreasing the excess of antigen and allowing the formation of inductive complexes.

Mitchison (1963) cited an experiment of Dresser in which cells from mice tolerant to bovine gamma globulin were transplanted into lethally irradiated isogeneic mice. The mice were then challenged with bovine gamma globulin and proved to be nontolerant. In this case the cells from the tolerant animal were not subjected to irradiation. Dresser suggested that termination of tolerance may have been the result of depletion of antigen in the donor cells, which had divided repeatedly in the host.

Other workers have been unable to break complete tolerance by whole-body irradiation (Denhardt and Owen, 1960; Weigle, 1964a; Brooke, 1966). Weigle (1964a) suggested that successful termination of partial tolerance by ionizing radiation may reflect the presence of a few immune cells in the partially tolerant hosts. Since immune or primed cells are more radiation-resistant than nonimmune cells (Taliaferro, 1957), they would be selected out for repopulation of the host after the nonimmune cells have been destroyed by irradiation.

The few examples of abrogation of tolerance by Freund's adjuvant or by whole-body irradiation do not clarify the problem of the nature of tolerance. The results can be rationalized in terms of either cellular or extracellular antigen recognition. In any case, the explanations suggested are quite speculative.

C. Adoptive and Passive Immunization

Transplantation of immunologically competent cells from an immune donor to a compatible recipient represents adoptive transfer of immunity (Mitchison, 1955). If the cells are from a normal, nonimmune donor, the phenomenon may be termed adoptive transfer of immune potential (Hašek,

1963). In either case, the procedure may result in the rejection of a graft previously tolerated by the adoptively immunized host (Mitchison, 1953; Billingham *et al.*, 1954). The effect may be very dramatic if the recipient is a heavily irradiated animal which has accepted and is subsisting on allogeneic bone marrow. The marrow is rejected and the animal dies in consequence (Trentin, 1958). Termination of tolerance to nonliving antigens has also been obtained with adoptive immunity. For example, immunological paralysis to pneumococcal polysaccharides was abrogated by lymphoid cells from immune mice (Brooke and Karnovsky, 1961; Neeper and Seastone, 1963a).

Termination of tolerance by adoptive immunity or by adoptive transfer of immune potential can be related initially to the immune reaction of the transferred cells against the tolerated antigen. Once the concentration of antigen has been reduced sufficiently, the host cells may also become immune (Hašek, 1963). Abrogation of tolerance by adoptive immunization is certainly compatible with the clonal selection view, which regards the tolerant animal as lacking cells genetically capable of responding to the tolerated antigen. Introduction of cells with such genetic capability would be expected to remove this deficiency. However, since the transferred cells are producing antibodies (immune or natural) against the tolerated antigen, they may act by raising the concentration of such antibodies so as to favor the formation of inductive complexes (Eisen and Karush, 1964). In these experiments, therefore, there is no basis for a decision between antigen recognition at the cellular and extracellular levels.

In contrast, abrogation of tolerance by passive immunization would be more readily explained by models which assign an antigen recognition role to antibodies, since transfer of cells is not involved in this procedure. Mitchison (1962) was successful in terminating tolerance to turkey erythrocytes in chickens by passive administration of immune serum. Hašek and Puza (1963) reported similar findings in ducks tolerant to erythrocytes of ducks of a different genus. In contrast, Weigle was unable to terminate tolerance to bovine serum albumin in rabbits by passive administration of antibody (1964a) or by injection of antigen-antibody precipitates formed at equivalence (1962). Friedman's (1965) attempts to terminate tolerance to *Shigella* soluble antigens in mice by passive immunization resulted only in a temporary suppression of the tolerant state. The mice reverted to the tolerant state as soon as the passively administered antibody was catabolized.

A phenomenon which may possibly be related to termination of tolerance by passive immunization has been described by Dray (1962), who reported suppression of the production of gamma globulin of the A5 allotype in A4/A5 heterozygous rabbits born of does homozygous for the A4 allotype and preimmunized against A5. The suppression of production of the A5 allotype lasted for 24 weeks, whereas passively acquired maternal anti-A5 antibody had disappeared by the eighth week of age. Although this possibility was not considered by Dray (1962), it seems conceivable that the maternally derived anti-A5 antibody prevented the establishment of A5 tolerance in the offspring. The latter may then have become immune to their own A5 allotype and rejected the cells producing it.

The successful abrogation of tolerance to erythrocytes by passive immunization (Mitchison, 1962; Hašek and Puza, 1963) and the failure to affect tolerance to soluble antigens by similar means (Weigle, 1964a; Friedman, 1965) may reflect a fundamental difference between the nature of tolerance to particulate antigens and that to soluble antigens. Possibly, intracellular erythrocytic antigens are rapidly degraded, and maintenance of tolerance to erythrocytes may depend exclusively on the persistence of circulating antigen. Soluble antigens may form intracellular depots not accessible to passively injected antibodies. However, the failure to terminate tolerance to soluble antigens by passive immunization may be related to failure to establish *in vivo* the quantitative relations between antigen and antibody most appropriate for the formation of inductive complexes. With this possibility in mind, Hemphill *et al.* (1966) attempted abrogation of tolerance by exposing spleen cells from tolerant mice to antigen-antibody complexes *in vitro*. The treated spleen cells were then injected to the mice from which they had been derived. Upon challenge with the antigen (human serum albumin in incomplete Freund's adjuvant) a significant number of the treated mice became immune. Mice whose spleen cells had been exposed *in vitro* to antigen only or to saline remained tolerant. These results indicated that the spleen of tolerant mice contained cells capable of responding to the tolerated antigen, provided that the antigen was presented in the appropriate complex with antibody. The results are consistent with the model of Eisen and Karush (1964) and indeed were predicted by the model.

D. Stimulation with Altered and Cross-Reacting Antigens

Occasional production of antibodies to the tolerated antigen has been observed in tolerant rabbits injected with an antigen related to that used to induce tolerance (Cinader and Dubert, 1955; Curtain, 1959). The most extensive experiments of this type, however, were described by Weigle (1961, 1962, 1963, 1964b, 1965a). He found that a high proportion of rabbits tolerant to bovine serum albumin lost their tolerance upon injection of human or horse serum albumin (Weigle, 1961). Serum albumins of porcine and ovine origin were less effective. The albumins with the least degree of cross-reaction with the tolerated antigen were more effective in breaking tolerance. Similar results were obtained with bovine serum albumin altered by conjugation with haptens (Weigle, 1962). Again, the greater the extent of alteration, the greater was the probability of breaking tolerance. Although the injection of soluble cross-reacting antigens resulted in loss of tolerance, the same antigens incorporated in complete Freund's adjuvant were more effective (Weigle, 1964b). Simultaneous injection of the tolerated antigen and the cross-reacting antigen interfered with the termination of tolerance (Weigle, 1964b; Nachtigal *et al.*, 1965).

The conclusions of Weigle have been amply confirmed (Nachtigal and Feldman, 1964; Nachtigal *et al.*, 1965; Yoshimura and Cinader, 1966) and extended to other systems (Schechter *et al.*, 1964; Weigle, 1965b; Nisonoff *et al.*, 1967). The finding of Azar (1966) that mice tolerant to soluble human gamma globulin were immunized by heat-aggregated human gamma globulin may perhaps be related to alterations of the antigen induced by heating.

Abrogation of tolerance by stimulation with cross-reacting antigens can be explained on the basis of models which assume that natural antibodies constitute the immunological recognition system. Eisen and Karush (1964) point out that serologic cross-reactions may occur when two antigens share one or more antigenic determinants as well as possessing distinctive determinants, or when two antigens possess determinants that are structurally similar, but not identical. In the first case, the cross-reacting antigen molecule would form inductive complexes with natural antibodies directed against the distinctive determinants and gain entrance into the immunologically competent cells. Since this model postulates that tolerant animals have cells capable of responding to the tolerated antigenic determinant, antibody to the tolerated antigen should arise. In the second case, the determinant structurally similar to, but not identical with, the tolerated determinant, should find enough natural antibody of adequate affinity for itself, but of low enough affinity for the tolerated determinant to have escaped elimination through formation of Ag_2Ab complexes with the tolerated antigen. Antibodies would then be formed against the cross-reacting determinant. When such antibodies have reached a sufficient concentration, they may interact with the tolerated determinant and loss of tolerance may result.

The model of Talmage and Pearlman (1963) can also explain abrogation of tolerance by stimulation with cross-reacting antigens. While the tolerated antigen should be unable to find enough antibody to its determinants to aggregate on its surface, an antigen possessing some determinants identical as well as some different from those of the tolerated antigen may accomplish aggregation of the antibodies directed against the dissimilar determinants, thus producing the nonspecific stimulus necessary for replication of the cells with potential for synthesizing antibody against the tolerated determinant. In this model, however, abrogation of tolerance by cross-reacting antigens should be a rare event, since it would depend on the presence of sufficient numbers of newly arisen mutant cells with potential for antibody synthesis.

The relevance of abrogation of tolerance by cross-reacting antigens for spontaneous autoimmune conditions lies in the possibility that self-components may become sufficiently altered by pathological processes to acquire the properties of cross-reacting antigens. This possibility is made less likely by the finding (Weigle, 1964b; Nachtigal *et al.*, 1965) that the presence of tolerated antigen interfered with the ability of the cross-reacting antigen to bring about termination of tolerance.

IV. CONCLUSIONS

The foregoing considerations have been directed to the problem of the nature and the location of the immunological recognition system. Only two alternatives have been discussed: (1) The immunological recognition system is made of cells genetically capable of responding to foreign antigens, whereas cells capable of responding to self-components are absent; (2) the immunological recognition system is made of natural antibodies capable of combining with foreign antigens, whereas antibodies specific for self-components are absent. Much evidence seems to support the second view. However, other alternatives have not been considered. Perhaps each immunologically competent cell has a site for tolerance and a site for immunity; or there may be a target cell for tolerance and a different target cell for immunity (Mitchison, 1966, 1967); or perhaps a recognition system is not needed at all: Autoimmunity is the rule and tolerance only represents a persistence of the lag phase of the immune response for as long as antigen persists in circulation (Haurowitz, 1965a,b). Although some of these views may be valid, they are difficult to discuss at the present time because experiments to substantiate or deny them have not been devised.

Meanwhile, we do have to contend with autoimmune conditions. Autoimmune processes in blood diseases and their pathogenetic role are discussed in Chapter 4 of this book. No attempt will be made to review the subject here. It is clear, however, that certain erythrocytic infections result in the production of antibodies reacting with both infected and uninfected erythrocytes of the infected animals and of other individuals of the same species. Of the four headings under which experimental abrogation of tolerance has been discussed in the preceding pages, two do not seem applicable to this situation. Adoptive or passive immunization and ionizing radiation can hardly be invoked as possible causes of autoimmunity in blood infections.

An adjuvantlike effect of the causative agent is possible, but there is no evidence for it. Furthermore, adjuvants of immunity are not highly effective in abrogation of immunological tolerance.

Either of the two other mechanisms discussed above, depletion of antigen and stimulation with cross-reacting antigens, may be relevant to autoimmunity in blood infections. Depletion of antigen need not be systemic in order to result in abrogation of self-tolerance. Conceivably, a local fall in antigen concentration at a site of antibody production could result in a local immune response. Because of the autocatalytic nature of antibody production envisioned by those models which view immunization as being dependent on antibody concentration (Jerne, 1955, 1960; Talmage and Pearlman, 1963; Eisen and Karush, 1964), a localized antibody response could trigger systemic immunization. In the case of blood infections, destruction of parasitized erythrocytes might conceivably lead to a localized decrease in concentration of erythrocytic antigens followed by autoimmunity.

Alternatively, infection of the erythrocytes may result in sufficient antigenic alteration to make them behave as cross-reacting antigens. Self-tolerance to unaltered erythrocytes might be abrogated under these conditions. The finding of Weigle (1964b) and Nachtigal et al. (1965) that the presence of native tolerated antigen interferes with the ability of the cross-reacting antigen to bring about termination of tolerance may argue against the likelihood of this mechanism in the pathogenesis of autoimmunity in blood diseases. However, Nisonoff et al. (1967) were able to induce production of antibodies against rabbit cytochrome c in rabbits injected with cytochromes c prepared from other species, or with rabbit cytochrome c conjugated to acetylated bovine gamma globulin.

The availability of genetically homogeneous strains of mice may afford an opportunity to test the hypothesis that autoimmunity in blood diseases results from alterations of the antigens of infected erythrocytes. If this hypothesis were correct, it should be possible to immunize uninfected mice with parasite-free preparations of erythrocytes, or erythrocytic stromata, from infected isogenic mice. Such treatment should result in the production of antibodies directed against the erythrocytes of the recipient mice.

The experimental situations in which immunological tolerance has been artificially terminated may be of value in understanding the pathogenesis of autoimmune diseases. Conversely, experimentation with autoimmune conditions may help evaluate current theories of immunity and tolerance. These are the dividends we have come to expect from the exchange of ideas among workers in different fields. Autoimmunity in blood infections may represent a very fruitful area of investigation in this respect.

REFERENCES

Alexander, J. (1931). *Protoplasma* **14,** 296.

Attardi, G., Cohn, M., Horibata, K., and Lennox, E. S. (1959). *Bacteriol. Rev.* **23,** 213.

Attardi, G., Cohn, M., Horibata, K., and Lennox, E. S. (1964a). *J. Immunol.* **92,** 335.

Attardi, G., Cohn, M., Horibata, K., and Lennox, E. S. (1964b). *J. Immunol.* **92,** 346.

Attardi, G., Cohn, M., Horibata, K., and Lennox, E. S. (1964c). *J. Immunol.* **93,** 94.

Azar, M. M. (1966). *Proc. Soc. Exptl. Biol. Med.* **123,** 571.

Bennet, J. C., Hood, L., Dreyer, W. J., and Potter, M. (1965). *J. Mol. Biol.* **12,** 81.

Billingham, R. E., Brent, L., and Medewar, P. B. (1953). *Nature* **172,** 603.

Billingham, R. E., Brent, L., and Medewar, P. B. (1954). *Proc. Roy. Soc.* **B143,** 58.

Brambell, F. W. R. (1958). *Biol. Rev.* **33,** 488.

Breinl, F., and Haurowitz, F. (1930). *Z. Physiol. Chem.* **192,** 45.

Brooke, M. S. (1966). *J. Immunol.* **96,** 364.

Brooke, M. S., and Karnovsky, M. J. (1961). *J. Immunol.* **87,** 205.

Burnet, F. M. (1957). *Australian J. Sci.* **20,** 66.

Burnet, F. M. (1958). *In* "Immunity and Virus Infection" (V. A. Najjar, ed.), pp. 1–17. Wiley, New York.

Burnet, F. M. (1959). "The Clonal Selection Theory of Acquired Immunity." Vanderbilt Univ. Press, Nashville, Tennessee.

Burnet, F. M. (1963). *In* "Conceptual Advances in Immunology and Oncology" (R. W. Cumley *et al.*, eds.), pp. 7–19. Harper (Hoeber), New York.

Burnet, F. M., and Fenner, F. (1949). "The Production of Antibodies." Macmillan, Melbourne, Australia.

Bussard, A., ed. (1963). "Tolérance acquise et tolérance naturelle à l'égard de substances antigéniques definies." C. N. R. S., Paris.

Celada, F., and Wigzell, H. (1966). *Immunology* **11,** 453.

Chase, M. W. (1946). *Proc. Soc. Exptl. Biol. Med.* **61,** 257.

Chase, M. W. (1959). *Ann. Rev. Microbiol.* **13,** 349.

Cinader, B., and Dubert, J. M. (1955). *Brit. J. Exptl. Pathol.* **36,** 515.

Cohen, S., and Porter, R. R. (1964). *Advanc. Immunol.* **4,** 287.

Coons, A. H. (1963). *In* "Tolérance acquise et tolérance naturelle à l'égard de substances antigéniques définies" (A. Bussard, ed.), pp. 121–132. C. N. R. S., Paris.

Curtain, C. C. (1959). *Brit. J. Exptl. Pathol.* **40,** 255.

Dawe, D. L., Segre, D., and Myers, W. L. (1965). *Science* **148,** 1345.

Denhardt, D. T., and Owen, R. D. (1960). *Transplant. Bull.* **7,** 394.

Dixon, F. J., and Maurer, P. H. (1955a). *J. Exptl. Med.* **101,** 233.

Dixon, F. J., and Maurer, P. H. (1955b). *J. Exptl. Med.* **101,** 245.

Dray, S. (1962). *Nature* **195,** 677.

Dreyer, W. J., and Bennet, J. C. (1965). *Proc. Natl. Acad. Sci. U.S.* **54,** 864.

Ehrlich, P. (1900). *Proc. Roy. Soc.* **B66,** 424.

Ehrlich, P., and Morgenroth, J. (1900). *Berlin. Klin. Wochschr.* **21,** 453.

Eisen, H. N., and Karush, F. (1964). *Nature* **202,** 677.

Feldman, M., and Mekori, T. (1966). *Immunology* **10,** 149.

Felton, L. D., and Ottinger, B. (1942). *J. Bacteriol.* **43,** 94.

Felton, L. D., Prescott, B., Kauffmann, G., and Ottinger, B. (1955.) *J. Immunol.* **74,** 205.

Fishman, M. (1961). *J. Exptl. Med.* **114,** 837.

Fishman, M., van Rood, J. J., and Adler, F. L. (1965). *In* "Molecular and Cellular Basis of Antibody Formation" (J. Šterzl *et al.*, eds.), pp. 491–498. Academic Press, New York.

Friedman, H. J. (1965). *J. Immunol.* **94,** 921
Gallily, R., and Feldman, M. (1967). *Immunology* **12,** 197.
Gray, W. R., Dreyer, W. J., and Hood, L. (1967). *Science* **155,** 465.
Hanan, R., and Oyama, J. (1954). *J. Immunol.* **73,** 49.
Hašek, M. (1953). *Cesk. Biol.* **2,** 265.
Hašek, M. (1963). *In* "Tolérance acquise et tolérance naturelle à l'égard de substances antigéniques définies" (A. Bussard, ed.), pp. 217–227. C. N. R. S., Paris.
Hašek, M., and Puza, A. (1963). *In* "Mechanisms of Immunological Tolerance" (M. Hašek, A. Lengerová, and M. Vojtíšková, eds.), pp. 257–265. Academic Press, New York.
Hašek, M., Lengerová, A., and Hraba, T. (1961). *Advan. Immunol.* **1,** 1.
Hašek, M., Lengerová, A., and Vojtíšková, M., eds. (1963). *In* "Mechanisms of Immunological Tolerance." Academic Press, New York.
Haurowitz, F. (1965a). *Ann. N. Y. Acad. Sci.* **124,** 50.
Haurowitz, F. (1965b). *Nature* **205,** 847.
Hemphill, F. E., Segre, D., and Myers, W. L. (1966). *Proc. Soc. Exptl. Biol. Med.* **123,** 265.
Hiramoto, R. N., and Hamlin, M. (1965). *J. Immunol.* **95,** 214.
Hood, L. E., Gray, W. R., and Dreyer, W. J. (1966). *Proc. Natl. Acad. Sci. U. S.* **55,** 826.
Hraba, T., and Hašek, M. (1956). *Cesk. Biol.* **5,** 89.
Humphrey, J. H. (1960). *In* "Mechanisms of Antibody Formation" (M. Holub and L. Jarošková, eds.), p. 353. Academic Press, New York.
Humphrey, J. H. (1964a). *Immunology* **7,** 449.
Humphrey, J. H. (1964b). *Immunology* **7,** 462.
Jerne, N. K. (1955). *Proc. Natl. Acad. Sci. U. S.* **41,** 849.
Jerne, N. K. (1960). *Ann. Rev. Microbiol.* **14,** 341.
Kaeberle, M. L., and Segre, D. (1964). *Am. J. Vet. Res.* **25,** 1103.
Kaplan, M. H., Coons, A. H., and Deane, H. W. (1950). *J. Exptl. Med.* **91,** 15.
Karush, F. (1961). *In* "Immunochemical Approaches to Problems in Microbiology" (M. Heidelberger and O. J. Plescia, eds.), pp. 368–376. Rutgers Univ. Press, New Brunswick, New Jersey
Karush, F. (1963). *In* "Tolérance acquise et tolérance naturelle à l'égard de substances antigéniques définies" (A. Bussard, ed.), pp. 451–469. C. N. R. S., Paris.
Kennedy, J. C., Till, J. E., Siminovitch, L., and McCulloch, E. A. (1966). *J. Immunol.* **96,** 973.
Kerman, R., Segre, D., and Myers, W. L. (1967). *Science* **156,** 1514.
Lederberg, J. (1959). *Science* **129,** 1649.
Locke, R. F., Segre, D., and Myers, W. L. (1964). *J. Immunol.* **93,** 576.
McBride, R. A., and Schierman, L. W. (1966). *Science* **154,** 655.
Mäkelä, O., and Nossal, G. J. V. (1962). *J. Immunol.* **88,** 613.
Mitchison, N. A. (1953). *Nature* **171,** 267.
Mitchison, N. A. (1955). *J. Exptl. Med.* **102,** 157.
Mitchison, N. A. (1962). *Immunology* **5,** 359.
Mitchison, N. A. (1963). *In* "Tolérance acquise et tolérance naturelle à l'égard de substances antigéniques définies" (A. Bussard, ed.), pp. 199–211. C. N. R. S., Paris.
Mitchison, N. A. (1966). *Progr. Biophys. Mol. Biol.* **16,** 3.
Mitchison, N. A. (1967). *In* "Regulation of the Antibody Response" (B. Cinader, ed.) Charles C Thomas, Springfield, Illinois (in press).

Mudd, S. (1932). *J. Immunol.* **23,** 292.
Myers, W. L., and Segre, D. (1963). *J. Immunol.* **91,** 697.
Nachtigal, D., and Feldman, M. (1964). *Immunology* **7,** 616.
Nachtigal, D., Eschel-Zussman, R., and Feldman, M. (1965). *Immunology* **9,** 543.
Nakano, M., and Braun, W. (1966). *Science* **151,** 338.
Neeper, C. A., and Seastone, C. V. (1963a). *J. Immunol.* **91,** 374.
Neeper, C. A., and Seastone, C. V. (1963b). *J. Immunol.* **91,** 378.
Nisonoff, A., Margoliash, E., and Reichlin, M. (1967). *Science* **155,** 1273.
Nossal, G. J. V., and Mäkelä, O. (1962). *J. Immunol.* **88,** 604.
Onoue, K., Yagi, Y., Grossberg, A. L., and Pressman, D. (1965). *Immunochemistry* **2,** 401.
Owen, R. D. (1945). *Science* **102,** 400.
Paraf, A., Fougereau, M., and Bussard, A. (1963a). *In* "Tolérance acquise et tolérance naturelle à l'égard de substances antigéniques définies" (A. Bussard, ed.), pp. 97–112. C. N. R. S., Paris.
Paraf, A., Fougereau, M., and Metzger, J. J. (1963b). *Compt. Rend.* **257,** 312.
Pauling, L. (1940). *J. Am. Chem. Soc.* **62,** 2643.
Potter, M., Dreyer, W. J., Kuff, E. L., and McIntire, K. R. (1964). *J. Mol. Biol.* **8,** 814.
Schechter, I., Bauminger, S., Sela, M., Nachtigal, D., and Feldman, M. (1964). *Immunochemistry* **1,** 249.
Segre, D. (1963). *In* "Tolérance acquise et tolérance naturelle à l'égard de substances antigéniques definies" (A. Bussard, ed.), pp. 417–427. C. N. R. S., Paris.
Segre, D., and Kaeberle, M. L. (1962a). *J. Immunol.* **89,** 782.
Segre, D., and Kaeberle, M. L. (1962b). *J. Immunol.* **89,** 790
Segre, D., and Myers, W. L. (1964). *Am. J. Vet. Res.* **25,** 413.
Smith, R. T. (1960). *In* "Mechanisms of Antibody Formation" (M. Holub and L. Jarošková, eds.), pp. 313–328. Academic Press, New York.
Smith, R. T. (1961). *Advanc. Immunol.* **1,** 67.
Smithies, O. (1965). *Science* **149,** 151.
Stark, O. K. (1955). *J. Immunol.* **74,** 130.
Sulzberger, M. B. (1929). *Arch. Dermatol. Syphilol.* **20,** 669.
Taliaferro, W. H. (1957). *Ann. N. Y. Acad. Sci.* **69,** 745.
Talmage, D. W. (1959a). *In* "A Symposium on Molecular Biology" (R. E. Zirkle, ed.), pp. 91–101. Univ. of Chicago Press, Chicago, Illinois.
Talmage, D. W. (1959b). *Science* **129,** 1643.
Talmage, D. W., and Claman, H. N. (1964). *In* "The Thymus in Immunobiology" (R. A. Good and A. E. Gabrielsen, eds.), pp. 49–67. Harper (Hoeber), New York.
Talmage, D. W., and Pearlman, D. S. (1963). *J. Theoret. Biol.* **5,** 321.
Terres, G., and Hughes, W. L. (1959). *J. Immunol.* **83,** 459.
Terres, G., and Wolins, W. (1959). *Proc. Soc. Exptl. Biol. Med.* **102,** 632.
Terres, G., and Wolins, W. (1961). *J. Immunol.* **86,** 361.
Thorbecke, G. J., Siskind, G. W., and Goldberger, N. (1961). *J. Immunol.* **87,** 147.
Till, J. E., and McCulloch, E. A. (1961). *Radiation Res.* **14,** 213.
Titani, K., Whitley, E., Avogadro, L., and Putnam, E. W. (1965). *Science* **149,** 1090.
Traub, E. (1936). *J. Exptl. Med.* **64,** 183.
Traub, E. (1938). *J. Exptl. Med.* **68,** 229.
Traub, E. (1939). *J. Exptl. Med.* **69,** 801.
Trentin, J. J. (1958). *Ann. N. Y. Acad. Sci.* **73,** 799.
Trentin, J. J., and Fahlberg, W. J. (1963). *In* "Conceptual Advances in Imunology and Oncology" (R. W. Cumley *et al.*, eds.), pp. 66–72. Harper (Hoeber), New York.

Triplett, E. L. (1962). *J. Immunol.* **89,** 505.

Waksman, B. H. (1959). *In* "Mechanisms of Hypersensitivity" (J. H. Shaffer, G. A. LoGrippo, and M. W. Chase, eds.), pp. 679–696. Little, Brown, Boston, Massachusetts.

Weigle, W. O. (1961). *J. Exptl. Med.* **114,** 111.

Weigle, W. O. (1962). *J. Exptl. Med.* **116,** 913.

Weigle, W. O. (1963). *In* "Tolérance acquise et tolérance naturelle à l'égard de substances antigéniques définies" (A. Bussard, ed.), pp. 233–245. C.N.R.S., Paris.

Weigle, W. O. (1964a). *J. Immunol.* **92,** 113.

Weigle, W. O. (1964b). *Immunology* **7,** 239.

Weigle, W. O. (1965a). *Ann. N. Y. Acad. Sci.* **124,** 133.

Weigle, W. O. (1965b). *J. Exptl. Med.* **121,** 289.

Wells, H. G., and Osborne, T. B. (1911). *J. Infect. Diseases* **8,** 77.

Yoshimura, M., and Cinader, B. (1966). *J. Immunol.* **97,** 959.

4

Autoimmune Response and Pathogenesis of Blood Parasite Diseases*

WILLIAM F. SCHROEDER AND MIODRAG RISTIC

I. THE IMMUNOLOGICAL BASIS OF ANEMIA IN ERYTHROCYTIC INFECTIONS

A. IMMUNOGENIC DISORDERS

Disease may result from an immune reaction to a latent infection or from antigen-antibody reactions *in vivo* (Magill, 1954; Dragstadt, 1941; Dixon, 1961, 1964; Wissler, 1962; McCormack *et al.*, 1963). Further, the injury resulting from the reaction to an infectious agent may be more severe than the injury produced by the agent itself (Steiner and Volpe, 1961; Waksman,

* Research was supported in part by research grant HE-10609 from the U.S. Public Health Service.

1959, 1960, 1962). These observations led to the development of a specialized study in immunopathology and to the recognition of so-called "immunological diseases." The acquired hemolytic anemias of man of autoimmunological origin are within this group of disorders (Dacie, 1962). More recently it has been suggested that anaplasmosis (Ristic, 1961; Kreier *et al.*, 1966), piroplasmosis (Ristic, 1966; Schroeder *et al.*, 1966), and malaria (McGhee, 1960; Zuckerman, 1964; Cox *et al.*, 1966) may be included in this group of diseases.

Current investigations are being directed toward utilizing immunochemical and biophysical techniques to detect and define autoantigens and autoantibodies as well as the site of reaction in these erythrocytic infections.

B. Anemia Not Commensurate with Parasitemia

An outstanding feature of erythrocytic infections is the frequent occurrence of anemia that is often not proportional to the prevailing parasitemia. Christophers and Bentley (1908) compared the anemia occurring in canine piroplasmosis with that occurring in humans with a sequela of *Plasmodium falciparum* malaria called "blackwater fever." They pointed out that the anemia of these two disorders occurred after the peak of parasitemia and that, in many instances in blackwater fever patients, the anemia occurred suddenly in the absence of observable parasitemia.

These observations were confirmed experimentally in *Babesia canis* infections of puppies by Maegraith *et al.* (1957). These investigators suggested that the anemia in these animals was due to two factors: one, the destruction of erythrocytes by the parasite per se and the other, the removal or destruction of erythrocytes by immunogenic factors. In animals with a high degree of parasitemia, destruction of erythrocytes per se was thought to exceed the anemia-producing effects of immunogenic factors. These latter factors only became evident in animals with a low degree of parasitemia. It has further been found that in *Babesia rodhaini*-infected rats (Schroeder *et al.*, 1966; Zuckerman, 1966) and *Anaplasma marginale*-infected cattle (Schroeder, 1967), infected animals fall into two groups: one with mild and the other with more severe parasitemia. In *P. berghei*-infected rats, maximum anemia followed peak parasitemia while peak anemia occurred between the ninth and eleventh day after inoculation in both *P. berghei*- and *B. rodhaini*-infected rats, irrespective of the degree of parasitemia. In *Anaplasma*-infected cows the anemia was as great in animals with 1% parasitemia as it was in animals with 8 to 10% parasitemia. The failure of George *et al.* (1966) to confirm Zuckerman's studies with *P. berghei*-infected rats may be partly due to the fact that parasitemias were reported to be high in all infected rats of this study. In other studies, investigators have used antiparasite drugs to control the incidence of parasitemia. McGhee (1963, 1964, 1965) found that when *P. lophurae* infection of ducks was controlled with quinine dihydrochloride marked anemias occurred although the degree of parasitemia remained low. Schroeder and Ristic (1965a) observed severe anemia in *Anaplasma*-infected calves despite the fact that parasitemia in these animals was controlled by continuous tetracycline treatment.

Marked anemia has often been observed following treatment of malaria patients with antiparasitic drugs. The hemolytic activity of the 8-amino-quinolines appears to be associated with a hereditary defect in the patients' mature erythrocytes. Other antimalarial drugs, such as quinine, quinidine, and quinacrine, appear to act as haptens. The anemia following treatment with stibophen has been shown to be due to the formation of a drug-antibody complex that nonspecifically absorbed to erythrocytes and caused their destruction (Dixon, 1966). It was not felt, however, that the anemia observed in experimental animals treated with antiparasitic drugs was of of this nature. The subject of drug-induced anemia in malaria has been discussed and reviewed by Zuckerman (1964).

C. IMMUNOGENIC FACTORS

1. Toxins and By-Products of Infection

Early investigators observing anemias not commensurate with the degree of parasitemia in canine babesiosis and blackwater fever concluded that parasite toxins or by-products of erythrocyte infection destroyed the erythrocytes and caused anemia. Experimental studies utilizing immuno-chemical and biophysical techniques have not revealed any evidence of parasite toxins in any of the erythrocyte infections. The observation that maximum anemia occurs after peak parasitemia does not fit into the concept that a parasite toxin is involved. The subject has been reviewed and discussed by Zuckerman (1964). Possible products of parasite infection in the form of peptides, some of which appear to be pharmaceutically active, have been reported in the blood and urine of *Babesia*-infected dogs and rats (Maegraith *et al.*, 1957; Goodman and Richards, 1960). The role of these peptides in the pathogenesis of infection or of anemia remains undetermined.

2. Soluble Antigens

Studies of soluble antigens in the serum or plasma of animals with erythrocytic infections has been complicated by the presence of host cell contaminants. It has been reported that the purest preparations of *Plasmodium lophurae* antigens contained considerable quantities of host hemoglobin due to extensive phagotrophy by *Plasmodium* organisms (Sherman, 1964).

Diggs (1964) emphasized, however, that contaminant antigens occurred as a part of the natural disease and must be considered when one studies the pathogenesis of anemia in infected animals. Antibodies were formed in *Plasmodium*-infected animals against host gamma-2 globulin, host hemoglobin, and three white blood cell antigens. Other studies by Eaton and Coggeshall (1939) and Zuckerman (1945) demonstrated that normal erythrocytes absorbed agglutinins and opsonins which are directed against *Plasmodium*-parasitized erythrocytes.

More recently, application of immunochemical and biophysical techniques has resulted in the isolation of soluble serum and plasma antigens from *Anaplasma*-infected cattle (Amerault and Roby, 1964), from *Babesia*-infected horses and dogs (Sibinovic, 1966; Sibinovic et al., 1967a,b), and from *Plasmodium gallinaceum*-infected chickens (Todorovic, 1967; Todorovic et al., 1967a,c.).

According to Sibinovic (1966), babesial soluble serum antigens contain two physically and serologically distinct components. These antigens are genus rather than species specific and they may be similar to antigens reported earlier in *P. knowlesi* infections by Eaton (1939) and *P. lophurae* infections by McGhee (1960). It is not known whether these antigens are entirely parasitic in origin since they may also be by-products of infection.

The use of enzymes to alter erythrocytes was introduced by Morton and Pickles (1951) and Morton (1962). This discovery led to the detection and characterization of a number of antigenic sites on the surface of the erythrocyte (Springer, 1963). It was subsequently determined that these exposed antigens reacted with antibodies not detectable by routine methods. Antisera from patients with acquired hemolytic anemia were shown to react with enzyme-treated erythrocytes (Henningsen, 1949; Mabry et al., 1956a,b). Following injection of autologous trypsin-treated erythrocytes in rabbits, Dodd et al. (1953) demonstrated hemagglutinins in the serum which reacted *in vitro* with the treated erythrocytes. Treatment of bovine erythrocytes with trypsin uncovered antigenic sites that reacted with certain of the isoantibodies of cattle, with the antibody from patients with infectious mononucleosis, and with free-serum autohemagglutinins of *Anaplasma*-infected calves (Burnet and Anderson, 1946; Stone and Miller, 1955; Tomsick and Baumann-Grace, 1960; Mann and Ristic, 1963a,b). Stone and Miller (1955) presented evidence that the presence of these antigens on bovine erythrocytes was genetically determined. In investigating erythrocytic antigens to which autohemagglutinins are formed, a soluble antigen precipitated with protamine sulfate (Ristic et al., 1963) was obtained from uninfected as well as *Anaplasma*-infected erythrocytes which reacted with free-serum autohemagglutinin (Schroeder and Ristic, 1965b). In other studies concerned with uncovering antigenic sites of erythrocytes, Athineos et al. (1962) revealed that desialization of mucoproteins (which are constituents of erythrocytic stroma) will greatly enhance the antigenicity of these antigenically poor complexes.

3. Antibodies

The globulin-antiglobulin technique of Coombs *et al.* (1945) and Dunsford and Grant (1959) and the fluorescent antibody technique of Coons *et al.* (1941) have been used to detect erythrocyte-bound globulins. In applying these techniques it was soon determined that they could be used to detect antibody globulin, antigen-antibody complexes, complement, and transferrins as well. Thus, it has been shown that the positive antiglobulin test obtained during the course of *Plasmodium berghei* infection or as a result of inducing anemia artificially in rats is due to the nonspecific adherence of transferrins to immature erythrocytes (George *et al.*, 1966). These workers applied an immunoelectrophoresis technique to obtain more purified preparations of antigamma globulin and were able to eliminate false positive antiglobulin reactions in these instances. Positive antiglobulin reactions were obtained by Mann and Ristic (1963a) during the course of *Anaplasma* infection in calves but the nature of the bound globulin was not determined. A bound antibody detected by a hemolytic test has also been demonstrated during the course of anaplasmosis in calves (Ristic, 1961).

An antigen-antibody complex has been detected during the course of *P. berghei* infection in mice and *Anaplasma* infection in cattle by means of the indirect fluorescent antibody technique (Kreier and Ristic, 1964; Ristic and White, 1960). Fluorescein-labeled rabbit antimouse gamma globulin stained the malaria parasite between 3 to 8 days after infection. This is not entirely unexpected, however, since host antigens have been demonstrated to occur within the *Plasmodium* parasite (Sherman, 1964). A similar phenomenon has been demonstrated during the course of *B. rodhaini* infection in mice (Cusick *et al.*, 1964) and *Anaplasma* infection of calves (Ristic and White, 1960).

An opsonin that stimulated erythrophagocytosis in an *in vitro* system was demonstrated in the sera of chickens hyperimmunized to *P. gallinaceum* and *P. lophurae* (Zuckerman, 1945), but the opsonin could not be demonstrated in the serums of naturally infected chickens. More recently, Schroeder (1966) demonstrated the presence of an opsonin in the serum of *P. berghei*- and *B. rodhaini*-infected rats and *Anaplasma*-infected calves* by means of an *in vitro* technique utilizing mouse peritoneal exudate cells. The opsonins sensitized autologous and homologous uninfected erythrocytes to phagocytosis by cells of the reticuloendothelial system (RES). The opsonin was readily eluted from rat erythrocytes but not from bovine erythrocytes. Studies are underway to define the nature of the opsonins.

* Schroeder and Ristic (1968).

Hansard and Foote (1959) demonstrated an erythropoiesis-depressing factor in *Anaplasma*-infected calves which appeared concurrently with increased erythrophagocytic activity in the spleen and bone marrow. These workers suggested that this factor acted as an opsonin *in vivo*.

II. CONCEPTS OF AUTOIMMUNE RESPONSE

Dixon (1966) has pointed out the similarity of basic antigen-antibody reactions in immunogenic disorders. The classification of an autoimmunogenic response depends primarily upon the demonstration that endogenous (autoantigens) are involved in the reactions. In reviewing the studies that revealed immunological factors in Section I, e.g., bound antibody, hemagglutinin for trypsin-treated erythrocytes, and opsonins, it is evident that these antibodies react with normal autologous erythrocytes in such a way that the erythrocytes are lysed or sensitized to phagocytosis. Thus many investigators have concluded that the factors associated with the anemia in erythrocytic parasite infections are part of an autoimmune reaction.

A. Exposure of Normal "Self-Antigen" to Immunologically Competent Cells

The well-established mechanisms in certain diseases of autoimmune nature, such as Hashimoto's thyroiditis (McMaster *et al.*, 1961; Rose *et al.*, 1961), iritis (Halbert *et al.*, 1964), and aspermatogenesis (Burnet, 1962), do not appear to apply to the erythrocytic infections. However, since erythrocytes circulate continuously throughout the body and are in close contact with the immunologically competent cells of the RES, there is continual breakdown of aged erythrocytes which provides for uninterrupted liberation of host antigens.

B. Development of "Immunologically Competent Cell Colonies" for "Self-Antigen"

Perhaps the best example of this type of immune mechanism is that of the lymphoproliferative disease entities as exemplified by the hereditary lymphomas in man (Kaplan and Smithers, 1959), macroglobinemia (Dameshek, 1964), and Runt disease in New Zealand black mice (Howie, 1964). The extensive lymphoid reaction in theileriosis of cattle may well be another example. In these disorders, it appears that the abnormal immunologically competent cells of the RES produce antibody against normal erythrocytes and induce anemia. It has been suggested that the autohemagglutinin observed in certain erythrocytic infections in which there is hypersplenism may represent an altered response of antibody-producing cells (Schroeder and Ristic, 1965a).

C. The Immune Response to Parasite-Altered Erythrocytes

Williams and Kunkel (1963) showed experimentally that circulating autologous protein does not stimulate autoantibody production unless altered in some manner. Tolerance to such proteins may be terminated however by alteration of the globulin (McClusky *et al.*, 1962). Thus a mechanism whereby altered erythrocytes could become antigenic is suggested. Studies of *Anaplasma*-infected erythrocytes indicate that the erythrocytes become altered. Dimopoullos and Bedell (1962) demonstrated a decrease in the total phospholipid concentration of the infected erythrocytes while Rao and Ristic (1963) have reported the production of neuraminidase by *Anaplasma*. Thus under the circumstances both the permeability as well as the antigenicity of the erythrocyte may be altered. Springer (1963) reported that neuraminidase action on erythrocytes may result in the release of sialic acid and alteration of the erythrocytic surface charge and stromal antigenicity. Alteration of the metabolic activity of *Plasmodium-* or *Babesia*-infected erythrocytes has been reported by Rao (1965). Electron microscope studies of the initial bodies of *Anaplasma* indicate that these bodies are embedded in an envelope which may be a mixture of parasite and host cell antigens (Ristic and Watrach, 1961; Ristic, 1967). Release of these complexes may introduce antigen into the circulation and induce formation of an autoantibody. Another possibility is that altered antigens may leak out of the infected erythrocyte due to its increased permeability. Thus the concept of the altered erythrocytic antigen appears to fit the situation most closely in erythrocyte infections.

III. IMMUNOPATHOLOGY

Many studies have been directed toward determining the role of immunogenic factors in the pathogenesis of anemia in the erythrocytic infections. Although there is disagreement as to whether the antigens and antibodies involved are autoantigens and autoantibodies there is general agreement that antigen-antibody complexes or reactions are responsible for the observed anemia, alterations in vascular endothelium, splenomegaly, and erthrophagocytosis and hematuria.

A. Role of Soluble Antigens in Producing Anemia

Studies utilizing soluble antigens obtained during the acute phase of *Babesia* or *Plasmodium* infections have shown that these substances induce anemia as well as resistance to infection (Sibinovic, 1966; Todorovic *et al.*, 1967c; Ferris *et al.*, 1967. It is possible that the protection from infection following injection of serum obtained from *P. lophurae* ducks was due to a similar factor (Sloan and McGhee, 1965). Dixon (1966) suggested that soluble antigen might combine with antibody and nonspecifically adhere to normal erythrocytes, thereby sensitizing them to lysis or phagocytosis. The time interval between appearance of soluble antigen and antibody to this antigen however does not coincide with the appearance of anemia in infected animals and the mechanism may be different in the infected animal than when antigen is injected into a normal animal. Fluorescent antibody studies in *Babesia*-infected dogs indicates that antigen adheres to normal erythrocytes (Sibinovic, 1966). These observations suggest that antibody is formed against altered erythrocytes rather than against the antigen alone in the infected animal.

B. Hemagglutinins for Altered Erythrocytes

The role of hemagglutinins for trypsin-treated erythrocytes is not clear. Hemagglutinins were found in the serum of many but not all animals infected with either *Babesia, Plasmodium,* or *Anaplasma* organisms (Mann and Ristic, 1963; Schroeder, 1966; Schroeder *et al.*, 1966; Kreier *et al.*, 1966; Cox *et al.*, 1966). When present the hemagglutinins were better related to the onset of anemia than with the incidence of parasitemia. It has also been demonstrated by the fluorescent antibody technique that these autohemagglutinins combine with trypsin-treated erythrocytes but not untreated erythrocytes. A relationship between a soluble antigen from normal and *Anaplasma*-infected bovine erythrocytes has been shown by absorption tests (Schroeder and Ristic, 1965b). It has been suggested that these hemagglutinins react with autoantigens that are present on the surface of erythrocytes of many species and which are exposed by treatment with enzymes or exposed following infection (Wiener, 1943; Springer, 1963; Schroeder and Ristic, 1965a). Stone and Miller (1955) found that the presence of such agglutinogens on bovine erythrocytes was genetically controlled; thus the failure of some animals to form autohemagglutinins may be due to a lack of antigen on their erythrocytes. Agglutinins for nontrypsinized mouse erythrocytes were formed independently of hemagglutinins for trypsin-treated erythrocytes in rats injected with *Babesia*-infected blood. Control rats injected with uninfected mouse erythrocytes formed hemagglutinins for normal mouse blood but not for trypsin-treated mouse or rat erythrocytes (Schroeder, 1966).

C. The Role of Erythrocyte-Bound Complexes

Radioisotope tagging techniques have been used to measure the survival rate of erythrocytes *in vivo* and indirectly indicate alteration of the erythrocyte. Reduced erythrocyte survival time was demonstrated by this technique when erythrocytes from blackwater fever patients were injected into normal individuals or when normal erythrocytes were injected into blackwater fever patients. Foy *et al.* (1945) interpreted this as indicating that the surface of the erythrocyte had been altered by a substance in the serum of patients with blackwater fever. Other studies conducted by Koret *et al.* (1955), Kuvin *et al.* (1962), and Stohlman *et al.* (1963) support this view. A recent report by George *et al.* (1966) is in contrast to these earlier findings. These workers were unable to demonstrate any decrease in the survival of normal rat erythrocytes when exposed to serum from *P. berghei*-infected rats and subsequently injected into normal rats. More extensive studies similar to those conducted in human beings with acquired hemolytic anemia are in order. Such studies should be accompanied by the use of the Coombs antiglobulin technique, hemolytic test, and fluorescent antibody technique.

D. Splenomegaly, Opsonins, and Erythrophagocytosis

Splenomegaly and accompanying erythrophagocytosis is a common finding in the erythrocytic parasite infections. On this basis, Taliaferro (1949) proposed a cellular basis for immunity to malaria. Taliaferro and Mulligan (1937) suggested that the initial phagocytic response was due to acquired immunity following exposure to erythrocytic infection. The literature on the significance of cellular immunity in both parasitic and bacterial infections has been reviewed by Suter and Ramsier (1964). Other investigators have suggested that splenomegaly in itself creates physical conditions that favor the stasis, trapping of erythrocytes, and subsequent phagocytosis or lysis (Jandl *et al.*, 1957; Craddock, 1962; George *et al.*, 1966). These investigators concluded as a result of applying radioisotope-tagging techniques to a study of erythrocyte survival in man and animals that the accompanying anemia was an expression of mechanical filtration and trapping of erythrocytes in the sinuses due to immune adherence. A similar condition could be induced by injection of inert particles.

Other investigators have pointed out that erythrophagocytosis is a normal physiological function of the RES which functions to remove aged or "effete" erythrocytes (Rous and Robertson, 1917; Rous, 1923; Bessis, 1962–1963). This phenomenon becomes markedly increased in erythrocytic infections and in the autoimmune hemolytic anemias of man. It has been suggested that infected erythrocytes or those with bound globulins are comparable to aged erythrocytes and are thus more rapidly removed. Phagocytosis of erythrocytes that had been "aged" *in vitro* or treated with trypsin was observed to occur more readily than phagocytosis of freshly collected erythrocytes (Miescher, 1957; Vaughan and Boyden, 1964; Mabry *et al.* 1956a,b). An increase in phagocytosis of apparently normal erythrocytes was observed in a number of bacterial infections, hemolytic diseases, and erythrocytic infections (Zinkham and Diamond, 1952; Miller and Perry-Pepper, 1916; Mallory, 1898; Hektoen, 1906; Wright, 1906; Kyes, 1915). Adler (1964) reported that splenic uptake of albumin-globulin complexes was more rapid than that of colloidal carbon. Simultaneous injection of carbon and of sensitized erythrocytes resulted in interference with erythrophagocytosis; however, phagocytosis of carbon particles was not interfered with when erythrocytes were injected first.

Erythrophagocytosis was studied in rats infected with *B. rodhaini* or *P. berghei* (Schroeder *et al.*, 1966; Cox *et al.*, 1966; Zuckerman, 1966). Phagocytosis of uninfected erythrocytes by free macrophages of the RES was generally observed to occur in the spleen of *B. rodhaini*- and *P. berghei*-infected rats as early as 7 days and persisted for 31 days after inoculation. Splenomegaly was a constant finding, however the size of the spleen was not always proportional to the degree of parasitemia or the extent of erythrophagocytosis. Ramakrishnan (1952) found that the increase in spleen size in rats infected with *P. berghei* was usually greater in those rats with low degree of parasitemia then in those with high degrees of parasitemia. Early appearance of histiocytes containing pigment was considered to represent a nonspecific natural immune response while the latter appearance of erythrocytes ingested by free macrophages was considered to represent an induced immune response. Zuckerman (1966) studied the extent of erythrophagocytosis in *P. berghei*-infected rats. She determined that significant numbers of erythrocytes were removed by macrophages to induce anemia. In her studies, she found that the size of the spleen was proportional to the extent of phagocytosis. Erythrophagocytosis has also been observed in the spleen and bone marrow of *Anaplasma*-infected calves and considered to contribute to the anemia (Ristic and Sippel, 1958; Hansard and Foote, 1959; Baker *et al.*, 1961; Seger and White, 1962; Kreier *et al.*, 1964). Sloan and McGhee (1965) demonstrated extensive phagocytosis of polychromatophil erythrocytes in chickens and ducks infected with *P. lophurae*.

In vitro systems consisting of the use of buffy coat and serum from patients with acquired hemolytic anemia or paroxysmal nocturnal hemoglobinuria revealed that complement enabled cold antibody to fix to erythrocytes and sensitized them to phagocytosis (Jordan *et al.*, 1952; Conway, 1953; Bonnin and Schwartz, 1954; Greendyke *et al.*, 1963). This technique has not proved effective in demonstrating opsonins in erythrocytic infections, however. By means of an *in vitro* tissue culture technique, Zuckerman (1945) demonstrated opsonins in the serum of chickens hyperimmunized to *P. gallinaceum* and *P. lophurae*. She was not able to demonstrate opsonins in the serum of naturally infected birds. More recently, Schroeder (1966) utilized an *in vitro* system consisting of rat and bovine erythrocytes, mouse peritoneal exudate cells, and serum from normal and infected animals. By this means it was demonstrated that opsonins occurred in the serum of *B. rodhaini-* and *P. berghei*-infected rats and *A. marginale*-infected calves. The opsonins were heat-stable and sensitized normal uninfected autologous and homologous erythrocytes to phagocytosis by macrophages of the RES. The appearance and titers of the opsonins were related to the onset, degree, and duration of anemia in infected animals. Studies are being conducted to identify the opsonin and determine its activity in uninfected animals.

E. ASSOCIATED PHENOMENA

Hematuria has been observed to occur in babesiosis and malaria. Evidence of nephrosis has been reported in dogs with *B. canis* infection by Maegraith *et al.* (1957) and Dorner (1965). Dixon (1966) reported that when the fluorescent antibody technique was used, deposits of host gamma globulin were detected on the glomerular basement membrane of patients with quartan malaria.

Increased capillary permeability has been observed in erythrocytic infections. Cell-bound antibodies have been demonstrated on the capillary endothelium in humans with thrombocytopenia of autoimmune origin (Waksman, 1960; McCormack *et al.*, 1963). In dogs infected with *B. canis* acute vascular collapse has been observed in the absence of detectable parasitemia (Popovic, 1965). Ristic and Watrach (1962) demonstrated *A. marginale* bodies within thrombocytes, and it may well be that immune mechanisms are involved in the increased capillary permeability observed in this disease, as has been demonstrated in thrombocytopenias (Harlington *et al.*, 1951; Waksman, 1960).

Delayed sexual maturity has been observed in cattle raised in areas where hematozoan infections are common. Prado (1966) has observed aspermatogenesis in bulls that had undergone infections with *Anaplasma* and *Babesia*. It remains to be demonstrated whether the aspermatogenesis is of the type believed to be due to autoimmune mechanisms (Fudenberg, 1962).

IV. PROTECTION RESULTING FROM THE AUTOIMMUNE RESPONSE

It would appear from the experimental data obtained at this time that autoimmune mechanisms of erythrocytic infections are complex. It has been suggested (Schroeder and Ristic, 1965a) that both altered antigen- and altered antibody-forming mechanisms may be responsible for the formation of autoantibody. Both mechanisms may be active concurrently in the same animal but may vary in importance or degree at different times during the infection. This would help in explaining the generally similar yet different in detail autoimmune response between animals and species studied.

It has been suggested that hemagglutinins and opsonins may be present in small quantities in healthy animals and that these aid in the disposal of aged erythrocytes. These antibodies are increased in quantity during the course of infection and contribute to the observed anemia. The same mechanism that may contribute to anemia during the acute phase may also provide protection from infection or reinfection. It is well established that a "state of premunition" is necessary for resistance to erythrocytic infections. When this state is terminated by treating the animals with chemotherapeutics, they become susceptible to reinfection. The mechanism whereby protection is maintained is not well understood; however, it has recently been demonstrated (Sibinovic, 1966; Todorovic, 1967; Ferris et al., 1967) that soluble antigens obtained from the plasma of animals in the acute phase of babesiosis and malaria not only induces anemia in normal uninfected animals but protects them from infection. Thus it is suggested that the maintenance of autoimmune mechanisms during the state of premunition may contribute to the resistance of the animal.

REFERENCES

Adler, F. L. (1964). *Progr. Allergy* 8, 41.
Amerault, T. E., and Roby, T. O. (1964). *Am. J. Vet. Res.* 25, 1642.
Athineos, E., Thornton, M., and Winzler, R. J. (1962). *Proc. Soc. Exptl. Biol. Med.* 111, 353.
Baker, N. F., Osebold, J. W., and Christensen, J. F. (1961). *Am. J. Vet. Res.* 22, 590.
Bessis, M. C. (1962–1963). *Harvey Lectures* 58, 125–156.
Bonnin, J. A., and Schwartz, L. (1954). *Blood* 9, 773.
Burnet, F. M. (1962). *Proc. Roy. Soc. Med.* 55, 619.
Burnet, F. M., and Anderson, S. G. (1946). *Australian J. Exptl. Biol. Med. Sci.* 25, 213
Christophers, S. R., and Bentley, C. (1908). *Sci. Mem. Med. Sanit. Dept. India* No. 35, pp. 1–179.
Conway, H. (1953). *J. Clin. Pathol.* 6, 208.
Coombs, R. R., Mourant, A. E., and Race, R. R. (1945). Lancet 249, 15.
Coons, A. H., Creech, H. J., and Jones, R. N. (1941). *Proc. Soc. Exptl. Biol. Med.* 47, 200.
Cox, H. W., Schroeder, W. F., and Ristic, M. (1966). *J. Parasitol.* 13, 327.
Craddock, C. G., Jr. (1962). *In* "Mechanisms of Anemia" (I. M. Weinstein and E. Beutler, eds.), Chapter 8. McGraw-Hill, New York.
Cusick, F., Ristic, M., and Sibinovic, S. (1964). Special Project, University of Illinois, Urbana, Illinois.
Dacie, J. V. (1962). "II. Autoimmune Hemolytic Anemias." Churchill, London.
Dameshek, W. (1964). *Ann. N. Y. Acad. Sci.* 124, 162–166.

Diggs, C. L. (1964). *Am. J. Trop. Med. Hyg.* **13,** Suppl., 217.

Dimopoullos, G. T., and Bedell, D. M. (1962). *Am. J. Vet. Res.* **23,** 813.

Dixon, F. J. (1961). *Proc. 2nd Intern. Symp. Immunopathol.*, pp. 72–87. Grune & Stratton, New York.

Dixon, F. G. (1964). *Ann. N. Y. Acad. Sci.* **124,** 162–166.

Dixon, F. G. (1966). *In* "Research in Malaria" (E. H. Sadun and H. S. Osborne, eds.), p. 1233. Walter Reed Army Inst. Res., Washington, D. C. (*Military Med.* **131,** Suppl.).

Dodd, M. C., Wright, C. S., Baxter, J. A., Bouroncle, B. A., Bunner, A. E., and Winn, H. J. (1953). *Blood* **8,** 64

Dorner, J. L. (1965). M.S. Thesis, University of Illinois, Urbana, Illinois.

Dragstadt, C. A. (1941). *Physiol. Rev.* **21,** 563.

Dunsford, I., and Grant, J. (1959). "The Antiglobulin (Coombs) Test in Laboratory Practice." Oliver & Boyd, Edinburgh and London.

Eaton, M. D. (1939). *J. Exptl. Med.* **69,** 517

Eaton, M. D., and Coggeshall, L. T. (1939). *J. Exptl. Med.* **70,** 131.

Ferris, D. H., Todorovic, R., and Ristic, M. (1967). *Z. Tropenmed. Parasitol.* (in press).

Foy, H., Kondi, A., Rebelo, A., and Soeiro, A. (1945). *Trans. Roy. Soc. Trop. Med. Hyg.* **35,** 119.

Fudenberg, H. (1962). *Proc. Conf. Immuno-Reproduction*, p. 63.

George, J. N., Stokes, E. F., Wicker, D. J., and Conrad, M. E. (1966). *In* "Research in Malaria" (E. H. Sadun and H. S. Osborne, eds.), p. 1217. Walter Reed Army Inst. Res., Washington, D. C. (*Military Med.* **131,** Suppl.).

Goodman, L. G., and Richards, W. H. G. (1960). *Brit. J. Pharmacol.* **15,** 152.

Greendyke, R. M., Brierty, R. E., and Swisher, S. N. (1963). *Blood* **22,** 295.

Halbert, S. P., Manski, W., and Ehrlich, S. (1964). *Ann. N. Y. Acad. Sci.* **124,** 332–351.

Hansard, S. L., and Foote, L. E. (1959). *Am. J. Physiol.* **197,** 711.

Harlington, W. J., Minnich, V., Hollingsworth, J. W., and Moore, C. V. (1951). *J. Lab. Clin. Med.* **38,** 1.

Hektoen, L. (1906). *J. Infect. Diseases* **3,** 721.

Henningsen, K. (1949). *Acta Pathol. Microbiol. Scand.* **26,** 339.

Howie, J. B. (1964). *Ann. N. Y. Acad. Sci.* **124,** 167–177.

Jandl, J. H., Richardson-Jones, A., and Castle, W. B. (1957). *J. Clin. Invest.* **36,** 1428.

Jordan, W. S., Prouty, R. L., Heinle, R. W., and Dingle, J. H. (1952). *Blood* **7,** 387.

Kaplan, H. S., and Smithers, D. W. (1959). *Lancet* **II,** 1.

Koret, D. E., Clatanoff, D. V., and Shilling, R. F. (1955). *Clin. Res. Proc.* **3,** 195.

Kreier, J. P., and Ristic, M. (1964). *Am. J. Trop. Med. Hyg.* **13,** 6.

Kreier, J. P., Ristic, M., and Schroeder, W. F. (1964). *Am. J. Vet. Res.* **25,** 343.

Kreier, J., Shapiro, H., Dilley, D., Szilvassy, I. P., and Ristic, M. (1966). *Exp. Parasitol.* **19,** 155.

Kuvin, S. F., Beye, H. K., Stohlman, F., Contacos, P. G., and Coatney, G. R. (1962). *Trans. Roy. Soc. Trop. Med. Hyg.* **56,** 371.

Kyes, P. (1915). *Anat. Record* **9,** 97.

Mabry, D. S., Bass, J. A. Dodd, M. C., Wallace, J. H., and Wright, C. S. (1956a). *J. Immunol.* **76,** 54.

Mabry, D. S., Wallace, J. H., Dodd, M. C., and Wright, G. S. (1956b). *J. Immunol.* **76,** 62.

Mann, D. K., and Ristic, M. (1963). *Am. J. Vet. Res.* **24,** 709.

McClusky, R. T., Miller, F., and Benacerraf, B. (1962). *J. Exptl. Med.* **115,** 253.

McCormack, F., O'Brient, D. J., and Oliver, R. A. M. (1963). *J. Clin. Pathol.* 16, 436.
McGhee, R. B. (1960). *J. Infect. Diseases* 107, 410.
McGhee, R. B. (1963). *Proc. 1st. Intern. Congr. Protozool., Prague, 1961* p. 470. Academic Press, New York.
McGhee, R. B. (1964). *Am. J. Trop. Med. Hyg.* 13, Suppl., 219.
McGhee, R. B. (1965). *Proc. 2nd. Intern. Congr. Protozool., London,* Intern. Congr. Ser. No. 91, p. 171. Excerpta Med. Found., Amsterdam.
McMaster, P. R. B., Lerner, B. M., and Exam, E. D. (1961). *J. Exptl. Med.* 113, 611.
Maegraith, B. G., Giles, H. M., and Devakul, K. (1957). *Z. Tropenmed. Parasitol.* 8, 485.
Magill, T. P. (1954). *J. Immunol.* 74, 1.
Mallory, F. B. (1898). *J. Exptl. Med.* 3, 611.
Mann, D. K., and Ristic, M. (1963a). *Am. J. Vet. Res.* 24, 703.
Mann, D. K., and Ristic, M. (1963b). *Am. J. Vet. Res.* 24, 709.
Miescher, P. (1957). "Physiopathology of the Reticuloendothelial System," p. 147. Blackwell, Oxford.
Miller, G. T., and Perry-Pepper, O. H. (1916). *J. Immunol.* 1, 383.
Morton, J. A. (1962). *Brit. J. Haematol.* 8, 134.
Morton, J. A., and Pickles, M. M. (1951). *J. Clin. Pathol.* 4, 189.
Popovic, N. (1965). M.S. Thesis, University of Illinois, Urbana, Illinois.
Prado, A. H. (1966). Facultad Veterinary Medicine, Maracay, Venezuela, personal communication.
Ramakrishnan, S. P. (1952). *Indian J. Malariol.* 6, 189.
Rao, P. J. (1965). Ph.D. Thesis, University of Illinois, Urbana, Illinois.
Rao, P. J., and Ristic, M. (1963). *Proc. Soc. Exptl. Biol. Med.* 114, 447.
Ristic, M. (1961). *Am. J. Vet. Res.* 22, 871.
Ristic, M. (1966). *In* "The Biology of Parasites" (E. J. L. Soulsby, ed.), p. 128. Academic Press, New York.
Ristic, M. (1967). *Am. J. Vet. Res.* 28, 63.
Ristic, M., Mann, D. K., and Kodras, R. (1963). *Am. J. Vet. Res.,* 24, 472.
Ristic, M., and Sippel, W. L. (1958). *Am. J. Vet. Res.* 19, 44.
Ristic, M., and Watrach, A. M. (1961). *Am. J. Vet. Res.* 22, 109.
Ristic, M., and Watrach, A. M. (1962). *Am. J. Vet. Res.* 23, 626.
Ristic, N., and White, F. H. (1960). *Science* 131, 987.
Rose, N. R., Kite, J. R., and Doebbler, T. K. (1961). *Proc. 2nd Intern. Symp. Immunopathol.,* p. 161. Grune & Stratton, New York.
Rous, P. (1923). *Physiol. Rev.* 3, 75.
Rous, P., and Robertson, O. H. (1917). *J. Exptl. Med.* 25, 651.
Schroeder, W. F. (1966). Ph.D. Thesis, University of Illinois, Urbana, Illinois.
Schroeder, W. F. (1967). Current research.
Schroeder, W. F., and Ristic, M. (1965a). *Am. J. Vet. Res.* 26, 239.
Schroeder, W. F., and Ristic M. (1965b). *Am. J. Vet. Res.* 26, 679.
Schroeder, W. F., and Ristic, M. (1968). *Am. J. Vet. Res.* (In press).
Schroeder, W. F., Cox, H. W., and Ristic, M. (1966). *Ann. Trop. Med. Parasitol.* 60, 31.
Seger, C. L., and White, D. (1962). *Proc. 4th. Natl. Anaplasma Conf., Reno, Nevada, 1961* pp. 26–28.
Sherman, I. (1964). *Am. J. Trop. Med. Hyg.* 13, Suppl., 214.
Sibinovic, K. H. (1966). Ph.D. Thesis, University of Illinois, Urbana, Illinois.
Sibinovic, K. H., McLeod, R., Ristic, M., Sibinovic, S., and Cox, H. W. (1967a). *J. Parasitol.* (in press).

Sibinovic, K. H., Sibinovic, S., Ristic, M., and Cox, H. W. (1967b). *J. Parasitol.* (in press).

Sloan, B. L., and McGhee, R. B. (1965). *J. Parasitol.* **51**, Suppl., 36.

Springer, G. F. (1963). *Bacteriol. Rev.* **27**, 191.

Steiner, J. W., and Volpe, R. (1961). *Can. Med. Assoc. J.* **84**, 1297.

Stohlman, F., Contacos, P. G., and Kuvin, S. F. (1963). *J. Am. Med. Assoc.* **184**, 1020.

Stone, W. H., and Miller, W. J. (1955). *Genetics* **40**, 599.

Suter, E., and Ramsier, H. (1964). *Advan. Immunol.* **4**, 117.

Taliaferro, W. H. (1949). *Ann. Rev. Microbiol.* **3**, 159.

Taliaferro, W. H., and Mulligan, H. W. (1937). *Indian Med. Res. Mem.* **29**, 1.

Todorovic, R. (1967). Ph.D. Thesis, University of Illinois, Urbana, Illinois.

Todorovic, R., Ristic, M., and Ferris, D. H. (1967a). *Trans. Roy. Soc. Trop. Med.* (In press).

Todorovic, R., Ristic, M., and Ferris, D. H. (1967b). *Ann. Trop. Med. Parasitol.* **67**, 117).

Todorovic, R., Ristic, M., and Ferris, D. H. (1967c). *Trans. Roy. Soc. Trop. Med.* (in press).

Tomsick, J., and Baumann-Grace, J. B. (1960). *Pathol. Microbiol.* **23**, 172.

Vaughan, R. B., and Boyden, S. V. (1964). *Immunology* **7**, 18.

Waksman, B. H. (1959). *Intern. Arch. Allergy Appl. Immunol.* **14**, 1.

Waksman, B. H. (1960). *Am. J. Pathol.* **37**, 673.

Waksman, B. H. (1962). *Medicine* **41**, 93.

Wiener, A. S. (1943). "Blood Groups and Transfusion," 3rd. ed., pp. 48–49. Thomas Springfield, Illinois.

Williams, R. G., and Kunkel, M. G. (1963). *Proc. Soc. Exptl. Biol. Med.* **112**, 554.

Wissler, R. W. (1962). *Ann. Rev. Microbiol.* **16**, 265.

Wright, A. E. (1906). *Brit. Med. J.* **I**, 143.

Zinkham, W. H., and Diamond, L. K. (1952). *Blood* **7**, 592.

Zuckerman, A. (1945). *J. Infect. Diseases* **77**, 28.

Zuckerman, A. (1964). *Exptl. Parasitol.* **15**, 138.

Zuckerman, A. (1966). *In* "Research in Malaria" (E. H. Sadun and H. S. Osborne, eds.), p. 1201. Walter Reed Army Inst. Res., Washington, D. C. (*Military Med.* **131**, Suppl.).

5

Blood Parasite Antigens and Antibodies

AVIVAH ZUCKERMAN AND MIODRAG RISTIC*

A generation ago scarcely a reference to immune responses to protozoan parasites was to be found in standard texts on immunology. Immunity to protozoa was assumed to differ in some essential but undefined way from immunity to other microbes. However, as data have accumulated, it has become increasingly clear that immunological phenomena relating to the protozoa are qualitatively similar to those governing other host-parasite combinations. Recent studies on protozoan antigens and antibodies therefore closely parallel those on other microbes, employing the same technical and analytical methods as in other branches of immunology.

* This study was supported by grants AI-02859, AI-03315, and HE-10609 from the United States Public Health Service.

I. PLASMODIA

A. ANTIGENS OF PLASMODIA

1. Techniques for Obtaining Plasmodial Material Free of Host Cells

The antigenic analysis of a parasitic organism presupposes its separation, in convenient quantities, from host contaminants or from constituents of culture media. Where microbes are readily cultured *in vitro* this problem is relatively easy to solve. Analysis of the parasitic blood protozoa, on the other hand, has been impeded by several technical problems which have been extensively attacked only in the past few years (Zuckerman, 1964a). It has been correctly stated that a major obstacle to the detection of malarial antibody has been the difficulty in obtaining sufficient quantities of malarial antigen (Tobie, 1964).

Some protozoa, including the plasmodia, are obligate intracellular organisms which have yet to be cultivated *in vitro* in indefinitely replicated serial subculture, either within or independently of their host cells. Obtaining antigen for analysis from such parasites thus currently entails (*a*) bleeding an infected host, and (*b*) separating the parasites in the infected blood from their host cells and from other blood constituents. Such separation must obviously be accomplished without injury to the antigenic structure of the parasites.

The bulk of the leukocytes are generally removed from infected blood before the parasites are freed from the erythrocytes. In cases where heavily infected blood is readily available, this may be done by removing the leukocytic buffy coat after centrifugation. However, many parasites are perforce discarded when this procedure is employed, since the infected cells concentrate just beneath the buffy coat layer. Where parasitemias are low (as in human infections with *Plasmodium falciparum*), this method has been considered too wasteful, and most of the leukocytes and thrombocytes have been separated from the erythrocytes by aggregating the latter with the aid of dextran (Spira and Zuckerman, 1966) whereupon they sediment out of the suspension. The use of Millipore filters to remove leukocytes has been advocated (Corradetti *et al.*, 1964a), and this method may be applied after the parasites are free from their host cells.

Methods devised for freeing plasmodia from erythrocyte hosts have included treating infected blood with distilled water (Stein and Desowitz, 1964; Chavin, 1966), with antierythrocyte antiserum (Bowman *et al.*, 1960), or with saponin (Sherman and Hull, 1960b; Spira and Zuckerman, 1962; Bray, 1964; Corradetti *et al.*, 1964a; Kreier *et al.*, 1965; Ward and Conran, 1966; Sodeman and Meuwissen, 1966). While the technique employing hemolytic antiserum is the more "physiological," the saponin method is recommended for its simplicity. The fact that freed parasites obtained by either technique are viable, albeit at a reduced rate of infectivity (Bowman *et al.*, 1960; Spira and Zuckerman, 1962), argues against antigenic degradation by either of the separatory procedures. Saponin-freed avian plasmodia have further been treated with deoxyribonuclease (DNase) to dispose of the erythrocyte nuclei (Sherman and Hull, 1960b). Parasites of *Plasmodium gallinaceum* remain viable after such enzyme treatment (Spira and Zuckerman, 1964). Plasmodia have also been separated from their host erythrocytes by the application of controlled pressure in a French press (D'Antonio *et al.*, 1966). Lower pressures burst the red cells, and the washed parasites were later disintegrated by applying higher pressures.

Residual leukocytes in plasmodial preparations may release deoxyribonucleic acid (DNA) during the later stage of the harvesting process. This forms a gummy, fibrous mass, trapping numerous parasites in its meshes. Most of this network may be dispersed with the aid of DNase, without detriment to the parasites (Spira and Zuckerman, 1966).

Washed, freed parasites have been disintegrated by freezing and thawing (Sherman and Hull, 1960b; Bray, 1964; Sherman, 1964; Sodeman and Meuwissen, 1966), by grinding in a Hughes press (Spira and Zuckerman, 1962), by grinding in a homogenizer (Stein and Desowitz, 1964; Corradetti *et al.*, 1964a; Banki and Bucci, 1964a; Chavin, 1966; Sodeman and Meuwissen, 1966), by sonication (Kreier *et al.*, 1965; Ward and Conran, 1966; Sodeman and Meuwissen, 1966; Todorovic, 1967; Todorovic *et al.*, 1967a), and by grinding in a French press (D'Antonio *et al.*, 1966). The antigenic mixture so obtained may be lyophilized and reconstituted without apparent change in antigenicity (Zuckerman, 1964a). Alternatively, the freed parasites may be lyophilized whole, and may later be disintegrated when the powder is reconstituted.

Methods devised to date for harvesting plasmodial antigen are still obviously imperfect. Improvements must still be sought, and procedures standardized. Nevertheless, the admittedly imperfect techniques now available are already capable of serving during the present exploratory phase in the antigenic analysis of plasmodia.

2. Analysis of Plasmodial Antigenic Structure

While cytochemical studies of plasmodial structure are not strictly within the scope of antigenic analysis, their close relationship is obvious. In such a cytochemical study (Bahr, 1966) it is suggested that ribonucleic acid (RNA)-rich *Plasmodium berghei* parasites "prompt a rapid uptake of metabolites, many of which would not ordinarily enter the red cell."

Attempts at separating plasmodial homogenates into constituent proteins by electrophoresis have yielded increasing numbers of components as techniques have been progressively refined. Thus, employing paper electrophoresis, Sherman and Hull (1960b) found that the proteins of *Plasmodium lophurae* moved as a single, relatively uniform entity. Immunoelectrophoresis of *Plasmodium vinckei* (Spira and Zuckerman, 1962) yielded 6 to 8 precipitin arcs; similarly *P. berghei* yielded 6 (Zuckerman and Spira, 1963) to 11 arcs (Banki and Bucci, 1964a); *Plasmodium cynomolgi* yielded 9 arcs (Banki and Bucci, 1964b); and *P. gallinaceum* yielded 10 to 12 arcs (Spira and Zuckerman, 1964). Disc electrophoresis, a technique capable of even greater resolution than immunoelectrophoresis, was employed in the study of *P. berghei* (Sodeman and Meuwissen, 1966; Chavin, 1966), and of a series of 7 rodent, simian, and avian plasmodia, including *P. berghei* (Spira and Zuckerman, 1966). As many as 16 protein discs attributable to the parasites were demonstrated. Since disc electrophoresis can be combined with gel diffusion techniques (Spira *et al.*, 1966), it should be feasible to identify the contaminating protein discs representing host cell factors. It is noteworthy that the disc patterns of *P. berghei* obtained by the two groups of investigators by whom comparative mobilities were calculated (Sodeman and Meuwissen, 1966; Spira and Zuckerman, 1966), are closely comparable, a fact which encourages the hope that such patterns may ultimately aid in identifying species and strains of plasmodia.

Plasmodial enzymes constitute a special group of parasite proteins which have been studied in some detail. Cook *et al.* (1961) distinguished between the proteinases of *P. berghei* and *Plasmodium knowlesi* and of their respective host erythrocytes. Two distinct parasitic proteinases, associated with plasmodial cytoplasm, are maximally active at pH 8 and 5, respectively. Similarly Sherman (1961, 1965) described a lactic dehydrogenase of *P. lophurae* which differs from that of the parasite's host erythrocyte, and a malic dehydrogenase from the same parasite (Sherman, 1965) which, again, is distinct from that of the host cell.

Work in the entire area of antigenic analysis of plasmodia is still initial, exploratory, and fragmentary. Fractionation of plasmodial homogenates in order to obtain pure preparations of single antigens for analysis has been carried out for *P. knowlesi* and *P. falciparum* antigens (Mahoney *et al.*, 1966) and should be pursued all along the line. Identification and and characterization of the protective antigen(s) of each plasmodium is an obvious desideratum. Studies of this sort should ultimately aid in defining plasmodial species and strains, and in comparing the antigenic structure of various phases in the life cycle of the same parasite. It has been suggested that relapse strains of plasmodia may differ antigenically from the strains inducing primary infection (Cox, 1959, 1962; K. N. Brown and Brown, 1965). Also, passage through tissue culture containing heterologous serum has been thought to modify the antigenicity of a strain of *P. berghei*

(Weiss and De Giusti, 1964). Such variations, if they occur, should also be capable of analysis by the methods outlined above.

B. ANTIPLASMODIAL ANTIBODY

1. Established and Recent Methods of Demonstration

Earlier work on malarial immunity was reviewed in 1949 by Taliaferro, who described cellular and humoral immune factors acting in unison in the living host. The review cites the demonstration of circulating antibody by complement fixation, precipitation, agglutination, opsonization, and by the passive transfer of protection. However, results were by no means uniform in the hands of different investigators. A recent review by Tobie (1964) covers methods for the detection of malarial antibody employed during the past half century, bringing into the picture both the established methods cited by Taliaferro, and also certain of the newer techniques, such as hemagglutination and immunofluorescence. To the latter may be added the techniques of electrophoresis and diffusion in gel.

The older immunological tools have not yet been entirely superseded by the newer ones. Thus, Schindler (1964) studied complement fixation in mice with *P. berghei*, and concluded, as had earlier workers, that complement-fixing antibody is not protective antibody. Ward and Conran (1966) included complement fixation in a battery of tests on the simian malarias *P. cynomolgi* and *Plasmodium fieldi*. The reaction became positive following immunization with sonicated parasites. Complement fixation tests were carried out with *P. knowlesi* and *P. falciparum* antigens by D'Antonio *et al.* (1966). Finally, titrations of complement components were done during *P. knowlesi*, *P. berghei*, and *P. gallinaceum* infections (Fogel *et al.*, 1966; Cooper and Fogel, 1966). The results suggest the fixation of complement by antigen-antibody complexes in the circulation.

The selective agglutination of *P. knowlesi* schizonts by anti-*knowlesi* antibody (Eaton, 1938) was confirmed anew by K. N. Brown and Brown (1965) in the course of their study on the antigenic heterogeneity of plasmodial relapse strains. Kreier *et al.* (1965) observed the agglutination of *P. gallinaceum* freed from red cells by plasma from infected chickens, recalling a similar study by Stauber *et al.* (1951) with *P. lophurae*.

Much fruitful work has followed the introduction of immunofluorescent techniques into malarial research, and the end is not yet in sight. Voller (1964b) has described the technique in detail, and has reviewed the rapidly evolving events in its application. Immunofluorescence was quickly adapted to use in rodent, avian, simian, and human malarias. Soon after its introduction it was determined that the fluorescent antibody technique (FAT) is more sensitive in detecting malarial antibody than is complement fixation (Tobie *et al.*, 1962). It can be adapted to field trials, using very small samples of finger prick blood as an antibody source (Bray, 1962).

Fluorescent antibody titers rise abruptly during initial infection with human plasmodia, paralleling the rise in serum gamma globulins (Kuvin *et al.*, 1962a,b, 1963; Tobie and Coatney, 1964; Voller, 1964b; Collins *et al.*, 1964b; Abele *et al.*, 1965; McGregor, 1965). Titers fall gradually following initial infection (Kuvin *et al.*, 1963), and are markedly reduced when persons formerly residing in an endemic region move to a nonmalarious area (Kuvin and Voller, 1963). Reinfection titers are significantly higher than those following primary infection (Collins *et al.*, 1964c). In contrast to unprotected human beings, those under drug protection in a hyperendemic area do not develop significant FAT titers (Voller and Wilson, 1964). Thus, repeated infection is apparently needed to maintain a high antibody level.

While much has been learned, we do not yet know what proportion of the gamma globulin increment observed during infection represents antimalarial antibody (Kuvin and Voller, 1963). Nor do we know what proportion of malarial antibody, as demonstrated by the FAT or by other methods, is protective antibody (Kuvin *et al.*, 1962b; Powell and Brewer, 1964; Voller and Wilson, 1964; McGregor, 1965; Targett and Voller, 1965), although McGregor *et al.* (1965) observed a reciprocal relationship between parasite density and FAT titers, which they interpreted as a direct correlation between FAT levels and effective immunity. It has been estimated that protective antibody may represent no more than a small percentage of the immune globulin in simian and human malarias (Cohen *et al.*, 1961; Targett and Fulton, 1965; Targett and Voller, 1965). We are in need of studies designed to differentiate between group-specific and species-specific protective antibody (Voller, 1964a,b; Collins *et al.*, 1965b). In addition to the protective mammalian antibody demonstrated in the above studies, passive transfer of protection has recently been demonstrated in ducklings with *P. lophurae* (Corwin *et al.*, 1965).

In studying the asexual blood forms of plasmodia with the aid of the FAT, it has been found that the cytoplasm and not the nucleus of the parasite fluoresces. In addition, such erythrocytic elements as the stromata of uninfected red cells, as well as Schüffner's dots and Maurer's clefts in mammalian malarias also fluoresce (Tobie *et al.*, 1961; Voller and Bray, 1962; Collins *et al.*, 1964a). Furthermore, malarial sporozoites (Sodeman and Jeffery, 1964; Corradetti *et al.*, 1964b) and exoerythrocytic stages (R. L. Ingram, 1963; R. L. Ingram and Carver, 1963; Voller and Taffs, 1963) can be demonstrated with the aid of the FAT; as can an antigen-antibody complex occurring *in vivo* (Kreier and Ristic, 1964).

Plasmodia freed of their red cell hosts and exposed to a FAT-positive antiserum specifically absorb the antibody and render the serum FAT-negative (Ward and Conran, 1966).

Malarial antibody can be demonstrated with the aid of the FAT in nephrotic urines (Kibukamusoke and Wilks, 1965a,b).

A natural, nonprotective antibody has been demonstrated with the aid of the FAT in mice but not in rats (Sodeman and Jeffery, 1965).

A hemagglutination technique for the detection of antiplasmodial antibody was described by Desowitz and Stein (1962) and Stein and Desowitz (1964). They coated formalinized, tanned sheep erythrocytes with antigen obtained from plasmodia freed from their host cells by lysis in distilled water. Hemagglutination studies were carried out with rodent, simian, and human plasmodia. Antibody was detected at high dilutions of serum from malarious subjects, and control serums from subjects never exposed to malaria were negative. Chemotherapy of *P. cynomolgi* infections led to a rapid fall in hemagglutination titer (Desowitz, 1965). Similarly, a program of intensive chemotherapy reduced the high hemagglutination titers of human subjects in an endemic area to practically nil (Desowitz and Saave, 1965). As in the case of the FAT studies, the conclusion was drawn that continuing antigenic stimulus is required to maintain detectable antiplasmodial titers. Desowitz *et al.* (1966) have reviewed work done on hemagglutination, and, as in the case of the FAT, have pointed out that the positive hemagglutination titers of malarial serums need not necessarily indicate an immune state.

In the hands of Desowitz' group formalinized, tanned sheep erythrocytes have been relatively stable, and once prepared and sensitized, could be stored for many weeks. Such relative stability is a *sine qua non* in the application of the hemagglutination test to malarial research on an extensive scale; the fact that the test has been applied by them under field conditions indicates that such stability of the prepared erythrocytes can be achieved. Other groups have unfortunately been less consistently successful in preparing stable, tanned erythrocytes for long-term studies; this technical difficulty has impeded the introduction of this desirable technique into other laboratories. Bray (1964) has suggested the freeing of parasites for antigen with the aid of saponin rather than of distilled water. A hemagglutination test using trypsinized erythrocytes was employed in studying *P. berghei* in the rat (Cox *et al.*, 1966).

A particle agglutination test which entirely bypasses the use of coated erythrocytes, and substitutes sensitized latex or bentonite particles as the antigen carrier, has been described by Todorovic *et al.* (1967b,c), using avian, rodent, simian, and human plasmodial antigens. Latex particles coated with serum antigens from infections with *P. gallinaceum* were used to detect antibody against the human plasmodia. (Fig. 1.)

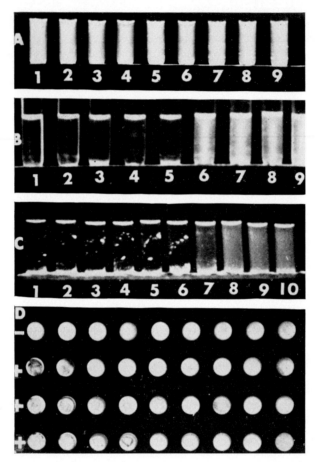

FIG. 1. Appearance of reactions involved in the tube-latex-agglutination (TLA) test. (A) With no reaction with serum from noninfected birds or mammals, tube contents appear uniformly opaque. (B) Reaction to a titer of 1:80 with serum from birds or mammals infected with *Plasmodium* species. In tubes where a reaction occurred the latex particles have settled to the bottom of the tube. (C) Reaction at a titer of 1:160 with serum from birds or mammals infected with *Plasmodium* species. Agitation of tubes causes the appearance of aggregates in tubes in which a reaction occurred. (D) Sedimentation patterns of latex particles. Row 1, no reaction; row 2, titer 1:40; row 3, titer 1:640; row 4, titer 1:160. (Courtesy: Todorovic, R., Ristic, M., Ferris, D., and the *Trans. Roy. Soc. Trop. Med. Hygiene*, 1967.)

Circulating precipitin has been demonstrated with the aid of gel diffusion techniques in malarias of rodents (Zuckerman *et al.*, 1965b; Chavin, 1966; Goberman, 1966; Todorovic *et al.*, 1967b), monkeys (K. N. Brown and Brown, 1965; Todorovic *et al.*, 1967b), and human beings (Tobie, 1965; McGregor, 1965; Mahoney *et al.*, 1966; Todorovic *et al.*, 1967b). Homologous systems were employed except in the study of Todorovic *et al.* (1967b), in which a soluble *P. gallinaceum* antigen reacted in gel with a series of heterologous mammalian antibodies.

2. Immunoglobulins Concerned in Immunity to Malaria

It has already been noted above that the serum gamma globulin level generally rises during malarial infection. This has been recorded for birds (Sherman and Hull, 1960a), rodents (Briggs *et al.*, 1960; Gail and Kretschmar, 1965; Tella and Maegraith, 1965) and human beings (McGregor *et al.*, 1956; Kuvin *et al.*, 1962a,b; Cohen and McGregor, 1963; Curtain *et al.*, 1964; Edozien, 1964; McGregor, 1965). Tella and Maegraith (1965) observed no rise in gamma globulins in monkeys with *P. knowlesi*. Where an increment in gamma globulin is found, it is not yet certain what proportion is due to malarial antibody. In hyperendemic malarial areas, the problem is complicated by the possible role of other concurrent infections in inducing the rise of gamma globulins in the human population (Cohen and McGregor, 1963; Curtain *et al.*, 1965).

Cohen and McGregor (1963) have discussed in detail the role of serum globulins in immunity to malaria. Their extensive studies in Gambia clearly point to the protective effect of the 7 S gamma globulin fraction of immune serum. This conclusion is confirmed for a Nigerian population by Edozien (1964) (see Section, I,B,3). Moreover, the antigen-binding capacity of 7 S gamma globulin from infected human beings in New Guinea was reported by Curtain *et al.* (1964). Protective antimalarial activity is also thought by some investigators to be correlated with a rise in beta as well as gamma globulins (Causse-Vaills *et al.*, 1961; Abele *et al.*, 1965; Tobie, 1965). Beta globulins rose in a transient manner in chicks with *P. lophurae* (Sherman and Hull, 1960a).

Haptoglobins are a group of serum proteins which combine with hemoglobin in the circulation. Low haptoglobin levels are observed in the populations of certain West African areas in which *P. falciparum* is endemic. Blumberg *et al.* (1963) have speculated on the possibility that the endemic malaria may be indirectly responsible for the low haptoglobinemia.

3. Passive Transfer of Immunity to Malaria

The fact that a measure of passive protection against malaria may be transmitted transplacentally from an immune mother to her offspring (reviewed by Bruce-Chwatt, 1963) has suggested that the protective factor must be a molecule small enough to traverse the transplacental barrier. Recent studies (Cohen and McGregor, 1963; Edozien, 1964) have shown that protection resides in the 7 S gamma globulin, which is, indeed, small enough to fulfill this condition (McGregor, 1964a, 1965). Such passive protection of human infants via the placenta is transient, and with its aid the infant exposed to infection is tided over only the initial months of its life (Edozien *et al.*, 1962; Bruce-Chwatt, 1963; Voller and Wilson, 1964). In rodents, protective antiplasmodial antibody can be transmitted from mother to offspring not only via the placenta, but also, and chiefly, via the maternal milk (Bruce-Chwatt, 1963; Adler and Foner, 1965).

Passive protection by the inoculation of immune serum was one of the earliest experimental methods employed in malarial immunology (Taliaferro, 1949). Recently, 7 S gamma globulins from immune adult Gambians were shown to protect Gambian infants against the local strain of *P. falciparum* (McGregor *et al.*, 1963; McGregor, 1964a). The fact that they also protected East African infants suggests that a protective antigen is shared by East and West African strains of *P. falciparum* (McGregor *et al.*, 1963; McGregor, 1964b). Neither 19 S gamma globulin from immune serum nor 7 S gamma globulin from normal serum was protective (Cohen *et al.*, 1961).

Immune gamma globulin from West African human beings protected splenectomized chimpanzees against blood-induced infection with *P. falciparum* from the same area, but not against a drug-resistant, southeast Asian strain (Sadun *et al.*, 1966).

The protective effect of the serum of hyperimmunized donors is proportional to the number of immunizing injections in monkey malaria (Coggeshall and Kumm, 1938) and in rodent malaria (Briggs *et al.*, 1966).

Passive immunization is thought to act specifically against the circulating, mature, asexual forms of the parasite (McGregor, 1964a).

C. Relationships among Plasmodia as Explored by the Previous Methods

Until quite recently, our knowledge of the relationships among plasmodia was based almost exclusively on cross-immunization. Where cross-immunization among strains or species occurred, it was assumed that some antigenic components were shared, and where it was absent, that none were shared. The degree of cross-immunization was further assumed to indicate the degree of antigenic congruence among the strains or species in question. Such cross-protection experiments are still being pursued as new plasmodia present themselves for examination (R. S. Nussenzweig *et al.*, 1966).

Certain of the newer techniques, such as immunofluorescence and hemagglutination, have clearly demonstrated that the assumption of shared antigenic components is correct. Furthermore, they have yielded important information as to the quantity of such congruence. Thus the FAT has revealed varying degrees of antigenic congruence among various plasmodia, even among phylogenetically distant species of the same genus [Voller (1964b) and Tobie (1964) review the literature to date; Collins *et al.*, 1965a,b; Diggs and Sadun, 1965]. Of particular interest is the fact that cross-reactions have occurred regularly among the human and simian plasmodia (Tobie *et al.*, 1963; Voller, 1964a).

One of the first serological demonstrations of an antigenic difference between geographically distant strains of the same plasmodial species was done with the aid of the FAT (Collins *et al.*, 1964c).

Varying degrees of antigenic congruence among plasmodial species have also recently been demonstrated by complement fixation (D'Antonio *et al.*, 1966), by hemagglutination (reviewed by Desowitz *et al.*, 1966), and by latex agglutination (Todorovic *et al.*, 1967c). Cross-reactions using this group of techniques are not, however, indicative of cross-protection (Voller, 1964a; Targett and Voller, 1965).

At the present state of our knowledge neither FAT nor particle agglutination is capable of identifying the shared components in the antigenic melanges of which plasmodia consist. This limitation may eventually be overcome when effective techniques are elaborated to fractionate plasmodial extracts so as to make it possible to analyze each component antigen separately.

On the other hand, by virtue of their reactions of identity and non-identity, gel diffusion techniques (Ouchterlony double diffusion in agar, immunoelectrophoresis) are capable of supplying information as to which precipitinogens are shared by different plasmodia (Zuckerman and Spira, 1963, 1965, 1966; Zuckerman, 1964a; Banki and Bucci, 1964b; Spira and Zuckerman, 1966). Species-specific precipitinogens in plasmodia which do not cross-immunize are of particular interest, since they are under suspicion of being the antigen(s) responsible for the production of specific, protective antibody (Zuckerman, 1964a). For obvious reasons we urgently need to isolate and characterize the protection-inducing antigens from among the spectrum of plasmodial antigens. Similarly we require the separation of protective antibody from the total product of antiplasmodial antibody, and its eventual analysis as an uncontaminated entity. We need more correlative studies like that of Mahoney *et al.* (1966) to compare the various serological methods in use and to determine whether they are demonstrating the presence of the same or different antibodies (Voller, 1964a; Desowitz *et al.*, 1966).

D. PLASMODIAL ANTIGENS AS VACCINES

The idea of using nonliving plasmodial antigen to induce active immunity against malaria is not new. Promising attempts have been made and set aside several times during the past few decades (reviewed by Powell and

Brewer, 1964; Zuckerman, 1964a). More recently, incomplete protection was recorded when monkeys were vaccinated with formalinized *P. knowlesi* (Targett and Fulton, 1965; Targett and Voller, 1965), and when rats were vaccinated with an extract of *P. berghei* (Hamburger, 1965; Zuckerman *et al.*, 1965a). In both studies, the use of Freund's adjuvant was optional. Complement fixing and immunofluorescent antibodies were observed in the serum of monkeys vaccinated with a sonicated preparation of *P. cynomolgi* (Ward and Conran, 1966). Similarly, anti-*knowlesi* precipitins were observed in monkeys vaccinated against *P. knowlesi* (Mahoney *et al.*, 1966).

If vaccination were similarly to prove feasible in human malaria, the induction of even partial active immunity might possibly aid the initially exposed subject (the indigenous infant in a hyperendemic region, or the "unsalted" adult moving into a malarious area for the first time) to withstand the clinical effects of the initial acute attack, while at the same time actively acquiring additional immunity as a result of the patent infection. The question should be raised and debated whether it is desirable to achieve such partial immunity, or whether total eradication should be our eventual goal. For example, Desowitz *et al.* (1966) have pointed out that a campaign of chemotherapy and antivector measures have resulted in the virtual disappearances of detectable antibody in children being reared in a Trobriand Island area. They question "whether it is judicious to deny immunologic protection to a population."

It is true that any project aimed at vaccinating human beings against malaria is shackled by the fact that obtaining human plasmodial antigen is exceedingly arduous (Spira and Zuckerman, 1966). Human plasmodial antigen has yet to be produced *in vitro* in bulk. While the propagation of certain human plasmodia in nonhuman primates has been reported, the amounts of antigen currently obtainable in this manner could scarcely supply a vaccination program aimed at millions of human beings at risk. Synthesis of the protective antigen(s) would, of course, be the ideal solution. However, if these antigens were already fully characterized (which they are not), their synthesis would constitute a problem insurmountable at this time. As an additional complication, the protective antigens probably vary from strain to strain, since geographically widely separated strains do not cross-immunize; and they possibly even vary within the same strain, from relapse to relapse (see Section I,A,2).

Nevertheless, the fact that antiplasmodial vaccination against mammalian plasmodia is technically feasible presents a challenge which cannot be ignored, even though we are indeed still far from being able to apply this fact to human malaria.

II. TRYPANOSOMES

A. ANTIGENS OF TRYPANOSOMES

1. Analysis of Trypanosomal Antigenic Structure

Since the beginning of the century, the immunological analysis of the trypanosomes has been constantly pursued. The extreme antigenic lability of this group of organisms has been repeatedly stressed. As in the case of the plasmodia, the introduction of new immunological techniques has quickly been followed by their application in the analysis of the trypanosomes. In fact, probably due to the relatively easier task of obtaining trypanosomal antigen, immune analysis of the trypanosomes is at certain points some paces ahead of that of the plasmodia. The older immunological literature on the trypanosomes has been reviewed by Taliaferro (1930), the newer, by K. N. Brown (1963) and Weitz (1963a).

An extensive cytochemical study of a group of mammalian trypanosomes was recently undertaken by Williamson and his co-workers (Williamson and Desowitz, 1961; K. N. Brown, 1961; K. N. Brown and Williamson, 1962a,b, 1964; Williamson, 1963; Williamson and Brown, 1964). In this series the reactions of trypanosome antigens to heat and to enzymatic degradation were studied by the diffusion of homologous antisera in gel against the trypanosome extracts following a battery of pretreatments.

In addition to standard cytochemical analysis, a major objective of this group of studies was the characterization of the trypanosomal antigens. Extracts of trypanosomes were obtained following lysis by water, sonication, or grinding in a Mickle disintegrator or in a Hughes press. Cell-free homogenates were studied by column fractionation, ultracentrifugation, various electrophoretic techniques, and gel diffusion.

Williamson's group concluded that their trypanosomal antigens were located chiefly in the cytoplasm, and were almost entirely protein and not carbohydrate, lipid, or nucleoprotein in nature. The DNA in trypanosome nuclei was shown by immunofluorescent analysis to be similar to that in mammalian tissue (Beck and Walker, 1964).

Although some cross-reactions have been recorded among trypanosome strains and species, indicating the sharing of antigenic components, antigenic differences among strains and relapse variants have been stressed (K. N. Brown, 1961, 1963; K. N. Brown and Williamson, 1962a). In addition to the weak common antigens observed in two strains of *Trypanosoma brucei* (Williamson and Brown, 1964), variant-specific antigens were recorded, and fell into two groups: (1) a major 4 S group located largely in the mitochondria, and (2) a minor, unlocalized 1 S group. While the relation of these antigens to the induction of protection is not yet known, absorption experiments suggest that the agglutinating, precipitating, and trypanocidal activities of immune serum are probably associated with the 4 S antigens (K. N. Brown and Williamson, 1964). It has been suggested that each precipitating arc in the gel diffusion studies may represent not one but several antigens. Thus, 4 S trypanosomal antigens described by K. N. Brown and Williamson (1962a) included at least four antigenically similar protein components which could, however, be distinguished by immunoelectrophoresis.

The electrophoresis of sonicated trypanosomes yielded reproducible patterns characteristic of the species and strain studied (Rosseau-Baelde and Misselyn, 1963). Similarly, Pautrizel *et al.* (1962a) have reported that they were able to differentiate the morphologically indistinguishable *Trypanosoma gambiense*, *Trypanosoma rhodesiense*, and *T. brucei* (the *brucei* group) on the basis of their antigenic structure with the aid of specific agglutinins in the serum of parasitized rabbits, but that human patients' serum did not permit such differentiation. Also using the agglutination test and rabbit antisera, Seed (1964) studied the antigenic structure of the culture forms of a series of mammalian trypanosomes. All species of the *brucei* group which he studied cross-reacted; he concluded that culture trypanosomes, unlike the blood-inhabiting forms, are antigenically stable and uniform.

The immunogenic properties of *T. brucei* were not destroyed by β-propiolactone, which inactivates the trypanosomes. In contrast, both formalin and phenol inactivated the trypanosomes and also destroyed their ability to induce the formation of protective antibody (Soltys, 1965).

Trypanosome enzymes, like those of plasmodia, have recently been studied, and are mentioned in the present context since enzymes, being proteins, are also antigens. The level of transaminase, and particularly of glutamic pyruvic transaminase, increases in the serum of cattle or sheep with *Trypanosoma vivax*. Gray (1963) showed that *T. vivax* homogenates contain large amounts of transaminase, and suggested that the high serum levels of the enzyme probably originated in destroyed, circulating trypanosomes. D'Alesandro and Sherman (1964) observed that the lactic dehydrogenase content of *Trypanosoma lewisi* fell at the time when ablastin inhibited the continued multiplication of trypanosomes, but that ablastin did not inhibit the activity of the residual enzyme. Hexokinases were described in *Trypanosoma cruzi* (Warren and Guevara, 1964) and in *T. gambiense* and *T. rhodesiense* (Seed and Baquero, 1965).

2. Antigenic Variation of Trypanosomes

The proverbial antigenic diversity of the trypanosomes is one of the outstanding features of this group of organisms, and is probably a major factor in their pathogenicity, since the immune response to a given strain is ineffective against a relapse strain having a different antigenic structure. However, the exact relationship between antigenic constitution and the ability to infect mammalian or arthropod hosts is still conjectural (Cunningham and Vickerman, 1961a). Most experiments are done with strains maintained in laboratory animals, since many trypanosomes have not yet been cultured (Weitz, 1963a). Possible differences in antigenic structure among forms occurring in mammalian blood, in the arthropod vector, or in culture must be borne in mind.

Under certain conditions underlying antigenic similarities are apparent. Thus, Desowitz (1961) has suggested that antigenic diversity within the *brucei* group may be associated with the polymorphic condition, since such diversity among the three species tends to disappear when the strains become monomorphic. Moreover, the culture forms of different species of the *brucei* group have an apparently stable serotype, and extensive cross-agglutination occurs among them (Seed, 1964).

Despite such similarities, antigenic dissimilarities have repeatedly been described among strains and species of trypanosomes. Thus the blood forms of members of the *brucei* group were distinguished by specific agglutinabilities (Pautrizel *et al.*, 1962a; Gray, 1962). Cunningham and Vickerman (1961a,b) compared 12 East African cattle (*T. brucei*) and human (*T. rhodesiense*) strains from the *brucei* group by the agglutination of frozen and thawed trypanosomes in infected mouse blood. They distinguished 6 distinct antigens, 1 to 5 per strain, some of which were shared by different cattle—and human—species and by successive relapse strains of the same species. Only a single antigen was detected in a clone strain of *T. rhodesiense*. The presence of both shared and variant-specific antigens in trypanosomes was also described by Weitz (1963a,b), who stated that the soluble antigens (see Section II,A,3) were variant-specific, as tested by gel diffusion and immunoelectrophoresis, whereas the bound antigens tended to be shared. Both shared and variant-specific antigens were also demonstrated by means of gel diffusion and immunoelectrophoresis (K. N. Brown and Williamson, 1962a; K. N. Brown, 1963; Williamson and Brown, 1964). The trypanolytic power of immune serum was found to be exclusively variant-specific (K. N. Brown and Williamson, 1962a).

Gray (1962) came to the conclusion that various strains of *T. brucei* are potentially capable of producing numerous antigens, and that the antigenic proportions in a strain population are affected by the presence of specific antibody against certain of the antigens, resulting in their suppression. His results further suggested that *T. brucei* strains may contain a major shared component and minor variant-specific components. In contrast, Watkins (1964) considered that variations in the antigenic constituents of trypanosome populations could be explained as due to mutation and selection rather than adaptation.

V. Nussenzweig et al. (1963a,b) compared a series of strains of *T. cruzi* isolated from man, from other mammalian hosts and from triatomid vectors. They employed agglutination and gel diffusion tests. Up to 10 precipitin arcs developed in the presence of rabbit antiserum. The authors concluded that the clinical diversity of Chagas' disease may be based on differences in *T. cruzi* serotypes.

As in the case of the plasmodia (see Section I,C), K. N. Brown (1963) has emphasized the need for the purification and fractionation of trypanosomal homogenates into separate antigenic components, to permit the analysis of each as a distinct entity. The unequivocal association of each of the biological and serological effects of a trypanosome on its host with a given antigenic component must await such fractionation.

3. Exoantigens of Trypanosomes

In addition to the cell-bound antigens discussed up to this point, soluble trypanosomal antigens (called "exoantigens" by Weitz, 1960a, 1963a) can be demonstrated in the plasma of infected hosts. Exoantigens have been found in infected rats (Weitz and Lee-Jones, 1961; Weitz, 1960a, 1963a; Gray, 1961b; Seed, 1963; Seed and Weinman, 1963; Gill, 1965c; Lanham, 1966), mice (Dodin et al., 1962; Weitz, 1963a), goats (Gray, 1961a,b), cattle (Weitz and Lee-Jones, 1961), and human beings (Toussaint et al., 1965), and in the culture media in which *T. cruzi* has been grown (Tarrant et al., 1965). The presence of exoantigens can be proved by inoculating serum suspected of containing them into clean homologous hosts. Thus, antisera against *T. brucei* exoantigen in rat serum was prepared by inoculating such serum into rats (Weitz, 1963a). No antibody was produced against the rat serum proteins, but precipitin was produced against the trypanosomal exoantigen.

Exoantigens, which are apparently excretions or secretions of the trypanosomes into the host's plasma (Weitz, 1960a), also occur on the surface of the organisms, but are not firmly bound at this site since washing removes them (Weitz, 1960b), rendering the trypanosome both noninfective and nonagglutinable. Exoantigen, in addition to being extruded into the plasma, is also an integral part of the intact trypanosome (Weitz, 1963a). Thus, it partially inhibits agglutinin produced in response to a homogenate of whole trypanosomes (Weitz and Lee-Jones, 1961). Gill (1965c) found that a *Trypanosoma evansi* homogenate contained three antigens, one of which was identical with its exoantigen. Gray (1961a,b) has reported that fewer precipitinogens occur in trypanosome homogenates than in the plasma of animals infected with the same organism.

Exoantigen from a single trypanosome may consist of several distinct components. Thus, Seed (1963) distinguished between an exoantigenic precipitinogen and a protective factor, separable from one another by elution.

The prior administration of exoantigens protects against homologous viable trypanosomes (Weitz, 1960a, 1963a; Weitz and Lee-Jones, 1961). The protective activity of exoantigen in the serum of mice with *T. gambiense* or *Trypanosoma congolense* resided in the gamma globulin fraction precipitated with 50% ammonium sulfate (Dodin *et al.*, 1962). *T. rhodesiense* exoantigen was similarly isolated by ammonium sulfate (Seed and Weinman, 1963). While exoantigen fails to take up protein stain (Weitz, 1960b), its close association with a protein component of the serum is attested by the above fractionation procedures. Dissociation from such protein may destroy its immunogenicity (Weitz, 1963a). *Trypanosoma brucei* exoantigen moves electrophoretically more slowly than gamma globulin (Weitz, 1960a,b). Lanham (1966), using column fractionation procedures, found that both the precipitinogenic and the protective factors in *T. brucei* exoantigen in mouse serum were eluted chiefly with the gamma globulins, but also in part with the alpha and beta globulins, while the albumin fraction contained some protective exoantigen but no precipitinogen. The major component of an exoantigen of *T. cruzi* grown *in vitro* is a glycoprotein (Tarrant *et al.*, 1965), against which detectable circulating antibody can be produced. A puzzling observation, in view of the well-documented protective effect of trypanosomal exoantigen, is the fact that, when added to an inoculum of viable trypanosomes, exoantigen enhanced the infectivity of *T. brucei* in mice (Weitz, 1963a).

Precipitin, protection, and immunofluorescent tests show that the exoantigens of *T. vivax* and the *brucei* group, in contrast to bound antigens, are species-specific (Weitz, 1963a,b). *Trypanosoma rhodesiense* exoantigens are even strain-specific, as judged by cross-absorption, gel diffusion, and protection tests (Seed and Weinman, 1963). The specificity of an *in vitro* exoantigen of *T. cruzi* was demonstrated by complement fixation (Tarrant *et al.*, 1965). Exoantigen regularly detected antibody in the serum of patients with Chagas' disease (Toussaint *et al.*, 1965).

B. ANTITRYPANOSOMAL ANTIBODY

1. Established and Recent Methods of Demonstration

A generation ago it was already possible to demonstrate antitrypanosomal antibody by means of protection, agglutination, complement fixation, adhesion phenomena, and the inhibition of reproduction. Each of these techniques has also been employed in more recent studies.

Protective antibody is produced in numerous trypanosomal infections, and can be passively transferred. Thus, Kagan and Norman (1962) harvested serum from surviving mice primed with an avirulent strain of *T. cruzi*, and challenged with a virulent one. Such serum partially protected mice exposed to the virulent strain by prolonging the infection and by raising the survival rate from 0 to 28%. K. N. Brown and Williamson (1964) found that the agglutinative and trypanocidal activity of *T. rhodesiense* antiserum are both associated with the 4 S group of antigens. Similarly Watkins (1964) concluded that the agglutinating and the protective antibody in mice with *T. brucei* are identical. Passive immunization of such mice delayed death by several days. Protection against relapse variants was specific for each variant strain. Pautrizel *et al.* (1965) found that rabbit antiserum against *T. gambiense* protected mice against this organism. As has already been pointed out in Section II,A,3, trypanosomal exoantigens induce the formation of protective antibody. Killed *T. congolense* and *T. cruzi* were capable of partially protecting mice against homologous viable organisms (Johnson *et al.*, 1963). Similarly, killed *T. evansi* protected mice and guinea pigs against viable parasites (Gill, 1965b).

The agglutination reaction was used by Cunningham and Vickerman (1961b) to study 12 strains of the *brucei* group. They attributed the enhanced agglutination which they observed to the fact that their antigen consisted of whole blood and not of washed trypanosomes, since agglutinogenic exoantigens on the surface of the trypanosomes are removed by washing. Agglutination reactions have been used to distinguish between trypanosomes of the *brucei* group and in studying *T. equiperdum* infections in rabbits (Pautrizel *et al.*, 1962c). Common antigens were detected by agglutination in a series of relapse strains of the *brucei* group (Seed, 1963, 1964). It is noteworthy that washed trypanosomes were used in this study. In contrast, Watkins (1964) found that strains of *T. brucei* agglutinated in homologous antiserum, but generally failed to agglutinate in antiserum against the preceding or the succeeding strain in the relapse series. McGhee and Hanson (1963) studied the interrelations among a group of *Crithidia* by agglutination tests in chicken antisera. *Crithidia* derived from Diptera were closely related to one another, as were those from Hemiptera, but there were few cross-reactions between the two groups.

An improved complement fixation test for the diagnosis of sleeping sickness was devised by Pautrizel *et al.* (1959), using *Trypanosoma equiperdum* trypanosomes freed of red cells, and therefore devoid of anticomplementary activity, as antigen. The use of culture forms of *T. cruzi* as antigen was advocated in a complement-fixing test for the diagnosis of Chagas' disease (Berrios and Zeledon, 1960). Complement-fixing antibody against *T. cruzi* appeared in the serum of a number of newborn infants of mothers with Chagas' disease. Their antibody titres became negative in a few months (Figallo Espinal, 1962).

Adhesion phenomena in trypanosomiasis, in which an antibody called "thrombocytobarin" or "adhesin" plays an essential role, have been reviewed by Zuckerman (1964b). The adhesion of several species of mammalian trypanosomes to the macrophages of infected rodent hosts was described in a study by Dodin et al. (1962) that was not included in the above review.

Ablastin, the reproduction-inhibiting antibody occurring in the serum of rats with T. lewisi (Taliaferro, 1923), has recently been restudied by Ormerod (1963), in whose view ablastin is not distinct from antitrypanosomal antibody, as it is considered by Taliaferro to be.

As in malaria, the newer techniques for the detection of antibody have also been introduced into research in trypanosomiasis. Information is still very inadequate as to whether the various established or new serological techniques are demonstrating the presence of the same or of distinct antibodies, although some studies in this direction have already been undertaken (e .g., Pautrizel et al., 1962c; Gill, 1965a).

Protracted trypanosomiasis leads to the production of precipitins, demonstrable by double diffusion in agar, in the serum of cattle with T. vivax and of human beings with T. gambiense (Gray, 1961a,b). While gel diffusion experiments demonstrated differences among the antigens of a stable strain of T. rhodesiense as compared with a relapse variant of the same strain (K. N. Brown, 1961), Williamson and Brown (1964) reported reactions of identity among T. brucei strains and their variants. Antiserum against T. vivax yielded both species-specific gel diffusion reactions and also cross-reactions with T. gambiense and T. brucei (Gray, 1961a,b). One of the methods of choice for the detection of trypanosomal exoantigens and for the investigation of their properties has been double diffusion in agar (Gray, 1961b; Weitz, 1963a; Seed, 1963; Seed and Weinman, 1963; Gill, 1965c). Gray (1961b) considers gel diffusion to be a more sensitive reaction in trypanosome research than agglutination. Autoimmune antiheart precipitins were demonstrated with the aid of gel diffusion by Kozma (1962) in guinea pigs and human beings with T. cruzi.

Immunofluorescent techniques have been employed in the diagnosis of Chagas' disease (Fife and Muschel, 1959; Sadun et al., 1963) and of sleeping sickness (Sadun et al., 1963). Immunofluorescent cross-reactions were observed between T. rhodesiense and T. gambiense (Sadun et al., 1963). Trypanosoma lewisi antigen also cross-reacted with T. rhodesiense and T. gambiense antisera (Williams et al., 1963). Fife and Muschel (1959) compared the FAT with complement fixation in the diagnosis of Chagas' disease, and found the FAT somewhat the more sensitive of the two.

Weitz (1963b) considered the FAT a potentially useful tool in trypanosome research, particularly because it can aid in locating the antigen(s) in the body of the trypanosome being studied. In investigating the exoantigens of *T. brucei* and *T. vivax*, he obtained FAT-positive results only against the homologous trypanosomes. Essenfeld and Fennell (1964) succeeded in demonstrating both the trypaniform and the leishmaniform stages of *T. cruzi* by means of the FAT, where double diffusion in agar failed.

The hemagglutination of tanned red cells sensitized with trypanosome homogenates or exoantigens has been explored by Weitz and Lee-Jones (1961). The reaction can yield both qualitative and quantitative information

Tanned erythrocytes coated with soluble *T. evansi* antigen were capable of detecting anti-*evansi* antibody in high dilution (Gill, 1964). Agglutination and hemagglutination tests were the methods of choice in detecting *T. evansi* antiserum, as they became positive earlier than other tests (precipitin, FAT, complement fixation; Gill, 1965a).

The immunoconglutinin reaction, which measures a specific anticomplement antibody, has been described in a series of trypanosomal infections (D. G. Ingram and Soltys, 1960, reviewed in Zuckerman, 1964b). Pautrizel *et al.* (1962b,c) found that conglutinin reactions paralleled the appearance of antitrypanosomal antibody in human patients and in experimental animals with sleeping sickness.

The use of disc electrophoresis in the analysis of certain of the lower Trypanosomatidae (Hill and Guttman, 1965) has again demonstrated the excellent resolving power of this technique by comparison with older electrophoretic methods (Williamson and Desowitz, 1961; K. N. Brown, 1961). It is to be anticipated that this technique will find extensive application in the analysis of trypanosomes.

2. Immunoglobulins Concerned in Immunity to Trypanosomiasis

In addition to the classic protective, precipitating, and agglutinating antibodies discussed in Section II,B,1, a beta-2 macroglobulin has been prominently featured in recent studies on the diagnosis of sleeping sickness. A strain of trypanosomes of the *brucei* group, transmitted to three cercopithecus monkeys, induced a marked and early rise in the titer of the beta-2 macroglobulins in their serum (Mattern *et al.*, 1963). This observation was quickly extended to human beings with trypanosomiasis at a West African site where sleeping sickness is endemic (Bentz, 1963). Elevated beta-2 macroglobulin titers were recorded in 97 % of several hundred cases (Bentz, 1963; Mattern, 1964), and the titer rose in some of the patients when they were still clinically healthy. While it was not considered certain that a rise in beta-2 macroglobulin was pathognomonic of trypanosomiasis, persons exhibiting such a rise were strongly suspected of being trypanosome carriers (Bentz and Macario, 1963; Bentz, 1964; Nicoli *et al.*, 1964). Attempts were made to standardize the technique used in demonstrating the presence of beta-2 macroglobulin (Bentz and Diallo, 1963; Bentz, 1964), and to define the ratio of beta-2 macroglobulin to gamma globulin in health and during infection (Bentz and Mattern, 1963). The technique as it is now practiced in the field (Mattern, 1964) permits diagnosis at an earlier stage in the disease than was heretofore possible. It consists of collecting finger-prick blood in heparinized capillary tubes, and developing a standard dilution of the plasma by double diffusion in agar against antiserum against the beta-2 macroglobulin fraction of horse serum. Precipitation arcs appear in this standardized system only when titers of beta-2 macroglobulin are significantly elevated. Cross-reactions with sera from patients with other diseases have not generally been noted, although a rise in beta-2 macroglobulins has been reported in human malaria (Abele *et al.*, 1965; Tobie, 1965). Hundreds of daily tests can conveniently be done at negligible expense by comparison with other titration procedures involving ultracentrifugation, chromatography, or immunoelectrophoresis (Bentz, 1964). Lumsden (1966) has suggested a modification of the precipitation in gel technique similar to that introduced into malaria research by Tobie (1965). He incorporates the antibeta-2 macroglobulin antibody in the gel, and titrates the beta-2 macroglobulin content of the examined serum by measuring the diameter of the zone of precipitation.

The accuracy of this diagnostic test is estimated at more than 90%, as compared with 30% accuracy attained by the microscopic examination of blood or lymphatic tissue (Mattern, 1964). Elevated titers of beta-2 macroglobulin in the cerebrospinal fluid indicate the onset of the meningo-encephalitic stage of the disease. Before this stage plasma titers may rise at a time when cerebrospinal titers are still normal. The early onset of a positive reaction, and the prompt return to normal titers following anti-trypanosomal treatment (Bentz and Macario, 1963) should aid materially in assessing the degree of endemicity of sleeping sickness in areas where the disease is established.

III. *THEILERIA*

Research on the theilerias is fragmentary due to the expense of using cattle as host animals, and to the fact that theilerias affecting small laboratory animals are unknown. Barnett (1963) has reviewed our scanty information on the antigenic structure of the theilerias and on their host-parasite relationships. What antigenic definition we have is based largely on cross-immunity, and antigenic analysis in the strict sense is still largely lacking for this important group of parasites.

Barnett lists three species considered valid since they do not cross-immunize: *Theileria mutans*, *T. annulata*, and *T. parva*. The schizogonic and the blood stages of *Theileria* may differ antigenically, since cattle recovered from the former were still susceptible to the latter (Neitz, 1957).

Several antigen-antibody systems have been used in attempts to develop a means for the serological diagnosis of theileriosis or to provide better criteria for differentiation among *Theileria* species. Bailey and Cowan (1961) were able to demonstrate a reaction in the agar gel diffusion system between an extract obtained from lymph nodes of *Theileria*-infected cattle and the serum of cattle recovered from clinical theileriosis. Schaeffler (1963) prepared an agglutinogen from *Theileria* species of deer based upon the technique used for preparation of agglutinating antigen from *Anaplasma marginale* (Ristic, 1962). Staining *Theileria* by the FAT gave no serological cross-reaction with *A. marginale* or *Eperythrozoon ovis* (Kreier et al., 1962).

In recent years considerable progress has been made in devising tissue culture systems for *in vitro* propagation of *Theileria* species. These systems consititute a simpler means for studying the immunoserological properties of *Theileria* (Hulliger *et al.*, 1966; Tsur and Adler, 1965). The effect of "immune" sera and of "immune" cells on multiplication of *Theileria* propagated in tissue cultures and on infectivity for cattle were investigated by Hulliger *et al.* (1966). Cells infected with *T. parva* were grown with and without "immune" sera secured from convalescent animals. Three animals which were inoculated with cells cultivated in "immune" sera developed more severe theileriosis than two other animals which were inoculated with tissue grown in a control serum obtained from susceptible animals. Also, no significant differences in multiplication rate of the cells were noted and no indication of an effect due to "immune" sera on the intracellular form of the parasite was found.

This and other experimental evidence concerning the immune mechanism of *Theileria* studied *in vitro* prompted Hulliger *et al.* (1966) to conclude that the immune factor associated with *T. parva* is formed within the parasitized cell and can apparently act directly on the parasite. The authors, however, feel that additional evidence will be needed to support this concept.

IV. *BABESIA*

Babesia is a genus of parasitic protozoa which invade and multiply in the erythrocytes of various domestic and wild animals, causing the disease known as babesiosis. Little is known about the host spectrum of *Babesia* which fact is best illustrated by a recently described case of babesiosis in man due to *Babesia bovis* (Skrabalo and Deanovic, 1957).

The causative agent of "Texas fever," *Babesia bigemina*, was the first protozoan parasite shown to be transmitted by an arthropod (Smith and Kilbourne, 1893). For more than half a century babesiosis was recognized as a specific infectious entity and a severe economic threat to the cattle industry in many areas of the world; however, the nature of the immune host-parasite interaction has remained little understood.

A characteristic feature of babesial infections is that animals which recover from an acute infection become carriers of the respective hemoparasite. These carrier animals cannot be clinically distinguished from noninfected animals and *Babesia* cannot be found in blood films by any of the contemporary staining methods. Thus, in order to identify carrier animals, it was necessary to develop serological methods to detect specific antibodies rather than the parasites themselves.

Recently some fundamental knowledge concerning the immunoserology of several babesial species has led to the development of serodiagnostic procedures which provided a means for studying the pathogenesis of babesiosis and the detection of animals with subclinical infections. The antigens used in these procedures originated from erythrocytes and serum of animals with acute babesiosis.

A. ERYTHROCYTIC ANTIGENS

1. Complement Fixation Tests

The antigen consisting of stromata of erythrocytes from acutely infected horses was used in a complement fixation (CF) test for diagnosing equine babesiosis (Hirato *et al.*, 1945). These authors found that antibodies became detectable by the test 11 to 15 days after the onset of parasitemia and persisted for at least 100 days.

Mahoney (1962) prepared an antigen from water-lysed erythrocytes of cattle acutely infected with *Babesia argentina* and *B. bigemina*. The antigen was shell-frozen in 1.0 ml quantities at −79°C and then preserved at −20°C. Homologous antibodies became detectable in the CF test within 7 to 21 days after infection and remained detectable during the acute and convalescent stages of infection. The antigens were apparently species-specific and reacted with heterologous sera at low titers. The author indicated the need for additional work in order to determine the usefulness of this test for studying the epidemiology of bovine babesiosis.

Schindler and Dennig (1962a,b) prepared a CF antigen from erythrocytes of dogs acutely infected with *Babesia canis* by means of immunohemolysis and, more recently, by exposing the erythrocytes to 0.3% saponin. The antibody became detectable in the CF test within 12 to 34 days following infection. No serological cross-reaction was observed between *Babesia rodhaini* and *B. canis* using the antigen of the latter parasite. Lyophilization was found to be an efficient means for long-term preservation of the antigen.

2. Gel Precipitation Test

By means of precipitation with protamine sulfate, Ristic and Sibinovic (1964) obtained an antigen from the lysate of erythrocytes of horses acutely infected with *Babesia caballi*. The antigen was termed the protamine sulfate (PS) antigen. An antibody was demonstrated in horses with babesiosis by bringing the PS antigen in gel into contact with sera of infected horses. The "positiveness" of equine sera in the gel precipitation (GP) test was correlated with the persistence of the latent infection as determined by transmitting infections from these carriers into susceptible horses. The specificity of the GP test was shown by the absence of reaction with sera of horses with various other infections, including viral infectious anemia.

3. A Polysaccharide (BPS) Antigen and Its Use in the Passive Hemagglutination (HA) Test

Exposure of babesial PS antigen to boiling water for 30 minutes resulted in the formation of (1) a serologically inactive precipitate and (2) a serologically reactive supernatant which was termed the boiled PS (BPS) antigen (S. Sibinovic *et al.*, 1966). In a precipitation test with immune babesial serum, the PS antigen formed two precipitin lines while the BPS antigen formed only one line; the two lines obtained with the PS antigen coalesced into a single BPS line at the point at which the two precipitin lines intersected.

The serological reactivity of the PS antigen was destroyed by Taka-Diastase and trypsin; however, only Taka-Diastase destroyed the serological reactivity of BPS antigen. Results from the Folin test for carbohydrates indicated that the BPS antigen contained as much as 300 μg of polysaccharide per ml. The sedimentation coefficient of this antigen was 4 S (S. Sibinovic *et al.*, 1966).

The BPS antigen was adsorbed on intact sheep erythrocytes and these cells were then used in a passive hemagglutination test for detection and titration of antibodies in serum of horses infected with *B. caballi* (S. Sibinovic *et al.*, 1967). Reactions were observed with convalescent sera of *B. caballi* but not with convalescent sera of *B. canis* or *B. rodhaini*.

B. SERUM ANTIGENS

1. Identification

A precipitation reaction in gel was observed when serum from a horse acutely infected with *B. caballi*, used as the antigen, was allowed to react with serum from a convalescent horse, used as antibody. The antigen disappeared from the serum of infected horses approximately 1 month after infection. Antibodies to the serum antigen were detectable at 3 months after infection and persisted at variable levels for an additional 7 months (K. H. Sibinovic *et al.*, 1965).

2. A Tube Latex Agglutination Test

The serum antigens isolated from dogs with acute babesiosis were adsorbed to latex particles having an average diameter of 0.81 μ, in accordance with the method described by Todorovic *et al.* (1967b,c) for adsorption of plasmodial antigens. The sensitized latex particles were used as agglutinogens in a tube latex agglutination (TLA) test for titration of antibodies in sera of dogs infected with *B. canis*. The earliest that antibodies were detected in the TLA test was approximately 15 days following infection; about 2 months later they reached a maximal titer of 1:640 and persisted at titers of 1:40 to 1:80 during the next 4 months. The persistence of serum agglutinins was generally correlated with the persistence of a carrier stage which, in some cases, terminated between 8 and 18 months postinfection and, in other cases, persisted as long as 3 years postinfection. Recently, the TLA test has also been successfully used to diagnose a second case of human babesiosis (Braff and Condit, 1967; Anonymous, 1967).

In this latter case, a 46-year-old Caucasian male resident of San Francisco, California, was hospitalized in June, 1966, with chills and fever. Malaria was suspected; however, microscopic examination of his blood revealed the presence of intraerythrocytic parasites of unknown nature. This patient's serum was also examined for malarial antibodies, using the fluorescent antibody (FA) technique, and found negative to 10 malarial antigens, including the 4 human species of malaria. The *Babesia* research team at the United States Department of Agriculture also examined blood films from this patient and found that the forms of the parasite in question resembled *Babesia caballi* and *Babesia equi*. We have examined the serum from this patient by the TLA test and found the following: the serum reacted at a titer of 1:80; positive dog serum reacted at 1:160 and 5 control human sera showed no reaction at a 1:2 dilution.

3. Biophysical and Biochemical Properties

K. H. Sibinovic (1966; K. H. Sibinovic *et al.*, 1967a,c) subjected serum antigens from dogs and horses with acute babesiosis to purification procedures which resulted in the concentration of the serologically reactive substances approximately 100-fold. Analytic ultracentrifugation studies showed that two components were present in both the dog and horse antigens and that the components possessed significantly different sedimentation coefficients. The coefficients of the dog components were 7 S and 20 S while those of the horse were 8 S and 23 S. The presence of two serologically reactive components was also confirmed by means of paper electrophoretic and by zonal density gradient centrifugation procedures. Each of the two antigenic components of equine and canine origin reacted in a gel precipitation test with sera of horses, dogs, and rats which had recovered from acute babesiosis.

Biophysical properties such as maximal absorption of ultraviolet light at 280 mμ, thermolability, denaturation by distilled water, and precipitation by ammonium or sodium sulfate and by low pH buffers, indicated that the antigens were of proteinaceous nature. Amino acid analysis showed that the composition of the antigens was different from that expected for serum proteins or for most other common proteins since proline was absent (K. H. Sibinovic, 1966; K. H. Sibinovic *et al.*, 1967a). Methionine, the amino acid known to be essential for the metabolism of malaria parasites (Trager, 1957), was absent in the 7–8 S components and histidine was absent in the 20–23 S components. Papain digestion of the antigens indicated the presence of peptide chains containing leucine or glycine, which are adjacent basic amino acids susceptible to hydrolysis by papain. Trypsin digestion of the antigens indicated the presence of peptide bonds between the carboxyl groups of arginine or lysine and the amino group of another basic amino acid susceptible to degradation by trypsin. The destruction of serological reactivity by lipase indicated the presence of lipid in both fractions and the destruction of similar reactivity by lecithinase indicated the presence of phosphatides. The effect of Taka-Diastase on the 20–23 S components suggested the presence of polysaccharides containing alpha-4-glucose linkages. Therefore, the data indicated that the antigens are of complex chemical structure consisting of peptides, lipids, phosphatides, and polysaccharides, all of which apparently were essential for serological reactivity.

4. Immunogenic Properties

The immunogenic activity of the serum antigens was tested extensively in rats and dogs (K. H. Sibinovic, 1966; K. H. Sibinovic *et al.*, 1967c). When homologous purified antigens were used, these animals were protected against challenge infections. The animals apparently became susceptible to challenge infection at the time that the TLA titer fell below 40. When rats were immunized with the antigen of dog and horse origin, respectively, partial protection was obtained when these rats were challenged with *B. rodhaini*. The degree of protection in the challenged rats was considered to be significant if the degree of parasitemia was the criterion for evaluation of protection; however, when anemia was considered the criterion, the protection was not significant. Antisera prepared against purified serum antigens of dog or rat origin produced complete protection for periods up to 45 days when administered subcutaneously in a 20-ml dose to dogs and in a 3-ml dose to rats.

Intravenous inoculation of purified *B. canis* or *B. rodhaini* serum antigens into normal rats resulted in anemia manifested by a progressive decrease of erythrocyte counts beginning within 2 hours after inoculation (K. H. Sibinovic *et al.*, 1967b). Recovery from anemia and the return of erythrocyte counts to preexperiment values occurred within 3 days after inoculation. An attempt was made to trace inoculated antigen by staining blood films from inoculated animals with the fluorescein-labeled anti-antigen-antibody. A "doughnutlike" halo effect around erythrocytes was first noted within 10 minutes after inoculation. Samples collected subsequent to this revealed intense fluorescence of the entire erythrocyte surface. There was no fluorescence of erythrocytes collected 8 to 10 hours after injection.

5. A Protective Antibody and Its Effect on the Organism and on the Course of Parasitemia

Since the appearance of an antibody to the serum antigen coincided with the beginning of latent infection and acquired resistance, it was suspected that it may be a protective antibody. Titration of the serum antigen and its antibody throughout the course of babesial infections revealed that its titer is inversely related to the severity of parasitemia and the concentration of serum antigen (Fig. 2). More exactly, it was ascertained that in animals with babesial infections, the disappearance of the antibody to serum antigen was followed by the appearance of parasitemia and serum antigen, which then resulted in the reappearance of antibody and disappearance of both parasitemia and the serum antigen (K. H. Sibinovic *et al.*, 1967c).

The exact mechanism by which the antibody to serum antigen may affect the parasite and its development is not clearly understood. Studies of the *in vitro* effects of immune globulins revealed that the primary effect was agglutination of the infected erythrocytes or of the parasites freed by

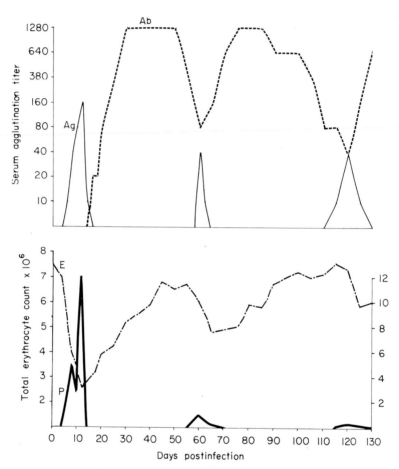

Fig. 2. Immunohematological abnormalities in a dog infected with *Babesia canis*. The upper graph shows the results of titrating serum antigen (Ag) and its corresponding antibody (Ab) by means of a tube latex agglutination (TLA) test. In each case latex particles were sensitized with the antibody and the antigen respectively. The lower graph shows the correlation between parasitemia (P) and erythrocyte counts (E). It is evident that the antibody titer is inversely related to the severity of parasitemia and the concentration of serum antigen. (Courtesy of Dr. Kyle H. Sibinovic.)

sonic oscillation (Fig. 3). Normal sera or globulins did not agglutinate infected erythrocytes or free parasites, neither were normal erythrocytes agglutinated by the immune globulins.

When washed erythrocytes from a dog acutely infected with *B. canis* were stained with the fluorescein-labeled anti-antigen-antibody, a diffuse fluorescence of the cytoplasm of infected erythrocytes was noted, indicating that the serum antigen is not an integral part of the parasite.

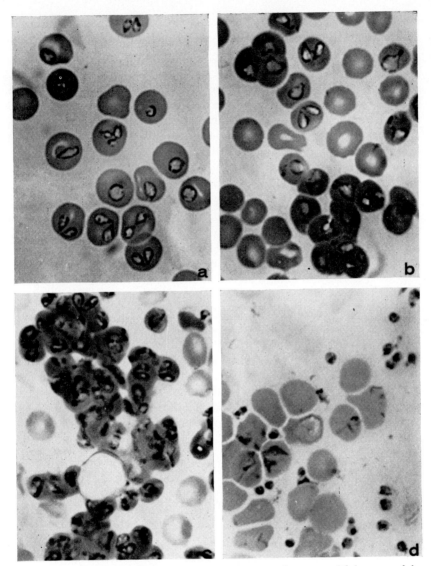

FIG. 3. Effect of treating *Babesia canis*-infected erythrocytes with immune globulins prepared against babesial serum antigen of dog origin. Time after adding immune globulin: (a) 0 minutes; (b) 30 minutes; (c) 60 minutes; (d) 120 minutes. Giemsa stain: × 950. (Courtesy of Dr. Kyle H. Sibinovic.)

Although additional work will be needed to ascertain more exactly the role of serum antigens in acquired immunity in babesiosis, it appears that immunogenic mechanisms known to occur in viral and bacterial infections are also operational in this hemosporidian infection. However, the antigens involved in the latter infection apparently represent a substance associated with the developmental cycle of the parasite rather than the parasite per se.

6. Fluorescent Antibody Techniques

Ristic *et al.* (1964) used the fluorescent antibody techniques (FAT) to study the major erythrocytic growth and developmental stages of *Babesia* isolated from horses in Florida. By use of this serological staining method rather than histochemical techniques it was hoped that all antigens capable of stimulating the production of an antibody in infected horses could be revealed. Considered on the basis of their size, mode of multiplication, and the number of parasites in the infected erythrocyte, two species of parasites were observed. *Babesia caballi* comprised the major part of the parasitic population. This species was larger than half the diameter of an erythrocyte, multiplied by binary fission, and appeared alone or in pairs in erythrocytes (Fig. 4A). The second form, which was typical of *B. equi*, was also observed. These parasites were less than half the diameter of the erythrocytes, multiplied by division into four daughter cells, occurred alone or in groups of two to four, and were irregularly arranged in the erythrocytes (Fig. 4B). Thus, it may be concluded that the equine blood specimen from Florida contained both *B. caballi* and *B. equi*.

The one-step fluorescein-labeled antibody inhibition test in which an antigen is reacted simultaneously with a mixture of fluorescein-labeled and unlabeled antibody was used by Ristic and Sibinovic (1964) as a means of detecting circulating anti-*Babesia* antibody. These authors concluded that a one-step fluorescein-labeled antibody inhibition test could be considered only as a research tool requiring further study and evaluation of its accuracy and specificity.

C. Autohemagglutinins, Opsonins, Erythrophagocytosis, and Anemia

Techniques used in studies of autoimmune processes in anaplasmosis (Mann and Ristic, 1963b; Ristic, 1961) have been applied to similar studies of the pathogenesis of anemia associated with *B. rodhaini* and *Plasmodium berghei* infections (Cox *et al.*, 1966; Schroeder *et al.*, 1966). In both infections, the appearance of autohemagglutinins was found to be associated with the appearance of anemia and erythrophagocytosis in the spleen and bone marrow.

Recently, by means of an *in vitro* technique, a heat-stable opsonin was demonstrated in the sera of rats infected with *B. rodhaini* or *P. berghei* (Schroeder, 1966). The opsonin sensitized normal autologous and homologous erythrocytes to phagocytosis by macrophages. The presence and reactivity of the opsonin in sera of infected animals coincided with the degree and persistence of anemia (Fig. 5). A factor that inhibited erythro-

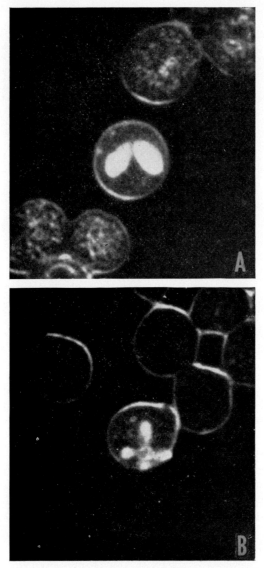

FIG. 4. *Babesia caballi* (A) and *Babesia equi* (B) stained by the fluorescein-labeled antibody technique. The "maltese cross" form of the developing parasite in B is shown. Fluorescent-antibody stain: × 1200.

phagocytosis *in vitro*, but apparently not *in vivo*, was demonstrated in sera from infected animals. This factor could be removed from the sera, without removing hemagglutinins or opsonins, by absorption with intact erythrocytes at 25°C.

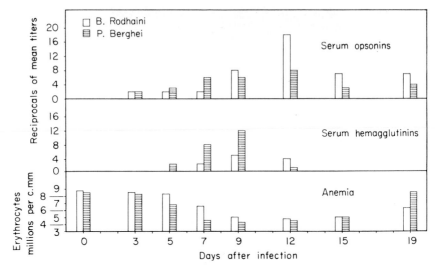

FIG. 5. Relationship of serum opsonic titers and hemagglutinin activity for trypsin-treated erythrocytes to the onset and duration of anemia in 15 *Babesia rodhaini-* and 15 *Plasmodium berghei*-infected rats. (Courtesy of Dr. W. F. Schroeder.)

It was suggested that the anemia observed in babesiosis and malaria may be a partial result of autoimmunization.

V. *ANAPLASMA*

Whereas all attempts have failed to isolate a whole-cell babesial antigen from infected erythrocytes, such *Anaplasma* antigens have been isolated, purified, and used as an agglutinogen for detection of anti-*Anaplasma* antibody (Ristic, 1962). The isolation of this *Anaplasma* antigen led to the development of a capillary tube agglutination (CA) test for diagnosing various forms of *Anaplasma* infection including the subclinical, or so-called "carrier," form the detection of which is apparently most essential from the standpoint of controlling field anaplasmosis (Ristic, 1962; Welter and Zuscheck, 1962; Welter, 1964).

The ease with which *Anaplasma* whole-cell antigen can be isolated and with which its serological reactivity can be preserved for long periods of time by storage at 4°C, underlies another basic biological difference between *Anaplasma* and *Babesia* species. These characteristics of *Anaplasma* are in accord with those known to exist in bacterial and rickettsial species and are also in agreement with findings concerning the structural (Ristic and Kreier, 1963; Ristic and Watrach, 1961) and biochemical properties (Pilger *et al.*, 1961) of *Anaplasma*, all of which point to the nonprotozoan nature of this parasite (Ristic, 1960, 1963).

A. ERYTHROCYTIC ANTIGEN

1. The Complement Fixation Test

The techniques used for the preparation of CF antigens of *Anaplasma* are in most instances similar to those described for preparation of malarial stroma antigens (Heidelberger and Mayer, 1944). The essential principle in these techniques was to retain the organisms, as antigens, in the washed stromata of osmotically lysed, infected erythrocytes. These antigens must be preserved in a freezer at −65°C or lower (Gates *et al.*, 1954). For many years, the CF test has been the only serological technique available for detecting *Anaplasma* carriers. Inherent variables and difficulties in the CF test have greatly hampered the development of an effective and intensified program for identification of *Anaplasma* carriers.

2. Capillary Tube Agglutination Test

Studies of biophysical properties of *Anaplasma* revealed that a technique could be devised to obtain a purified whole-cell *Anaplasma* antigen from infected erythrocytes and that such an antigen could then be employed in a simple test, such as agglutination, for detection of serum antibodies in carrier animals. Employing differences in rates of sedimentation by centrifugation of disrupted *Anaplasma*-infected bovine erythrocytes, Ristic and co-workers (Ristic *et al.*, 1963) obtained two *Anaplasma* antigens. The sedimentable antigen was designated as "S" and the nonsedimentable one was termed "NS". The S antigen, which is considered the causative agent of anaplasmosis, was further purified and then used in a capillary tube agglutination (CA) test for detection of a serum antibody which develops in *Anaplasma*-infected animals (Ristic, 1962; Ristic and Kreier, 1963). The development of the CA test provided a simple, yet highly specific means of detecting the carrier state of anaplasmosis (Hibbs *et al.*, 1966; Jatkar *et al.*, 1966; Kuttler, 1963; Welter and Zuscheck, 1962; Welter, 1964). The popularity of the CA test increased greatly and the test is now also being used in foreign countries where anaplasmosis is a major economic problem (Brown, C. G. D., 1963; Castillo and Chavez, 1963–1965; Castillo *et al.*, 1966; Chavez and Castillo, 1963–1964; Schindler *et al.*, 1966; Viterbo, 1964).

3. Gel Precipitation Test

The isolation of soluble antigens which can be used in a gel precipitation (GP) test was accomplished by precipitating a supernatant lysate of erythrocytes from acutely infected cattle with protamine sulfate (PS antigen) or with dilute hydrochloric acid (HCl antigen) (Ristic *et al.*, 1963). In *Anaplasma*-infected animals the antibody detected in the GP test ap-

peared at least 1 week after the appearance of the agglutinating CA antibody. By means of cross-absorption and cross-precipitation tests, it was
determined that the soluble PS and HCl antigens were antigenically different from the particulate *Anaplasma* CA antigen. The chemical resemblance of the PS antigen to a similar preparation made from uninfected
erythrocytes suggested that it may possibly be the antigen responsible
for the autoimmune processes observed in anaplasmosis.

B. Serum Antigens

The only report describing the presence of a serum antigen in acute
Anaplasma infections is that of Amerault and Roby (1964). By means of
a gel-diffusion system it was shown that an antigen present in serum of
acutely infected animals will react with an antibody found in the serum
of animals convalescing from anaplasmosis. The presence of this antigen
in the serum was closely correlated with the time of occurrence of maximal
parasitemia. This antigen, which was described as being an "exoantigen,"
is believed to originate from the infected erythrocytes of acutely infected
animals and its presence in the serum was thought to indicate the dis-
integration of erythrocytes *in vivo*.

C. Autohemagglutinins and Opsonins: Their Role in the Pathogenesis of Anemia

In studying the anemia associated with bovine anaplasmosis, serological
evidence of erythrocyte-bound autoantibodies (Ristic, 1961) and of a free-
serum autohemagglutinin (Mann and Ristic, 1963a) has been demonstrated.
Further study of the possible role of autoimmune mechanisms in the pathogenesis of the anemia showed that anemia was not commensurate with
parasitemia in the convalescent phase of the disease and that a free-serum
autohemagglutinin was associated with anemia (Kreier *et al.*, 1964; Schroeder and Ristic, 1965) and erythrophagocytosis. The autohemagglutinins
were isolated from the euglobulin fraction of serum and identified as beta-
2 macroglobulin (Mann and Ristic, 1963b). It was necessary to trypsinize
the test erythrocytes in order to demonstrate the hemagglutinating activity
of this antibody. The appearance of autohemagglutinins in the sera of infected cattle was associated with the onset of erythrophagocytosis in the
bone marrow. Other changes in the cellular elements of the bone marrow
did not indicate that erythropoietic depression played a role in the anemia
of acute anaplasmosis (Kreier *et al.*, 1964).

In order to gain better insight into the mechanism by which erythrocytic
autoantibodies produce anemia in anaplasmosis, Rao (1965) studied the
effect of these autoantibodies on the fragility of erythrocytes suspended in
autologous serum acidified to pH 7.0. Erythrocyte fragility increased during

the course of infection. It was revealed that erythrocytes from infected animals maintained their increased susceptibility to hemolysis by acidified serum for several weeks after parasitemia had subsided. Fragility of erythrocytes in anaplasmosis was found to be more closely related to the titers of free-serum and erythrocyte-bound autoantibodies than to the per cent of parasitemia. Thus, it was proposed that hemagglutinins might opsonize erythrocytes and that the subsequent phagocytosis would contribute to the development of anemia.

Recently, by means of an *in vitro* test, Schroeder (1966) demonstrated the presence of opsonins in the serum of *Anaplasma*-infected cattle. These opsonins sensitized normal autologous and homologous erythrocytes to phagocytosis by cells of the reticuloendothelial system. The performance of the test, the nature of opsonins, and their role in the pathogenesis of anemia associated with *Anaplasma* infection will be discussed in greater detail in Volume II, Chapter 23, "Anaplasmosis."

Rao and Ristic (1963) produced indirect evidence which suggested that *Anaplasma* possesses the ability to desialize erythrocytic mucoproteins. In view of the fact that desialization of mucoproteins may result in enhanced antigenicity of these antigenically poor complexes, the operation of such a process in anaplasmosis may help explain the mechanisms of autoantigenicity (Rao and Ristic, 1963).

D. Fluorescent Antibody Techniques

Ristic and co-workers (1957) were the first to use fluorescein-labeled anti-*Anaplasma* globulin as a means of detecting marginal inclusions in erythrocytes of cattle with acute anaplasmosis. This specific staining identified the *Anaplasma* inclusions as the agents responsible for producing such an antibody and eliminated the possibility that they were nonspecific products of the host.

In a test which Ristic and White (1960) termed a "self-labeling indirect fluorescent antibody test," they were able to demonstrate that the *Anaplasma*-antibody reaction was a natural process occurring within the blood vascular system of an infected animal. Thus, in order to demonstrate this reaction, the only manipulation required was to expose washed erythrocytes from suspected animals to fluorescein-labeled antibovine globulin. With this method, it was possible to observe most of the *Anaplasma* growth forms including marginal inclusions and initial bodies.

Madden (1962) and Kreier and Ristic (1963b,c) employed the direct fluorescent antibody technique in an effort to reveal the serological relationship between *A. marginale* and the *Anaplasma*-like organism described by España *et al.* (1959) and named *Paranaplasma caudata* by Kreier and Ristic (1963d). On the basis of cross-absorption of immune fluorescein-labeled sera, Kreier and Ristic (1963c) demonstrated that the antigen found in the tail-like appendage extending from the marginal body of *P. caudata* cannot be serologically stained by an antibody against *A. marginale*. Recently, on the basis of electron microscopic studies on ultra-thin sections of erythrocytes infected with *P. caudata*, Simpson *et al.* (1965) concluded that the tail-like structure of this parasite is not an integral part of the parasite itself but rather a condensation of a protein-like mass derived from the marginal inclusion, from erythrocytes, or from both of them.

E. Localization of Serologically Reactive Antibodies in Bovine Serum

Dimopoullos *et al.* (1960) reported that early in infection changes of serum proteins in *Anaplasma*-infected animals were predominantly associated with alpha and beta globulins, and that the gamma globulin fraction was involved in the convalescent phase. Subsequent studies (Rogers and Dimopoullos, 1962) revealed that during the acute stage of infection, the CF antibody was localized in the alpha and beta globulins of lower mobility and in the gamma globulins of higher mobility. In convalescent sera, the CF antibody was associated with the beta and gamma globulins. Mann (1964) found that the major portion of convalescent *Anaplasma* antibody, demonstrated by the capillary tube agglutination (CA) test, was contained in the gamma globulin fraction.

Murphy *et al.* (1966) followed the sequences and duration of synthesis of antibodies in *Anaplasma*-infected animals with ion exchange chromatography and found that the antibodies demonstrable by the CA and CF tests were contained in the 19-S globulin fraction during the acute phase of the infection. Subsequent to this phase of infection the localization of these antibodies progressively changed from the 19-S globulin to the 6.2-S globulin fraction. This change occurred faster with the CF antibody than with the CA antibody.

VI. *EPERYTHROZOON* and *HAEMOBARTONELLA*

Wigand (1956a,b) described a complement-fixation (CF) test suitable for detecting antibody to *Haemobartonella muris* and *Eperythrozoon coccoides*. The antigen consisted of stromata of lysed erythrocytes obtained from acutely infected mice. The antibody was first detected by the CF test 5 days after infection and reached a maximum titer within 3 to 5 weeks. Treatment with neoarsphenamine caused the antibody titers to subside and fall below detectable levels.

A complement-fixing (CF) antigen from erythrocytes of pigs infected with *Eperythrozoon suis* was prepared by Splitter (1958) using a technique for preparation of the *Anaplasma* CF antigen (Gates *et al.*, 1954). The CF antibody was detected approximately 3 days after the onset of *Eperythrozoon* infections in pigs. It was not possible to detect subclinical infections by this test. *Eperythrozoon* sera also reacted in the CF test for bovine anaplasmosis.

Kreier and Ristic (1963a) obtained immune gamma globulin from cattle hyperimmunized with *Eperythrozoon wenyoni* and conjugated it to fluorescein isothiocyanate. By means of the FAT, serological cross-reaction was shown between *Eperythrozoon ovis*, *E. wenyoni*, and *A. marginale*. The fluorescent *Eperythrozoon* bodies were morphologically similar to those seen in Giemsa-stained blood films. A serological relationship between *Eperythrozoon* and *Anaplasma* was further ascertained by reacting the sera from animals infected with *E. ovis* and *E. wenyoni* in a CF test using *Anaplasma* antigen.

Small and Ristic (1967) studied *Haemobartonella felis* by means of Giemsa, acridine orange, and fluorescent-labeled antibody staining methods and by electron microscopy using shadow cast technique and ultrathin sections. The fluorescent *Haemobartonella* bodies were predominantly oval to round and measured approximately 0.3 to 0.4 μ in diameter. On the basis of acridine orange staining, it was concluded that *H. felis* contained both DNA and RNA. *Haemobartonella felis* bodies were accurately identified in film samples of peripheral blood when the acridine orange and FA techniques were used, while the Giemsa staining method was found to be less than 50% reliable for detecting the presence of such bodies.

REFERENCES

Abele, D. C., Tobie, J. E., Hill, G. J., Contacos, P. G., and Evans, C. B. (1965). *Am. J. Trop. Med. Hyg.* 14, 191–197.

Adler, S., and Foner, A. (1965). *Israel J. Med. Sci.* 1, 988–993.

Amerault, T. E., and Roby, T. O. (1964). *Am. J. Vet. Res.* 25, 1642–1649.

Anonymous. (1967). "Human Babesiosis," p. 4. California Communicable Disease Center, Vet. Public Health Notes.

Bahr, G. F. (1966). *In* "Research in Malaria" (E. H. Sadun and H. S. Osborne, eds.), pp. 1064–1070. Walter Reed Army Inst. Res., Washington, D. C. (*Military Med.* 131, Suppl.).

Bailey, K. P., and Cowan, K. (1961). Personal communication as cited in Barnett (1963, p. 193).

Banki, G., and Bucci, A. (1964a). *Parassitologia* 6, 251–257.

Banki, G., and Bucci, A. (1964b). *Parassitologia* 6, 269–274.

Barnett, S. F. (1963). *In* "Immunity to Protozoa" (P. C. C. Garnham, A. E. Pierce, and I. Roitt, eds.), pp. 180–195. Blackwell, Oxford.

Beck, J. S., and Walker, P. J. (1964). *Nature* 204, 194–195.

Bentz, M. (1963). *Proc. 7th Intern. Congr. Trop. Med. Malariol., 1963* Vol. 2, pp. 223–224. Rio de Janeiro.

Bentz, M. (1964). *Rev. Franc. Etudes Clin. Biol.* 9, 657–661.

Bentz, M., and Diallo, I. (1963). *Bull. Soc. Pathol. Exotique* 56, 434–437.

Bentz, M., and Macario, C. (1963). *Bull. Soc. Pathol. Exotique* 56, 416–421.

Bentz, M., and Mattern, P. (1963). *Bull. Soc. Pathol. Exotique* **56,** 862–866.
Berrios, A., and Zeledon, R. (1960). *Rev. Biol. Trop., Univ. Costa Rica* **8,** 225–231; seen in *Trop. Diseases Bull.* **58,** 1224 (1961).
Blumberg, B. S., Kuvin, S. F., Robinson, J. C., Teitelbaum, J. M., and Contacos, P. G. (1963). *J. Am. Med. Assoc.* **184,** 1021–1023.
Bowman, I. B. R., Grant, P. T., and Kermack, W. O. (1960). *Exptl. Parasitol.* **9,** 131–136.
Braff, E., and Condit, P. (1967). *Morbidity Mortality Weekly Rept.* **16,** 8.
Bray, R. S. (1962). *Trans. Roy. Soc. Trop. Med. Hyg.* **56,** 436.
Bray, R. S. (1964). *In* "Colloque Internationale sur le *Plasmodium berghei*," pp. 1–8. Inst. Med. Trop. Prince Leopold.
Briggs, N. T., Garza, B. L., and Box, E. D. (1960). *Exptl. Parasitol.* **10,** 21–27.
Briggs, N. T., Wellde, B. T., and Sadun, E. H. (1966). *In* "Research in Malaria" (E. H. Sadun, and H. S. Osborne, eds.), pp. 1243–1249. Walter Reed Army Inst. Res., Washington, D. C. (*Military Med.* **131,** Suppl.).
Brown, C. G. D. (1963). Report No. 3 to the Research Department, Diamond Laboratories, Inc., Des Moines, Iowa, U.S.A.; Wellcome Research Laboratories, Kabete, Kenya, East Africa.
Brown, K. N. (1961). *Ann. Trop. Med. Parasitol.* **55,** 143–144.
Brown, K. N. (1963). *In* "Immunity to Protozoa" (P. C. C. Garnham, A. E. Pierce, and I. Roitt, eds.), pp. 204–212. Blackwell, Oxford.
Brown, K. N., and Brown, I. N. (1965). *Nature* **208,** 1286–1288.
Brown, K. N., and Williamson, J. (1962a). *Nature* **194,** 1253–1255.
Brown, K. N., and Williamson, J. (1962b). *Trans. Roy. Soc. Trop. Med. Hyg.* **56,** 12.
Brown, K. N., and Williamson, J. (1964). *Exptl. Parasitol.* **15,** 69–86.
Bruce-Chwatt, L. J. (1963). *In* "Immunity to Protozoa" (P. C. C. Garnham, A. E. Pierce, and I. Roitt, eds.), pp. 89–108. Blackwell, Oxford.
Castillo, A., and Chavez, C. E. (1963–1965). Separata de la Revista de la Facultad de Medicina Veterinaria, Vols. 18–20. Universidad Nacional Mayor de San Marcos, Lima, Peru.
Castillo, A., Chavez, C. E., and La Rosa, V. (1966). Separata de la Revista de la Facultad de Medicina Veterinaria, Vols. 18–20. Universidad Nacional Mayor de San Marcos, Lima, Peru.
Causse-Vaills, C., Orfila, J., and Fabiani, M. G. (1961). *Ann. Inst. Pasteur* **100,** 232–242.
Chavez, C. E., and Castillo, A. (1963–1964). Separata de la Revista de la Facultad de Medicina Veterinaria, Vol. 18–19, pp. 1–12. Universidad Nacional Mayor de San Marcos, Lima, Peru.
Chavin, S. I. (1966). *In* "Research in Malaria" (E. H. Sadun and H. S. Osborne, eds.), pp. 1124–1136. Walter Reed Army Inst. Res., Washington, D. C. (*Military Med.* **131,** Suppl.).
Coggeshall, L. T., and Kumm, H. W. (1938). *J. Exptl. Med.* **68,** 17–27.
Cohen, S., and McGregor, I. A. (1963). *In* "Immunity to Protozoa" (P. C. C. Garnham, A. E. Pierce, and I. Roitt, eds.), pp. 123–159. Blackwell, Oxford.
Cohen, S., McGregor, I. A., and Carrington, S. (1961). *Nature* **192,** 733–737.
Collins, W. E., Jeffery, G. M., and Skinner, J. C. (1964a). *Am. J. Trop. Med. Hyg.* **13,** 1–5.
Collins, W. E., Jeffery, G. M., and Skinner, J. C. (1964b). *Am. J. Trop. Med. Hyg.* **13,** 256–260.
Collins, W. E., Jeffery, G. M., and Skinner, J. C. (1964c). *Am. J. Trop. Med. Hyg.* **13,** 777–782.

Collins, W. E., Skinner, J. C., and Guinn, E. G. (1965a). *Proc. 2nd Intern. Conf. Protozool., London, 1965* Intern. Congr. Ser. No. 91, p. 175. Excerpta Med. Found., Amsterdam.

Collins, W. E., Skinner, J. C., Guinn, E. G., Dobrovolny, C. G., and Jones, F. E. (1965b). *J. Parasitol.* **51**, 81–84.

Cook, L., Grant, P. T., and Kermack, W. O. (1961). *Exptl. Parasitol.* **11**, 372–379.

Cooper, N. R., and Fogel, B. J. (1966). *In* "Research in Malaria" (E. H. Sadun and H. S. Osborne, eds.), pp. 1180–1190. Walter Reed Army Inst. Res., Washington, D. C. (*Military Med.* **131**, Suppl.).

Corradetti, A., Verolini, F., and Ilardi, A. (1964a). *Parassitologia* **6**, 279–281.

Corradetti, A., Verolini, F., Sebastiani, A., Proietti, A. M., and Amati, L. (1964b). *Bull. World Health Organ.* **30**, 747–750.

Corwin, R. M., McGhee, R. B., and Sloan, B. L. (1965). *J. Parasitol.* **51**, Sect. 2, 36.

Cox, H. W. (1959). *J. Immunol.* **82**, 209–214.

Cox, H. W. (1962). *J. Protozool.* **9**, 114–118.

Cox, H. W., Schroeder, W. F., and Ristic, M. (1966). *J. Protozool.* **13**, 327–332.

Cunningham, M. P., and Vickerman, K. (1961a). *Ann. Trop. Med. Parasitol.* **55**, 142–143.

Cunningham, M. P., and Vickerman, K. (1961b). *Trans. Roy. Soc. Trop. Med. Hyg.* **55**, 12.

Curtain, C. C., Kidson, C., Champness, D. L., and Gorman, J. G. (1964). *Nature* **203**, 1366–1367.

Curtain, C. C., Gorman, J. G., and Kidson, C. (1965). *Trans. Roy. Soc. Trop. Med. Hyg.* **59**, 42–45.

D'Alesandro, P. A., and Sherman, I. W. (1964). *Exptl. Parasitol.* **15**, 430–438.

D'Antonio, L. E., von Doenhoff, A. E., Jr., and Fife, E. H., Jr. (1966). *In* "Research in Malaria" (E. H. Sadun and H. S. Osborne, eds.), pp. 1152–1156. Walter Reed Army Inst. Res., Washington, D. C. (*Military Med.* **131**, Suppl.).

Desowitz, R. S. (1961). *J. Immunol.* **86**, 69–72.

Desowitz, R. S. (1965). *Proc. 2nd Intern. Conf. Protozool., London, 1965* Intern. Congr. Ser. No. 91., p. 169. Excerpta Med. Found., Amsterdam.

Desowitz, R. S., and Saave, J. J. (1965). *Bull. World Health Organ., Tech. Rept. Ser.* **32**, 149–159.

Desowitz, R. S., and Stein, B. (1962). *Trans. Roy. Soc. Trop. Med. Hyg.* **56**, 257.

Desowitz, R. S., Saave, J. J., and Stein, B. (1966). *In* "Research in Malaria" (E. H. Sadun and H. S. Osborne, eds.), pp. 1157–1166. Walter Reed Army Inst. Res., Washington, D. C. (*Military Med.* **131**, Suppl.).

Diggs, C. L., and Sadun, E. H. (1965). *Exptl. Parasitol.* **16**, 217–223.

Dimopoullos, G. T., Schrader, G. T., and Foote, L. E. (1960). *Am. J. Vet. Res.* **21**, 222–225.

Dodin, A., Fromentin, H., and Gleye, M. (1962). *Bull. Soc. Pathol. Exotique* **55**, 291–299.

Eaton, M. B. (1938). *J. Exptl. Med.* **67**, 857–870.

Edozien, J. C. (1964). *Am. J. Trop. Med. Hyg.* **13**, Suppl., 233–234.

Edozien, J. C., Gilles, H. M., and Udeozo, I. O. K. (1962). *Lancet* **2**, 951–955.

España, C., España, E. M., and Gonzales, D. (1959). *Am. J. Vet. Res.* **20**, 795–805.

Essenfeld, E., and Fennell, R. H. (1964). *Proc. Soc. Exptl. Biol. Med.* **116**, 728–730.

Fife, E. H., and Muschel, L. H. (1959). *Proc. Soc. Exptl. Biol. Med.* **101**, 540–543.

Figallo Espinal, L. (1962). *Arch. Venezolanos Med. Trop. Parasitol. Med.* **4**, 243–264; seen in *Trop. Diseases Bull.* **61**, 1130 (1964).

Fogel, B. J., von Doenhoff, A. E., Jr., Cooper, N. R., and Fife, E. H. Jr. (1966). *In*

"Research in Malaria" (E. H. Sadun and H. S. Osborne, eds.), pp. 1173–1179. Walter Reed Army Inst. Res., Washington, D. C. (*Military Med.* 131, Suppl.).

Gail, K., and Kretschmar, W. (1965). *Naturwissenschaften* 16, 480.

Gates, D. W., Mohler, W. M., Mott, L. O., Poelma, L. T., Price, K. E., and Mitchell, J. (1954). *Proc. 58th Ann. Meeting U.S. Livestock Sanit. Assoc.* pp. 105–114.

Gill, B. S. (1964). *Ann. Trop. Med. Parasitol.* 58, 473–480.

Gill, B. S. (1965a). *J. Comp. Pathol. Therap.* 75, 175–183.

Gill, B. S. (1965b). *J. Comp. Pathol. Therap.* 75, 233–240.

Gill, B. S. (1965c). *J. Gen. Microbiol.* 38, 357–361.

Goberman, V. (1966). M. S. Thesis, Hebrew University, Jerusalem, Israel.

Gray, A. R. (1961a). *Ann. Trop. Med. Parasitol.* 55, 142.

Gray, A. R. (1961b). *Immunology* 4, 253–261.

Gray, A. R. (1962). *Ann. Trop. Med. Parasitol.* 56, 4–13.

Gray, A. R. (1963). *Exptl. Parasitol.* 14, 374–381.

Hamburger, Y. (1965) M. S. Thesis, Hebrew University, Jerusalem, Israel.

Heidelberger, M., and Mayer, M. M. (1944). *Science* 100, 359–360.

Hibbs, C. M., Weide, K. D., and Marshall, M. (1966). *J. Am. Vet. Med. Assoc.* 148, 545–546.

Hill, G. C., and Guttman, H. W. (1965). *Proc. 2nd Intern. Conf. Protozool., London, 1965* Intern. Congr. Ser. No. 91., pp. 132–133. Excerpta Med. Found., Amsterdam.

Hirato, K., Ninomiya, Y., Uwano, T., and Kutii, T. (1945). *Japan. J. Vet. Sci.* 7, 197–205.

Hulliger, L. (1965). *J. Protozool.* 12, 649–655.

Hulliger, L., Brown, C. G. D., and Wilde, J. K. H. (1966). *Nature* 211, 328–329.

Ingram, D. G., and Soltys, M. A. (1960). *Parasitology* 50, 231–239.

Ingram, R. L. (1963). *Proc. 7th Intern. Congr. Trop. Med. Malariol., Rio de Janeiro, 1963* Vol. 5, pp. 23–24.

Ingram, R. L., and Carver, R. K. (1963). *Science* 139, 405.

Jatkar, P. R., Kreier, J. P., Akin, E. L., and Tharp, V. (1966). *Am. J. Vet. Res.* 27, 372–374.

Johnson, P., Neal, R. A., and Gall, D. (1963). *Nature* 200, 83.

Kagan, I. G., and Norman, L. (1962). *J. Parasitol.* 48, 584–588.

Kibukamusoke, J. W., and Wilks, N. E. (1965a). *E. African Med. J.* 42, 203–206.

Kibukamusoke, J. W., and Wilks, N. E. (1965b). *Lancet* 1, 301–302.

Kozma, C. (1962). *Z. Tropenmed. Parasitol.* 13, 175–180.

Kreier, J. P., and Ristic, M. (1963a). *Am. J. Vet. Res.* 24, 488–500.

Kreier, J. P., and Ristic, M. (1963b). *Am. J. Vet. Res.* 24, 676–687.

Kreier, J. P., and Ristic, M. (1963c). *Am. J. Vet. Res.* 24, 688–696.

Kreier, J. P., and Ristic, M. (1963d). *Am. J. Vet. Res.* 24, 697–702.

Kreier, J. P., and Ristic, M. (1964). *Am. J. Trop. Med. Hyg.* 13, 6–10.

Kreier, J. P., Ristic, M., and Watrach, A. M. (1962). *Am. J. Vet. Res.* 23, 657–662.

Kreier, J. P., Ristic, M., and Schroeder, W. F. (1964) *Am. J. Vet. Res.* 25, 343–352.

Kreier, J. P., Pearson, G. L., and Stillwell, D. (1965). *Am. J. Trop. Med. Hyg.* 14, 529–532.

Kuttler, K. L. (1963). *J. Am. Vet. Med. Assoc.* 143, 729–733.

Kuvin, S. F., and Voller, A. (1963). *Brit. Med. J.* 2, 477–479.

Kuvin, S. F., Beye, H. K., Stohlman, F., Contacos, P. G., and Coatney, G. R. (1962a). *Trans. Roy. Soc. Trop. Med. Hyg.* 56, 371–378.

Kuvin, S. F., Tobie, J. E., Evans, C. B., Coatney, G. R., and Contacos, P. G. (1962b). *Am. J. Trop. Med. Hyg.* 11, 429–436.

Kuvin, S. F., Tobie, J. E., Evans, C. B., Coatney, G. R., and Contacos, P. G. (1963). *J. Am. Med. Assoc.* **184**, 943–945.

Lanham, S. M. (1966). *Trans. Roy. Soc. Trop. Med. Hyg.* **60**, 125–126.

Lumsden, W. H. R. (1966). *Trans. Roy. Soc. Trop. Med. Hyg.* **60**, 125.

McGhee, R. B., and Hanson, W. L. (1963). *J. Protozool.* **10**, 239–243.

McGregor, I. A. (1964a). *Am. J. Trop. Med. Hyg.* **13**, 237–239.

McGregor, I. A. (1964b). *Trans. Roy. Soc. Trop. Med. Hyg.* **58**, 80–92.

McGregor, I. A. (1965). *W. African Med. J.* [N.S.] **14**, 6–9.

McGregor, I. A., Gilles, H. M., Walters, J. H., Davies, A. H., and Pearson, F. A. (1956). *Brit. Med. J.* **II**, 686–692.

McGregor, I. A., Carrington, S. P., and Cohen, S. (1963). *Trans. Roy. Soc. Trop. Med. Hyg.* **57**, 170–175.

McGregor, I. A., Williams, K., Voller, A., and Billewicz, W. Z. (1965). *Trans. Roy. Soc. Trop. Med. Hyg.* **59**, 395–414.

Madden, P. A. (1962). *Am. J. Vet. Res.* **23**, 921–924.

Mahoney, D. F. (1962). *Australian Vet. J.* **38**, 48–52.

Mahoney, D. F., Redington, B. C., and Schoenbechler, M. (1966). *In* "Research in Malaria" (E. H. Sadun and H. S. Osborne, eds.), pp. 1141–1151. Walter Reed Army Inst. Res., Washington, D. C. (*Military Med.* **131**, Suppl.).

Mann, D. K. (1964). M. S. Thesis, University of Illinois, Urbana, Illinois.

Mann, D. K., and Ristic, M. (1963a). *Am. J. Vet. Res.* **24**, 703–708.

Mann, D. K., and Ristic, M. (1963b). *Am. J. Vet. Res.* **24**, 709–712.

Mattern, P. (1964). *Ann. Inst. Pasteur* **107**, 415–421.

Mattern, P., Fromentin, H., and Pilo-Moron, E. (1963). *Bull. Soc. Pathol. Exotique* **56**, 301–305.

Murphy, F. A., Osebold, J. W., and Aalund, O. (1966). *J. Infect. Diseases* **116**, 99–111.

Neitz, W. O. (1957). *Onderstepoort J. Vet. Res.* **27**, 275–430.

Nicoli, J., Acker, P., and Demarchi, J. (1964). *Ann. Inst. Pasteur* **107**, 232–244.

Nussenzweig, R. S., Yoeli, M., and Most, H. (1966). *In* "Research in Malaria" (E. H. Sadun and H. S. Osborne, eds.), pp. 1237–1242. Walter Reed Army Inst. Res., Washington, D. C. (*Military Med.* **131**, Suppl.).

Nussenzweig, V., Deane, L. M., and Kloetzel, J. (1963a). *Exptl. Parasitol.* **14**, 221–232.

Nussenzweig, V., Kloetzel, J., and Deane, L. M. (1963b). *Exptl. Parasitol.* **14**, 233–239.

Ormerod, W. E. (1963). *In* "Immunity to Protozoa" (P. C. C. Garnham, A. E. Pierce, and I. Roitt, eds.), pp. 213–227. Blackwell, Oxford.

Pautrizel, R., Lafaye, A., and Duret, J. (1959). *Bull. Soc. Pathol. Exotique* **52**, 318–331.

Pautrizel, R., Duret, J., Tribouley, J., and Ripert, C. (1962a). *Bull. Soc. Pathol. Exotique* **55**, 383–390.

Pautrizel, R., Duret, J., Tribouley, J., and Ripert, C. (1962b). *Bull. Soc. Pathol. Exotique* **55**, 391–397.

Pautrizel, R., Duret, J., Tribouley, J., and Ripert, C. (1962c). *Rev. Immunol.* **26**, 157–166.

Pautrizel, R., Duret-Tribouley, J., and Tribouley, J. (1965). *Proc. 2nd Intern. Conf. Protozool., London, 1965* Intern. Congr. Ser. No. 91, pp. 139–141. Excerpta Med. Found., Amsterdam.

Pilger, K. S., Wu, W. G., and Muth, O. H. (1961). *Am. J. Vet. Res.* **22**, 298–307.

Powell, R. D., and Brewer, G. J. (1964). *Am. J. Trop. Med. Hyg.* **13**, Suppl., 228–232.

Rao, P. J. (1965). Ph.D. Thesis, University of Illinois, Urbana, Illinois.

Rao, P. J., and Ristic, M. (1963). *Proc. Soc. Exptl. Biol. Med.* **114**, 447–452.

Ristic, M. (1960). *Advan. Vet. Sci.* **6**, 111–192.
Ristic, M. (1961). *Am. J. Vet. Res.* **22**, 871–876.
Ristic, M. (1962). *J. Am. Vet. Med. Assoc.* **114**, 588–594.
Ristic, M. (1963). *Proc. 17th World Vet. Congr., Hannover, Germany*, Vol. 1, pp. 815–819.
Ristic, M., and Kreier, J. P. (1963). *Am. J. Vet. Res.* **24**, 985–992.
Ristic, M., and Sibinovic, S. (1964). *Am. J. Vet. Res.* **25**, 1519–1526.
Ristic, M., and Watrach, A. M. (1961). *Am. J. Vet. Res.* **22**, 109–116.
Ristic, M., and White, F. H. (1960). *Science* **131**, 987–988.
Ristic, M., White, F. H., and Sanders, D. A. (1957). *Am. J. Vet. Res.* **18**, 924–928.
Ristic, M., Mann, D. K., and Kodras, R. (1963). *Am. J. Vet. Res.* **24**, 472–477.
Ristic, M., Oppermann, J., Sibinovic, S., and Phillips, T. N. (1964). *Am. J. Vet. Res.* **25**, 15-23.
Rogers, T. E., and Dimopoullos, G. T. (1962). *Proc. Soc. Exptl. Biol. Med.* **110**, 359–362.
Rosseau-Baelde, M., and Misselyn, G. (1963). *Compt. Rend. Soc. Biol.* **157**, 1854–1857.
Sadun, E. H., Duxbury, R. E., Williams, J. S., and Anderson, R. I. (1963). *J. Parasitol.* **49**, 385–388.
Sadun, E. H., Hickman, R. L., Wellde, B. T., Moon, A. P., and Udeozo, I. O. K. (1966). *In* "Research in Malaria" (E. H. Sadun and H. S. Osborne, eds.), pp. 1250–1262. Walter Reed Army Inst. Res., Washington, D. C. (*Military Med.* **131**, Suppl.).
Schaeffler, W. F. (1963). *Am. J. Vet. Res.* **24**, 784–791.
Schindler, R. (1964). *In* "Colloque Internationale sur le *Plasmodium berghei*," pp. 1–9. Inst. Med. Trop. Prince Leopold.
Schindler, R., and Dennig, H. K. (1962a). *Berlin Muench. Tierarztl. Wochschr.* **75**, 111–112.
Schindler, R., and Dennig, H. K. (1962b). *Z. Tropenmed. Parasitol.* **13**, 480–488.
Schindler, R., Ristic, M., and Wokatsch, R. (1966). *Z. Tropenmed. Parasitol.* **17**, 337–360.
Schroeder, W. F. (1966). Ph.D. Thesis, University of Illinois, Urbana, Illinois.
Schroeder, W. F., and Ristic, M. (1965). *Am. J. Vet. Res.*, **26**, 239–245.
Schroeder, W. F., Cox, H. W., and Ristic, M. (1966). *Ann. Trop. Med. Parasitol.* **60**, 31–38.
Seed, J. R. (1963). *J. Protozool.* **10**, 380–389.
Seed, J. R. (1964). *Parasitology* **54**, 593–596.
Seed, J. R., and Baquero, M. A. (1965). *J. Protozool.* **12**, 427–432.
Seed, J. R., and Weinman, D. (1963). *Nature* **198**, 197–198.
Sherman, I. W. (1961). *J. Exptl. Med.* **114**, 1049–1062.
Sherman, I. W. (1964). *J. Protozool.* **11**, 409–417.
Sherman, I. W. (1965). *Proc. 2nd Intern. Conf. Protozool., London, 1965* Intern. Congr. Ser. No. 91, pp. 170–171. Excerpta Med. Found., Amsterdam.
Sherman, I. W., and Hull, R. W. (1960a). *J. Protozool.* **7**, 171–176.
Sherman, I. W., and Hull, R. W. (1960b). *J. Protozool.* **7**, 409–416.
Sibinovic, K. H. (1966). Ph.D. Thesis, University of Illinois, Urbana, Illinois.
Sibinovic, K. H., Ristic, M., Sibinovic, S., and Phillips, T. N., (1965). *Am. J. Vet. Res.* **110**, 147–153.
Sibinovic, K. H., MacLeod, R., Ristic, M., Sibinovic, S., and Cox, H. W. (1967a). *J. Parasitol.* **53**, 919–923.
Sibinovic, K. H., Milar, R., Ristic, M., and Cox, H. W. (1967b). In preparation.

Sibinovic, K. H., Sibinovic, S., Ristic, M., and Cox, H. W. (1967c). *J. Parasitol.* (in press).

Sibinovic, S., Sibinovic, K. H., Ristic, M., and Cox, H. W. (1966). *J. Protozool.* **13,** 551–552.

Sibinovic, S., Sibinovic, K. H., and Ristic, M. (1967). In preparation.

Simpson, C. F., Kling, J. M., and Neal, F. C. (1965). *J. Cell Biol.* **27,** 225–235.

Skrabalo, Z., and Deanovic, Z. (1957). *Doc. Med. Geograph. Trop.* **9,** 11–16.

Small, E., and Ristic, M. (1967). *Am. J. Vet. Res.* **28,** 845–851.

Smith, T., and Kilbourne, F. L. (1893). *U.S. Dept. Agri., Bur. Animal Ind., Bull.* **1,** 177.

Sodeman, W. A., and Jeffery, G. M. (1964). *J. Parasitol.* **50,** 477–478.

Sodeman, W. A., and Jeffery, G. M. (1965). *Am. J. Trop. Med. Hyg.* **14,** 187–190.

Sodeman, W. A., and Meuwissen, J. H. E. T. (1966). *J. Parasitol.* **52,** 23–25.

Soltys, M. A. (1965). *Proc. 2nd Intern. Conf. Protozool., London, 1965* Intern. Congr. Ser. No. 91, pp. 138–139. Excerpta Med. Found., Amsterdam.

Spira, D., and Zuckerman, A. (1962). *Science* **137,** 536–537.

Spira, D., and Zuckerman, A. (1964). *J. Protozool.* **11,** Suppl., 43–44.

Spira, D., and Zuckerman, A. (1966). *In* "Research in Malaria" (E. H. Sadun and H. S. Osborne, eds.), pp. 1117–1123. Walter Reed Army Inst. Res. Washington, D.C. *(Military Med.* **131,** Suppl.).

Spira, D., Hamburger, Y., and Zuckerman, A. (1966). Unpublished data.

Splitter, E. J. (1958). *Am. J. Vet. Res.* **132,** 47–49.

Stauber, L. A., Walker, H. A., and Richardson, A. P. (1951). *J. Infect. Diseases* **89,** 31–34.

Stein, B., and Desowitz, R. S. (1964). *Bull. World Health Organ.* **30,** 45–49.

Taliaferro, W. H. (1923). *Am. J. Hyg.* **3,** 204–205.

Taliaferro, W. H. (1930). *In* "The Immunology of Parasitic Infections." John Bale, Sons and Danielsson, Ltd., London.

Taliaferro, W. H. (1949). *In* "Malariology" (M. F. Boyd, ed.), Vol. 2, pp. 935–965. Saunders, Philadelphia, Pennsylvania.

Targett, G. A. T., and Fulton, J. D. (1965). *Exptl. Parasitol.* **17,** 180–193.

Targett, G. A. T., and Voller, A. (1965). *Brit. Med. J.* **II,** 1104–1106.

Tarrant, C. J., Fife, E. H., and Anderson, R. I. (1965). *J. Parasitol.* **51,** 277–285.

Tella, A., and Maegraith, B. C. (1965). *Ann. Trop. Med. Parasitol.* **59,** 153–158.

Tobie, J. E. (1964). *Am. J. Trop. Med. Hyg.* **13,** Suppl., 195–203.

Tobie, J. E. (1965). *Proc. 2nd Intern. Conf. Protozool., London, 1965* Intern. Congr. Ser. No. 91, pp. 164–165. Excerpta Med. Found., Amsterdam.

Tobie, J. E., and Coatney, G. R. (1964). *Am. J. Trop. Med. Hyg.* **13,** 786–789.

Tobie, J. E., Coatney, G. R., and Evans, C. B. (1961). *Exptl. Parasitol.* **11,** 128–132.

Tobie, J. E., Kuvin, S. F., Contacos, P. G., Coatney, G. R., and Evans, C. B. (1962). *Am. J. Trop. Med. Hyg.* **11,** 589–596.

Tobie, J. E., Kuvin, S. F., Contacos, P. G., Coatney, G. R., and Evans, C. B. (1963). *J. Am. Med. Assoc.* **184,** 945–947.

Todorovic, R. (1967). Ph.D. Thesis, Univ. of Illinois, Urbana, Illinois.

Todorovic, R., Ferris, D. H., and Ristic, M. (1967a). *Ann. Trop. Med. Parasitol.* **61,** 117–124.

Todorovic, R., Ristic, M., and Ferris, D. H., (1967b). *Trans. Roy. Soc. Trop. Med. Hyg.* (in press).

Todorovic, R., Ristic, M., and Ferris, D. H. (1967c). *Trans. Roy. Soc. Trop. Med. Hyg.* (in press).

Toussaint, A. J., Tarrant, C. J., and Anderson, R. I. (1965). *J. Parasitol.* **51**, Sect. 2, 29.

Trager, W. (1957). *Acta Trop.* **14**, 289–301.

Tsur, I., and Adler, S. (1965). *Proc. 2nd Intern. Conf. Protozool., London, 1965* Intern. Congr. Ser. No. 91, pp. 37–38. Excerpta Med. Found., Amsterdam.

Viterbo, G. H. (1964). M.S. Thesis, University of the Laguna, The Philipines.

Voller, A. (1962). *Bull. World Health Organ.* **27**, 283–287.

Voller, A. (1964a). *Am. J. Trop. Med. Hyg.* **13**, Suppl., 204–208.

Voller, A. (1964b). *Bull. World Health Organ.* **30**, 343–354.

Voller, A., and Bray, R. S. (1962). *Proc. Soc. Exptl. Biol. Med.* **110**, 907–910.

Voller, A., and Taffs, L. F. (1963). *Trans. Roy. Soc. Trop. Med. Hyg.* **57**, 32–33.

Voller, A., and Wilson, H. (1964). *Brit. Med. J.* **II**, 551–552.

Ward, P. A., and Conran, P. (1966). *In* "Research in Malaria" (E. H. Sadun and H. S. Osborne, eds.), pp. 1225–1232. Walter Reed Army Inst. Res., Washington, D. C. (*Military Med.* **131**, Suppl.).

Warren, L. G., and Guevara, A. (1964). *J. Protozool.* **11**, 107–108.

Watkins, J. F. (1964). *J. Hyg.* **62**, 69–80.

Weiss, M. L., and De Giusti, D. L. (1964). *Nature* **201**, 731–732.

Weitz, B. (1960a). *Nature* **185**, 788–789.

Weitz, B. (1960b). *J. Gen. Microbiol.* **23**, 589–600.

Weitz, B. (1963a). *In* "Immunity to Protozoa" (P. C. C. Garnham, A. E. Pierce, and I. Roitt, eds.), pp. 196–203. Blackwell, Oxford.

Weitz, B. (1963b). *J. Gen. Microbiol.* **32**, 145–149.

Weitz, B., and Lee-Jones, F. (1961). *Ann. Trop. Med. Parasitol.* **55**, 141–142.

Welter, C. J. (1964). *Am. J. Vet. Res.* **25**, 1058–1061.

Welter, C. J. and Zuscheck, T. (1962). *J. Am. Vet. Med. Assoc.* **141**, 595–599.

Wigand, R. (1956a). *Z. Tropenmed. Parasitol.* **7**, 322–340.

Wigand, R. (1956b). *Nature* **178**, 1288–1289.

Williams, J. S., Duxbury, R. E., Anderson, R. I., and Sadun, E. H. (1963). *J. Parasitol.* **49**, 380-384.

Williamson, J. (1963). *Exptl. Parasitol.* **13**, 348–366.

Williamson, J., and Brown, K. N. (1964). *Exptl. Parasitol.* **15**, 44–68.

Williamson, J., and Desowitz, R. S. (1961). *Exptl. Parasitol.* **11**, 161–175.

Zuckerman, A. (1964a). *Am. J. Trop. Med. Hyg.* **13**, Suppl., 209–213.

Zuckerman, A. (1964b). *Exptl. Parasitol.* **15**, 138–183.

Zuckerman, A., and Spira, D. (1963). *J. Protozool.* **10**, Suppl., 34.

Zuckerman, A., and Spira, D. (1965). *World Health Organ., Malariol. Ser.* **497.65**, 1–10.

Zuckerman, A., and Spira, D. (1966). *Proc. 1st Intern. Congr. Parasitol., Rome, 1964* p. 130. Pergamon Press.

Zuckerman, A., Hamburger, Y., and Spira, D. (1965a). *Proc. 2nd. Intern. Conf. Protozool., London, 1965* Intern. Congr. Ser. No. 91, pp. 50–51. Excerpta Med. Found., Amsterdam.

Zuckerman, A., Ron, N., and Spira, D. (1965b). *Proc. 2nd Intern. Conf. Protozool., London, 1965* Intern. Congr. Ser. No. 91, pp. 167–168. Excerpta Med. Found., Amsterdam.

6

Zoonotic Potential of Blood Parasites

R. S. BRAY

I. INTRODUCTION

The knowledge that a disease affecting man has an animal reservoir is of considerable importance to public health authorities who are responsible for the control or eradication of such a disease. The species of animal, its habitat and habit, and the mode of transmission of the infection from animals to man will profoundly affect not only the geographic and social distribution of the disease in man, but also the virulence and form of the infection and the rate of spread through populations. The existence of animal reservoirs has called into being new applications of ecology such as landscape epidemiology and the study of natural foci. These aspects of ecology have been discussed in greater detail in Chapter 1.

Biological disciplines other than microbiology have been employed in the study of an infecting organism and have included mammology, ornithology, and vertebrate and arthropod ecology. It is perhaps unfortunate that the existence of animal reservoirs of infection complicates the epidemiology and natural history of infection geometrically. If one can understand the epidemiology of an infection of man caused by a parasite transmitted by an arthropod, by knowing three salient facts about each agent then there exist 27 circumstances to be juggled; add one other agent —the animal reservoir—and we have 81 circumstances with which to play. It is indeed fortunate for the epidemiologist that the inherent biological interest of such a natural history is also increased geometrically.

II. DEFINITIONS

It is hopeless to approach the problem of animal reservoirs of infection— that is, the condition of a zoonosis—without some definition of terms. For the purposes of this chapter, the "unashamedly anthropocentric" definition of a zoonosis (Nelson, 1960) is accepted pro tempore and Hoare's (1960) view of *"Homo sapiens* as just another animal species" is temporarily set aside and a zoonosis is defined as "infections naturally transmitted between vertebrate animals and man" (WHO, 1959).

Obviously, this condition must be subdivided into particular forms of zoonoses for an easier understanding of the implications inherent in each zoonotic situation. Our understanding of these situations is built up on a series of biological facts made apparent from field and laboratory studies. With a World Health Organization (WHO) expert committee on zoonoses pending, no attempts will be made here to promulgate laws or even to attempt definitions beyond those of Nelson (1960) which suit the present purpose well enough. The three categories of zoonoses which concern us, then, according to Nelson (1960), are:

Type I: Infections of man naturally acquired from other vertebrates where the maintenance host is animal, and man is an accidental host.

Type II: Infections of vertebrates naturally acquired from man where the maintenance host is man and animals are incidental hosts.

Type III: Infections naturally transmitted between man and other vertebrates, the infection being maintained by either man or animals.

Type I infections will tend to a sort of centripetal concentration around foci (Meyer, 1942; Pavlovsky, undated, A) of animal infection, and type II around similar foci of human infection. In relation to leishmaniasis Garnham (1965) asserts that the disease in man, when of zoonotic source, is relatively mild whereas repeated interhuman transmission increases the virulence of the disease. This observation obviously has considerable general application.

Hoare (1962) believes that the infection in the maintenance animal host will always tend to be less virulent or asymptomatic as compared with the symptomatic infection in the incidental human host. This seems a rather more doubtful proposition in the light of the similar lesions on rodents and man due to *Leishmania tropica* and *L. mexicana* or the asymptomatic infections in marsupials, dogs, and man due to *Trypanosoma rangeli.* In any case, when thinking in terms of virulence one must always remember

Garnham's (1959, 1965) evolutionary sequence where domestic animals are frequently intermediate between feral and human hosts; so virulence in a dog does not, necessarily, mean failure of adaption but may mean that the domestic dog is almost as recent a host as man and may suffer considerably as in kala azar.

One other law may be dimly perceived from the fact of the great zoonotic potential inherent in certain animals. Anyone interested in zoonoses must be struck by the importance in most zoonoses of domesticated animals and animals, such as rodents, which are capable of living with man.

One could postulate that a successful reservoir animal is one that can live with man by serving him or resisting him. An unsuccessful reservoir animal is one which retreats from man, or attempts to compete with him for living space. The obvious zoonotic potential of the great apes who share with man so many of his diseases is almost totally vitiated by their lack of contact with man. The bushbuck host of *Trypanosoma rhodesiense* may eventually suffer from man's need for grazing land for his animals. Even the tastiness of roast armadillo *garni* may play its part eventually in the epidemiology of *T. cruzi*. The unstable "ecotones" of Heisch (1956) where man and animals are in uneasy relationship will tend always to become stable areas of human domination unless the animal concerned can serve man or occupy a niche unwanted by man (see also Pavlovsky, 1966).

III. DISEASES CAUSED BY BACTERIA AND SPIROCHETES

A. BARTONELLOSIS

Of the infections which come within the compass of this book we can dispose of only one which has no known animal reservoir. *Bartonella bacilliformis* has not been known to affect any animal in nature apart from man; this, however, is not to say that an animal reservoir does not exist.

B. RELAPSING FEVER

Borrelia recurrentis or epidemic louse-borne relapsing fever is not known to have an animal reservoir and the host specificity of the lice concerned would seem to rule out recourse to an animal reservoir during epidemics. However, the suddenness of the great epidemic outbreaks, in widely different areas, have set epidemiologists a problem. The recurrent form of the disease in man and newer knowledge of relapse variants among protista obviously provide an explanation for these outbreaks if one postulates that man is a reservoir. Nicolle and Anderson (1927) were the first to suggest another explanation—that louse-borne relapsing fever commenced as tick-borne relapsing fever in rodent reservoirs and accidentally became adapted to man and then to the louse. Baltazard *et al.* (1947), Heisch and Garnham (1948), and Boiron *et al.* (1948) showed that tick-borne *Borrelia* could be transmitted in the laboratory by lice. Heisch (1949) found lice infected with *Borrelia duttoni* (a tick-borne form) in nature in Kenya. More recently, however, Heisch (1965) has thrown some cold water on these ideas by showing that lice require high numbers of infectious units of tick-borne *Borrelia* (*B. duttoni*) before they can transmit. The rest is speculation.

Tick-borne relapsing fever is undoubtedly a type I and III zoonosis, but the extent of transmission between animals and man appears to vary greatly from one place to another. In North and Northwest Africa the *Borrelia* spp. from rodents are transmitted by ticks (*Ornithodorus*) to man when he inhabits the cave habitats of the tick (see Mooser, 1958; Garnham, 1959). In tropical Africa, ticks biting man have become largely domesticated, and in human beings, *B. duttoni* is transmitted from man to man without the intervention of an animal reservoir. *Borrelia duttoni* may now be incapable of infecting feral rodents in nature (Geigy and Aeschliman, 1957); however, *B. duttoni* of man is almost certainly the descendant of a *Borrelia* of rodents such as *B. dipodilli* which infects man in the laboratory (Heisch, 1954, 1956). On the other hand, Zumpt (1959) makes a case for the involvement of peridomestic multimammate rats (*Mastomys*) in tick-borne relapsing fever of man in South Africa. In the Americas the involvement of rodents and other small mammals in the epidemiology of relapsing fever is certain (Clark, 1942; Beck, 1937; Bohls, 1942) and in Asia, small mammals have been incriminated (See Mooser, 1958; Pavlovsky, undated, B). It would seem that in many parts of the world natural foci occur where *Borrelia* spp., such as *B. venezuelensis*, *B. dipodilli*, *B. crocidurae*, *B. sogdiana*, and *B. latychevi*, capable of infecting man are found in small mammals associated with *Ornithodorus* ticks. If man enters such foci and is bitten by ticks he will contract the infection. The disease cycle may there cease from man's point of view unless his own habitat is infested with man-biting *Ornithodorus*. In such a situation an endemic of relapsing fever may occur and continue in the absence of a reservoir host other than man (Garnham, 1959; Pavlovsky, undated, B).

IV. TRYPANOSOMIASIS

A. AFRICAN TRYPANOSOMIASIS

There exists no known animal reservoir of *Trypanosoma gambiense** nor does its known epidemiology lead one to postulate the necessary intervention of an animal host in the disease cycle. The disease is one of riverine or lakeside people in reasonable numbers and in settled conditions. The

* However, see Volume II, Chapter 17, "The Human Trypanosomiases."

disease does not suddenly appear well beyond known foci except along traveled communication routes to new riverine foci. Within foci the disease is dependent upon a very high level of contact between individual tsetse flies and man, and will disappear where an adequate supply of preferred hosts (e.g., crocodile) are present within the confines of the fly habitat to divert flies from contact with man. Such conditions would seem to rule out the existence of a feral reservoir host. Not even the occasional epidemic situation (such as leads an epidemiologist to think about reservoir hosts) in the Chad riverine system led Duggan (1962) to postulate a zoonosis of *T. gambiense* but rather to suggest, as one alternative, the presence of a *T. rhodesiense* zoonosis. Ashcroft (1959a) thought of *T. gambiense* as being "poorly adapted in virulence to infect wild animals which are the hosts to tsetse flies"—a remarkable comment on the poor ability of *T. gambiense* to infect tsetse flies in the light of the more usual thoughts made on virulence in feral maintenance hosts. Despite all these arguments, the persistence of *T. gambiense* sleeping sickness in the face of widespread pentamidine prophylaxis in parts of the former French West Africa led Willett (1963) to say "the possibility of the existence of a reservoir other than man should, therefore, not be dismissed as unimportant" (see also Hoare, 1962). If such a reservoir exists it is almost certain to be domestic and the favourite candidates are obviously the domestic pig and goat (van Hoof, 1947; Watson, 1962).

It is a well-established and well-known fact that endemic rhodesian sleeping sickness due to *T. rhodesiense* is a zoonosis. With minor exceptions, its epidemiology is typical of a type I zoonosis consisting of sporadic and widely scattered cases. Typical is the River Ugala area of the Tabora District, Tanzania, investigated by Apted (1956) and Apted *et al.* (1963). Infection is contracted when hunting, fishing, or beeswax gathering in an almost unpopulated area abounding in game and *Glossina morsitans*.

It would seem that man becomes infected only when crossing the line of the vector flight between animal hosts and thus falling a chance prey to the vector. Such an endemic disease is rarely transmitted from man to man since the vectors are oriented toward animals, and there are no known reports by sleeping sickness epidemiologists about such interhuman endemics. The classic Tinde experiment (Fairbairn and Burtt, 1946; Willett and Fairbairn, 1955; Ashcroft, 1959b) showed both the susceptibility of a number of domestic and wild animals to infection with *T. rhodesiense* and also the continued virulence of the parasite for man after maintenance in animals by fly transmission for 23 years.

However, for many years this zoonosis remained an epidemiological concept rather than a proven fact. *Trypanosoma rhodesiense* exists as a species because of its ability to infect man; it is indeed *T. brucei* of man.

What remained to be done was to discover a trypanosome of the *T. brucei* group in a wild animal which would infect man. The work of Weitz and Glasgow (1956) pointed to the bushbuck (*Tragelaphus scriptus*) as the main host of *Glossina pallidipes* and MacKichan (1944) and work at the East African Trypanosomiasis Research Organization (Eatro, 1956, 1958) had shown that *G. pallidipes* harbored *T. rhodesiense*. Heisch and his co-workers (1958) isolated a number of strains of trypanosomes from antelopes in rats and inoculated these into human volunteers. One volunteer succumbed a week later to infection by a trypanosome originating in a bushbuck. This was the final proof of the zoonotic character of endemic rhodesian sleeping sickness.

The ubiquity of the bushbuck in tropical Africa might be thought of concern enough, but Onyango *et al.* (1966) have recently shown that of three *T. brucei*-like trypanosome strains isolated from domestic cattle in Uganda, one infected man and was therefore *T. rhodesiense*. Bushbuck, reedbuck, and domestic cattle, all ubiquitous, must be under constant suspicion, nor perhaps can warthog and bushpig be entirely exonerated as yet, despite lower densities in experimental infections (Ashcroft *et al.*, 1959). The role of larger game animals such as elephant, giraffe, gnu, etc., remains unknown, but they seem unlikely reservoirs in the view of the studies by Ashcroft *et al.* (1959). In the opinion of Apted *et al.* (1963), given the habits of these various animals "bushbuck is the more likely reservoir," though one must now include domestic cattle. With the incrimination of domestic cattle perhaps one would be relieved to learn that the fate of African wild game may not now be decided by any overenthusiastic eradication schemes aimed at the elimination of sleeping sickness.

The recent work of Gray (1965a,b), Brown (1963), and Cunningham and Vickerman (1962) on antigen variation among single lines or clones of *T. brucei* might lead to speculation on the origin of *T. rhodesiense* as merely an antigenic variant of *T. brucei* turning up now and again and capable of infecting man. According to Gray (1965a,b), such a variant would return to the basic variant No. 1 on inoculation back into animals, by syringe or tsetse flies, in at least a proportion of cases. The Tinde experiment shows that *T. rhodesiense* retains its integrity in animals even after innumerable animal-to-animal passages; consequently, unless the variant of *T. brucei* which is *T. rhodesiense* "fixes" in a way unknown among other variants, then antigenic variation of *T. brucei* does not explain the appearance of *T. rhodesiense*. Also, areas are known such as Zululand, where *T. brucei* occurs in the absence of human sleeping sickness.

It should be noted that so far only endemic sleeping sickness of the rhodesian type has been considered. MacKichan (1944) described an epidemic situation involving *T. rhodesiense* and *G. pallidipes* in the Busoga area of Uganda and Willett (1965) described a similar situation involving *T. rhodesiense* and *Glossina fuscipes* in the Alego area, Central Nyanza, Kenya. Both authors state, or imply, that the epidemic proportions of the outbreaks were due to man-to-man transmission by the tsetse fly and that the last stage of Garnham's evolutionary sequence in zoonotic events had arrived in the case of rhodesian sleeping sickness. It would seem that epidemics due to transmission of *T. rhodesiense*, from man to man do occur although they are fairly rare and likely to die out when the conditions causing the intimate man-to-fly contact disappear.

B. AMERICAN TRYPANOSOMIASIS

Chagas' disease, caused by *T. cruzi*, is a typical type III zoonosis. The infection is transmitted from man to man by domestic triatomid bugs such as *Triatoma infestans* and *Panstrongylus megistus* which lay eggs in the mud walls of houses and feed upon man and his dogs or cats. At the same time the organism occurs in the blood of many wild animals and is transmitted from animal to animal by a variety of triatomids of greater or lesser host restriction (Deane and Deane, 1957; Barretto, 1963). It is not proposed to list here all the animals in which *T. cruzi* or "*T. cruzi*-like trypanosomes" have been found. These are described fully and excellently by Barretto (1964).

The full gamut of epidemiological possibilities are run in *T. cruzi* infections. This can be expected when one examines the long list of vertebrate hosts which includes forest and savannah animals; diurnal and nocturnal animals; wild animals and the domestic dog, cat, and guinea pig; and primates and rodents. Obviously infection can, from the anthropocentric view, pass from man to man, domestic animal to man, wild animal to man, wild animal to domestic animal to man. Given a degree of indiscriminate voracity, as is normal for triatomid bugs, all these pathways are perfectly possible and all probably occur in many South American localities (Dias and Chandler, 1949; Pessôa, 1958). Animals such as opossums (*Didelphys* spp.) and wood rats (*Neotoma* spp.) which can, and do, occupy areas cultivated by man, are of more obvious danger to man than those which may prefer to avoid man and are hunted by him, such as armadillos (*Dasypus* spp.). It should be stressed, however, that the main reservoir of infection in man with *T. cruzi* is man himself, as the last stage of Garnham's evolutionary sequence of zoonoses.

However, the frequency of infection of animals and triatomid bugs in the United States and the relative lack of human infection still remains a puzzle. Barretto (1964) lists 13 species of animals, largely wood rats, as infected with *T. cruzi* in southern United States from California to Maryland. Wood (1943), Yaeger (1959), Wood and Wood (1961), Olsen *et al.*

(1964), and Ryckman and Ryckman (1965) list 14 species and subspecies of *Triatoma* in the United States they found infected with *T. cruzi*. Despite these records, only two indigenous human infections have been described in the United States (Woody and Woody, 1955; Yaeger, 1959). One possibility is that the strains of *T. cruzi* in the United States are less virulent than South American strains. This is partially contradicted by the work of Packchanian (1943), who achieved an infection with a United States strain in man which was manifested by appreciable parasitemia and clinical symptoms similar to those occurring in natural infections. Nonetheless, inapparent and relatively asymptomatic infections may occur and this argument has been lent support by the recent work of Woody *et al.* (1961, 1965). These workers showed by complement fixation tests that antibodies to *T. cruzi* occurred in about 2.5 % of persons bitten by triatomid bugs who were living in rural Texas in association with *Triatoma gerstaeckeri*, opossums, and wood rats.

Turning for a moment to the vectors, Dias and Chandler (1949) and Woody *et al.* (1965) have suggested that there is a tendency for certain South American triatomid bugs to defecate during a meal, while United States triatomid bugs usually defecate after completion of the blood meal when they have moved away from the host. Obviously in these conditions the chances of the *T. cruzi* metacyclic forms in the feces contaminating the wound or mucous membranes is less in the case of United States triatomid bugs. Contact between normally zoophilic triatomid bugs and man is also of great importance, as discussed by Wood (1958) and Wood and Wood (1961) in relation to *Triatoma gerstaeckeri* in the Carlsbad caves.

The other South and Central American form of trypanosomiasis is caused by *Trypanosoma rangeli*. This disease is obviously a type I zoonosis and is a natural infection of wild animals, notably opossums of the genera *Metachirops* and *Didelphis* (Deane, 1958; Torrealba *et al.*, 1951; Zeledon, 1954). The dog is also involved and presumably is infected by the feral reservoir. The infection in all vertebrate hosts is asymptomatic while the infection in the triatomid vector *Rhodnius prolixus* is frequently harmful to the bug, at least in the laboratory (Grewal, 1957).

One must also mention the report of *Trypanosoma lewisi* of rodents in the blood of man in Malaya (Johnson, 1933). There is a considerable resemblance between *T. rangeli*, *T. lewisi*, *T. conorhini*, and even *T. primatum* of chimpanzees. The trypanosome found in a child in Malaya may have been a *T. lewisi-* or a *T. rangeli*-like trypanosome such as *T. conorhini* of triatomid bugs or even an unknown simian trypanosome. In the same vein, Macfie (1917) and Lavier (1927) reported *T. vivax* in man. A photograph of Macfie's trypanosome preserved at the London School of Hygiene and Tropical Medicine leaves this author in no doubt that the trypanosome is *T. vivax*, and it must be assumed that occasionally *T. vivax* can produce a transient infection in man.

V. MALARIA

Malaria is a disease transmitted from man to man by anopheline mosquitoes, but it is also an infection capable of being a type I, II or III zoonosis. For example, a number of simian malaria parasites can infect man in the laboratory; these parasites include *Plasmodium vivax schwetzi* (Rodhain and Dellaert, 1955) and *P. malariae* of chimpanzees (Rodhain, 1940); *P. cynomolgi cynomolgi* (Schmidt *et al.*, 1961; Coatney *et al.*, 1961), *P. cynomolgi bastianellii* (Eyles *et al.*, 1960), *P. cynomolgi* spp. (Bennett and Warren, 1965), *P. knowlesi* (Knowles and Das Gupta, 1932; Chin *et al.*, 1965), and *P. inui* (Das Gupta, 1938) of macaques and finally *P. brasilianum* (Contacos *et al.*, 1963) of New World Cebidae. The points at which and the conditions under which these parasites might infect man in nature have been fully discussed (Bray, 1963, 1965; Coatney, 1963; Contacos and Coatney, 1963). The susceptibility of man to these parasites establishes the possibility of a type I zoonosis. The existence of such a zoonosis has been established recently by the work of Dr. G. Robert Coatney's group at the National Institutes of Health in Washington, D.C. (Chin *et al.*, 1965). The events described in this milestone of malaria research were as follows. An American attached to the U.S. Army, and working in Malaya fell ill in Bangkok but returned to the United States. A physician treated him for upper respiratory infection following a history of sore throat, chills, and fever. The man went home where his private physician diagnosed malaria and admitted him to the Clinical Centre of the National Institutes of Health. Here an infection with *P. malariae* was diagnosed and blood was sent to the Laboratory of Parasite Chemotherapy where it was inoculated into a volunteer. Following this infection the parasite was inoculated into more volunteers and into rhesus monkeys. All monkeys died with a fulminating infection due to *P. knowlesi*. Thus it was shown that a simian malaria parasite had infected man during a 4-week stay in Malaya.

The question in this case is whether the occurrence of this zoonotic event is an isolated or a common occurrence in forest, or one which could and has led to the establishment of an endemic of simian malaria in man in rural villages. Obviously it is at least an isolated event, so individuals in Malaya and by inference, Thailand and Cambodia too, can be infected with *P. knowlesi*, probably in the forest. We have no real way of knowing whether it is a common event in Malayan forests. The known vector of *P. knowlesi* is *Anopheles hackeri* (Wharton and Eyles, 1961) which has never been taken biting man (Wharton *et al.*, 1963), but in inland forest, members of the *A. leucosphyrus* group are involved in the transmission of simian malaria (Wharton *et al.*, 1962; Eyles *et al.*, 1963) and the two species involved, *A. leucosphyrus* and *A. balabacensis introlatus*, will come to bite man, particularly in the forest. Moreover, *A. balabacensis balabacensis* in Cambodia and Thailand is thought to be a vector of simian and human malaria (Eyles *et al.*, 1964).

On the other hand, it might be assumed that zoonotic events are rare since there is a definite lack of descriptions or follow-ups of odd-seeming malaria parasites despite the presence in Malaya of malariologists such as Dr. John Field and of workers such as the late Dr. Don Eyles and his team from the National Institutes of Health.

However, we know the event can occur and to know whether such an isolated infected individual could set up a human endemic of malaria of simian origin requires very comprehensive knowledge of the epidemiology of Southeast Asian malaria and of the behavior of simian malaria parasites in local populations. Epidemiological evidence partially supports the possibility of a type III zoonosis. *Anopheles maculatus*, the chief vector of malaria in Malaya, is highly susceptible to monkey malaria parasites of the *Plasmodium cynomolgi* group (Warren and Wharton, 1963) and the natural infection of *Anopheles balabacensis introlatus* with *P. cynomolgi* might lead one to suppose that *A. balabacensis balabacensis*, now the chief vector of malaria in some areas of Thailand and Cambodia, could also transmit this simian parasite. Similarly, *Anopheles leucosphyrus* is a vector of *Plasmodium inui* (Wharton *et al.*, 1963) in Malaya and a major human malaria vector in Sarawak and Sumatra where *P. inui* is to be found (Eyles, 1963).

Two possible situations arise from these observations. The first is that a vector such as *Anopheles balabacensis balabacensis* may feed indiscriminately on peridomestic monkeys and village inhabitants setting up a type I and possibly a type III zoonosis. The second is that a primarily domestic and anthropophilic *Anopheles*, such as *A. maculatus*, could pick up the chance simian malaria infection in man (a forest hunter perhaps) and transmit it cyclically from man to man, setting up an endemic of simian malaria side-by-side with a human malaria endemic. The first possibility has the tacit support of Eyles *et al.* (1964). Yet there is evidence accruing that divergence within *Anopheles* species may be tied to host restriction. Such a case is *A. gambiae*-type, species A feeding on man, and *A. gambiae*-type, species C feeding on animals (Paterson *et al.*, 1963). There is the strong possibility that the village-visiting and man-biting *A. balabacensis balabacensis* is not the same mosquito which feeds on monkeys in the forest canopy and occasionally bites man in the forest. After all, *A. balabacensis balabacensis* is now only one of 13 forms formerly recognized as a single species—*A. leucosphyrus* (Colless, 1956, 1957; Reid, 1949).

With reference to the second possible situation, it is obvious that *A. maculatus* could transmit simian malaria from man to man provided the donor possessed a gametocytemia sufficient to infect the mosquito and an endemic could ensue provided such gametocytemias were maintained. Man in this case is not the nonimmune of the laboratory but the semi-immune of the Southeast Asian village. The infection would be achieved by the bite of one lightly infected *Anopheles*, not necessarily by the heavy laboratory infections following some possible adaptation to man.

Before speculation can be particularly useful, we must know more about the behavior of simian malaria parasites in local peoples, particularly children, following the inoculation of low numbers of sporozoites. The fact remains that simian malaria parasites have not yet been recovered from village inhabitants in Southeast Asia but intensive attempts to do so are in progress.

We have primarily considered the situation in Southeast Asia where infected monkeys exist close to human habitation. Most authorities agree that the possibilities of a type I zoonosis involving the malaria parasites of chimpanzees in Africa are remote because there is little contact between man and chimpanzee (Bray, 1960, 1963, 1965; Hoare, 1962) and because of the refractivity of *A. gambiae* to chimpanzee malaria parasites (Bray, 1957, 1958, 1960). The situation in South America is too obscure at present for comment to be useful. *Plasmodium brasilianum* of the Cebidae is infective to man (Contacos *et al.*, 1962) but little is known of its vectors in nature, though Dr. L. Deane and his colleagues may soon rectify this.

The possibilities of a type II zoonosis should be briefly mentioned, first in relation to *P. malariae* of man which can infect chimpanzees. Bray (1960) was never able to produce gametocytes of human *P. malariae* in chimpanzees and the contact between man and the anthropophilic mosquitoes on the one hand and chimpanzees on the other was slight. I believe this possibility can be dismissed. More than a mile into the Liberian forests, one of the haunts of the chimpanzee, we never caught man-biting *Anopheles* and were never bitten by *Anopheles* at any time or altitude. In Central America the howler monkey (*Alouatta*) is susceptible to *P. falciparum* blood infections but gametocytes were not produced (Taliaferro and Taliaferro, 1934). Very recently we have learned that the night monkey (*Aotus*) and the titi marmoset (*Saguinus*) are susceptible to *P. vivax* (Porter and Young, 1966), but the full implications of this discovery remain to be assessed.

VI. BABESIOSIS

The last zoonosis to be mentioned is the remarkable fatal infection of man with *Babesia bovis* reported in Yugoslavia by Škrabalo and Deanović

(1957). The man in question had been splenectomized 11 years previously and was living on a tick-infested farm where the cattle were infected with *Babesia*. Garnham and Bray (1959) and Garnham and Voller (1965) were later to show that splenectomized chimpanzees and rhesus monkeys but not intact animals were susceptible to bovine babesiosis but not to bovine theileriosis (Bray and Garnham, 1961). A second case of human babesiosis has been discovered recently which was serologically diagnosed by the tube latex agglutination (TLA) test (Morbidity and Mortality Weekly Report, Vol. 16, January 7, 1967, page 8, Communicable Diseases Center, U.S.P.H.S., Atlanta, Georgia) (see Chapter 5). It would seem that man becomes susceptible at least to the smaller bovine *Babesia* spp. following splenectomy.

Acknowledgments

I should like to acknowledge the critical help given in the preparation of this chapter by Professor P. C. C. Garnham, Drs. F. I. C. Apted, A. J. Duggan, R. B. Heisch, C. A. Hoare, M. P. Hutchinson, and G. S. Nelson.

REFERENCES

Apted, F. I. C. (1956). Sleeping Sickness Service Annual Report for 1955, Tanganyika Medical Department.
Apted, F. I. C. Ormerod, W. E., Smyly, D. P., Stronach, B. W., and Szlamp, E. L. (1963). *J. Trop. Med. Hyg.* **66,** 1.
Ashcroft, M. T. (1959a). *Trop. Diseases Bull.* **56,** 1073.
Ashcroft, M. T. (1959b). *Ann. Trop. Med. Parasitol.* **53,** 137.
Ashcroft, M. T., Burtt, E., and Fairbairn, H. (1959). *Ann. Trop. Med. Parasitol.* **53,** 147.
Baltazard, M., Mofidi, C., and Bahmanyax, M. (1947). *Ann. Inst. Pasteur* **73,** 1066.
Barretto, M. P. (1963). *Arquiv. Hig. Saude Publ. (Sao Paulo)* **28,** 43.
Barretto, M. P. (1964). *Rev. Brasil. Malariol. Doencas Trop., Publ. Avulsas* **16,** 527.
Beck, M. D. (1937). *J. Infect. Diseases* **60,** 64.
Bennett, G. F., and Warren, M. (1965). *J. Parasitol.* **51,** 79.
Bohls, S. W. (1942). *In* "Symposium on Relapsing Fever in the Americas," Publ. No. 18, pp. 125–130. Amer. Assoc. Advan. Sci., Washington, D. C.
Boiron, H., Koerber, R., and Carronier, B. (1948). *Bull. Soc. Pathol. Exotique* **41,** 81.
Bray, R. S. (1957). *Ann. Soc. Belge Med. Trop.* **37,** 169.
Bray, R. S. (1958). *J. Parasitol.* **44,** 46.
Bray, R. S. (1960). *Am. J. Trop. Med. Hyg.* **9,** 455.
Bray, R. S. (1963). *Ergeb. Mikrobiol., Immunitaetsforsch. Exptl. Therap.* **36,** 168.
Bray, R. S. (1965). *Advan. Sci.* p. 25.
Bray, R. S., and Garnham, P. C. C. (1961). *J. Parasitol.* **47,** 538.
Brown, K. N. (1963). *In* "Immunity to Protozoa" (P. C. C. Garnham, A. E. Pierce, and I. Roitt, eds.), pp. 204–212. Blackwell, Oxford.
Chin, W., Contacos, P. G., Coatney, G. R., and Kimball, H. R. (1965). *Science* **149,** 865.
Clark, H. C. (1942). *In* "Symposium on Relapsing Fever in the Americas," Publ. No. 18, pp. 29–34. Am. Assoc. Advance. Sci., Washington, D.C.

Coatney, G. R. (1963). *J. Am. Med. Assoc.* **184**, 876.
Coatney, G. R., Elder, H. A., Contacos, P. G., Getz, M. G., Greenland, R., Rossan, R. N., and Schmidt, L. H. (1961). *Am. J. Trop. Med. Hyg.* **10**, 541.
Colless, D. H. (1956). *Trans. Roy. Entomol. Soc. (London)* **108**, 37.
Colless, D. H. (1957). *Proc. Roy. Entomol. Soc. (London)* **B-26**, 131.
Contacos, P. G., and Coatney, G. R. (1963). *J. Parasitol.* **49**, 912.
Contacos, P. G., Elder, H. A., Coatney, G. R., and Genther, C. (1962). *Am. J. Trop. Med. Hyg.* **11**, 186.
Contacos, P. G., Lunn, J. S., Coatney, G. R., Kilpatrick, J. W., and Jones, F. E. (1963). *Science* **142**, 676.
Cunningham, M. P., and Vickerman, K. (1962). *Trans. Roy. Soc. Trop. Med. Hyg.* **56**, 48.
Das Gupta, B. M. (1938). *Proc. Natl. Inst. Sci. India* **4**, 241.
Deane, L. M. (1958). *Rev. Brasil. Malariol. Doenças Trop., Publ. Avulsas* **10**, 531.
Deane, L. M., and Deane, M. P. (1957). *Rev. Brasil. Malariol. Doenças Trop., Publ. Avulsas* **9**, 577.
Dias, E., and Chandler, A. C., (1949). *Mem. Inst. Oswaldo Cruz* **47**, 403.
Duggan, A. J. (1962). *Trans. Roy. Soc. Trop. Med. Hyg.* **56**, 439.
EATRO. (1956). East African Trypanosomiasis Research Organization, Annual Report, 1955–1956.
EATRO. (1958). East African Trypanosomiasis Research Organization, Annual Report, 1956–1957.
Eyles, D. E. (1963). *J. Parasitol.* **49**, 866.
Eyles, D. E., Coatney, G. R., and Getz, M. G. (1960). *Science* **131**, 1812.
Eyles, D. E., Warren, M., Guinn, E., Wharton, R. H., and Ramachandran, C. P. (1963). *Bull. World Health Organ.* **28**, 134.
Eyles, D. E., Wharton, R. H., Cheong, W. H., and Warren, M. (1964). *Bull. World Health Organ.* **30**, 7.
Fairbairn, H., and Burtt, E. (1946). *Ann. Trop. Med. Parasitol.* **40**, 270.
Garnham, P. C. C. (1959). *Med. Press* **142**, 251.
Garnham, P. C. C. (1965). *Am. Zoologist* **5**, 141.
Garnham, P. C. C., and Bray, R. S. (1959). *J. Protozool.* **6**, 352.
Garnham, P. C. C., and Voller, A. (1965). *Acta Protozool. (Warsaw)* **3**, 183.
Geigy, R., and Aeschliman, A. (1957). *Z. Tropenmed. Parasitol.* **8**, 96.
Gray, A. R. (1965a). *Ann. Trop. Med. Parasitol.* **59**, 27.
Gray, A. R. (1965b). *J. Gen. Microbiol.* **41**, 195.
Grewal, M. S. (1957). *Exptl. Parasitol.* **6**, 123.
Heisch, R. B. (1949). *Brit. Med. J.* **1**, 17.
Heisch, R. B. (1954). *Ann. Trop. Med. Parasitol.* **48**, 28.
Heisch, R. B. (1956). *Brit. Med. J.* **2**, 669.
Heisch, R. B. (1965). *E. African Med. J.* **42**, 31.
Heisch, R. B., and Garnham, P. C. C. (1948). *Parasitology* **38**, 247.
Heisch, R. B., McMahon, J. P., and Manson-Bahr, P. E. C. (1958). *Brit. Med. J.* **3**, 1203.
Hoare, C. A. (1960). *Trans. Roy. Soc. Trop. Med. Hyg.* **54**, 318.
Hoare, C. A. (1962). *Acta Trop.* **19**, 281.
Johnson, P. D. (1933). *Trans. Roy. Soc. Trop. Med. Hyg.* **26**, 467.
Knowles, R., and Das Gupta, B. M. (1932). *Indian Med. Gaz.* **67**, 301.
Lavier, G. (1927). "Interim Report of the League of Nations International Commission on Human Trypanosomiasis," CH536, p. 141. League of Nations, Geneva.
Macfie, J. W. S. (1917). *Brit. Med. J.* **1**, 12.

MacKichan, I. W. (1944). *Trans. Roy. Soc. Trop. Med. Hyg.* **38,** 49.

Meyer, K. F. (1942). *Medicine* **21,** 143.

Mooser, H. (1958). *Ergeb. Mikrobiol., Immunforsch. Exptl. Therap.* **31,** 184.

Nelson, G. S. (1960). *Trans. Roy. Soc. Trop. Med. Hyg.* **54,** 301.

Nicolle, G., and Anderson, C. (1927). *Arch. Inst. Pasteur Tunis* **16,** 123.

Olsen, P. F., Shoemaker, J. P., Turner, H. F., and Hays, K. L. (1964). *J. Parasitol.* **50,** 599.

Onyango, R. J., Van Hoeve, K., and De Raadt, P. (1966). *Trans. Roy. Soc. Trop. Med. Hyg.* **60,** 175.

Packchanian, A. (1943). *Am. J. Trop. Med.* **23,** 309.

Paterson, H. E., Paterson, J. S., and Van Eeden, G. J. (1963). *Med. Proc.* **9,** 414.

Pavlovsky, E. N. (1966). *Proc. 1st Intern. Congr. Parasitol., Rome, 1964* Vol. 1, p. 146.

Pavlovsky, Y. N., ed. (undated, A). "Human Diseases with Natural Foci," pp. 9–44. Foreign Language Publ. House, Moscow.

Pavlovsky, Y. N., ed. (undated, B). "Human Diseases with Natural Foci," pp. 138–184. Foreign Language Publ. House, Moscow.

Pessôa, S. B. (1958). *Rev. Goiana Med.* **4,** 83.

Porter, J. A., and Young, M. D. (1966). *Military Med., 131:* (No. 9, Suppl.), 952.

Reid, J. A. (1949). *Proc. Roy. Entomol. Soc. (London)* **B-18,** 42.

Rodhain, J. (1940). *Compt. Rend. Soc. Biol.* **133,** 276.

Rodhain, J., and Dellaert, R. (1955). *Ann. Soc. Belge Med. Trop.* **35,** 757.

Ryckman, R. E., and Ryckman, A. E. (1965). *J. Med. Entomol.* **2,** 87.

Schmidt, L. H., Greenland, R., and Genther, C. S. (1961). *Am. J. Trop. Med. Hyg.* **10,** 679.

Škrabalo, Z., and Deanović, Ž. (1957). *Doc. Med. Geograph Trop.* **9,** 11.

Taliaferro, W. H., and Taliaferro, L. G. (1934). *Am. J. Hyg.* **19,** 318.

Torrealba, J. F., Pifano, F., and Romer, M. (1951). *Gaceta Med. Caracas* **116,** 111.

van Hoof, L. (1947). *Trans. Roy. Soc. Trop. Med. Hyg.* **40,** 728.

Warren, M., and Wharton, R. H. (1963). *J. Parasitol.* **49,** 892.

Watson, H. J. C. (1962). Unpublished document quoted by Willett (1963).

Weitz, B., and Glasgow, J. P. (1956). *Trans. Roy. Soc. Trop. Med. Hyg.* **50,** 593.

Wharton, R. H., and Eyles, D. E. (1961). *Science* **134,** 279.

Wharton, R. H., Eyles, D. E., Warren, M., and Moorhouse, D. E. (1962). *Science* **137,** 758.

Wharton, R. H., Eyles, D. E., and Warren, M. (1963). *Ann. Trop. Med. Parasitol.* **57,** 32.

WHO. (1959). *World Health Organ., Tech. Rept. Ser.* **169,** 6.

Willett, K. C. (1963). *Bull. World Health Organ.* **28,** 645.

Willett, K. C. (1965). *Trans. Roy. Soc. Trop. Med. Hyg.* **59,** 374.

Willett, K. C., and Fairbairn, H. (1955). *Ann. Trop. Med. Parasitol.* **49,** 278.

Wood, S. F. (1943). *Am. J. Trop. Med.* **23,** 315.

Wood, S. F. (1958). *Bull. S. Calif. Acad. Sci.* **57,** 113.

Wood, S. F., and Wood, F. D. (1961). *Am. J. Trop. Med. Hyg.* **10,** 155.

Woody, N. C., and Woody, H. B. (1955). *J. Am. Med. Assoc.* **159,** 676.

Woody, N. C., Dedianous, N., and Woody, H. B. (1961). *J. Pediat.* **58,** 738.

Woody, N. C., Hernandes, A., and Suchow, B. (1965). *J. Pediat.* **66,** 107.

Yaeger, R. G. (1959). *Rev. Goiana Med.* **5,** 461.

Zeledon, R. (1954). *Rev. Biol. Trop., Univ. Costa Rica* **2,** 231.

Zumpt, F. (1959). *Nature* **184,** 793.

II

GENERAL CHARACTERISTICS OF THE BLOOD PROTISTS

7

Definition and Classification

JOHN O. CORLISS

The microbial world, hardly known to exist before the nineteenth century when advances in microscopy first allowed revelation of many of its members, poses many difficult problems today with respect to acceptable schemes of classification which accommodate the great diversity of forms generally included in it. Although the primary concern of this brief chapter is to discuss the taxonomic affinities (or lack thereof) of only a few representatives of the world of microorganisms—those which happen to possess in common the same relatively restricted ecological niche (blood of vertebrates)—it may be helpful to mention first some of the controversial aspects of classification of the so-called lower forms of life.

I. THE CONCEPT OF PROTISTA

Exactly a century ago, the great German biologist Ernst Haeckel proposed a third kingdom, the Protista, to contain all microorganisms, thus segregating them from the typically larger, multicellular, more highly differentiated forms of life, the Plantae and Animalia proper. Since the original separation was based primarily on the rather poor (if used alone) criterion of size and left unanswered the major question of the ancestry of the obviously plant or animals forms, Haeckel's proposition, in modern times, has often been considered to represent a compromise of limited value and one of considerable artificiality. Nevertheless, if additional criteria are applied, if the great diversity existing *within* this third group is recognized, and if the various pitfalls of drawing phylogenetic conclusions without adequate facts or fossils are kept clearly in mind, then the concept "Protista" still has some merit today—in fact, perhaps more value, in its refined form, then it originally possessed.

Tentative though any detailed scheme of classification of the Protista must be, in light of our very fragmentary data concerning their evolution and possible phylogenetic interrelationships, microbiologists, bacteriologists, mycologists, phycologists, and protozoologists generally agree that there are—at the cellular level (thus excluding, for the moment, the viruses or phage organisms)—two principal subdivisions within the Protista. The "lower" protists or Monera, though a heterogeneous group among themselves, include principally the bacteria (including the rickettsiae) and the blue-green algae. They comprise organisms typically small in size (but not less than 0.2–2.0 μ in the smallest diameter), possessing a primitively constructed nuclear apparatus (exhibiting the so-called "procaryotic" condition) and a unique system of genetic transfer of heritable information (transduction or type transformation), without true plastids, and exhibiting flagella not structurally homologous with the locomotor organelles of any other kinds of organisms. The "higher" protists, showing a tremendous diversity of forms and surely long separated evolutionarily from each other, embrace the various major kinds or phyla of algae (above the level of the blue-greens), the protozoa *sensu lato*, and the fungi (including alleged relatives). They are often larger microorganisms (sometimes tremendously larger) than the lower protists, their nuclear material is organized into the "true" nucleus characteristic of still "higher" plant and animal forms (thus termed "eucaryotic"), their cytoplasm generally contains the same subcellular organelles recognizable in the cells of the tissues and organ systems of plants and animals proper, and vibratile organelles (cilia, flagella), when present, are basically identical in structure with the locomotor organelles of higher organisms. The ability to carry out photosynthesis runs throughout both lower and higher protists, present in some subgroups, absent in others; differences are great even here, however, with respect to location and kinds of pigments involved, etc.

Viruses, mentioned briefly above, are not considered to be cellular organisms at all; thus most authors exclude them entirely from the category Protista. A viral particle consists of a nucleic acid molecule, either ribonucleic acid (RNA) or deoxyribonucleic acid (DNA), enclosed in a protein coat; multiplication can occur only within a host cell, under direction of the host's enzymatic machinery. Today no transitional forms between bacteria and viruses are thought to exist, though the rickettsiae, being small obligate intracellular parasites, were once so viewed. If Protista *sensu lato* are considered to include viruses, it must be with the clear understanding that no direct-line interrelationship between such dependent and acellular organisms and any other protistan group is being postulated.

II. PROBLEMS OF PROTISTAN TAXONOMY

A good scheme of classification must be not only a convenient and easily understandable one but also one possessing high predictive value. Proposed affinities among groups or subgroups must be based on sound data available in considerable quantity, if the taxonomic arrangements suggested are to stand the test of even a short time. Internal consistency should be a major objective of any practical scheme. But protistologists face a particularly difficult situation with their organisms, when it comes to classifying them, because of the very nature of the material with which they have to work. The usual absence of multicellular organization (if present, only at a low level of differentiation), the general lack of fossils (when known, as in the foraminifera, they are of no evolutionary significance with respect to relationship with *other* major taxa), the cosmopolitan distribution of many species, the widespread lack of "overt" sexuality—such factors represent major drawbacks to understanding possible phylogenetic interrelationships among protistan groups.

At the lower taxonomic levels, these same deficiencies make confident recognition of specific or even generic limits most difficult. In fact, the question "What is a species?" for the vast majority of forms which can never be tested by standard genetic procedures has yet to be answered satisfactorily.

Thus, often left with gaps and guesses, protistologists should not be surprised that classification of their organisms remains an area of active controversy. Taxonomy or systematics is far from a static science, however, so change based on new data resulting, for example, from application of new techniques, is welcomed; and there is hope that in some decade a fairly stable, widely acceptable scheme of classification for most protistan groups will emerge.

New or relatively new techniques include phase, electron, and other kinds of specialized microscopy, refined cytochemical tests, more sophisticated biochemical and physiological methods, and comparative analyses of specific differences in DNA base composition of nucleic acids. Encouraging also is the renewed attention to the importance of gathering knowledge concerning the entire life cycle of the organism (whether it be a free-living or parasitic form). Such knowledge would involve data on its ecology (including factors of host-parasite interrelationships) as well as the occurrence of polymorphism or other possibly unsuspected features of its ontogenetic morphogenesis. Since any investigator is obliged to work solely with extant, present-day species, however, a word of caution is in order concerning the real possibilities of both morphological and physiological convergence of forms sharing a more or less common ecological niche. An outstanding example may be found among four actually quite different kinds of protistan organisms—in this instance the same general type of disease in certain mammals is caused by all four of them. After invading the red blood cells of the host, they cause anemia and fever, and remain in the host long after symptoms of the disease disappear; and they are all carried by arthropod vectors. But the genera of parasites involved belong, according to other more appropriate characteristics including ultrastructural and full life cycle data, to four very widely separated protistan (*sensu lato*) groups, since one is a hemotropic virus; another, a rickettsial bacterium (lower protist); another, a sarcodinid protozoon (higher protist); and the last, a sporozoan protozoon (a different group of higher protist).

III. RULES OF NOMENCLATURE

Although nomenclature should be the handmaiden and not the master of taxonomy, there are innumerable problems of importance in the area of naming organisms and groups of organisms which cannot be totally ignored. A special complexity with respect to the Protista arises from the fact that its members fall, nomenclaturally, under three different sets of rules or codes: the bacteriological, the botanical, and the zoological. It is sufficient here merely to call attention to the importance of names and the rules governing their formation, usage, priority, etc.; various appropriate references cited in the annotated bibliography (see Section VI) may be consulted for detailed information.

It may be worthwhile to emphasize that there are few regulations in the international codes of nomenclature concerned with names of the higher (i.e., suprafamilial) taxa. Thus no legal attempt has been made to set up or control taxonomy at the level of orders, classes, phyla, and the like. As long as biologists do not unwisely treat the flexibility thus permitted, it is sensible to have such freedom of choice nomenclaturally at these levels.

IV. DEFINITION OF BLOOD PROTISTS

Blood protists, in a definition appropriate for the present volume, may be said to include all microorganisms known to date to pass a major portion of their life in the erythrocytes, leukocytes, or plasma of the blood of vertebrate animals. The hosts, here limited primarily to man and other mammals, may be the sole host or one of several in the full life history of the parasite. The infective organism may be an intra- or extracellular parasite in the blood tissue of the vertebrate host. The reaction(s) of the host and the degree of severity of the disease may vary in accordance with many factors, including the specific identity of the parasite. The reader is referred to appropriate chapters for details concerning host-parasite interrelationships, etc. (see generic names in the Subject Index as convenient guides).

V. SKELETAL SCHEME OF CLASSIFICATION

It may be instructive to conclude the present brief chapter with a skeletal classification of the Protista, including highly abbreviated definitions or characterizations of major groups and lists of genera whose species are treated elsewhere in this volume. Taxonomic allocation is based on the latest information available that seems sound enough to warrant such a position for the genera whether or not they have been traditionally so located. As intimated above, however, almost all schemes of classification for protistan forms must be considered tentative at best. Yet this very uncertainty of protistologists concerned with systematics represents a healthy outlook for the future. Flexibility is to be preferred over rigidity at a time when we still have so much to learn about the fascinating world of microorganisms.

For details of classification of any one organism or relatively restricted group of organisms the reader is referred to appropriate chapters of this volume prepared by specialists on the group (use Subject Index for names of desired taxa) or to the several pertinent works listed in the annotated bibliography (Section VI) of this chapter.

PROTISTA *sensu lato*

I. *Viruses*

Acellular, totally host-dependent, minute, intracellular parasites. (Hemotropic forms from the bloodstream of various vertebrates are known, but none is included in the present volume.)

II. *Lower Protists*

Greatly simplified cell structure; procaryotic; small organisms; primitive locomotor organelles, if any; free-living and parasitic forms.

A. *Blue-green algae.* None in vertebrate blood.

B. *Bacteria.* It may be interesting to note that many of the organisms listed here were for a long time thought to be protozoa.

1. *Bartonella,* found in blood of man. (Other, nonvector-transmitted bacterial diseases are not included in the present volume.)

2. Various microorganisms, in blood of numerous vertebrates (those marked with asterisk are found in man): *Anaplasma, Chlamydia,* * *Colesiota, Colettsia, Cowdria, Coxiella,* * *Ehrlichia, Eperythrozoon, Grahamella, Haemobartonella, Miyagawanella,* * *Neorickettsia, Rickettsia,* * *Ricolesia.*

3. Various spirochaetes, with species in man or other vertebrates: *Borrelia, Cristispira, Leptospira, Treponema.*

III. *Higher Protists*

Cell structure often quite complex; if multicellular, low degree of differentiation; eucaryotic; cilia or flagella, if present, showing structure typical of those of higher plants and animals; relatively large microscopic forms; free-living and parasitic species.

A. *All algae above blue-green level.* Six major phylogenetic groups are standardly recognized. No species occur in vertebrate blood.

B. *Fungi and relatives.* Ascomycetes. (Not included in the present volume.)

C. *Protozoa sensu lato.* (Classification used below follows that of the recent Honigberg Report: see bibliography for reference.)

   1. Flagellates, order Kinetoplastida: *Trypanosoma*, a polymorphic flagellate found in the circulating blood of many vertebrates—fishes, amphibians, reptiles, birds, and mammals, including man.

   2. Sarcodines, order Piroplasmida: *Aegyptianella, Babesia, Cytauxzoon,* and *Theileria,* all in blood of mammals other than man except for *Aegyptianella,* whose species are found only in birds.

   3. Sporozoa, order Eucoccida (suborder Haemosporina): *Haemoproteus, Hepatocystis, Leucocytozoon, Nycteria, Plasmodium, Polychromophilus,* in the blood of various vertebrates, especially mammals and birds, with only *Plasmodium* containing species found in man. Order Toxoplasmida: *Besnoitia, Encephalitozoon* (possibly a cnidosporan, order Microsporida!), *Sarcocystis, Toxoplasma,* in various vertebrates, occasionally including man.

   4. Ciliates, order Hymenostomatida: *Tetrahymena,* a small, practically ubiquitous ciliate, at least one species of which is found in the blood vessels of injured fishes or certain amphibians, as a facultative parasite. (Not included elsewhere in this volume, however.)

VI. ANNOTATED BIBLIOGRAPHY

The editors have granted me permission to include an appendix here of references pertinent to the substance of the material covered in preceding pages of this chapter, rather than making direct—and more limited—citation therein; and to cite the references in full, including complete title and pagination (so often indispensable in taxonomic work); and, finally, to append comments to each citation in order to guide the interested reader to some of the most helpful or most appropriate major modern papers, monographs, or books to be found in the tremendous accumulation of literature on the subject of systematics of lower organisms. It may be recalled that the term protista, taken in its broadest sense, covers organisms conventionally assigned to the allegedly vastly separated kingdoms or groups of microbes, plants, and animals, small and unicellular though most of them be. Thus the brief list of references which follows is very far from being an exhaustive one.

Ainsworth, G. C., and Sneath, P. H. A. (1962). Microbial Classification. *Symp. Soc. Gen. Microbiol.* **12,** 1–483. (Includes discussion of many current problems in the taxonomy of microorganisms, with contributions from some 23 authorities on special groups.)

Breed, R. S., Murray, E. G. D., and Smith, N. R. (1957). "Bergey's Manual of Determinative Bacteriology," 7th ed., 1,094 pp. Williams & Wilkins, Baltimore, Maryland. (The latest edition of a time-honored and invaluable comprehensive key to all of the multitudinous species of bacteria.)

Copeland, H. F. (1956). "The Classification of Lower Organisms," 302 pp. Pacific Books, Palo Alto, California. (A detailed treatment of the taxonomy of protistan forms, including unconventional groupings at the higher levels and a careful review of nomenclatural aspects of their classification.)

Cheissin, E. M., and Poljansky, G. I. (1963). On the taxonomic system of protozoa. *Acta Protozool. (Warsaw)* **1,** 327–352. (A presentation of the classification system favored for the protozoa by one modern school of thought, a system which, however, is rather strongly traditional in a number of respects.)

Corliss, J. O. (1960). *Tetrahymena chironomi* sp. nov., a ciliate from midge larvae, and the current status of facultative parasitism in the genus *Tetrahymena*. *Parasitology* **50,** 111–153. (A review of species of the nearly ubiquitous hymenostome genus *Tetrahymena*, with emphasis on those found in parasitic association with other organisms.)

Corliss, J. O. (1962a). Taxonomic procedures in classification of protozoa. *Symp. Soc. Gen. Microbiol.* **12,** 36–67. (A consideration of the general problems and controversies encountered in such work, the bases used in protozoan systematics, and some specific problems related to hierarchical levels, nomenclature, etc.)

Corliss, J. O. (1962b). Taxonomic-nomenclatural practices in protozoology and the new International Code of Zoological Nomenclature. *J. Protozool.* **9,** 307–324. (Discussion and interpretation of provisions in the new Code which particularly affect protozoan taxonomy in its nomenclatural aspects, including an encouragement to parasitologists and protozoologists to make more and better use of the Code in their systematic work.)

Corliss, J. O. (1963). Application of modern techniques to problems in the systematics of the Protozoa. *Proc. 16th Intern. Congr. Zool., Washington, D. C., 1963,* Vol. 4, pp. 97–102. (A brief consideration of some new or relatively new technical approaches which should yield data useful in revising older schemes of classification of these organisms.)

Corliss, J. O. (1967). Systematics of the phylum Protozoa. *In* "Chemical Zoology" (M. Florkin and B. T. Scheer, eds.), Vol. I, pp. 1–20. Academic Press, New York. (An annotated scheme of classification of the protozoa which endorses fully that published in 1964 as "The Honigberg Report"; see Honigberg *et al.,* 1964.)

Dogiel, V. A. (1965). "General Protozoology," 747 pp. Oxford Univ. Press, London and New York. (An English translation of the 1962 2nd edition of this general textbook, a revision which was written by Poljansky and Cheissin, the late Professor Dogiel's two most productive students.)

Dougherty, E. C., and Allen, M. B. (1960). Is pigmentation a clue to protistan phylogeny? *In* "Comparative Biochemistry of Photoreactive Systems" (M. B. Allen, ed.), pp. 129–144. (Consideration of the important problem of the possible role of the pigments in elucidating phylogenetic interrelationships among relatively primitive forms of life.)

Garnham, P. C. C. (1966). "Malaria Parasites and Other Haemosporidia," 1,114 pp. Blackwell, Oxford. [A modern authoritative account of the biology and taxonomy of a very important group of blood (and other tissue) protists among the sporozoan Protozoa.]

Grassé, P. P., ed. (1952–1953). "Traité de Zoologie. Anatomie, Systématique, Biologie," Vol. I, Parts 1 and 2, 2,231 pp. Masson, Paris. [These two important publications include treatment of (animal) phylogeny and of all major groups of the Protozoa except the ciliates.]

Haeckel, E. (1866). "Generelle Morphologie der Organismen. II. Allgemeine Entwicklungschichte der Organismen." G. Reimer, Berlin. (The classic work which includes the first use and definition of the term Protista.)

Hoare, C. A. (1966). The classification of mammalian trypanosomes. *Ergeb. Mikrobiol., Inmmunitaetsforsch. Exptl. Therapie* **39,** 43–57. (An up-to-date treatment of the taxonomy of a very important group of blood parasites by an expert in the field.)

Honigberg, B. M., Balamuth, W., Bovee, E. C., Corliss, J. O., Gojdics, M., Hall, R. P., Kudo, R. R., Levine, N. D., Loeblich, A. R., Jr., Weiser, J., and Wenrich, D. H. (1964). A revised classification of the phylum Protozoa. *J. Protozool.* **11,** 7–20. (A modern classification of the phylum Protozoa at suprafamilial levels, including definitions or descriptions of some 140 major taxonomic categories: the result of years of cooperative effort of an international group of specialists and consultants.)

Hutner, S. H., and Provasoli, L. (1955). Comparative biochemistry of flagellates. *In* "Biochemistry and Physiology of Protozoa" (S. H. Hutner and A. Lwoff, eds.), Vol. 2, pp. 17–43. Academic Press, New York. (Includes consideration of possible phylogenetic interrelationships among protozoan and algal forms.)

International Code of Botanical Nomenclature adopted by the 10th International Botanical Congress, Edinburgh, August 1964. (1966). Utrecht. 372 pp. (The rules of nomenclature for all plant organisms, exclusive of the bacteria.)

International Code of Nomenclature of Bacteria and Viruses adopted by the International Committee on Bacteriological Nomenclature. (1958). Iowa State College Press, Ames, Iowa. 186 pp. (The rules of nomenclature generally followed by all microbiologists concerned with taxonomic problems of the bacteria.)

International Code of Zoological Nomenclature adopted by the 15th International Congress of Zoology, London, July 1958. (1961). London. 176 pp. (The rules of nomenclature for all animal organisms.)

Jawetz, E., Melnick, J. L., and Adelberg, E. A. (1966). "Review of Medical Microbiology," 7th ed., 494 pp. Lange Med. Publ., Los Altos, California. (A standard review of various aspects, including some systematics, of medical microbiology — now in its 7th edition.)

Kudo, R. R. (1966). "Protozoology," 5th ed., 1,174 pp. Thomas, Springfield, Illinois. (The latest edition of the long-time standard American textbook of protozoology.)

Levine, N. D. (1961). "Protozoan Parasites of Domestic Animals and of Man," 412 pp. Burgess, Minneapolis, Minnesota. (A most useful manual, comprehensive, modern, well-documented.)

Levine, N. D., ed. (1966). Discussion of the Classification of Protozoa at the Second International Conference on Protozoology, London, England. *J. Protozool.* 13, 189–195. (Edited remarks of some 12–15 participants made at a lively, informal meeting in London at which recent schemes of classification of the Protozoa were discussed from a comparative point of view.)

Lwoff, A. (1962). "Biological Order," 101 pp. M.I.T. Press, Cambridge, Massachusetts. (A modern commentary on molecules and organisms.)

Lwoff, A., and Tournier, P. (1966). The classification of viruses. *Ann. Rev. Microbiol.* *20:* 45–74. (The most recent treatment of this important topic.)

Marmur, J., Falkow, S., and Mandel, M. (1963). New approaches to bacterial taxonomy. *Ann. Rev. Microbiol.* *17:* 329–372. (Important treatment of very new techniques.)

Mayr, E., Linsley, E. G., and Usinger, R. L. (1968). "Methods and Principles of Systematic Zoology," 2nd ed. McGraw-Hill, New York. (A book destined to become authoritative in the field of animal taxonomy.)

Pringsheim, E. G. (1949). The relationships between bacteria and myxophyceae. *Bacteriol. Rev.* 13, 47–98. (An authoritative discussion on interrelationships of two major groups of the lower protists.)

Raabe, Z. (1964). Remarks on the principles and outline of the system of Protozoa. *Acta Protozool.* 2, 1–18. (A presentation of a classification system for the protozoa differing in certain major respects from other modern schemes.)

Simpson, G. G. (1961). "Principles of Animal Taxonomy." 247 pp. Columbia Univ. Press, New York. (An important book by a recognized authority, emphasizing the phylogenetic approach to systematics.)

Smith, G. M. (1955). "Cryptogamic Botany," 2nd ed., 2 vols. McGraw-Hill, New York. (Latest edition of a classic treatise on the lower plants.)

Sokal, R. R., and Sneath, P. H. A. (1963). "Principles of Numerical Taxonomy," 359 pp. Freeman, San Francisco, California. (A thesis in defense of Neo-Adansonian classification, emphasizing, in particular, the computerized use of many characters all equally weighted: an approach especially feasible for various protistan groups.)

Sonneborn, T. M. (1957). Breeding systems, reproductive methods, and species problems in Protozoa. *In* "The Species Problem," Publ. No. 50, pp. 155–324. Am. Assoc. Advance. Sci., Washington, D.C. (Including a modern treatment of the problem of the definition of a species among forms many of which do not exhibit sexuality.)

Stanier, R. Y. (1959). Introduction to the Protista. *In* "Ward and Whipple's Fresh-Water Biology" (W. T. Edmondson, ed.), 2nd ed., pp. 7–15. Wiley, New York. (A succinct but cogent account of the groups of organisms comprising the Protista.)

Stanier, R. Y., Doudoroff, M., and Adelberg, E. A. (1963). "The Microbial World," 2nd ed., 753 pp. Prentice-Hall, Englewood Cliffs, New Jersey. (A modern treatise rapidly becoming an authoritative standard volume on the subject.)

8

Cultivation and Nutritional Requirements

WILLIAM TRAGER

I. INTRODUCTION

The cultivation *in vitro* of parasitic and in particular of disease-producing organisms serves a number of useful purposes. It may aid in the diagnosis of the infection or in the production of an effective vaccine. Perhaps its most important outcome, however, is the opening of avenues leading to an understanding of the physiological nature of the host-parasite relationship and ultimately to rational methods of treatment and chemotherapy.

It must be admitted at the outset that cultivation of the more important pathogenic protozoa parasitic in the blood of vertebrates has not progressed to a sufficiently high level of development. Some of these parasites, of which malaria is the most notable example, are obligate intracellular forms and have been maintained *in vitro* only within their living host cells (with one exception). But even with species such as the trypanosomes, which are generally not intracellular, the stages occurring in the blood of the vertebrate host have not yet been grown in culture (again with a few exceptions). Furthermore, of the blood protozoa which have been cultured axenically only a few species have been grown in a defined medium. Hence little is known about the nutritional requirements of most species.

Since the cultural requirements of parasites may reasonably be supposed to be related to the conditions under which they develop in nature, we shall consider, in the discussion which follows, first those protozoa living in the blood plasma and then those developing inside erythrocytes or leukocytes and related cells. Blood parasites of bacterial nature, such as the spirochetes and Bartonellaceae, and *Anaplasma* will not be treated in this chapter. *Anaplasma* has not yet been grown even in cell cultures or erythrocyte suspensions (see Chapter 23, by Ristic). Of the Bartonellaceae only the human pathogen *Bartonella* has been cultivated but little is known concerning its nutrition (see Peters and Wigand, 1955; Volume II, Chapter 15). Among the spirochetes, species of *Leptospira* are readily cultured and much progress has been made with nutritional studies (Geiman, 1952; Schlossberger and Brandeis, 1954; Johnson and Rogers, 1964), but the relapsing fever spirochetes of the genus *Borrelia* continue to present difficulties (see Chapter 18).

II. PROTOZOA DEVELOPING PRIMARILY IN THE PLASMA: THE HEMOFLAGELLATES

The flagellate protozoa of the family Trypanosomatidae represent probably the most extensive group of parasites of the blood of vertebrates. Typically these organisms are transmitted by insects or other invertebrate vectors, within which they undergo a reproductive cycle characterized by stages of a different morphology from the stages in the vertebrate blood. The stages in the invertebrate vector resemble related species of flagellates parasitic exclusively in invertebrates.

Many of these parasites of invertebrates, especially those from the alimentary tract of various insects, have been cultured axenically in the blood agar medium first used by Novy and MacNeal (1903) and by Nicolle (1908) to obtain cultures of trypanosomes from infected mammalian hosts. It is a remarkable fact that the bloodstream trypanosomes give rise in culture to forms morphologically and physiologically equivalent to the stages found in the invertebrate vector. Cultures reproducing as the bloodstream form had not been obtained until very recent successes with two species (see below). Only recently, too, have small beginnings been made toward understanding the conditions responsible for differentiation of the invertebrate form into the form found in the vertebrate blood. As a result, most of the work on cultivation and nutritional requirements of hemoflagellates has been done with species parasitic only in invertebrates (insects especially), much less has been done with the invertebrate stages of species parasitic in vertebrates, and hardly anything with the vertebrate stages of the latter.

A. HEMOFLAGELLATES OF INVERTEBRATES AND STAGES FROM INVERTEBRATE VECTORS

1. Cultivation

Whole blood, usually with complement inactivated, supplies the needed factors for cultivation of these protozoa. Strains have been kept in blood agar, at temperatures of 23°–28°C, for many years. Their infectivity for the invertebrate host has rarely been tested, but seems usually to have been retained (Wenyon, 1926). Infectivity for the vertebrate host depends on several factors, especially the species and the kind of life cycle it undergoes (see below). Utilization of such cultures for diagnosis and for attempts at vaccination is best discussed in the chapters concerned with the specific pathogenic hemoflagellates. Here we shall emphasize use of the cultures for determining nutritional requirements and for studies on morphogenesis.

Table I shows the composition of several media generally useful for the primary isolation and for the maintenance of strains of hemoflagellates. A partial list of culture strains of hemoflagellates available to investigators (as of 1958) may be found in the Catalogue of Laboratory Strains of Free-

TABLE I

MEDIA FOR CULTIVATION OF HEMOFLAGELLATES

1. NNN. Agar 14 gm, NaCl 6 gm, dist. water 900 ml. Heat to boiling, tube, autoclave 15 min., 15 lb. To use liquefy in boiling water, cool to 48°, add per tube ⅓ volume of sterile defibrinated rabbit blood. Mix by rotating and let solidify in slanted position. Many modifications of this medium are used, as with more, or less, blood, human blood (outdated from blood banks), or other blood in place of rabbit blood, etc. Often a small amount (as 0.5 ml per tube) of fluid overlay (as Locke's solution) is added after the slant has solidified.

2. SNB-9. Broth overlay on blood agar base. Broth: NaCl 0.6 gm, Neopeptone (Difco) 2 gm, dist. water 100 ml. Base: agar 2 gm, above broth 100 ml. Both autoclaved 20 min. at 16 lb. To mix liquefy agar base, cool to 50°, add ¼ volume sterile defibrinated rabbit blood, let cool at slant. To each tube holding 2–3 ml medium add 0.5 ml sterile broth overlay (Diamond and Herman, 1954).

3. Senekjie's medium. Bactobeef (Difco) 50 gm, Neopeptone (Difco) 20 gm, agar 20 gm, NaCl 5 gm, dist. water 1000 ml. Heat to dissolve agar, sterilize by autoclaving. For use liquefy, cool to 50°, add 10% defibrinated rabbit blood, let cool at slant. For overlay use Locke's solution: NaCl 8 gm, KCl 0.2 gm, CaCl₂ 0.2 gm, KH₂PO₄ 0.3 gm, dextrose 2.5 gm, dist. water 1000 ml (Senekjie, 1943).

4. Brain-heart infusion. Add 10% blood to Difco brain-heart infusion broth, heat to boiling, filter through paper, and sterilize by autoclaving (Zeledon, 1959).

5. Tobie's medium. Solid phase: 1.5 gm Bactobeef, 2.5 gm Bactopeptone, 4.0 gm NaCl, 7.5 gm agar in 500 ml dist. water. Adjust to pH 7.2–7.4 with NaOH. Autoclave 20 min. at 15 lb. Cool to about 45°. Add whole rabbit blood (citrated, inactivated at 56° for 30 min.) in proportion of 25 ml blood to 75 ml base. Liquid phase: Sterile Locke's solution (as for medium 3). To slant with 5 ml base add 2 ml Locke's. To flask with 25 ml base add 10–15 ml Locke's (Tobie et al., 1950).

6. Weinman's medium. (a) Autoclaved base of Difco nutrient agar 31 gm plus Bacto-agar 5 gm in 1000 ml dist. water. (b) Reconstituted human blood prepared from: citrated human plasma 200 ml, heated 30 min. at 56°, plus 200 ml human red cells washed 3 times in 3 volumes 0.9% NaCl solution. Cool base to 45°, mix 1 part blood to 3 parts base, cool, and slant (Weinman, 1960). Initial cultures are more often successful if infected blood, drawn with polyvinyl sulfuric acid as anticoagulant, is added (see Chapter 17).

living and Parasitic Protozoa (Committee on Cultures, 1958). A similar up-dated catalogue is in process of preparation by the Society of Protozo-ologists.

2. Defined Media

The first defined medium for a hemoflagellate was that of Cowperthwaite *et al.* (1953) for the insect parasite *Crithidia fasciculata* (see Table II). Several strains of this species, a few other species of *Crithidia* parasitic in insects, and a strain of *Blastocrithidia culicis*, also a parasite of the insect gut, have since been grown in this medium or slight modifications of it (Guttman and Wallace, 1964). Kidder and Dutta (1958) could get only light growth of *C. fasciculata* in this medium as originally described. They developed a much improved medium (Table II) and used this to determine many of the nutritional requirements of this flagellate (see Section II, A,3,a).

A hemoflagellate of lizards, *Leishmania tarentolae*, was the first species parasitic in a vertebrate to be cultured in a defined medium (Table II) (Trager, 1957a). The leptomonads of this *Leishmania* have more complex requirements than the *Crithidia*. Whereas *L. tarentolae* will not grow in media suitable for *Crithidia*, the latter do grow abundantly in the *L. tarentolae* medium. Guttman (1966a,b) has recently reported growth in defined medium for several species of *Leptomonas* from insects and for a second species of hemoflagellate parasitic in a cold-blooded vertebrate, *Trypanosoma ranarum* of the frog.

No successful defined medium has yet been reported for the invertebrate stages of any of the numerous species of hemoflagellates parasitic in mam-mals. The closest approach was that of Citri and Grossowicz (1955a), who grew a strain of *Trypanosoma cruzi* (cause of Chagas' disease) and 36 different strains of *Leishmania tropica* (cause of Oriental sore) in a partially defined medium. They were, however, unable to culture in this medium *L. infantum*, *L. donovani*, *L. brasiliense* (parasites of man), or the lizard parasite *L. agamae* (Citri and Grossowicz, 1955b).

Leishmania mexicana and *L. tropica minor* likewise did not grow in it (Strejan, 1963), nor did strains of *T. cruzi* other than those originally used (Boné and Parent, 1963). Such results are not surprising. With the more exacting parasites, strains may differ considerably in ease of cultivability even in rich natural media. Thus only some strains of *T. cruzi* will grow in the brain-heart infusion medium of Zeledon (1959); these strains are also less virulent for the mammalian host.

In the defined medium for *L. tarentolae*, *L. donovani* leptomonads (pro-mastigotes; see Hoare and Wallace, 1966) do not give rise to cultures. It is of interest, however, that the leishmanial (amastigote) forms of this

TABLE II
DEFINED MEDIA FOR HEMOFLAGELLATES

I. Medium of Nathan and Cowperthwaite for *Crithidia fasciculata*, as modified (see Guttman and Wallace, 1964) (also used for other species of *Crithidia* and for *Blastocrithidia culicis*). If supplemented with glycine at 100 μg/ml this supports *Leptomonas mirabilis*. If supplemented with both glycine (100 μg/ml) and choline·Cl at 1.5 μg/ml it supports *L. collosoma* and *L.* sp. from *Dysdercus* (Guttman, 1966a).

II. Medium of Kidder and Dutta (1958) for *C. fasciculata*.

III. Medium of Trager (1957a) for *Leishmania tarentolae*. An absolute requirement or a marked stimulatory effect has been demonstrated for each of the organic components, with the following exceptions in medium III: (*a*) The block of purines and pyrimidines is required but no attempt has been made to see whether adenine alone would satisfy this requirement for *L. tarentolae* as it does for *Crithidia*. (*b*) Among the vitamins, inositol and *p*-aminobenzoic acid are probably not essential. Pyridoxine and choline could both be omitted if pyridoxal or pyridoxamine were supplied. (*c*) Among the amino acids one, two, or all of the group alanine, aspartic acid, and glycine are required. This group was not dissected into its components.

Component	Concentration (μg/ml)		
	I	II	III
L-Arginine-HCl	500	430	300
L-Histidine-HCl	300	210	150
DL-Isoleucine	100	630	600[a]
L-Leucine	100[b]	970	1500
L-Lysine-HCl	400	760	1250
DL-Methionine	100	340	300
DL-Phenylalanine	60	500[c]	400
DL-Threonine	0	440	500
DL-Tryptophane	80	120[d]	200
DL-Valine	50	660	500
DL-Alanine	0	0	700
DL-Aspartic acid	0	0	1200
L-Glutamic acid	1000	0	1900
Glycine	0	0	100
L-Proline	0	0	500
DL-Serine	0	0	400
L-Tyrosine	60	200	400
p-Aminobenzoic acid	0	0	3.0
Biotin	0.01	0.2	0.2
Choline-Cl	0	3.0	3.0
Inositol	0	0	3.0
Nicotinamide	3.0[e]	5.0	5.0
Pyridoxine-HCl	0	0	2.0
Pyridoxal-HCl	0	2.0	2.0

TABLE II—*Continued*

Component	Concentration (μg/ml)		
	I	II	III
Pyridoxamine-2HCl	1.0	2.0	2.0
Calcium pantothenate	3.0	8.0	8.0
Riboflavin	0.6	2.0	2.0
Thiamine	6.0	2.0	2.0
Folic acid	1.0	2.0	1.6
Adenine	10.0	50	1.7
Guanine	0	0	1.7
Xanthine	0	0	1.7
Uracil	0	0	1.7
Cytidylic acid	0	0	0.5
Hemin	25	25	25
Triethanolamine	5000	50	0
Glucose	0	10,000	5000
Sucrose	15,000	0	0
Tween 80	0	5000	0
Ethylenediaminetetraacetic acid	600	700	0
NaCl	0	4000	2000
Na_2HPO_4	0	1250	1250
KH_2PO_4	0	500	500
K_3PO_4	150	0	0
$MgSO_4 \cdot 7H_2O$	650	400	1000
$CaCl_2$	0.9	7.3	26
$FeSO_4 \cdot 7H_2O$	0	4.9	10
$Fe(NH_4)_2(SO_4)_2 \cdot 6H_2O$	7.0	0	0
$ZnSO_4 \cdot 7H_2O$	221	68	22
$MnSO_4 \cdot H_2O$	430	77	3
$CoSO_4 \cdot 7H_2O$	12	2.4	0.44
$CuSO_4$	10	0.6	0.5
H_3BO_3	3	0.7	0.11
KI	0	0.03	0
$Na_2MoO_4 \cdot 2H_2O$	0	10.1	0
pH	7.8–8.0	7.8–8.0	7.8–8.0

[a] As 600 μg L-isoleucine.
[b] As 100 μg DL-leucine.
[c] As 500 μg L-phenylalanine.
[d] As 120 μg L-tryptophane.
[e] As nicotinic acid.

parasite removed from the spleen of an infected hamster will transform
into promastigotes in this medium (L. Simpson, 1965), thereby permitting
detailed study of the nutritional requirements for transformation. The
L. tarentolae medium, when supplemented with stearate, has also supported
growth of one strain of *T. cruzi* and of a strain of *T. vespertilionis* through
three weekly transfers but not beyond (Krassner, 1964).

It seems clear that the invertebrate stages of hemoflagellates parasitic
in mammals require at least the nutrients present in the medium for *L.
tarentolae* plus others, perhaps especially certain lipids. Still more complex
are the requirements of the vertebrate stages of these parasites. These will
be discussed later, together with the conditions needed for the transformation from one stage to another.

3. Nutritional Requirements for Substances of Known Chemical Nature

These are indicated in a general way by the composition of the defined
media. The following requirements have been given special study.

a. Hemin. This was the first specific substance shown to be essential for
growth of hemoflagellates. A. Lwoff (1933) and M. Lwoff (1933) found
that the growth of *Crithidia fasciculata* in a peptone medium was proportional to the concentration of added blood in the range from 1:200,000
down to 1:600,000. Blood could be replaced by hematin, protohemin, or
protoporphyrin, but not by other related porphyrins.

With the development of defined media an absolute requirement of
C. fasciculata for hemin has been demonstrated (Cowperthwaite *et al.*,
1953; Kidder and Dutta, 1958). The same has been done for *Leishmania
tarentolae* (L. Simpson, 1966). For the latter the minimal concentration
required for continuous growth was 200 mμg/ml. The concentration of
blood of 1:200,000 to 1:600,000 found by A. Lwoff (1933) to affect growth
of *C. fasciculata* would correspond roughly to about 750 to 250 mμg hemoglobin or 30 to 10 mμg hemin per ml. Although *L. tarentolae* does respond
to a hemin concentration of 20 mμg/ml it cannot be subcultured in defined
medium with this low concentration. The two species may differ somewhat
in their hemin requirement, or it might be that hemoglobin is more efficiently utilized as a hemin source than is hemin itself.

In view of the need for blood in media for hemoflagellates it seems likely
that all require an external source of hemin. One apparent exception has
been *Crithidia oncopelti* (Newton, 1957; Guttman and Wallace, 1964),
a flagellate which has been grown in a defined medium much simpler than
that for *C. fasciculata*. It is clear, however, that *C. oncopelti* contains
endosymbiotic bacteria (Gill and Vogel, 1963).

b. Amino Acids. Species of *Crithidia* require the 10 amino acids essential for rats, including threonine (Kidder and Dutta, 1958). In the

TABLE III

CONCENTRATIONS OF AMINO ACIDS FOR HALF-MAXIMAL AND OPTIMAL GROWTH OF
Crithidia fasciculata[a, b]

Amino acid	Concentration (μg/ml)	
	Half maximal	Maximal
L-Histidine-HCl	20	80
L-Phenylalanine	16	80
DL-Isoleucine	119	320
DL-Valine	85	200
L-Leucine	80	160
L-Lysine	32	160
L-Arginine-HCl	56	350
L-Tyrosine	19	60
DL-Methionine	40	180
L-Tryptophan	7.3	24
DL-Threonine	c	80

[a] From Kidder and Dutta (1958).

[b] Organisms grown in medium II (Table II) without the amino acid being tested
and with the following in μg/ml: DL-alanine, 550; L-aspartic acid, 610; glycine, 50;
L-glutamic acid, 1165; L-proline, 770; DL-serine, 440.

[c] Slow, suboptimal growth without threonine.

presence of glutamic acid threonine could be omitted, as in medium I
(Table II). The *Crithidia* also require tyrosine. Species of *Leptomonas*
must in addition have glycine (Guttman, 1966a). The requirements of
Leishmania tarentolae are still more complex. They include, besides the
ten essential for rats, glutamic acid, tyrosine, serine, proline, and one or
more of the group glycine, alanine, and aspartic acid (Trager, 1957a).

Quantitative requirements for amino acids have been reported for *C.
fasiculata* grown in a medium with 17 amino acids (Kidder and Dutta,
1958). They are shown in Table III.

c. Purines and Pyrimidines. As shown in Table II, adenine alone satisfies
these requirements for species of *Crithidia* and *Leptomonas*. Whether this
is also true for *Leishmania tarentolae* has not been investigated. For *C.
fasciculata* adenine can be replaced by guanine, xanthine, hypoxanthine, or
inosine, or by their nucleosides or nucleotides. All of these compounds are
about equally active (Kidder and Dutta, 1958).

Trypanosoma cruzi synthesizes purine nucleotides chiefly from preformed
purines (Fernandes and Castellani, 1959). The synthesis is inhibited by
certain analogs which have some trypanocidal activity (Castellani and
Fernandes, 1962).

d. Carbohydrates. This requirement can be fulfilled by either glucose or
sucrose, and probably by a variety of other sugars. It has not been thoroughly investigated with defined media.

e. Minerals. A need for Ca^{++}, Mg^{++}, and the trace metals mixture could be shown for *L. tarentolae* by repeated subculture in defined medium (Trager, 1957a). Much more refined experiments would be needed to study the metal requirements in detail. The requirements for Fe might be of special interest in relation to the hemin requirement and its possible replacement by protoporphyrin.

f. Vitamins. Thiamine, riboflavin, pyridoxal or pyridoxamine, pantothenic acid, biotin, nicotinamide, and folic acid are all essential for both *L. tarentolae* and *C. fasciculata*. *Crithidia oncopelti* when it has associated symbionts requires several vitamins including thiamine. Since thiamine is a heat-labile compound, a requirement for the complete thiamine molecule was demonstrated years ago before defined media for hemoflagellates had been developed (M. Lwoff, 1938). A concentration as low as 1 mμg per ml permitted abundant growth of *C. oncopelti* in a heated peptone medium. The defined media evidently supply thiamine in a many-fold higher concentration (see Table II). This is probably also true of other vitamins. Minimal requirements for *C. fasciculata* are shown in Table IV.

The minimal requirment of *L. tarentolae* for riboflavin would depend in part on one's definition. To get cultures which can be subcultured indefinitely, with peak populations of 20–30 $\times$ 10^6 organisms per ml, it is necessary to have only 10–20 mμg riboflavin per ml (Trager and Rudzinska, 1964). At 10 mμg/ml very few of the flagellates had a flagellum and none were motile. At 20 mμg/ml most had a short flagellum and there was some

TABLE IV

VITAMIN REQUIREMENTS FOR *Crithidia fasciculata*[a, b]

Vitamin	Concentration for half-maximal growth (μg/ml)
Thiamin	0.011
Riboflavin	0.015
Ca pantothenate	0.046
Nicotinamide[c]	0.065
Nicotinic acid[c]	0.295
Pyridoxal-HCl[d]	0.019
Pyridoxamine-2HCl[d]	0.0035
Folic acid[e]	0.0004
Biopterin[f]	0.0001

[a] As determined in the same medium used for amino acid tests (see Table III).
[b] From Kidder and Dutta (1958).
[c] Nicotinic acid or nicotinamide active alone.
[d] Pyridoxal or pyridoxamine active alone.
[e] Tested in presence of 0.005 μg/ml biopterin [2-amino-4-hydroxy (1-*threo*)-1',2'-dihydroxypropyl pteridine].
[f] Tested in presence of 0.03 μg/ml folic acid.

motility. At 200 mμg/ml, however, many of the organisms had a moderate to long flagellum and were actively motile, resembling those seen in a blood agar culture. The still higher concentration of 2000 mμg/ml, as used in the stock defined medium, had no further favorable effect. Hence 10 mμg/ml is a minimal value if we consider only population density, but about 200 mμg/ml is minimal if one also considers flagellar development and motility.

The flagellum may well be especially sensitive to a variety of vitamin deficiencies. Thus for *L. tarentolae* choline is an essential nutrient if pyridoxamine is replaced by pyridoxine (Trager, 1957a). A concentration of $0.4 \times 10^{-5}\ M$ choline sufficed for growth, but the flagellates were rounded and aflagellate. At a concentration of $2.0 \times 10^{-5}\ M$ choline, optimal growth of flagellated motile leptomonads was obtained. The interchangeable requirement of *L. tarentolae* for either choline plus pyridoxine or pyridoxal (or pyridoxamine) suggests that choline may be essential for the formation of pyridoxal from pyridoxine. *Leptomonas collosoma* and a *Leptomonas* sp. from *Dysdercus* require choline even in the presence of pyridoxal (Guttman, 1966a).

Also of special interest is the folic acid requirement of hemoflagellates. As first found for *C. fasiculata* this seemed to be inordinately high. It was then shown (Nathan and Cowperthwaite, 1955) that the high level of folic acid was serving as a source not only of folic but also of a new growth factor, a pteridine later isolated as biopterin [2-amino-4-hydroxy-6-(1-*threo*)-1′,2′-dihydroxypropyl pteridine] (Patterson *et al.*, 1955). In the presence of this material, or the related neopterin [2-amino-4-hydroxy-6-(L-erythro-1,2,3-trihydroxypropyl)-pteridine] (Rembold and Buschmann, 1963), folic acid was required at only the more usual level of about 0.1 mμg/ml. The possible role of pteridine in the nutrition of hemoflagellates other than *Crithidia* has not been investigated.

Responses of *C. fasciculata* to biopterin are shown in Fig. 1. Evidently this organism can cleave a conjugated pteridine, though only very inefficiently, but it cannot synthesize folic acid. Folic acid, but not the pteridine requirement, can be bypassed by supplying thymine at about 25 μg/ml (Nathan *et al.*, 1956), provided methionine is present.

Since both cytidine and lipids (see below) have a sparing effect on the pteridine requirement (Kidder and Dewey, 1963) it is probable that pteridine is involved in the biosynthesis of these materials.

g. Lipids. It is likely that some hemoflagellates require lipid nutrients, but little definitive information is as yet available. Stearate stimulated growth of *T. cruzi* in a Bactotryptone medium (Boné and Parent, 1963) and also of one strain capable of growth through three subcultures in the defined medium for *L. tarentolae* (Krassner, 1964).

Lipids have been found to have a sparing effect on the pteridine require-

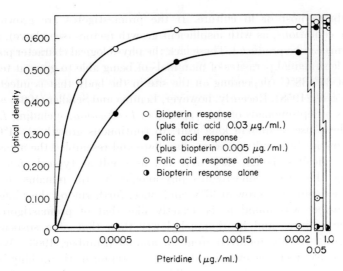

FIG. 1. Dose response of *Crithidia fasciculata* to biopterin and to folic acid. Inoculum from 10-day culture in low (0.001 μg/ml) folic acid and no biopterin. (From Kidder and Dutta, 1958.)

ment of *C. fasciculata* (Dewey and Kidder, 1966). In these experiments the medium II (Table II) was used with Tween 80 replaced by Triton WR 1339 (a *p*-iso octyl polyoxyethylene phenol polymer containing no free acid) at 0.5–1.0 mg/ml, and with a supplement of cytidine (40 μg/ml) which also has a pteridine-sparing effect. With pteridine at a concentration of only 0.04 mμg/ml, a level by itself supporting no growth of suitably depleted organisms, the addition of oleic acid at 10 μg/ml gave good growth. Arachidonic and linoleic acids were equally active, but α-linolenic was less so. Palmitoleic acid was inhibitory; its effect could be reversed by either L-neopterin or oleic acid. Dicarboxylic acids with 16 or 20 C atoms also had a pteridine-sparing effect.

B. CULTURE OF THE VERTEBRATE STAGES

Two main groups of hemoflagellates parasitize vertebrates: the members of the genus *Leishmania* and those of the genus *Trypanosoma* and their close allies. The stages characteristic of the vertebrate host have rarely been obtained in culture. These stages when placed *in vitro* either fail to grow or grow as organisms having the morphological and physiological properties of the stages occurring in the invertebrate vector.

1. Leishmania

These are typically small aflagellate intracellular forms in the vertebrate host (amastigotes), but flagellated leptomonads (promastigotes) in the invertebrate host or in culture. If the promastigotes are grown under adverse conditions, as with insufficient growth factors (see above), aflagellate forms can be produced. These lack the physiological character possessed by the leishmanial parasites of mammals of being able to grow at temperatures of 34°–38°C (depending on the site in the body that is infected; see Pereira *et al.*, 1958). Recently, however, Lemma and Schiller (1964) adapted cultures of leptomonads of three species of *Leishmania*, including *L. donovani*, the cause of human kala azar, to continuous growth at 34°C. The organisms in these cultures were aflagellate and resembled the amastigotes morphologically, though they were not as small as typical amastigotes. Whether they were amastigotes physiologically still remains in doubt, since they could not grow at 37°C and since, furthermore, their infectivity for hamsters was found to be exactly like that of promastigotes from ordinary cultures and hence much lower than that of a suspension of amastigotes freshly removed from an animal (Stauber, 1963). With the recent development of several additional criteria for distinguishing between promastigotes and amastigotes it will be of interest to find how such temperature-adapted strains behave. Thus promastigotes but not amastigotes give a strong positive cytochemical test for porphyrin (Krassner, 1966). The oxygen uptake per cell is 2 to 3 times higher for promastigotes than for amastigotes (L. Simpson, 1966). Mitochondria are more highly developed in promastigotes than in amastigotes (Rudzinska *et al.*, 1964). Gradual adaptation to higher temperatures does not seem likely to be the natural mechanism whereby promastigotes transform to amastigotes, since promastigotes placed at 37°C in tissue cultures, for example, can infect cells and multiply in them without having been first exposed to intermediate temperatures (Tchernomoretz, 1946; Hawking, 1948; Lamy *et al.*, 1964). Nevertheless, temperature must be one of the factors and certainly it is one of the most important in the reverse transformation from amastigote to promastigote (see below).

It is significant that *L. donovani* placed in a living tissue culture at 37°C can exist not only in the intracellular amastigote form but also as extracellular motile promastigotes (Weinman, 1939; Herman, 1966), whereas in acellular culture media 32° is the upper temperature limit (except for specially adapted cultures). The need for additional growth factors at higher temperatures is a fairly general phenomenon (Hutner *et al.*, 1957). It has been found that *L. tarentolae*, though incapable of growth at 33°C in a defined medium can grow well at 33° if the medium is supplemented with a low concentration of red cell extract, and could grow even at 37°C if further supplemented with supernatant from a tissue culture (Krassner, 1965).

Initial multiplication of *L. donovani* at 37°C over a 4-day period was obtained by inoculating amastigotes from spleen into a medium containing hamster serum and a concentrated extract of human erythrocytes (Trager, 1953). The organisms grew chiefly as aflagellate forms of size and morphology intermediate between promastigotes and amastigotes. Their positive fluorescence test for porphyrins would place them physiologically with the promastigote forms (Krassner, 1966).

2. Trypanosomes

These show in their vertebrate hosts a much wider range of types, physiologically as well as morphologically, than do the *Leishmania*. They range from the more primitive species in the section Stercoraria (subgenera *Megatrypanum, Herpetomonas*) (Hoare, 1964) to the highly specialized forms in the subgenus *Trypanozoon*. The former group have well-developed mitochondria in the bloodstream as well as in the culture forms, whereas in the bloodstream forms of the latter the mitochondria are reduced to a hollow tube (see Trager, 1964a; Rudzinska and Vickerman, Chapter 10, this volume).

The blood forms of two species of the Stercoraria group have been grown *in vitro* at 37°C: *T. theileri* in a red cell extract medium (Ristic and Trager, 1958; C. F. Simpson and Green, 1959), and *T. conorhini* (Deane and Kirchner, 1963; Desowitz, 1963). Cultures of the latter organism are of special interest. At 28°C this trypanosome multiplies as the crithidial stage (epimastigote; Hoare and Wallace, 1966) in either of two rich media: liver infusion-tryptone-hemoglobin-serum or Hanks' solution-hematin-lactalbumin hydrolyzate-serum. At 37° it fails to grow in the first medium, but it does multiply in initial culture in the second, where it grows as the typical bloodstream trypanosome form (trypomastigote). Thus, though temperature is the determining factor, an appropriate medium is also essential. At present no guess can be made as to why the bloodstream forms of *T. conorhini* cannot be subcultured. A contributing factor may be the much lower population of organisms obtained at 37° as compared with cultures kept at 28°C.

Other species of trypanosomes can be kept multiplying as bloodstream forms indefinitely by syringe passage from animal to animal, or, in nature, by venereal transmission (*T. equiperdum*) or by mechanical transmission by biting flies (*T. evansi, T. equinum*). Yet so far these bloodstream forms, even of species of the Stercoraria, have not been cultured *in vitro*. The closest approach is the initial multiplication of *T. (Herpetomonas) lewisi* bloodstream forms at 37° during the first 24 hours in a medium of Hanks' solution, rat serum, rat blood lysate, yeast extract, and lactalbumin hydrolyzate (D'Alesandro, 1962).

When trypanosomes of the group Salivaria were placed in tissue cultures they could not grow at 36°C, and at lower temperatures they grew as the insect forms, just as in NNN medium (Demarchi and Nicoli, 1960; Fromentin, 1961). On the other hand *Trypanosoma (Schizotrypanum) cruzi*, which has in the vertebrate host an intracellular multiplicative amastigote phase, grows in this phase within appropriate tissue culture cells at 37° (Kofoid *et al.*, 1937; Meyer and Chagas, 1950; Neva *et al.*, 1961; Trejos *et al.*, 1963).

C. MORPHOGENESIS IN HEMOFLAGELLATES

1. Change from Vertebrate to Invertebrate Stage

It is not surprising that temperature should be a major morphogenetic factor for hemoflagellates having stages parasitic in invertebrates and in warm-blooded vertebrate hosts. The changes occurring when such a hemoflagellate is transferred from its vertebrate to its invertebrate host can generally be easily duplicated *in vitro*. It suffices to place bloodstream trypanosomes or the tissue forms of *Leishmania* in an appropriate medium at a temperature (~28°C) likely to be encountered in the invertebrate vector in nature. Although temperature is the conspicuous environmental change, the matter of an appropriate medium is of considerable importance. Blood agar is adequate, but in most instances no attempt has been made to define the substances required for the transformation from vertebrate to invertebrate stage. For *Trypanosoma cruzi* hemoglobin and at least a second factor present in red cells, nutrient broth, meat infusion, peptone, and blood extract were essential for the change from blood trypomastigotes to epimastigotes (Muniz and Freitas, 1945). Similar factors seemed to play a role also for *T. conorhini* (Deane and Kirchner, 1963).

A more precise analysis of the factors concerned has been made for *Leishmania donovani* (L. Simpson, 1966). The transformation of the intracellular amastigote forms from the mammalian host into promastigotes requires not only a lowering of temperature but also the presence of an energy source (glucose) and of certain specific amino acids. It is interesting that although transformation occurs in a defined medium (Table II), this medium does not support growth of the promastigotes of *L. donovani*.

With *L. donovani*, up to 80 % of a suspension of amastigotes may transform into promastigotes, but this percentage varies greatly from preparation to preparation under the same environmental conditions, suggesting that intrinsic differences may exist among amastigotes in their ability to give rise to promastigotes. That such intrinsic differences do exist for trypanosomes of the Salivaria group has recently been strongly indicated by Vickerman's (1965) finding that "stumpy" blood forms of *T. gambiense* have a well-developed mitochondrial system (similar to, though less extensive than, that of culture forms) which is lacking in "slender" blood forms. The "stumpy" trypanosomes, like culture forms, can retain their motility with only α-ketoglutarate as energy source, showing that the switch in type of metabolism has already occurred. The "stumpy" forms are preadapted for life in the insect vector. What might bring about such a change within the mammalian bloodstream is not known. Methods for *in vitro* maintenance of the blood forms might provide some clues; such methods, unfortunately, are lacking.

2. Change from Invertebrate to Vertebrate Stage

Just as for the reverse change already considered, temperature, specific substances, and intrinsic factors are all involved. This change is however much more difficult to bring about *in vitro*. Thus merely transferring a culture of a mammalian trypanosome or *Leishmania* to a temperature of 37° usually results only in the death of the culture (see however *T. conorhini*, Section II, B,2 above). Actually two problems are involved in attempting to duplicate *in vitro* the entire life cycle of a two-host hemoflagellate. The first is the formation in culture of forms infective to the vertebrate host. The second is the propagation of the vertebrate forms.

The appearance of infective forms of hemoflagellates in the usual culture media depends on the type of organism. Cultures of promastigotes of *Leishmania* ordinarily are infective, but they produce in the inoculated host only 10% of the number of amastigotes produced by inoculation of a comparable dose of amastigotes (Stauber, 1963). This suggests that only some leptomonads in a culture are competent to transform into leishmanias. In cultures of *Trypanosoma cruzi*, where the metacyclic forms can be distinguished morphologically, these do not appear until after the log phase of growth of the culture has finished (Camargo, 1964).

The metacyclic or infective forms of trypanosomes of the Salivaria group have been obtained *in vitro* only exceptionally. When hanging drop cultures of *Trypanosoma vivax* growing in the presence of living tsetse fly tissue were exposed overnight to 38°C and then inoculated to a sheep 2 out of 6 cultures proved to be infective (Trager, 1959). It was of interest that these were also the two cultures which showed large numbers of very active trypanosomes after the overnight exposure to high temperature, suggesting that these cultures may have contained, for reasons not now understood, trypanosomes preadapted to life in the mammalian host. Thus Weinman (1957) (see Vol. II, Chap. 17) obtained a few infections from cultures of *T. rhodesiense* and thought this to be associated with the presence of trehalose. Later Geigy and Kauffman (1964) thought other sugars might be responsible (this aspect of the problem is considered in greater detail in Chapter 17), but most recently it has been claimed that the age of the culture is the chief determining factor. It would seem that a subculture of *T. brucei* on blood agar is likely to have infective forms after 18 days at 25°, but not before or after (Amrein *et al.*, 1965). Confirmation and extension of these results will be of great interest.

Meanwhile there is one species of trypanosome for which the transformation from invertebrate to vertebrate form can be controlled. Cultures of *T. mega*, a parasite of amphibia, grow ordinarily as epimastigotes typical of the invertebrate host. Steinert (1958) found that exposure of a culture to urea at the physiological level of 0.5–1.0 mg/ml was followed by up to 10 % of the organisms assuming the large trypanosome form characteristic of the vertebrate host. Urea was effective only if added to cultures past the logarithmic phase of growth; hence only such cultures were "competent." Furthermore, anaerobiosis and reducing agents increased the per cent of transformation obtained in presence of urea, as did also a high pH (Steinert, 1965). These results suggest that urea may modify the secondary and tertiary structures of a high-molecular-weight component involved in transformation. Further studies along these lines with *T. mega* and with *Leishmania* (L. Simpson, 1966) promise results of broad significance with regard to regulatory mechanisms in cellular differentiation. In the differentiations undergone by hemoflagellates the characteristic DNA-containing organelle, the kinetoplast, may play a special role (Trager, 1965).

III. ORGANISMS DEVELOPING WITHIN THE BLOOD CELLS

The axenic cultivation of an intracellular parasitic protozoan has yet to be achieved. The closest approach so far has been obtained with an avian malaria parasite, *Plasmodium lophurae*, which has been kept alive and developing extracellularly *in vitro* for periods up to 5 days (see below).

A number of such parasites, like viruses or rickettsiae, including species parasitizing blood cells, have been cultured in cultures of appropriate host cells.

A. Cultivation in Tissue Culture Systems

1. Theileria

Tissue stages of *T. annulata* developed in explants of infected spleen or lymph node (Brocklesby and Hawking, 1948) but the infection could not be transmitted to uninfected tissue cultures. More recently Hulliger (1965) obtained long-term cultivation of *Theileria parva*, *T. annulata*, and *T. lawrencei*. The intralymphocytic theilerial particles grew within bovine lymphocytes maintained in culture in the presence of a line of baby hamster kidney cells. The latter, necessary for maintenance of the bovine lymphocytes, did not themselves become infected. In such cultures the theilerias multiplied at about the same rate as their host cells. Although the division by binary fission of the theilerial chromatin particles was not synchronous with that of the host cell, the theilerial bodies present in a cell undergoing mitosis became closely associated with the spindle fibers and were pulled apart and distributed to both daughter cells in late anaphase.

The usefulness of such cultures in studying the pathology of the infection has been beautifully demonstrated in experiments where temperature again has been shown to have a morphogenetic effect (Hulliger *et al.*, 1966). Whereas at 37° there appeared only the small theilerial particles, at 40° the large forms characteristically found in the internal organs of infected cattle developed.

2. Malaria: Exoerythrocytic Stages

These are the stages which develop first in the vertebrate host following the inoculation of sporozoites. In bird malaria these stages occur in cells of the reticuloendothelial system (Huff, 1957); in primate and rodent malaria they occur in hepatic cells (Shortt *et al.*, 1951; Yoeli *et al.*, 1966). The latter have not been grown in tissue culture, perhaps because of the difficulty of maintaining differentiated hepatic cells *in vitro*. But with bird malaria parasites both the infection of tissue cultures from sporozoites and maintenance of the exoerythrocytic forms have been obtained (see Pipkin and Jensen, 1958).

Continuous maintenance in tissue culture of the exoerythrocytic forms of *Plasmodium gallinaceum* was first described by Meyer (1949). Huff *et al.* (1960) used initial cultures for detailed studies of the morphology of the exoerythrocytic forms of *P. gallinaceum* and also *P. fallax*, at first by light microscopy and more recently by electron microscopy (Hepler *et al.*, 1966). *Plasmodium fallax* has proved especially amenable to growth in tissue culture and methods for large-scale production of the exoerythrocytic forms of this parasite in turkey embryo brain tissue cultures have now been developed (Davis *et al.*, 1966). The best medium consisted of equal parts modified Medium 199 (to contain 10 % by volume fetal calf serum and 1 % folinic acid solution) and a Diploid Growth Medium of the following composition: Eagle's basal medium +, by volume, 10 % fetal calf serum, 1 % folinic acid solution, 1 % glutamine solution, 1 % penicillin-streptomycin solution (5000 units each), 1 % MEM nonessential amino acid (100 ×), 1 % $NaHCO_3$ solution (7.5 %). About 50 % of the cell sheet became infected, with an average of 6 parasites per infected cell. Indeed, a main difficulty now is that the parasites develop so well that the host cells are destroyed, resulting secondarily in death of the parasites.

It seems certain that such cultures will prove valuable in biochemical and physiological studies of exoerythrocytic forms—studies all of which are yet to be done.

3. Malaria: Erythrocytic Stages

Good intraerythrocytic development of these stages of malaria parasites in red cell suspensions incubated *in vitro* was first obtained with two species: the bird malaria *Plasmodium lophurae* (Trager, 1943, 1947) and the monkey malaria *P. knowlesi* (Geiman *et al.*, 1946). The method consisted essentially of mixing infected blood with an appropriate diluent containing salts, glucose, amino acids, vitamins, etc., and incubating the mixture (at 40° for the bird parasite and 37° for the monkey parasites) under an atmosphere of 5 % CO_2 in air and on a gently rocking platform which kept the cells in even suspension. Under these conditions parasites initially present completed development to segmenters and some of the merozoites so produced invaded uninfected erythrocytes. Since the erythrocytes did not remain in good condition for more than 2 days, to get prolonged cultivation it was necessary to add fresh uninfected cells daily or at least every other day. When this was done *P. lophurae* was maintained for up to 8 days (however, with a decrease in parasite numbers because their multiplication was exceeded by the dilution with fresh blood) and *P. knowlesi* for 6 days. *P. gallinaceum* of the chicken was propagated for 10 days in chicken erythrocytes suspended in a medium of chicken erythrocyte extract in chicken serum supplemented with glucose (Anderson, 1953). Fresh, uninfected red cells were added every other day. The parasites multiplied to such an extent that the count per 1000 red cells was the same at the end of the 10-day period as it had been initially, despite a 20,000-fold dilution with normal erythrocytes. This exceptionally good result has never been equaled since, either with *P. gallinaceum* or with any other species.

The cultures of this type have found their primary usefulness for short-term studies extending over 1 or 2 days, a period in which one could feel reasonably certain that both the parasites and their host cells remained in good condition. The method was found applicable to a number of species in addition to the three already mentioned: *P. hexamerium* (Nydegger and Manwell, 1962), *P. falciparum* (Geiman, 1948; Trager, 1958, 1966), *P. coatneyi* (Trager, 1966), and *P. cynomolgi* (Geiman *et al.*, 1966b).

From these studies something was learned concerning the nutritional requirements of the parasite-host cell complex. Glucose was a prime requirement. There was a need for a source of purines and pyrimidines and of B vitamins. Plasma could not be dispensed with, perhaps because it helped to maintain the integrity of the host cells. Two specific growth factor requirements were demonstrated: calcium pantothenate for *P. lophurae* (Trager, 1943, 1966) and *p*-aminobenzoic acid for *P. knowlesi* (Geiman *et al.*, 1946).

The penetration of erythrocytes by *P. lophurae in vitro*, an essential step for multiplication in such cell suspension systems, turned out to be very sensitive to factors in the medium. It was easy to find changes which reduced the extent of penetration (Sherman, 1966), among them being addition of bovine albumin or of inosine, high concentrations of glutathione, and aged vitamin solutions. Much more work is needed along these lines.

In keeping with the requirement of the *P. lophurae*-host cell complex for pantothenate was the inhibitory effect of antipantothenate on *P. lophurae* developing intracellularly in duck erythrocyte suspension (Table V) (Trager, 1966). Antipantothenate also inhibited development of *P. coatneyi* in rhesus monkey erythrocyte suspensions (Table VI) and *P. falciparum* in chimpanzee or human erythrocyte suspensions (Trager, 1966), indicating that pantothenic acid is required for the intracellular development of these parasites.

Improved development of *P. knowlesi* in monkey erythrocyte suspensions has recently been obtained by two small modifications in the methods earlier used by Geiman *et al.*: use of a glycyl-glycine buffer and dilution of the infected blood with 6 volumes rather than 3 volumes of medium (Geiman *et al.*, 1966b). Whole monkey plasma could be replaced with a rather concentrated solution of human plasma fraction IV-4 (alpha-globulin) (Geiman *et al.*, 1966a), or by stearate (Siddiqui *et al.*, 1967).

TABLE V

EFFECT OF ANTIPANTOTHENATE ON *Plasmodium lophurae* DEVELOPING WITHIN DUCK ERYTHROCYTES MAINTAINED *in Vitro*[a]

Flask No.	Calcium pantothenate (µg/ml)	Antipantothenate[b] (µg/ml)	Percent parasites[c] after 1 day with	
			2–4 nuclei	>4 nuclei
1	0	0	64	19
2			65	17
3	0	130	54	4
4			47	6
5	10	0	69	15
6			68	18
7	10	130	62	9
8			63	13

[a] From Trager (1966).

[b] Antipantothenate: d-(2-pantoylamino)ethylsulfono-4-chloroanilide.

[c] At 0 time there were 5% with 2–4 nuclei and 2% with over 4 nuclei.

TABLE VI

EFFECT OF ANTIPANTOTHENATE ON *Plasmodium coatneyi* IN MONKEY ERYTHROCYTES
in Vitro[a, b]

Flask No.	Antipanto- thenate (μg/ml)	Calcium pantothenate (μg/ml)	Parasites[c] per 10,000 RBC after 2 days			
			Total	R	T	S
1	0	0	100	86	0	14
2			78	70	0	8
3	130	0	8	2	2	4
4			18	0	4	14
5	0	10	98	86	0	12
6			114	108	0	6
7	130	10	26	14	0	12
8			28	4	4	20

[a] From Trager (1966).

[b] See Table V.

[c] R = rings; T = uninucleate trophozoites; S = forms with over 1 nucleus. At 0 time there were 170 parasites per 10,000 RBC, of which 32 were rings, 124 were trophozoites, and 14 were multinucleate.

At day 1 all flasks had about 150 parasites per 10,000 RBC, of which the great majority were multinucleate. Segmentation was more advanced in the flasks without antipantothenate. After centrifugation, resuspension in fresh medium and mixing with 0.5 ml of fresh heparinized uninfected monkey blood, the counts were about 100 parasites per 10,000 RBC.

B. EXTRACELLULAR DEVELOPMENT *in Vitro* OF *Plasmodium lophurae*

This avian malaria parsite has been removed from its host duck erythrocytes by means of hemolytic antiserum and kept alive and developing extracellularly *in vitro* up to 5 days (see Trager, 1957b, 1964b). The proportion of surviving parasites then dropped precipitously despite the fact that the cultures continued to be provided with fresh medium and fresh supplements every 12 hours.

The medium is an extract of duck erythrocytes prepared in a nutrient solution containing gelatin and supplemented with a number of cofactors of proved favorable effect (Table VII). The original papers should be consulted for details of preparation (Trager, 1957b, 1958, 1966). It will be sufficient here merely to summarize the principal results obtained.

For best development over a 4-day period a so-called full strength duck red cell extract is essential. This is the extract obtained when 1 volume of frozen-thawed red cells is mixed with 1 $\frac{1}{3}$ volume of diluent and centrifuged 50 minutes at 3000 rpm. The clear, deep red extract contains about 14 %

TABLE VII

COMPOSITION OF THE ERYTHROCYTE EXTRACT

A. Nutrient medium used to prepare the extract
      High K$^+$ content
      Also the following ions: Na$^+$, Ca^{++}, Mg^{++}, Mn^{++}, Cl$^-$, PO$_4^{---}$, HCO$_3^-$, SO$_4^{--}$
      Lactalbumin hydrolyzate
      Glutathione
      Water-soluble vitamins
      Adenine, guanine, xanthine, uracil, cytidylic acid
      Bovine plasma fraction V
      Gelatin
      Duck serum
B. One volume frozen-thawed duck erythrocytes in 1⅓ volumes of A, centrifuged
      50 min. at 2000 g.
      Clear, deep red supernatant = extract, pH to 6.9.
C. To each 10 ml extract added 0.2 ml of a solution to give concentrations of: yeast
      adenylic acid, 1.4; L-malic acid, 6.0; NAD, 0.15 mM per liter.
D. Supplementary additions to give following final concentrations of added

ATP	2 mM/liter
Na pyruvate	5 mM/liter
CoA	0.02 mM/liter
Leucovorin	0.012 mM/liter

hemoglobin. It also contains other nondialyzable materials shown to be favorable to extracellular survival and development of *P. lophurae* (Trager, 1958).

In the presence of full strength duck erythrocyte extract good survival to the second day was obtained only with added adenosine triphosphate (ATP), to the third day only with further addition of malate and coenzyme A(CoA), and to the fourth day only with the still further addition of folinic acid. Since the requirements for CoA and ATP could themselves account for the obligate intracellular parasitism, they have been given special study. The favorable effect of each of these factors can be shown in 1-day tests if a dilute red cell extract is used (Tables VIII and IX). It is of interest to note that whereas ATP increases the proportion of parasites becoming multinucleate during a 1-day period of incubation, the effect of CoA is to reduce the proportion of degenerate parasites.

In keeping with the requirement for an external source of ATP is the finding of Brewer and Powell (1965) that *P. falciparum* developed faster in individuals whose red cells had a relatively higher ATP content.

The enzymatic defect in the parasites which might account for their need for externally supplied ATP remains to be found. The parasites contain both pyruvic and 3-phosphoglyceric kinases (Trager, 1967) and can therefore make ATP from adenosine diphosphate (ADP) in the glycolytic

TABLE VIII

EFFECT OF ADENOSINE DIPHOSPHATE (ADP) WITH PHOSPHOENOLPYRUVATE (PEP) AND OF ADENOSINE TRIPHOSPHATE (ATP) WITH PYRUVATE ON EXTRACELLULAR DEVELOPMENT OF *Plasmodium lophurae*[a, b]

Flask No.	Addition	Average percent parasites after 1 day[c]		
		With over 1 nucleus	With over 2 nuclei	Degenerate
1–3	None	17	4	12
4–6	PEP, 1.5 mM	22	7	13
7–9	PEP, 1.5 mM ADP, 5 mM	39	13	1
10–12	Na pyruvate, 4 mM ATP, 10 mM	52	18	0.3

[a] From Trager (1967).

[b] All flasks contained dilute red cell extract (0.36 ×) plus all supplements except ATP and pyruvate.

[c] At 0 time there were 7% with over 1 nucleus and 2% with over 2.

TABLE IX

EFFECT OF DIFFERENT SOURCES OF CoA AND LACK OF EFFECT OF PANTETHINE AND AN ANTIPANTOTHENATE ON EXTRACELLULAR DEVELOPMENT OF *Plasmodium lophurae* IN PRESENCE OF 0.36 × DUCK ERYTHROCYTE EXTRACT AND ALL OTHER SUPPLEMENTS[a]

Expt. No.	Flask No.	Addition	Concn. (µg/ml)	Average percent parasites after 1 day[b]	
				Multinucleate	Degenerate
A	1–3	None	—	21	7
	4–6	CoA	40	24	2
	7–9	Acetyl CoA (70%)	57	24	3
	10–12	Liver concentrate	570	25	2
B	1–3	None	—	29	9
	4–6	Pantethine	40	32	7
	7–9	CoA-(B)	40	28	3
	10–12	CoA-(P)	40	31	2
C	1–3	None	—	24	13
	4–6	Antipantothenate[c]	100	24	11
	7–9	CoA	40	25	4
	10–12	CoA	40	25	3
		Antipantothenate[c]	100		

[a] From Trager (1966).

[b] At 0 time there were, in Expt. A, 7% multinucleate parasites; in Expt. B, 7% multinucleate parasites; in Expt. C, 8% multinucleate parasites.

[c] Antipantothenate: d-(2-pantoylamino)ethylsulfono-4-chloroanilide.

cycle. They might, however, be unable to form adenylic acid from adenine, an activity of which mature human erythrocytes are capable (Tsuboi, 1965).

Coenzyme A plays a major role in the metabolism of malaria parasites (see Moulder, 1962). It now appears that the requirement for an external source of this cofactor rests on a deficiency of the parasite in pantothenate kinase, the first enzyme in the biosynthetic pathway of CoA (Bennett and Trager, 1967). This enzymatic activity was readily demonstrated in extracts prepared in three different ways of normal duck erythrocytes, and in somewhat less amount, of infected duck erythrocytes. It could not be found, however, in the free parasites of *P. lophurae*.

It may well be that malaria parasites lack still other key enzymes, of the nature of which we have as yet no suspicion. Provision of these enzymes or perhaps better of their required products (as in the case of CoA) might then permit continuous extracellular, i.e., axenic, culture of *P. lophurae*. The situation with regard to extracellular cultivation may, however, be much more complex. The importance of the physicochemical environment is indicated by the favorable effect of gelatin, an effect also found with bacterial protoplasts and with chelate-requiring mycobacteria (see Hanks, 1966, for a pertinent review of these and related problems). There may furthermore be a much more intimate relationship between the parasite and its host erythrocytes than has been hitherto suspected. Thus electron micrographs of human erythrocytes infected with *P. falciparum* or of monkey erythrocytes with *P. coatneyi* show in the host cell characteristic clefts bounded on each side by two unit membranes (Trager *et al.*, 1967). There is reason to suspect that these membranes may be extensions of the double membrane bounding the parasite itself. If so, their biochemical functions would probably have a bearing on attempts at extracellular cultivation of the parasites.

REFERENCES

Amrein, Y. U., Geigy, R., and Kauffmann, M. (1965). *Acta Trop.* 22, 193.
Anderson, C. R. (1953). *Am. J. Trop. Med. Hyg.* 2, 234.
Bennett, T. P., and Trager, W. (1967). *J. Protozool.* 14, 214.
Boné, G. J., and Parent, G. (1963). *J. Gen. Microbiol.* 31, 261.
Brewer, G. J., and Powell, R. D. (1965). *Proc. Natl. Acad. Sci. U. S.* 54, 741.
Brocklesby, D. W., and Hawking, F. (1948). *Trans. Roy. Soc. Trop. Med. Hyg.* 52, 414.
Camargo, E. P. (1964). *Rev. Inst. Med. Trop. Sao Paulo* 6, 93.
Castellani, O., and Fernandes, J. F. (1962). *Exptl. Parasitol.* 12, 52.
Citri, N., and Grossowicz, N. (1955a). *J. Gen. Microbiol.* 13, 273.
Citri, N., and Grossowicz, N. (1955b). *Trans. Roy. Soc. Trop. Med. Hyg.* 49, 603.
Committee on Cultures, Society of Protozoologists. (1958). *J. Protozool.* 5, 1–38.
Cowperthwaite, J., Wever, M. M., Packer, L., and Hutner, S. H. (1953). *Ann. N. Y. Acad. Sci.* 56, 972.

D'Alesandro, P. A. (1962). *J. Protozool.* **9,** 351.

Davis, A. G., Huff, C. G., and Palmer, T. T. (1966). *Exptl. Parasitol.* **19,** 1.

Deane, M. P., and Kirchner, E. (1963). *J. Protozool.* **10,** 391.

Demarchi, J., and Nicoli, J. (1960). *Ann. Inst. Pasteur* **99,** 120.

Desowitz, R. W. (1963). *J. Protozool.* **10,** 390.

Dewey, V. C., and Kidder, G. W. (1966). *Arch. Biochem. Biophys.* **115,** 401.

Diamond, L. S., and Herman, C. M. (1954). *J. Parasitol.* **40,** 195.

Fernandes, J. F., and Castellani, O. (1959). *Exptl. Parasitol.* **8,** 480.

Fromentin, H. (1961). *Bull. Soc. Pathol. Exotique* **54,** 1046.

Geigy, R., and Kauffmann, M. (1964). *Acta Trop.* **21,** 169.

Geiman, Q. M. (1948). *Dept. State Publ.* **3246.**

Geiman, Q. M. (1952). *Ann. Rev. Microbiol.* **6,** 299.

Geiman, Q. M., Anfinsen, C. B., McKee, R. W., Ormsbee, R. A., and Ball, E. G. (1946). *J. Exptl. Med.* **84,** 583.

Geiman, Q. M., Siddiqui, W. A., and Schnell, J. V. (1966a). *Science* **153,** 1129.

Geiman, Q. M., Siddiqui, W. A., and Schnell, J. V. (1966b). *Military Med.* **131,** Suppl., 1015.

Gill, J. W., and Vogel, H. J. (1963). *J. Protozool.* **10,** 149.

Guttman, H. N. (1966a). *J. Protozool.* **13,** 390.

Guttman, H. N. (1966b). *J. Protozool.* **13,** Suppl., 18.

Guttman, H. N., and Wallace, F. G. (1964). *Biochem. Physiol. Protozoa* **3,** 459–494.

Hanks, J. H. (1966). *Bacteriol. Rev.* **30,** 114.

Hawking, F. (1948). *Trans. Roy. Soc. Trop. Med. Hyg.* **41,** 545.

Hepler, P. K., Huff, C. G., and Sprinz, H. (1966). *J. Cell Biol.* **30,** 333.

Herman, R. (1966). *J. Protozool.* **13,** 408.

Hoare, C. A. (1964). *J. Protozool.* **11,** 200.

Hoare, C. A., and Wallace, F. G. (1966). *Nature* **212,** 1385.

Huff, C. G. (1957). *Exptl. Parasitol.* **6,** 143.

Huff, C. G., Pipkin, A. C., Weathersby, A. B., and Jensen, D. V. (1960). *J. Biochem. Biophys. Cytol.* **7,** 93.

Hulliger, L. (1965). *J. Protozool.* **12,** 649.

Hulliger, L., Brown, C. G. D., and Wilde, J. K. H. (1966). *Nature* **211,** 328.

Hutner, S. H., Baker, H., Aaronson, S., Nathan, H. A., Rodriquez, E., Lockwood, S., Sanders, M., and Peterson, R. A. (1957). *J. Protozool.* **4,** 259.

Johnson, R. C., and Rogers, P. (1964). *Arch. Biochem. Biophys.* **107,** 459.

Kidder, G. W., and Dewey, V. C. (1963). *Biochem. Biophys. Res. Commun.* **12,** 280.

Kidder, G. W., and Dutta, B. N. (1958). *J. Gen. Microbiol.* **18,** 621.

Kofoid, C. A., Wood, F. D., and McNeil, E. (1937). *Univ. Calif. (Berkeley) Publ. Zool.* **41,** 23.

Krassner, S. M. (1964). Unpublished data.

Krassner, S. M. (1965). *J. Protozool.* **12,** 73.

Krassner, S. M. (1966). *J. Protozool.* **13,** 286.

Lamy, L., Samso, H., and Lamy, H. (1964). *Bull. Soc. Pathol. Exotique* **57,** 16.

Lemma, A., and Schiller, E. L. (1964). *Exptl. Parasitol.* **15,** 503.

Lwoff, A. (1933). *Compt. Rend. Soc. Biol.* **113,** 231.

Lwoff, M. (1933). *Ann. Inst. Pasteur* **51,** 707.

Lwoff, M. (1938). *Compt. Rend. Soc. Biol.* **128,** 241.

Meyer, H. (1949). *Rev. Brasil. Biol.* **9,** 211.

Meyer, H., and Chagas, S. (1950). *Anais. Acad. Brasil. Cienc.* **32,** 175.

Moulder, J. W. (1962). "The Biochemistry of Intracellular Parasitism." Univ. of Chicago Press, Chicago, Illinois.

Muniz, J., and Freitas, G. (1945). *Rev. Brasil. Med. (Rio de Janeiro)* **2**, 995.

Nathan, H. A., and Cowperthwaite, J. (1955). *J. Protozool.* **2**, 37.

Nathan, H. A., Hutner, S. H., and Levin, H. L. (1956). *Nature* **178**, 741.

Neva, F. A., Malone, M. F., and Myers, B. R. (1961). *Am. J. Trop. Med. Hyg.* **10**, 140.

Newton, B. A. (1957). *J. Gen. Microbiol.* **17**, 708.

Nicolle, C. (1908). *Compt. Rend.* **146**, 842.

Novy, F. G., and MacNeal, W. J. (1903). *J. Am. Med. Assoc.* **41**, 1266.

Nydegger, L., and Manwell, R. D. (1962). *J. Parasitol.* **48**, 142.

Patterson, E. L., Broquist, H. P., Albrecht, A. M., von Saltza, M. H., and Stokstad, E. L. R. (1955). *J. Am. Chem. Soc.* **77**, 3167.

Pereira, C., de Castro, M. P., and de Mello, D. (1958). *Arg. Inst. Biol. (San Pablo)* **25**, 21.

Peters, D., and Wigand, R. (1955). *Bacteriol. Rev.* **19**, 150.

Pipkin, A. C., and Jensen, D. V. (1958). *Exptl. Parasitol.* **7**, 491.

Rembold, H., and Buschmann, L. (1963). *Ann. Chem.* **662**, 72.

Ristic, M., and Trager, W. (1958). *J. Protozool.* **5**, 146.

Rudzinska, M. A., D'Alesandro, P. A., and Trager, W. (1964). *J. Protozool.* **11**, 166.

Schlossberger, H., and Brandeis, H. (1954). *Ann. Rev. Microbiol.* **8**, 133.

Senekjie, H. A. (1943). *Am. J. Trop. Med.* **23**, 523.

Sherman, I. W. (1966). *J. Protozool.* **52**, 17.

Shortt, H. E., Fairley, N. H., Covell, E. G., Shute, P. G., and Garnham, P. C. C. (1951). *Trans. Roy. Soc. Trop. Med. Hyg.* **44**, 405.

Siddiqui, W. A., Schnell, J. V., and Geiman, Q. M. (1967). *Science* **156**, 1623.

Simpson, C. F., and Green, J. H. (1959). *Cornell Vet.* **49**, 192.

Simpson, L. (1965). *Proc. 2nd Intern. Congr. Protozool., London, 1965.* Intern. Congr. Ser. No. 91, pp. 41–42. Excerpta Med. Found., Amsterdam.

Simpson, L. (1966). Unpublished data.

Stauber, L. A. (1963). *Proc. 16th Intern. Congr. Zool., Washington, D. C., 1963,* Vol. 4, p. 198.

Steinert, M. (1958). *Exptl. Cell Res.* **15**, 560.

Steinert, M. (1965). *Proc. 2nd Intern. Congr. Protozool., London, 1965.* Intern. Congr. Ser. No. 91, p. 40. Excerpta Med. Found., Amsterdam.

Strejan, G. (1963). *Israel J. Exptl. Med.* **11**, 21.

Tchernomoretz, L. (1946). *Harefuah* **30**, 87.

Tobie, E. J. (1964). *J. Protozool.* **11**, 418.

Tobie, E. J., von Brand, T., and Mehlman, B. (1950). *J. Parasitol.* **36**, 48.

Trager, W. (1943). *J. Exptl. Med.* **77**, 411.

Trager, W. (1947). *J. Parasitol.* **33**, 345.

Trager, W. (1953). *J. Exptl. Med.* **97**, 177.

Trager, W. (1957a). *J. Protozool.* **4**, 269.

Trager, W. (1957b). *Acta Trop.* **14**, 289.

Trager, W. (1958). *J. Exptl. Med.* **108**, 753.

Trager, W. (1959). *Ann. Trop. Med. Parasitol.* **53**, 473.

Trager, W. (1964a). *In* "The Cell" (J. Brachet and A. E. Mirsky, eds.), Vol. 6, pp. 81–137. Academic Press, New York.

Trager, W. (1964b). *Am. J. Trop. Med. Hyg.* **13**, 162.

Trager, W. (1965). *Am. Naturalist* **49**, 255.

Trager, W. (1966). *Trans. N.Y. Acad. Sci.* [2] **28**, 1044.

Trager, W. (1967). *J. Protozool.* **14**, 110.

Trager, W., and Rudzinska, M. A. (1964). *J. Protozool.* **11**, 133.

Trager, W., Rudzinska, M. A., and Bradbury, P. C. (1967). *Bull. World Health Organ.* **35**, 883.

Trejos, A., Godoy, G. A., Greenblatt, C., and Cedillos, R. (1963). *Exptl. Parasitol.* **13**, 211.

Tsuboi, K. K. (1965). *J. Biol. Chem.* **240**, 582.

Vickerman, K. (1965). *Trans. Roy. Soc. Trop. Med. Hyg.* **59**, 372.

Weinman, D. (1939). *Parasitology* **31**, 185.

Weinman, D. (1957). *Trans. Roy. Soc. Trop. Med. Hyg.* **51**, 560.

Weinman, D. (1960). *Trans. Roy. Soc. Trop. Med. Hyg.* **54**, 180.

Wenyon, C. M. (1926). "Protozoology," Vol. 1, p. 396. William Wood & Co., New York.

Yoeli, M., Upmanis, R. S., Vanderberg, J., and Most, H. (1966). *Military Med.* **131**, Suppl., 900.

Zeledon, R. (1959). *J. Parasitol.* **45**, 652.

9

Some Biological Leads to Chemotherapy of Blood Protista, Especially Trypanosomatidae

S. H. HUTNER, HUGUETTE FROMENTIN, AND KATHLEEN M. O'CONNELL

I. INTRODUCTION: NEED FOR BETTER SCREENING METHODS FOR ANTIPROTOZOAL AGENTS

> *Any scholar knows thousands of facts, but the prophet must choose just the right ones and lay them end to end*
>
> Don Marquis, "The Almost Perfect State"

The age of molecular enlightenment has dawned but no safe drug is available for Chagas' disease; mortality of *treated* African trypanosomiasis is not less than 14 % (Duggan and Hutchinson, 1966); drug-resistant strains of *Plasmodium* emerge from Brazil and Colombia to Cambodia and Viet Nam; and the Pan-American highway and feeder roads are being driven through jungles rife with Chagas' disease, malaria, and mucocutaneous leishmaniasis (Galloway, 1967). Moreover, the Peace Corps, its counterparts in other countries, and wars are increasing exposure to these diseases for which drugs are unsatisfactory. This is also the jet age, facilitating global swappings of hosts and parasites—a straw or anopheline in the wind—holidaying British school children acquired malaria at Dakar's airport (Shute, 1965).

Drugs are unsatisfactory for other blood diseases; Hawking (1963) adds theileriasis and anaplasmosis. The dearth of leads to new antiprotozoal drugs has been mentioned (Goodwin, 1964; Goble, 1967). Rollo (1964) recommended a period of consolidation of knowledge of antimalarials on hand since new ones were not needed; events soon proved this view premature.

Other authors in this volume have covered morphology, nutrition, and immunology of blood protozoa thoroughly, setting the stage for a discussion of the chemotherapeutic situation. Unease at the slowness of therapeutic progress despite heavy investments of labor and materials is voiced by Chain (1967): We are "... living on inherited capital [of fundamental biological knowledge], a very large capital to be sure, but not inexhaustable." He urges broadening of biological research, including the use of unusual microorganisms.

To put his thesis in the context of blood parasites, we rephrase the taunt, "If you're so smart, why aren't you rich?" to: "If you know so much about blood, why can't you cultivate so many blood parasites?" And if you know so much about blood parasites, why can't you cultivate them, let alone cure the ills they cause? Since nutrition of parasites is discussed by Trager in this volume (see Chapter 8), we shall touch on nutrition only as related to leads to chemotherapy. The lines of investigation sketched here aim at enlarging the therapeutically relevant stock of information. We sermonize uneasily, aware of the humblingly small quotient for the number of questions a wise man can answer, divided with those asked him; we are asking. One of us once remarked on a similar occasion that the prophet's mantle is apt to turn into a dunce cap.

Chemotherapy and much of its pharmacological background are covered by the volumes edited by Schnitzer and Hawking (1963–1966) and the volume on acridines by Albert (1966); by the series edited by Goldin and Hawking (1964, 1965); and by other reviews: Elslager (1966) and Thompson (1967). Resistance to antiprotozoal agents is covered by Schnitzer (1966) and McLoughlin (1967); and resistance to drugs in general by Emmelot (1964).

The scarcity of leads to new antiprotozoal agents is inferable from symposia edited by Hobby (e.g., 1967) and reports of congresses; the bulky report edited by Kuemmerle and Preziosi (1964) has only one lead—not wholly new at that—to new antiflagellate agents.

Chapters in DiPalma (1965) and Goodman and Gilman (1965) recount shortcomings of antiprotozoal agents, but tend to vagueness about the toxic manifestations that limit applicability. Toxicity information is often anecdotal; e.g., Busey (1966) regrets having to abandon the antileishmanial stilbamidine for treating North American blastomycosis because of neurotoxicity, with subsequent recourse to the perhaps less effective hydroxystilbamidine. Presumably new editions of Moser (1964) and pharmacology texts will have clearer warnings of hazards.

An official of a major American pharmaceutical firm estimates that, on average, a successful parasiticide takes 3 to 5 years and 3 to 5 million dollars to develop. This figure may be much higher if failures are fully taken into account. Pitfalls in the use of laboratory animals for drug testing have been reviewed (Durbin and Robens, 1964); reliable laboratory animals are not cheap. Such research outlays exceed the resources of underdeveloped countries, most of which are tropical or subtropical and need improved drugs. Soaring costs, reflecting, among other things, heavy reliance on experimentally infected animals, followed up by extensive field testing, fixes a heavy responsibility on biologists to provide the background information that will cut these costs. We shall therefore discuss measures to make *in vitro* screens more realistic so as to minimize recourse to the far more expensive *in vivo* tests. Chain's opinion spurs us:

Screening technics are indeed essential and sometimes crucial . . . but their development, particularly in the biologic field, is far from simple and dull, but taxes ingenuity and knowledge of the investigators to the limit. . . . To develop an adequate screening system . . . is to some degree equivalent to asking the right question in scientific research, and this often means that half the battle is won. Therefore, let no one talk disparagingly of screening methods as of a lower form of scientific activity: they are just as important and respectable a scientific tool as any other methodology.

Evidently time's onrush has not challenged the opinion of Woodruff and McDaniel (1958) that "Successful screening programmes are not the work of technicians but of trained scientists"—sentiments echoed in Calam's notes (1964) on the vagaries of antibiotic screening. Herrmann (1965) in discussing screening for anticancer agents likewise emphasizes that *in vitro* tests permit rapid screening of many more test substances—and many times faster—than animal tests would permit. Foster (1964) has bewailed the shortage of talent in industrial microbiology. We therefore have sought to emphasize aspects of chemotherapy that join practicality, where we have little first-hand experience, and theoretical allurements, where we can at least register susceptibilities.

As our interests in blood parasites center on trypanosomatid pathogens of mammals, and on the biochemistry of trypanosomatids of invertebrates not known to infect vertebrates, the following pages will deal mainly on how the interplay of research on these two kinds of parasites may be put to use in chemotherapeutic research. We shall refer only incidentally to the aforementioned reviews on chemotherapy.

II. PITFALLS IN SCREENING FOR ANTIPROTOZOAL AGENTS

To have to learn the same thing over and over again wastes the time of a race

Clarence Day, "This Simian World"

A. Problems in Recognizing Known Cytotoxic Compounds

The vast screening programs for antiviral and antitumor agents have uncovered many interesting compounds, mainly extracted from fermentation beers. Distinguishing known from new agents becomes ever more complicated. Also, new compounds are often accompanied by old ones. Preparing enough of fractions for *in vitro* identification tests and preliminary animal tests is expensive, even before matters reach the pilot-plant scale. Hence, *in vitro* methods of identification have been highly developed by industry. The following exposition focuses on how to detect potential antiprotozoal compounds in antibiotic beers; the principles are applicable to synthetic organic chemicals.

Discovery of new antibacterial antibiotics is speeded by use of sensitive and resistant pairs of microorganisms for "fingerprinting" (Woodruff and McDaniel, 1958; Woodruff, 1966) and elaborate paper chromatography with several solvent systems and bioautography (Calam, 1964). Presumably, the sensitive microorganisms for the purpose are those originally selected for sensitivity and amenability to routine assay procedures. Assay bacteria for antibacterial antibiotics as tabulated (Kavanagh, 1963) include species of *Sarcina, Micrococcus, Bacillus,* and *Klebsiella*; all grow well on simple agar media and are easy to maintain. Some yeasts meet these specifications: Ehrlich *et al.* (1965) identified the cycloheximide family of antibiotics with *Kloeckera brevis* as the resistant and *K. africana* the sensitive twin. One procedure is to streak an agar plate containing the antibiotic with various resistant strains along with appropriate controls of sensitives. Inhibition of a sensitive but not of its resistant counterpart identifies the antibiotic. The need for quick identification is underscored by the finding that in one series of 2500 *Streptomyces*-producing antibiotics, " . . . all but

250 made streptothricin-like antibiotics, about 125 produced streptomycin, 40 produced tetracyclines, 55 produced other previously described antibiotics and 30 produced new antibiotics" (Woodruff and McDaniel, 1958)— and that was a decade ago.

Each company making antibiotics has a collection of fingerprinting organisms. The catalogs of the American Type Culture Collection list many bacteria employed for antibiotic assay, but it does not list induced resistant cultures; hence, one could not immediately assemble a kit of resistant cultures for antibiotic identification without using considerable variety of different naturally resistant microorganisms, many of them demanding different conditions. It therefore seems more practical, for guiding the search for antiprotozoal antibiotics, to develop a reference collection of drug-resistant protozoa based, at least at the start, on the same protozoan. Cross-resistance will be discussed later, as well as use of different flagellates to develop an antibiotic spectrum.

Why bother with fermentation beers and their myriad impurities while organic chemists abound? In response, microorganisms, notably actino-mycetes and molds, are fantastically more adept than man at concocting antibiotic molecules, many embodying linkages whose existence would otherwise seem but a chemist's demented doodlings. Even penicillin contains a strained ring that required the utmost ingenuity to duplicate (Chain, 1964).

B. Advantages of Insect Trypanosomatids as Screening Organisms, Especially *Crithidia fasciculata*

Realism urges that the pathogens themselves be used for screening. But they are as yet somewhat hard to handle in a routine way. Some, like *Trypanosoma cruzi* are dangerous since they may infect if splashed on a mucous membrane; there is an instructive literature on laboratory infections. Hence the present choice might well be a harmless protozoan closely related to the pathogens, as easily manipulable as a bacterium or yeast, and growing so densely in simple media as to permit selection of even rare types of spontaneous resistant mutants. These specifications seem well met by strains of *Crithidia* belonging to several species. Perhaps members of other trypanosomatid genera, e.g., *Leptomonas*, *Blastocrithidia*, and *Herpetomonas*, will be advantageous; this is discussed later. Although these flagellates are already biochemical workhorses (see Chapters 8 and 10) and *Crithidia fasciculata* is a routine assay organism for biopterin, they have not been used as fingerprint organisms. Recent developments enhance their suitability. They grow quickly on agar, forming distinct colonies; hence, they are easy to clone (Happold and Stephenson, 1935; West et al., 1962) even though the plating efficiency may turn out to be low with present methods.

They tolerate high concentrations of natural crudes such as peptones, liver and yeast extracts, blood, and the like—common ingredients of practical antibiotic-production media; thus, they are likely to be carried over into crude antibiotic preparations. They are osmotically sensitive, yet thanks to the remarkable elasticity of their cell walls—at least in *Crithidia fasciculata*—they withstand remarkable extremes in osmolarity (Cosgrove and Kessel, 1958).

This osmotic hardihood, also evidenced by their resistance to drying (Wallace, 1966a) as well as being manifested by colony formation on rather dry agar, accounts for their surviving freezing if grown in media containing $\sim$10% glycerol or somewhat less ethylene glycol; such cultures have stayed alive at least a year at $-20°C$ in an household freezer (see Chapter 12) (O'Connell *et al.*, 1967). Unpublished experiments indicate that rates of cooling and thawing are not critical, and that they withstand high concentrations of some of the lipid-soluble cheap commercial antioxidants, e.g., butylated hydroxytoluene, which was studied by Epstein *et al.* (1967) for toxicity to *Tetrahymena* and inhibition of the photodynamic test for polybenzenoid carcinogens. Since autoxidation shortens the life of frozen cultures, inclusion of antioxidants in conservation media would probably extend their longevity. Use of antioxidants would probably also enhance the genetic stability of normal and induced resistant lines: Accumulation of free radicals in frozen cultures of bacteria correlates well with death of cultures (Heckly *et al.*, 1963), and free radicals are mutagenic (Freese and Freese, 1966). We also tested the survival of *C. fasciculata* at 30°C. In media containing 12–14% glycerol to slow growth, cultures survived for 6 months.

Another advantage of the simplicity of freeze-storage methods for *C. fasciculata* is that it will facilitate experiments in which inocula must be precisely controlled. In vitamin assays, for instance, it can be tricky to obtain healthy yet adequately depleted inocula. This situation prevails with several vitamins, notoriously *p*-aminobenzoic acid and biotin generally, and B_{12} for *Euglena*. This problem arises wherever metabolites that must be supplied exogenously are also targets for drugs. *Crithidia*, as noted later, requires a wide assortment of potential targets, especially for drugs with aromatic rings. This feature of *Crithidia* may make it the method of choice for large-scale vitamin assays, as in surveys of the nutritional status of populations. Critical indices include riboflavin, thiamine, and nicotinic acid (H. Baker *et al.*, 1967)—all needed by *C. fasciculata* and perhaps by all trypanosomatids. This may prove to be a strong argument for concentrating research effort on it.

C. Detection of Carcinogenicity

The intensity of research on oncogenic viruses has created overlappings of antiviral and antitumor screening programs. As intimated earlier, these programs might well be extended so as not to miss antiprotozoal agents. But as protozoa resemble man more than bacteria do—a view in keeping with current emphasis on the gulf between procaryotes and eucaryotes—the danger is substantial that initially promising antiprotozoal agents will prove too toxic to the host. That antitumor therapy is a two-edged sword, carrying the risk of carcinogenesis, is well recognized; it goes back to the experience with irradiation therapy, alkylating agents, and, most recently, radiomimetic antitumor agents from fermentation beers, e.g., mitomycin (review: Stock, 1966a). Antiviral screening is a good source of cytotoxic agents—toxic at least to cell cultures—as is clear from Buthala's statement (1965) that drugs active against viruses in cell cultures kill cells if the concentration is high enough. The odds are not really so intimidatingly unfavorable that cytotoxins so detected will not be useful against protozoa: Buthala's dictum was derived from cell cultures; metazoan detoxication mechanisms alter the odds. While *in vitro* activity is, as is all too well known, a poor predictor of practical success, the converse may also be true: seemingly excessive toxicity *in vitro* may not be so *in vivo*. The need in any event is to detect novel cytotoxic agents.

The risk of discouragement is real, as witnessed by the dashed hopes for puromycin: immense sums were spent on it, to the eventual benefit of biochemistry but not of therapy. One must learn to discount the fact that protozoa are likely to be misleadingly sensitive to many cytotoxic antibiotics finding practical use only in cancer therapy. For example, *Trypanosoma equiperdum* is highly sensitive *in vitro* to actinomycin, mitomycin, and profiromycin (Jaffe, 1965); *T. cruzi* to mitomycin and actinomycin (Fernandes et al., 1966); and *T. mega* to actinomycin (Steinert, 1965).

The dilemmas in antiprotozoal therapy, especially in advanced disease, are likely to be as harrowing as in anticancer therapy, where even a therapeutic index only slightly above 1.0 may warrant administering a drug. The organic arsenicals pose this dilemma. They carry a definite if unexplored carcinogenic hazard (Hueper, 1967); even less is known about the antimonials. Carcinogenicity of natural products, including antibiotics, has been briefly reviewed (Schoental, 1967).

The vast increase in toxological investigations, spurred by the thalidomide disaster, has led to appreciation of the hazards attending long-term administration of a drug, especially to pregnant women and infants, superimposed on the long-recognized hazard of drug sensitization. All this adds expense to developing drugs. Elimination of all carcinogenic hazard is now recognized as an impractical ideal—or else smoggy megalopolises would be deserted en masse by doctor's orders. This dilemma even comes up for penicillin, for like other 4-membered β-lactones, it definitely increased incidence of tumors on repeated subcutaneous injection into rats (Dickens, 1967); Dickens states that obviously other β-lactam antibiotics urgently need similar test.

The antifungal antibiotic griseofulvin brings the problem of carcinogenicity closer to antiprotozoal therapy. This remarkably potent and specific antidermatophyte, active when given by mouth, has cured many otherwise intractable, near-intolerable ringworm infections, even of the fingernails. One oral dose in rats, as expected, did not increase the incidence of tumors (Griswold et al., 1966), but it was (1) cocarcinogenic with methylcholanthrene on oral administration to mice (Barich et al., 1962); (2) it induced hepatic cancer in mice (Hurst and Paget, 1963); and (3) on repeated injection, 10 out of 10 male mice and 2 out of 10 female mice showed macroscopic liver tumors (De Matteis et al., 1966).

Granted that carcinogenic hazards must be estimated in every screening program, can trypanosomatids themselves help detect these hazards, thus effecting an economy of means? Sensitivity, cost, and accuracy of in vivo versus in vitro methods of detecting carcinogenicity have to be considered, as in estimating antimicrobial activity. Weisburger and Weisburger (1967), in an exhaustive review of animal tests for carcinogenicity, dismissed in vitro tests from consideration by the assertion that such tests were not generally applicable. True or not, this seems conservative—it brushes aside practical exigencies in screening. Use of newborn mice (neonates) substantially increased the sensitivity of tests for carcinogens (Epstein et al., 1966), presumably because detoxication enzymes are still undeveloped. From such evidence newborn or fetal animals better predict danger to the human fetus or fetuses of other animals than do weanling-weanling or adult-adult comparisons. Carcinogenicity of air samples here correlated well with carcinogenicity inferred from chemical analysis and from photodynamic tests with ciliates. The photodynamic test is restricted to ultraviolet-absorbing polycyclic benzenoid hydrocarbons such as benzpyrene and related N-heterocycles.

Lysogeny, i.e., induced lysis in bacteria carrying an ordinarily harmless (temperate) phage, has been used by the Bristol Laboratories group (Heinemann and Howard, 1965; Price et al., 1965), by Endo et al. (1963), and Gause (1966) to detect antitumor agents and possible carcinogenicity. The limitations of this method are much the same ones raised in dealing with the possible carcinogenicity of antitumor agents. Thus aminopterin, presumably noncarcinogenic, the extremely carcinogenic 4-nitroquinoline N-oxide (NQO), and ultraviolet (UV) light were all active. Only a few lysogenic systems have been examined for sensitivity; the much-studied

Escherichia coli K12(λ) system so far seems as good as any. The *E. coli* and *Staphylococcus aureus* phage systems are sensitive to ~0.06 µg/ml of strong inducers (Legator, 1966). Since some of the antitumor agents, e.g., mitomycin, and to some extent NQO, are activated by reduction (unmasking of an alkylating group for mitomycin, formation of a hydroxyamino derivative for NQO), it may be necessary to examine the different reductases of *Crithidia* so as not to miss agents of this sort. A start has been made with some glucose, lactic, and ethanol dehydrogenases in suspensions of several trypanosomatids under anaerobic conditions (Schwartz, 1961), and α-glycerophosphate dehydrogenase in blood and culture forms of *T. rhodesiense*, whose demonstration requires vitamin K₃ and a tetrazolium salt under aerobic but not under anaerobic conditions (Ryley, 1966). Other dehydrogenases of trypanosomatids are reviewed by Danforth (1967).

Another prospect opens with the discovery that two very different carcinogens, 4-nitroquinoline *N*-oxide and ethionine, induced complete cross-resistance in *Ochromonas danica* and *Euglena gracilis* (Hutner *et al.*, 1967). NQO enters mainly via tryptophan transport (Zahalsky *et al.*, 1963; Zahalsky and Marcus, 1965), and ethionine via methionine transport. This pattern suggests that their point of attack is an intracellular one, not the cell membrane. This target is unknown for ethionine, whose fate is unknown aside from its conversion to the *S*-adenosyl derivative, presumably an ethylating agent. NQO appears to affect phosphorylation site II of the mitochondrion and the corresponding photophosphorylating site in the chloroplast (Hutner *et al.*, 1967). *Ochromonas danica* and *E. gracilis* for these studies had been chosen because, among other reasons, both grow well at pH 4.0, *Euglena* down to pH 2.5, and NQO is much more stable in acid. The limited permeability of *Euglena* (Hutner *et al.*, 1968) and the difficulty of cloning *O. danica* make *C. fasciculata* more suitable for such cross-resistance studies now that efficient acidic media are in sight. It becomes useful to prepare carcinogen-resistant strains of *Crithidia*. Judged from the cross-resistance in *E. coli* between radiomimetic agents (e.g., mitomycin), UV-irradiation, proflavin (Karrer and Greenberg, 1964), and nitrofuran derivatives (McCalla, 1965), the categories of carcinogens should not become intimidatingly extensive; in any event cross-resistance would help reveal unexpected types of carcinogens.

III. THE ANTIMETABOLITE APPROACH: WAYS TO IDENTIFY NEW METABOLITES AS POTENTIAL CHEMOTHERAPEUTIC TARGETS

A. GENERAL CONSIDERATIONS

Inspired by the *p*-aminobenzoic sulfonamide relationship, authors galore have mused that, to modify a metabolite so that parasite more than host will be deceived by this fraudulent metabolite, this "spanner in the works", is a rational path to drug design. But in Chain's judgment (1967) the antimetabolite gambit has been disappointing, for in "nearly all" cases the antimetabolite has proved too toxic to the host to be useful in eliminating the parasite, an exception being the coccidiostat amprolium. Gale (1966) also notes that the antimetabolite approach has been disappointing for bacteria. Where antiprotozoal chemotherapy is concerned, such a picture is too gloomy. Chain does not mention the highly successful antimalarial pyrimethamine (Daraprim), originally intended to be an antiviral agent by hindering pyrimidine or purine metabolism. (The idea was that since viruses tend to be stripped down to their nucleic acids, then analogs of nucleic acid bases storm their genetic citadel.) Bergel (1967) points out that the antimetabolite principle yielded some of the most useful agents against leukemias and tumors, e.g., the antifolics. In the pyrimethamine work, *Lactobacillus casei* served as test organism. And now a methyltetrahydrofolic homolog shows promise as an antimalarial (Kisliuk *et al.*, 1967). Goodwin and Rollo (1955) outlined how pyrimethamine came to be tried as an antimalarial. The happenstance that pyrimethamine has still not found employment as an antiviral agent points up a witticism long bandied about in chemotherapeutic circles: "We have the cure; what we need is the disease." A slogan increasing in popularity is: "If anything acts against anything, try it on everything—viruses, bacteria, fungi, weeds, helminths, snails, lampreys, rats, tumors—and protozoa."

In summary, one may assert "Nothing more true than not to believe your senses. But then where are your other evidences?" (Byron: "Don Juan") can be paralleled with: "Nothing more impractical than to put your trust in antimetabolites, but then where are your other evidences?"

Investigators of the growth requirements of pathogenic trypanosomatids must gage the likelihood of meeting requirements for metabolites new to biochemistry. The initial goals may be to devise *any* defined media; then defined media supporting growth at body temperature; then identification of factors accounting for localization in organs, e.g., *Leishmania donovani* in monocytes; *T. cruzi* in heart muscle; *T. gambiense* in the central nervous system. The subtle factors for maintenance of infectivity and virulence may not be beyond reach; there are successes, e.g., the probable identification of erythritol as the tissue-localization (although probably not the intracellular-localization) factor for *Brucella* (Williams *et al.*, 1964).

Effective antimetabolites may diverge widely from the metabolite. Why, is formulated by Richmond (1966): Efficient inhibitors combine structural analogy to the metabolite with an additional reactive group to form a relatively stable inhibitor-enzyme complex. It was therefore naive to expect easy continual successes on the sulfonamide order. Screening, accordingly, remains a necessary organized opportunism or enlightened empiricism, as workers have variously expressed it. The metabolite must still be identified to supply a lead.

Identification of the target metabolite of a drug has another value not exploited in practical antiprotozoal therapy, yet it holds promise of supplying the metabolite along with the drug to exploit differences between host and parasite in protection by the metabolite. The classic, long-known example is the synergism between sulfa drugs and low-molecular antifolics of the pyrimethamine-cycloguanil type (general review: Jukes and Broquist, 1963; malaria: Chin et al., 1966). That the toxicity of this combination was lowered by folinic acid without affecting therapeutic activity—i.e., the therapeutic index was raised—has been known since 1953 (Goodwin and Rollo, 1955) and demonstrated also for Toxoplasma infections (Hitchings, 1962). Folinic acid in consequence has been recommended as an adjuvant to the pyrimethamine treatment of ocular toxoplasmosis since folinic also protected against the toxicity of pyrimethamine used by itself (Giles et al., 1964). With the development of long-acting repository preparations, e.g., antimalarials (Thompson et al., 1963), it may be practical to include protective metabolites in drug formulations. This puts a premium on identifying the target metabolites for every drug.

Since progress in these directions could improve screening and therapy, a pertinacious attack on these problems, operationally nutritional, is in order. Since the technical obstacles are formidable and not all obvious, various approaches will be outlined.

B. Direct Nutritional Approaches; Some Pitfalls with Pathogens

Direct analysis of the growth requirements of the pathogens seems bogged down in a morass of protein-bound factors, some of them lipid, e.g., Greenblatt and Lincicome (1966) traced growth factors in rat serum for Trypanosoma lewisi to the globulins. Perhaps dissociation of a growth factor-protein complex accounts for the effect of urea as a morphogenetic factor for T. mega (Steinert, 1958). If some of the trypanosome growth factors in blood resemble B vitamins, they will be bound to the globulins (H. Baker et al., 1962). One must then find enzymes which will liberate the factors while also supplying all likely nutrients available as pure chemicals. This approach, too, has encountered difficulties: After an initial success in using Parker's 199 tissue-culture medium supplemented with various blood preparations for T. gambiense (Dodin and Fromentin, 1962), hints of trouble ahead came from the observation that blood from glucose-6-dehydrogenase negative bloods were unsuitable (Fromentin, 1964). Recourse was then had to a perhaps less exacting trypanosome from the chameleon (Fromentin, 1966). Another approach to elucidating the favorable effect of erythrocytes is to regard them as a source of enzymes liberating the growth factors in serum, perhaps also contained in themselves. In Parker-hemin medium supplemented with glucose 6-phosphate, fructose 6-phosphate, and fructose-1,6-diphosphate, further supplementation with alkaline phosphatase supported growth of *T. rotatorium;* this combination of supplements replaced washings of rat blood cells (Fromentin, 1967). A preliminary note by Boné *et al.* (1966) mentions that the trypanosome of amphibia, *T. mega,* requires richer defined media than for *Crithidia* (5 to 10 times more amino acids, and high folic acid); and that the mammalian parasites (*Leishmania tropica, L. enrietii, T. cruzi,* and a bat trypanosome) will multiply in this autoclavable medium on supplementation with extra folic acid, thiamine, and stearic acid. To grow African trypanosomes, Pittam (1966) supplemented elaborate defined media with heated whole human blood filtrates.

Blood fractions may be contaminated with *Mycoplasma* (PPLO's), as suggested by the voluminous literature on *Mycoplasma* contamination of cell cultures (Foley and Epstein, 1964; Bang, 1966; Porterfield, 1967). Sterility tests might well be chosen to demonstrate mycoplasmas, and the behavior of cultures tested before and after treatment with macrolide antibiotics (e.g., tylosin or erythromycin) and tetracyclines; these eliminate mycoplasmas from cell cultures (Arai *et al.*, 1966; Friend *et al.*, 1966; Jao, 1967) or infections (Layton *et al.*, 1966); sometimes other antibiotics are necessary (Rahman *et al.*, 1967).

Perhaps, soon, media for *T. mega* and *T. ranarum* will be standardized: not only for growth in defined media, as noted preliminarily by Guttman (1966), but also the factors inducing the blood trypanosome form will be near identification. Drugs inhibiting the trypanosomal phase of the pathogen much more than the crithidial phase would clearly be ones to study further. As noted later, *T. conorhini* may be an especially strategic organism.

C. TEMPERATURE FACTORS; INORGANIC FACTORS

In vitro testing of drugs against blood parasites would be nearer *in vivo* conditions if conducted at blood temperatures. The remarkable structural-metabolic transformations induced in polymorphic blood trypanosomes by passage between arthropod vector and mammalian host (discussed in this volume by Rudzinska and Vickerman; see Chapter 10) is at least partly induced by the temperature difference between vector and host. Since nutritional aspects of host-parasite interactions have been reviewed in this volume by Trager (see Chapter 8), it seems appropriate to try to pin down what impedes knowledge of how to grow the parasites not merely in defined media at 28°C or so, but at blood heat, and to mention developments which suggest that a more powerful attack on this problem can now be mounted.

Inorganic requirements, especially trace elements, may be crucial. For *Euglena* and some yeasts, metal requirements rose steeply with temperature; for *Euglena* the interval between 35.9 and 36.7°C was critical (Hutner *et al.*, 1958). This work was abandoned because difficulties in pushing temperatures higher aroused suspicion (Hutner *et al.*, 1960) that the roster of elements essential for life was grossly incomplete, and that concoction of "nonbiological" culture media was unexpectedly difficult. Most commercial Fe salts are prepared from hematite deposits—fossilized bog bacteria—and likewise Ca and Mg salts came from limestones, salt deposits, or seawater, rich in the skeletons of marine animals (the chalk cliffs of Dover being a famous example). Later work strengthened this conviction: growth of *Tetrahymena* in acidic media (down to pH 4.0) required mixtures of salts of Al, Ni, Pb, Br, and other biologically unfamiliar elements (Hutner *et al.*, 1965). It would be going too far afield in describing developments that may unravel this tangle other than to state that evidence accumulates from other laboratories for the general essentiality of Ni, Cr, and As; and that the crop red seaweed *Porphyra tenera* in defined media in preliminary work responded quantitatively to Br, Sr, Rb, and Li, among other elements (Iwasaki, 1967).

The danger in ignoring inorganics as growth factors—above all, temperature factors—is of laboring to run down an "organic" factor, only to have it turn out to be a trace element supplied in a peculiarly favorable concentration or favorable balance with other elements; loss of activity on ashing is not a dependable distinction between organic and inorganic factors. At least one well-documented cautionary instance adorns the protozoological literature of a growth-factor hunt, ending with the identification of a familiar trace element. In preliminary experiments with acidic media for *Crithidia*, well-grown cultures did not undergo the drastic acidification seen with the older media, where the pH may drop from 7.4–8.0 down to 4.5. Such pH shifts befog interpretation of results with ionized drugs, since entrance into the cell of drug and metabolite may be unequally influenced by pH. Precipitation of heavy metals is much less troublesome below pH 7.0. Media can have more Ca and Mg; these are favorable for counteracting various toxicities, presumably by strengthening the cell membrane.

Attempts to follow some recipes for cultivating exacting protozoa, or to adapt tissue culture media to protozoological uses, reveal that many of these recipes are not fully designed for reproducibility or for storage as frozen powders. Specification of Ca as $CaCl_2$ or $CaCl_2 \cdot 2H_2O$, etc.; Mg as $MgCl_2 \cdot 6H_2O$, Fe as $FeSO_4$; and so on, presages poor reproducibility; such salts vary in composition with humidity, and tend to become anhydrous—effloresce—in the freezer, inadvertently raising the concentrations of the salts when the medium is reconstituted. Only analysis at time of use guarantees that the ultrahygroscopic $CaCl_2$ is not $CaCl_2 \cdot 6H_2O$ or even $\cdot 12H_2O$ as weighed out. A way around these difficulties is to supply Ca and Mg as carbonates, and to include a fixed acid, e.g., succinic, in the dry mix. If the pH is brought below 6.0 for the reconstituted medium, heating completely expels CO_2 as the ingredients dissolve. The medium is then alkalinized as necessary with a predetermined concentration of a solid, stable, noncorrosive alkali such as tris or by the amine, e.g., triethanolamine, used as solvent for the heme (supplied as hemin), which is seemingly required by all Trypanosomatidae. *Leptomonas oncopelti* (*Crithidia, Strigomonas*) is an apparent exception but, as mentioned later, it may harbor a symbiont. NaOH and KOH tend to vary in compositions and present hazards in handling.

Acidic dry-mix media are simple to reconstitute because they dissolve fast; they also are less ballasted with alkali. Some vitamins are more stable in acidic media, e.g., thiamine, folic acid, and cyanocobalamin. Control of Ca and Mg concentrations, which entails avoidance of precipitates, is as necessary as control of trace elements, for although Ca and Mg chelate more weakly than do Fe, Mn, Zn, Cu, and other transitional elements, Ca and Mg have to be supplied in higher concentration, and so compete effectively for chelating agents. Dry-mix techniques for tissue culture are described by Evans and Bryant (1965) and for *Euglena*, including trace-element mixtures, by Hutner et al. (1967).

These interactions, whether in protozoan or cell cultures, usually only become clear when crudes such as peptones and serum are replaced with pure chemicals. Very puzzling results may then ensue, sometimes leading, as noted, to wild-goose chases after "organic" factors. If only for insurance, the trace elements should be supplied as such from the outset, and in well-chelated (metal-buffered) media. Perhaps some of the difficulties in replacing crudes for tissue cultures bedevil similar experiments with trypanosomatids.

Quite likely, temperature factors overlap with the folic- and biopterin-sparing factors for *C. fasciculata* described in the next section. Temperature experiments with *Ochromonas malhamensis* may hint as to what to expect with trypanosomatids. The free-living, weakly photosynthetic *O. malhamensis* has a vitamin B_{12} requirement like man's. Its only other absolute organic requirements besides the carbon energy source are for thiamine and biotin. When grown above 35°C, its B_{12} requirement rose steeply; this enhanced B_{12} requirement was spared by folic acid. The conclusion was that, superficially at least, temperature acted like adding sulfanilamide to the cultures (Hutner et al., 1957). Arguing for applicability of such findings to trypanosomatid situations is that a very different flagellate, the marine dinoflagellate *Gyrodinium cohnii*, has an induced B_{12}-folic requirement when grown above 35°C (Gold and Baren, 1966); it may therefore be a good model system for elucidating temperature-sensitive B_{12}-folic relationships, perhaps mainly serving as a supplement with controllable folic-B_{12} content for media for trypanosomatids.

Guttman (1963) outlined the increased requirements of *C. fasciculata* as incubation temperatures rose from 22°–27° to 33.6°–34°C. Requirements increased for amino acids; Ca, Fe, and Cu; for additional carbohydrate; and then for choline and inositol; and for lipid factors supplied as crude lecithin and egg yolk. Will these temperature factors coincide with leishmania-trypanosome factors, especially when these pathogens are grown at blood heat? The factors listed by Krassner (1965) for growth of *L. tarentolae* at 37°C in a defined basal medium included a lysate of human red cells and a tissue-culture medium conditioned by the growth of fibroblasts; these temperature factors must be of formidable intricacy.

The ease of preserving frozen inocula of *C. fasciculata* may here pay dividends: Should growth at elevated temperatures involve adaptation, one could store adapted strains and compare them with inocula of similarly frozen but unadapted strains. Use of inocula from frozen cultures would avoid the imminent danger of loss of adaptation to elevated temperatures if inoculum cultures are grown at ordinary temperatures.

Temperature factors concerned in growth and infectivity are conspicuous in *Leishmania* (Greenblatt and Glaser, 1966). Bray and Munford (1967) noted that *L. braziliensis* from normal cutaneous leishmaniasis in Brazil or Costa Rica will not grow at temperatures above 30°C, but that strains of *L. mexicana* and *L. braziliensis sensu lato* from "diffusa" leishmaniasis will grow at 37.5°C. If, because of sheer difficulty, it should prove wise to try flanking, rather than frontal, tactics in deciphering the determinants of form and infectivity of the blood trypanosomes, the various leishmanias might fall into a neat series of stepwise increases in intimacy with the warm-blooded host, paralleling the probably more complicated, corresponding stages in the trypanosomes. Not only is there the evolutionary transition from cold-blooded to warm-blooded host, but also from the cooler parts of the mammalian body to the interior, e.g., with stages in monocytes, as in *L. donovani*. The leishmanias, perhaps easier to grow on the whole, might be usefully interpolated between the trypanosomes and the model systems represented by the insect trypanosomatids, perhaps themselves falling into distinctly different patterns of increasing complexity as the temperature factors are charted for *Crithidia*, *Blastocrithidia*, and *Herpetomonas*. The leishmanias themselves pose formidable problems, e.g., some of the *L. braziliensis* strains that Bray and Munford (1967) could not maintain on the usual blood agars were, in unpublished experiments by

Fromentin and Vaucel (1965), grown through many passages in NNN agar at 26°C, but required great care at first. A key organism in the entire panorama may be *Trypanosoma conorhini*. This parasite of reduviids grows well in artificial media, but differs from *T. cruzi* in that infections in ordinary laboratory rodents are light and tend to die out on passage (Deane and Deane, 1961), and it has no intracellular phase in vertebrates. It has the virtue of readily yielding "blood" forms *in vitro* (Milder and Deane, 1967). It might be a good guide to what it is in lactalbumin hydrolyzate that promotes the appearance of metacyclic forms in *T. cruzi* cultures (Castellani *et al.*, 1967).

D. VITAMIN SPARERS, RELATION TO TEMPERATURE FACTORS

This technique has uncovered extraordinarily interesting metabolites (some new to biochemistry) for *Crithidia fasciculata*. Since some are unstable, they may contribute to difficulties in rearing blood parasites, assuming that the nutrition of the blood parasites comprises additions overlaying the simplicities of the insect parasites. Biopterin was discovered as a folic-sparing factor for *C. fasciculata*, simultaneously with its identification as a *Drosophila* eye pigment (review: Hutner *et al.*, 1959). Biopterin (or rather, related unconjugated pteridines in reduction states analogous to tetrahydro forms of folic acid) serves as an enzyme cofactor for hydroxylation of aromatic rings (Kaufman, 1967). To supply biopterin-catalyzed products is to supply biopterin-sparing factors. These biopterin sparers include serotonin (Janakidevi *et al.*, 1966a); norepinephrine (Janakidevi *et al.*, 1966b); oleate (linoleic and arachidonic acids less active); hexadecanedioate or eicosanedioate (octadecanediote has not been tried) (Dewey and Kidder, 1966); and the new 19-methyl C_{20}-sphingosine extracted from *Crithidia* bodies where it occurs as ceramide (Carter *et al.*, 1966).

These findings are of extreme interest; they bring forth new models around which to design antimetabolites. Participation of norepinephrine and serotonin in trypanosomatid metabolism brings trypanosomatids into the ambit of the stupendous research on psychotropic drugs. Conceivably, physicians may have to administer antiprotozoal agents inducing temporary psychosis. Patients in late stages of trypanosomiasis with involvement of the central nervous system have a high incidence of psychoses (Duggan and Hutchinson, 1966); there would appear to be little to lose in cases where the central nervous system is in danger of irreparable damage.

Identification of metabolites by permitting stepwise diminution of concentration of the growth factor required for their synthesis is an old method. Shive (1950) thereby independently isolated vitamin B_{12} as a factor annulling toxicity of sulfanilamide toward *Escherichia coli*. His procedure was to hold sulfanilamide constant; metabolite after metabolite was identified by activity in enabling less and less *p*-aminobenzoic acid to restore growth. A hierarchy of products of 1-carbon metabolism was uncovered, and the concentration of *p*-aminobenzoic acid required to restore growth became vanishingly small—indeed probably close to the experimental error entering from general contamination of the environment by *p*-aminobenzoic acid. This situation for trypanosomatids is a set of Chinese boxes since, as noted, biopterin had itself had come to light as a folic-sparer.

This line has not been pursued in protozoa except for the instances just cited and for a few situations of analytical interest, e.g., sparing by methionine of the vitamin B_{12} requirement of *Ochromonas malhamensis*. Provocative findings in bacteria or in protozoa, their counterparts, have not been followed up. *Crithidia fasciculata*, thanks to its vitamin requirements, is suited for such work.

Perhaps the most intriguing is vitamin B_6 because it enters into the myriad reactions involving an amino group and formation and opening of *N*-heterocycles. In some streptococci vitamin B_6 is completely replaceable by D-alanine. The B_6 requirement of *Tetrahymena* was neither replaced nor spared by D-alanine (H. Baker *et al.*, 1966). The mysterious situation is the bacterial, not the protozoan one; since B_6 catalyzes many synthetic reactions, replacing B_6 should require many more compounds besides D-alanine. Nothing is known about B_6 sparing in *Crithidia;* would its B_6 requirement be spared by intact or hydrolyzed streptococcus bodies grown on D-alanine? Interpretation would be aided by assay of the streptococcal preparations with *Tetrahymena*. A positive B_6 response would require confirmation by chemical means to rule out the possibility that the response was not a response to B_6-sparing or B_6-replacing factors. Whether this line will reveal a host of new metabolites is unpredictable. An experimental obstacle in all such experiments is the desirability of inactivating the growth factor as it exists in crudes. The growth factors eligible for sparing studies would seem at first glance to be restricted to those with a catalytic function. But much recent data on the interchangeability of constituents of membranes would appear to broaden the range of applicability.

Biotin-replacing factors are discussed later. Other *Crithidia* growth factors are discussed later in connection with minimal media.

Biosyntheses catalyzed by enzymes with vitamin prosthetic groups are likely to be deranged by supraoptimal temperatures. "Supraoptimal" requires an operational definition; e.g., in *Leishmania*, the temperature for maintaining the leptomonad form is lower than that for maintaining

Leishmania as intracellular leishmania bodies. Sparing factors might emerge as temperature factors by impairment of the function of the primary growth factor. The extensive literature in genetics on temperature-sensitive, "conditional" lethal mutations suggests this possibility, for many of these "reparable" mutants (i.e., lethality at the heightened temperature prevented by supplementation of the growth medium) require known growth factors, and such mutants may affect an appreciable proportion of the genome (see review by Wagner and Mitchell, 1965).

Among 400 temperature mutants of *Saccharomyces cerevisiae* (growing at 23°C, but not in enriched media at 30°C), the mutant types included loss of ability to synthesize protein, ribonucleic acid (RNA), deoxyribonucleic acid (DNA), and cell walls, and to carry out cell division (Hartwell, 1967). The complex nutritional supplements used were autoclaved and were water-soluble. Materials to be used as sources of trypanosomatid temperature factors would have to be treated more gently since they might have to include autoxidizable lipids and easily oxidized compounds such as the aforementioned serotonin and norepinephrine. The yeast results, limited as they are, indicate that the causes of temperature-induced damage may include every conceivable metabolic distortion and structural damage. Perhaps when the trypanosomatids used in screening are grown at or near blood temperature, many metabolic targets may be accessible, since they might be very stable or concealed by the stability of alternate pathways at, say, 30°C. This in turn might spell not only greater sensitivity but also greater specificity in screening compounds, and provide a biochemical basis for the assumed greater realism of drug tests carried out at blood heat.

E. Flank Support from Cell and Tissue Culture and Metazoan Nutrition

1. Cell Culture

Ignorance about the makeup of blood and of the growth requirements of most metazoan cells is evidenced by present ability to grow only a limited variety of cells. Some difficulties have been mentioned in preceding sections. Few specialized cells, blood cells included, can be cultivated outside the fibroblast series [review of metazoan culture in (a) the context of comparative nutrition: Hutner and Provasoli, 1965; (b) other general reviews: Evans and Bryant, 1965; Waymouth, 1965; Paul, 1967; Swim, 1967; and invertebrate cells: Jones, 1966; (c) use of cell cultures in virology: Porterfield, 1967; (d) cell cultures in pharmacology: Rosenoer and Jacobson, 1966, and Dawson and Dryden, 1967]. As emphasized by Fell (1965), freshly isolated cells cannot be grown; a period of adaptation (or selection) is necessary. The prevalence of aneuploidy, and the frequent, poorly understood appearance of carcinomatous processes in cell lines (Evans and Bryant, 1965), underscore these limitations. Rather than the protozoologist deriving aid from cell and tissue culture, the reverse may well be obtained: The pathogenic trypanosomatids are a powerful means of exploring the properties of blood and tissue fluids, especially if trypanosomatids have as many lipid and hormonal growth requirments as do metazoan cells, but the responses of trypanosomatids are clearer. The special factors for infectivity as distinguished from mere growth may be of a subtlety beyond anything yet experienced in whole-cell microbiology.

2. Hints from the Metabolism of Insects and Other Metazoa

If pathogenic trypanosomatids have had a long evolutionary sojourn in invertebrates, especially arthropods, it would not be surprising to find them bearing imprints of arthropod—especially insect—metabolism. Several developments in insect physiology foster this thought. Cleveland (1957) found that gametogenesis in certain flagellates of the wood cockroach, *Cryptocercus*, followed upon hormonal stimuli from the host. Other bits of circumstantial evidence are that *Nosema*, a microsporidian, causes abnormal growth and metamorphosis of lepidoptera and the flour beetle *Tribolium;* extracts of *Nosema* spores were rich in material with juvenile hormone activity (Fisher and Sanborn, 1962). Juvenile hormone is widely distributed in metazoa (Gabe *et al.*, 1964).

The colonization of many latex plants by trypanosomatids presents fascinating parallels to blood. These flagellates even invade the hemocoel of the hemipteran vector (McGhee and Hanson, 1964). But unfortunately no latex flagellate has been cultivated in artificial media. Nevertheless, even if one dismisses as highly unlikely the possibility that some pathogenic trypanosomatids once infected plants, study of the latex flagellates might provide clues. That it is time to try again is suggested by information on how some plant-feeding insects depend on compounds, provided by the host plant, which were unidentified when the original negative experiments were done; many of these experiments used media containing latex. Perhaps latex is quantitatively deficient in certain plant hormones or other characteristic plant metabolites. Provocatively, the pollen factor for the honeybee is replaceable by gibberellic acid (Nation and Robinson, 1966), and silkworms require chlorogenic acid or the related gallic acid (Hamamura *et al.*, 1966; Kato and Yamada, 1966). The hatching factor for the potato nematode *Heterodera rostochiensis* is replaceable by picrolonic acid or anhydrotetronic acid. The ascorbic acid requirement of various insects and mammals, the plant sterol requirement of paramecia and the guinea pig, and carotenoid requirements—all hint that the metabolic nexus between metazoa and metaphyta may be intimate indeed.

Another set of fastidious trypanosomatids may be represented by *Leptomonas* (*Crithidia, Strigomonas*) *oncopelti*, harbored by the common milkweed-feeding hemipteran *Oncepeltus fasciatus*. *L. oncopelti* contains a "bipolar" body, presumably endosymbiotic, since the nutritional requirements of *L. oncopelti* are far simpler than that of any other trypanosomatid, e.g., they do not need heme. Earlier reports that this bipolar body can be eliminated by antibiotic treatment or rapid transfer in rich media and that the "cured" flagellate can then be cultivated, have been placed in doubt (Newton and Gutteridge, 1967). Success here might provide clues as to how to cultivate the latex forms; this knowledge in turn would help define the biochemical common denominator of the trypanosomatid line.

F. MINIMAL MEDIA AND IDENTIFICATION OF TARGET METABOLITES OF KNOWN DRUGS; CROSS-RESISTANCE

1. Growth Requirements and Cross-Resistance

Use of minimal media to detect metabolites as drug targets started in early sulfanilamide days. Lean media—mainly dilute peptone-serum—served to demonstrate that sulfanilamide was bactericidal *in vitro* (review: Neipp, 1964), paving the way for the discovery that *p*-aminobenzoic acid was the main target metabolite. This means of uncovering metabolites has not yet been exploited systematically despite the profusion of new cytotoxic drugs active *in vitro*, i.e., drugs not requiring activation by the host's metabolism. The need for metabolite identification as drug targets is especially evident in antitumor screening and may hold lessons for protozoology. Davenport (1962), after testing 17,000 broths, concluded that many antitumor agents had no antimicrobial activity whatever. Recourse to transplantable tumors may offer no particular improvement, for Schabel and Pittillo (1961), reviewing their work on detecting and identifying antitumor agents by means of a spectrum of resistant bacteria, emphasized that false negative results, i.e., insensitivity, was so great a problem in using transplantable tumors as to have dictated their recourse to microorganisms. If detection of antitumor agents is an index of how efficiently *Crithidia fasciculata* detects cytotoxic agents, the *Crithidia* methods need improvement. For example, Price *et al.* (1962) tested 35 tumor-active and 26 tumor-inactive antibiotics. With 1 μg/ml the arbitrary cutoff point for activity, HeLa cells detected 63%, *Tetrahymena pyriformis* 20%, *Ochromonas malhamensis* 9%, and *C. fasciculata* only 9% of them. Nevertheless, Rosenoer (1966), in an extensive review of methods for evaluating antitumor agents, noted the value of bioautograph methods to detect cytotoxic substances in fermentation beers, and since *C. fascicu-lata* grows on agar, it would be suitable once sensitivity improved.

The reasons for poor sensitivity and failure to exploit known drugs as a means of detecting antimetabolites is clear enough; success depends on quite precisely designed media and properly depleted, rigorously standard-ized inocula. When development of microbiological assays for vitamins and amino acids was in its heyday, microbiologists were accustomed to expend much effort on improving culture media and procedures to obtain the desired precision and accuracy. They worked almost entirely with bacteria and yeasts; suitable protozoa were not at hand. With the advent of chromatographic and tracer methods, many of these workers abandoned microbiological for biochemical methods. Needs have arisen again for reproducible media and standardized procedures.

Knowledge of mechanisms by which antitumor and antiprotozoal agents enter the cell goes far to explain the low sensitivity of most *in vitro* screens employing protozoa:

1. Some of the protozoa used have surplus capacity to synthesize target metabolites and, hence, are protected internally.

2. With organisms with limited synthetic capacity, rich culture media were needed. They tended to be overrich in nutrients counteracting many agents. This probably applied to most cells and cultures of parasitic proto-zoa.

3. Some organisms used have a limited repertory of transport mecha-nisms and so do not respond to many cytotoxic compounds since these compounds do not get in at all. Compounds acting on the cell membrane, e.g., the polyene antibiotics, are not subject to this limitation.

The scene is changing; media and techniques are improving, and proto-zoa are available that are likely to be well suited for screening purposes; possibly better ones are under scrutiny. At this time *Crithidia fasciculata* or one of its acidophilic related species (McGhee *et al.*, 1967) impress us as being potentially valuable. Media of Kidder and Dutta (1958) and several detailed by Guttman (1963) and Guttman and Wallace (1964), modified as considered in this chapter, are a point of departure for developing standardized minimal media. *Crithidia fasciculata* fortunately requires some of the growth factors whose transport accounts for the entrance of many important drugs; endogenous synthesis of these metabolites hence is not a complication. Riboflavin (required by *C. fasciculata*) is a remark-able nutrient because its transport mechanism appears to facilitate entry of a remarkable diversity of agents. Toxicity counteraction by riboflavin appears to depend on: (1) aromatic rings (up to at least four), tending toward quinone or quinonoid configurations; and (2) a hydrophilic side chain whose length approximates that of the ribityl side chain. Thus quinacrine, which combines these attributes except for the quinone configuration, is a fairly reliable inhibitor of flavoproteins *in vitro*, which far from explains its antimalarial action. Riboflavin antagonists may be single ring as shown by its counteraction of alloxan toxicity for *Neurospora* (Garnjobst and Tatum, 1956). Two-ring compounds include 8-azaguanine (duckweeds: Thimann and Radner, 1958), and the extremely cytotoxic carcinogen 4-nitroquinoline N-oxide (Zahalsky *et al.*, 1963). Three-ring compounds besides the aforementioned quinacrine are other acridine dyes and phenazines. The acridine relation is celebrated for trypanosomatids: Extra riboflavin permitted the appearance of a high proportion of acriflavine-induced dyskinetoplastic forms, which were still nonviable, however (Trager and Rudzinska, 1964). Stuart and Hanson (1967), adding riboflavin to the acriflavine-containing medium, obtained up to 86% dyskinetoplasty compared with the 48% obtained by Cosgrove (1966), who did not add riboflavin. Tetracyclines are examples of four-ring compounds whose antibiotic action is opposed by riboflavin (Foster and Pittillo, 1953). Vincristine toxicity for *Tetrahymena* is opposed by riboflavin (Slotnick *et al.*, 1966); the vincristine molecule contains a bipartite multiple-ring system, some rings in each part saturated; one portion of the molecule has the planar indolyl nucleus.

In animal cells the permeases for riboflavin apparently overlap with those for conjugated pteridines, for riboflavin and folinic acid inhibited the influx of amethopterin into Sarcoma 180 cells *in vitro* (Hakala, 1965). A clear-cut difference between the permeability of protozoa and at least some gram-negative bacteria emerged from the many unsuccessful efforts to prepare riboflavinless mutants of *E. coli* (Wilson and Pardee, 1962); riboflavin does not enter *E. coli*.

The variety of agents that interfere with riboflavin suggests that riboflavin is strategic in many kinds of cross-resistance. The wide-ranging cross-resistance seen in protozoa extends to some other organisms, perhaps making *in vitro* screening with protozoa more widely applicable. With activity of the *Hfr* sex factor of *E. coli* K12 as one indicator, the resistance to acriflavine accompanied resistance to basic dyes generally (Nakamura, 1965). *Aspergillus nidulans* showed cross-resistance between acriflavine and triphenylmethane dyes (Ball and Roper, 1966; Warr and Roper, 1965); the latter authors showed that riboflavin opposed inhibition of *A. nidulans* by acriflavine and malachite green.

Whether inhibition by quinacrine of the emergence of resistant strains of *Staphylococcus aureus* (Heller and Sevag, 1966) applies to protozoa, and how riboflavin affects this, are unknown. Riboflavin analogs with changes in the ribityl side chain rather than the ring system have been reviewed

(Robinson, 1966); the scant data on their effects on protozoa suggest that they were tried and found wanting.

Fortunately for studies of drug action and cross-resistance, *C. fasciculata* requires many aromatic compounds besides riboflavin. These include phenylalanine, tryptophan, tyrosine, a purine, heme, and, as folic sparers, thymine and biopterin. With minimal media *C. fasciculata* might be a versatile detector of those cytotoxic agents whose penetration depends on subversion of transport systems for aromatic-ring metabolites. If B_{12}, as likely, should emerge as a temperature factor, it would join the array of target metabolites. Anti-B_{12}'s have not received much attention aside from minor modifications in the amide side chains and the aminopropanol moiety. The properties of the transport system for heme (and the also active protoporphyrin) is unexplored. Perhaps some antiprotozoal drugs have an antiheme component in their action; but as heme is in excess in media for trypanosomatids that have been grown in defined media, and the heme supply of blood media is essentially uncontrolled, is heme ever a limiting factor in blood media for higher trypanosomatids? This possibility has not been tested.

Some practical considerations enter into the design of the all-purpose minimal media intended for: (1) detection of cytotoxic agents; (2) a baseline for detecting temperature-induced metabolic distortions; (3) induction of drug-resistant lines. The medium should be as physiological as possible, imposing minimal metabolic strain on the organism. It should not be so rich in any metabolite as to mask the activity of agents affecting that metabolite, yet not so poor as to cause slow or erratic growth or necessitate a period of adaptation. To meet these specifications, minimal means minimal total weights of metabolites, rather than the minimal number of constituents. To illustrate this point: In the absence of biopterin, *C. fasciculata* requires an excessively high (unphysiological) concentration of folic acid. Biopterin occurs in body fluids. A "physiological" minimal medium should thus include biopterin and, as an additional ubiquitous sparing factor, thymine. By this reasoning it would be inadvisable to supply only the essential amino acids and only a single purine to satisfy the purine requirement.

Such a medium may provide a good way to identify temperature factors. By plotting the requirements for metabolites as the temperature is raised, one might be able to extrapolate to higher temperatures, and discern which metabolites became crucial; it might then be possible to relieve the temperature-induced metabolic strain by changing a nutritional balance, e.g., of amino acids, or supplying hitherto omitted nutrients, or supplying more of the sparing factors to relieve the strain on a crippled biosynthetic catalyst. Such procedures might permit getting around otherwise insuperable-seeming, sharp temperature cutoffs.

One has to steer between too few metabolites on one hand and an impracticably lengthy roster of metabolites on the other. Improvements in compounding media as dry mixes, and their deployment of the constituents in blocks, as in investigations of cell-culture media, make it feasible to assemble complicated media. Another encouragement is the commercial availability of biopterin, even if of low ($\sim$7%) purity. Some algae may prove to be rich sources of pure biopterin; mass culture methods for them are being improved.

2. Antimetabolites Revealed by Drug Studies

Perhaps for the reason noted—only sporadic use of minimal media—there are only a few instances in which drug studies with trypanosomatids have led to identification of target metabolites. A striking general situation besides the sulfanilamide-p-aminobenzoic pair, worth exploring in trypanosomatids, is that of the actinomycins, whose toxicity is lowered by guanine, and in some instances, as in lactobacilli, strikingly by deoxyguanosine (review: Stock, 1966b). Binding of actinomycin to the cell membrane, as inferred from this growth antagonism, resembles remarkably closely its postulated guanine binding sites on DNA; had guanine been unknown it could have been discovered as an actinomycin antagonist. Guanine was long ago recognized as an antagonist of proflavine for *Staphylococcus aureus* (Martin and Fisher, 1944).

Crithidia fasciculata, interestingly, resembles the mammal in sensitivity to the anticonvulsant primidone, widely used in the treatment of epilepsy. Some patients develop a megaloblastic anemia. Growth of the bacteria *Ochromonas malhamensis* and *Tetrahymena pyriformis* were not much inhibited by primidone; growth of *C. fasciculata* was sensitive; this inhibition was annulled by folic acid (H. Baker *et al.*, 1962). Later clinical trials (Reynolds, 1967) revealed that the anemia was indeed a folate deficiency, as suspected on other grounds.

Do anticonvulsants of the primidone type, among them phenobarbital and diphenylhydantoin, act by being antifolics in the central nervous system, the megaloblastic anemia being a spilling over of toxicity, much as in vitamin B_{12} deficiency? Injury to the central nervous system, in some animal species or humans, may precede the anemia. These anticonvulsants may be regarded as small-molecule antifolics, thus primidone is 2-deoxyphenobarbital or 5-ethyl-5-phenylhexahydropyrimidine-4,6-dione. One wonders what happens in epileptics so treated who have trypanosomatid infections.

The unique conspicuousness of a biopterin requirement in *C. fasciculata* (other trypanosomatids have not been explored in this respect) may also be reflected in a special sensitivity to small-molecular antifolics of the pyrimethamine type, as observed by Nathan and Cowperthwaite (1954) and confirmed by Rembold (1964) in showing that 4,5-diamino-6-di-methylamino-2-methylpyrimidine, its 6-chloro derivatives, and pyrimeth-amine were strongly inhibitory and competitive with biopterin. Perhaps the designation "antifolic" for antimalarials of the pyrimethamine type should be broadened to "antipteridine." Biopterin is an oxidation artifact in the sense that its coenzyme forms are tetrahydro-reduced forms, as in the folic (conjugated pteridine) series. *C. fasciculata* would then correspond to folic-requirers like *Streptococcus faecalis* and man in possessing reductases for folic acid; indeed Jaffe (1967) has demonstrated folic reductase in *T. equiperdum*. The synergism of the pyrimethamine-sulfonamide pair might thus reside in simultaneous inhibition of both kinds of pteridines.

Flexibility in procedure in inducing resistant strains may unearth other leads to chemotherapy. Impairment of permeases may be readily detecta-ble by requirement for the target metabolite, thus obviating the need to prepare labeled drugs for measuring uptake. Many intracellular changes may cause resistance. Resistant strains might be induced in minimal media with and without supplements to explore the possibility that some modes of resistance involve acquisition of dependence on metabolites not in the minimal medium but present in a supplement such as a liver digest or other crudes. Collateral sensitivity, i.e., gain in resistance to one drug along with increased sensitivity to another, might also be looked for. The cardinal technique for making such studies feasible may be, as noted earlier, media with high plating efficiencies. Good plating media would simplify detection of drug antagonisms and synergisms by the paper-disc methods routine in work with antibiotics. Cross-feeding techniques would be useful in determining whether resistance was based on overproduction of the target metabolite. Biochemical genetics abounds in ingenious papers on plate methods for identifying metabolic alterations, and so this subject is not discussed further.

Since protein synthesis in *C. fasciculata* remains highly active in extracts of cells stored frozen in 10% glycerol (Kahan *et al.*, 1967; Kahan *et al.*, 1967a), it may be feasible to stockpile frozen mitochondria and ribosomes (protected with antioxidants) from resistant strains, especially where resistance does not reflect an altered cell membrane. Drugs which poison mitochondria or ribosomes may thus be studied conveniently. Ribosom-ologists are accumulating a vast literature, enabling finer resolution of sites vulnerable to ribosomal poisons; at least 30 separate proteins occur in ribosomes. Superficially similar agents may thus eventually be dis-tinguished *in vitro*. The need seems particularly acute for aminoacyl antibiotics, e.g., puromycin. Does the antitrypanosomal nucleocidin (very toxic to mammals) which inhibits protein synthesis at a stage after formation of an aminoacyl transfer RNA (Florini *et al.*, 1966), like puromycin (Fox *et al.*, 1966; Monro *et al.*, 1967), differ enough from puromycin to warrant animal trial?

IV. SEDUCTIONS OF MOLECULAR BIOLOGY; DNA BINDING

> *I believe that in our modern biochemical approach we have concentrated too much on the study of isolated enzyme systems and cell fractions and have neglected the study of the biochemistry of the intact animal.*

Chain, 1967

This thesis is amply documented for antiprotozoal agents. Interaction of antiprotozoal agents with DNA, transfer RNA, messenger RNA, and ribosomes may explain arrest of protein synthesis and growth—and death. This line of investigation, dismayingly repetitious, still has little predictive value. If a molecule has a flat portion, and carries a positively charged group, it will attach to DNA (at least to extracted DNA) at the PO_4 groups of the nucleotides, and may also plaster itself decalcomania-style onto the plane of the bases. Drug binding may depend on (*a*) intercalation between the bases; (*b*) orientation vertical to the plane of the bases; (*c*) its insertion in the minor and major grooves of the DNA double helix (binding may depend on the DNA being double-stranded); and (*d*) base composition: (adenine + thymine)/(guanine + cytosine) ratio or, as now generally expressed, GC mole % or merely GC %.

These drug-macromolecular interactions are being studied by the fifteen or so physical techniques devised for the study of macromolecules, e.g., those applied to synthetic polynucleotides (reviewed by Michelson *et al.*, 1967). Superimposed are aminating, phosphorylating, depolymerizing, and other degradative or modifying synthetic enzymes, ranging from snake venom phosphodiesterase to initiating enzymes for polypeptide synthesis, e.g., *N*-formylmethionine-attaching and -detaching enzymes. There are also systems that may control gene derepression and repression, perhaps embodied in nucleotide-methylating and -demethylating enzymes. Moreover, additional geometric features of DNA have to be considered, e.g., possible penetration of the drug within the helix, thereby interfering with base pairing. Such DNA model building, reviewed by Fuller (1967) for drugs generally, and by Reich (1966) for actinomycin, has inspired vast efforts for acridine dyes, especially proflavine and quinacrine, likewise for chloroquine, with ethidium not far behind (Waring, 1966; LePecq and Paoletti, 1967). If the compound has cytological as well as biochemical uses, its literature can reach stupefying dimensions, as witnessed by the acridines (Kasten, 1967).

Another question: *Which* DNA is decisive for a given drug? DNA occurs in extranuclear sites—mitochondria, chloroplasts, centrioles (reviews: Borst *et al.*, 1967; Granick and Gibor, 1967)—accompanied by RNA's, including ribosomes, multiplying the number of potential drug targets. This issue is raised by reports such as that by Kitagawa *et al.* (1966) that aminofluorene and 2-acetylaminofluorenes bind to the liver microsomes of normal rats.

A listing, far from complete, of recent papers on drug-DNA binding indicates something of the resources applied to this line of investigation: ethidium (Eron and McAuslen, 1966; Le Pecq and Paoletti, 1967); stilbamidine (Truhaut *et al.*, 1966); chloroquine (O'Brien *et al.*, 1966a,b). As noted, studies abound of binding by acridine dyes, including quinacrine.

To gage the present utility of this information it is instructive to scan the corresponding, vaster literature on carcinogens and antitumor agents and, in the context of the earlier discussion, the literature on carcinogens. Carcinogens, especially as they approximate the active form, bind to DNA. Again to select only a few recent papers: aflatoxin (Black and Jirgensons, 1967; Clifford *et al.*, 1967), 4-nitroquinoline *N*-oxide (Malkin and Zahalsky, 1966; Nagata *et al.*, 1966; Paul *et al.*, 1967), butter-yellow metabolite (Warwick and Roberts, 1967), *Vinca* alkaloids (Creasey, 1966), and nogalomycin (Zeleznick and Sweeney, 1967). The literature on standard carcinogens, e.g., aminofluorene, appears to be developing along similar lines.

The near ubiquity of riboflavin as a protectant against aromatic biocides may reside not only in the broad specificity of its transport system, but also in its binding strongly to DNA (Warr and Roper, 1965).

Present biophysical criteria fall far short of foretelling whether the drug is antibacterial, antiviral, or antiprotozoal; it could even be a vitamin like riboflavin. Thanks to sensitive methods of detecting free radicals, one might guess whether the compound were eligible to be immediately carcinogenic and so might also have antitumor action. Anyway, strong DNA binding, whatever the chemotherapeutic specificity, spells intense *in vitro* cytotoxicity. How DNA binding in different organelles affects the organism may emerge when organelle DNA is further localized. Thus, cytochrome *b* in the mitochondria (and the corresponding site in the chloroplast), whose function NQO paralyzes (Hutner *et al.*, 1967), strongly binds DNA (Burgoyne and Symons, 1966; Symons and Ellery, 1967). Perhaps this will lead to a practical *in vitro* test for proximate carcinogens.

Integration between molecular and organismal approaches may be prefaced by cytological studies. Simard (1966), Simard and Bernhard (1966), and others have described nucleolar damage in cell cultures by strong DNA binders, e.g., actinomycin, ethidium, proflavine, mitomycin, aflatoxin, and NQO. The effect of these compounds on the kinetoplast may prove to be a useful indicator of DNA binding. Acriflavine did not affect the nucleolus at concentrations eliciting kinetoplast damage in *Leishmania tarentolae* (Trager and Rudzinska, 1964) or *C. fasciculata* (Kusel *et al.*, 1967); carcinogens and antitumor agents seem not yet tested.

Two opinions summarize the low predictive power of present knowledge. Despite the prodigious labors lavished on actinomycin, the role of the peptide part of the molecule is unexplained, as pointed out by Stock (1966b), and ". . . there is no reason to suspect that nucleic acid-bound acridines are significant in the growth-inhibitory effects of acridines" (Silver, 1966). These remarks might apply equally well to antimalarials.

V. ENERGY METABOLISM AND CHEMOTHERAPEUTIC TARGETS

CARBOHYDRATE METABOLISM, ACRIFLAVINE, AND MITOCHONDRIA; DYSKINETOPLASTIC STRAINS; ANAEROBIC METABOLISM

Rudzinska and Vickerman outline in this volume (see Chapter 10) how acriflavine and some other dyes destroy most of the DNA-rich substance of the trypanosomatid kinetoplast. Flagellates so altered are not viable, except occasionally for blood forms of trypanosomes. Some trypanosomes are naturally akinetoplastic (probably, better *dyskinetoplastic*), e.g., some strains of *T. evansi* and *T. equinum* lack kinetoplasts (Hoare, 1954). The following key observations (reviewed in Chapter 10, this volume) have been made: (1) Trypanosomes whose mitochondria have cristae do not give rise to viable dyskinetoplastic strains; and (2) *T. brucei*, artificially dyskinetoplastic, cannot develop cyclically in the tsetse fly. Blood must contain quite special growth factors for dyskinetoplastic trypanosomes, perhaps new to protozoology; but consideration of how acriflavine acts suggests another possibility: Blood somehow rebalances an extreme nutritional imbalance reflecting a particular kind of mutation; rebalancing need not demand new factors. Crick *et al.* (1961), from studies of mutations in phage induced by acriflavine, anticipated that the genetic code was triplet, from the observation that only mutants of a like class which contained three or six mutations yielded the wild phenotype. Hence acriflavine is now thought to cause single-nucleotide additions and deletions, shifting the frame of reading, like missing the right sprockets in a film strip where the sprockets have to be engaged three at a time to yield a picture. These are *frameshift* mutations. Experiments on the readout of the amino acids in proteins, e.g., lysozyme, in bacteria infected with phage T4 (Streisinger *et al.*, 1967) have provided unequivocal proof of amino-acid shifts. Ames and Whitfield (1967) warn that quinacrine is itself a weak mutagen as shown by reversion of a frameshift mutant, thus raising "... the possibility that the standard antimalarials chloroquine, quinine, and quinacrine, which are known to bind to DNA strongly, are causing frameshift mutations in the human population." The very limited and inconclusive experiments on the carcinogenicity of acridine dyes, quinacrine included, have been reviewed (Truhaut, 1967). The inadvertent experiment represented by use of acriflavine and other acridine dyes in World War I (Albert, 1966) and of quinacrine (Atabrine, mepacrine) in World War II indicates that their carcinogenicity, if any, must be near the limit of statistical detection.

Characteristic of blood forms of trypanosomes is their intense O_2 consumption and aerobic behavior along with cytochromes absent, mitochondria rudimentary or absent, intense glycolysis, and almost complete lack of incorporation of carbohydrate carbon. This focused attention on their active α-glycerophosphate dehydrogenase and α-glycerophosphate oxidase. This shuttle, which equilibrates reduced nicotinamide adenine dinucleotide (NADH) between mitochondrial and extramitochondrial compartments, is highly active in insect, e.g., housefly muscle (Sacktor and Dick, 1965) and the lens of the mammalian eye (Griffiths, 1966). In these organs it is coupled to an O_2 uptake mediated by the usual cytochromes. The α-glycerophosphate cycle in blood forms of trypanosomes has aroused much interest (Baernstein, 1963; Grant, 1966). Racker (1965) thinks that the cycle in general has to do with the rapid production of energy, helping to maintain steady-state conditions by competition for lactic dehydrogenase, thus regulating pyruvate generation. Testing the meaning of this cycle in blood trypanosomes, while following up a possible chemotherapeutic lead, might be by testing inhibitors of this cycle on growth; p-nitrocinnamic acid strongly inhibits the insect muscle α-glycerophosphate oxidase system (O'Brien et al., 1965; Sacktor and Dick, 1965).

Other tests of the meaning of trypanosome O_2 consumption are needed. Can heme be eliminated from growth media for C. fasciculata, thus forcing its metabolism into the pattern of the blood-form trypanosomes? Do blood forms need heme at all? This idea comes from Krassner's observation (1966) that the leptomonad but not the leishmania form of L. donovani has heme, and from the existence of latex flagellates. If latex flagellates have a heme requirement, presumably it is satisfied from the debris of cytochromes and heme-containing peroxidases such as tryptophan pyrrolase.

Trypanosoma gambiense can be grown anaerobically (Weinman, 1953). Perhaps the blood forms have some parallels with "petites" in yeast smallcolony formers which have lost most of their mitochondrial apparatus, including cytochromes a and b (Slonimsky et al., 1965). Petites are inducible by anaerobiasis; with high efficiency by propamidine (Lindegren, 1959, 1961); by growth at 40°–42°C (Parks and Starr, 1963); and most widely used of all, by acridine dyes (Nagai et al., 1961). Ability to grow anaerobically correlates with ability to form petites (Bulder, 1964). While anaerobically grown yeast lacks morphological mitochondria (Linnane, 1965), DNA-carrying particles having several cytochrome c-coupled succinate oxidoreductases persist (Schatz, 1965). *Saccharomyces cerevisiae*, prone to form petites, can be grown anaerobically when the medium is supplemented with ergosterol and oleate (Andreasen and Stier, 1956); their synthesis has an O_2-mediated step. How this fits in with the previously cited observations that biopterin sparing of *C. fasciculata* involves oleate is puzzling, since biopterin is not known to catalyze desaturation of long-chain fatty acids. On the other hand, stearate has been claimed indispensable for *T. mega* (Boné et al., 1966). Oleate and aspartate spare biotin for many biotin auxotrophs, but biotin sparing in protozoa may require a diversity of lipids, including the cyclopropane C_{19} fatty acid made by several *Crithidia* species (Meyer and Holz, 1966); cyclopropane fatty acids have hitherto been found only in higher plants and bacteria. As in yeasts, the main sterol of *C. fasciculata* is ergosterol (Kusel and Weber, 1965; Halevy, 1966; Meyer and Holz, 1966). It would be valuable to know which lipids promote anaerobic growth (if obtainable) of *Crithidia* and other trypanosomatids. The experiment thus shaping up is simultaneous sparing of heme, biopterin, and biotin as a means of arriving at an *in vitro* approximation of the metabolic state of trypanosomes in blood: Success would prove that O_2 functions in biosynthesis and not in adenosine triphosphate (ATP) production.

The distinction between carbon sources and energy sources revealed by the very limited uptake of carbohydrate carbon by trypanosomatids (review: Danforth, 1967), may be nearer solution with the preliminary report (Krassner, 1967) that *L. tarentolae* in a defined medium did not need glucose when given proline. If proline turns out to be a reason for the requirement for high concentration of amino acids by trypanosomes, preliminarily reported by Boné et al. (1966) and Guttman (1966), more weight would be lent the idea of an imprint of the metabolism of lower metazoa on trypanosomatid metabolism, for the blood of insects is high in amino acids (Gilbert and Schneiderman, 1961). Also perhaps pertinently, the hemolymph of the triatomid *Rhodnius* is high in proline (Ormerod, 1967) as is also insect flight muscle whose mitochondria briskly oxidize proline (Sacktor and Childress, 1967).

New directions in chemotherapy are suggested by the discovery that trypanosomatids use the diaminopimelic pathway for lysine synthesis (Gutteridge, 1966; Guttman, 1967), hitherto seen in bacteria, water molds, and vascular plants; higher fungi, some phycomycetes, and *Euglena* use the aminoadipic pathway. The status of metazoa is unclear because they need preformed lysine and it is not known whether diaminopimelic acid can satisfy the lysine requirement. This apparent kinship of trypanosoma-tids to bacteria suggests trial of bacterial-type ATP-generating systems, e.g., the Stickland reaction, oxidation of one amino acid by another, found in clostridia.

One hope thus aroused is that media can be devised that will push the *in vitro* metabolism of the flagellates toward the way it is in the mammalian host. A compound inhibiting growth in the mammalian-type medium, but not in the ordinary medium, might have specificity, i.e., low general toxicity. If high enough blood levels could be maintained, one would have a practical drug.

VI. PROSPECTS

. . . i never took much stock in being scared of hypodermic propositions or hypothetical injections

Don Marquis, "archy and mehitabel"

When protozoologists are urged to plunge into molecular biology to come up with chemotherapeutic treasures, it brings to mind the traditional congratulation of a New Yorker who confronts an overproud possessor of a certificate of academic achievement: "With your diploma—and a subway token—you can ride any subway in New York!" (With molecular biology—*and* a good screening program. . . .) Yet, some very successful pharmaceutical companies are erecting institutes of molecular biology; those companies also are experienced in screening. The hope, presumably, is that some day molecular biology will contribute to therapy. In part, this review tries to see how it can contribute.

Most protozoologists were trained in parasitology and zoology; biochemical protozoology came later. The medical tradition was embodied in Paul Ehrlich, who combined stupendous ingenuity in framing *ad hoc* chemotherapeutic and immunological hypotheses with acumen in directing a team of chemists and animal experimentalists along the road from methylene blue and quinine, sulfonated dyes, eventually (with his successors) to quinacrine, chloroquine, suramin, tryparsamide, and later compounds selected by screening (Hawking, 1963). Chain's worry, we take it, is that this lesson may be lost on newcomers to biology and chemotherapy from chemistry and physics. Simpson (1967), quoting Barry Commoner that life is the secret of DNA more than DNA is the secret of life, contends that contributions of DNA studies to the mainstream of biology are still largely unachieved, and that "The next breakthrough . . . the one now in progress

... is between molecular and evolutionary-organismal biology, to the great benefit of each." This view seems general among microbiologists, e.g., two eminent mycologists (Emerson and Fuller, 1966) complain that few molecular biologists, in dealing with fungi, stray beyond the *Neurospora-Saccharomyces-Aspergillus* axis, and that, on the other hand, mycologists have resisted the molecule more strongly than many of their contemporaries. Antiprotozoal and antifungal drugs so overlap that weakness in research on one group of protists weakens the other.

Investigations on chemotherapy are exercises in applied phylogeny (Aaronson and Hutner, 1966). In this light it is deplorable that little is known about bodonid metabolism. Like the trypanosomatids they have a kinetoplast; being free-living, they may be primitive. They are common, and bacterized cultures are available. Other genera in the Kinetoplastida are unexplored biochemically, e.g., the fish parasites *Costia* and *Cryptobia*. If the Trypanosomatida are monophyletic, an ancestor-like flagellate—perhaps a nicely archaic *Bodo* or a hardy *Leptomonas*—could be valuable as a common-denominator screening organism. This is a bad situation—hardly any biological capital to exhaust.

Scientists loathe pure empiricism. But seeing screening as unmitigated trial-and-error is seeing a bogeyman; screening contributes to pure biochemistry by uncovering drugs like puromycin, antimycin, and actinomycin which become indispensable research tools even though therapeutically useless—but they do warn of unproductive directions. The screening operation is continuously modified as results come in; to be successful, it must be, as intimated earlier, large-scale; there is a critical level of activity. Hutchison (1963), reviewing cross-resistance and collateral sensitivity in cancer chemotherapy, complained of too many systems (she listed 352) and too few data. The proponents of a new screening system must bear the burden of proof of its value.

Much information is available on why different animals respond so differently to drugs. The repertory of drug-metabolizing enzymes for each laboratory animal is being catalogued (Buyske and Dvornik, 1966; Gillette, 1966). One may eventually be able to foretell, for each type of compound, which laboratory animals yield realistic results for farm animals and man, and how the sum of absorption, detoxification, and excretion determines drug levels in blood and organs. An outstanding example (Berberian *et al.*, 1967) of the value of this approach is identification of the active form of schistosomicides of the lucanthone (Miracil) series as hycanthone, originating from metabolic conversion of a methyl group to a hydroxymethyl. Animals differed in speed of this conversion, thus accounting for differences in effectiveness from one species to another.

The problem remains of why the side chain for antimalarial activity in compounds having two or three rings should be (with numerous although minor variations on this theme) aminoalkyl-aminoalkyl with about five carbons separating the nitrogens; for schistosomicides, as in the lucanthone series, it is two carbons; and for flagellates the side chain seems irrelevant, with the geometry of more or less complex heterocyclic systems becoming the overriding consideration. This would appear to be a good arena for testing the weapons of molecular biology. This should make efficient *in vitro* testing even more important since animal testing will become less of a bottleneck.

The poor correlation between cytotoxicity to cells and whole animals noted from tests with 36 compounds (Smith *et al.*, 1963) is one of many such reports. Another constraint has been pointed out (Roe and Ambrose, 1966) in detection of anticancer agents: the screens have selected for agents with antigrowth or antimitotic activities, hence for activity against rapidly growing tumors, and they damage the cells most mitotically active.

One can moralize from cordycepin (Williamson, 1966) that it is perilous to dismiss even a field as much picked over as the antipurines, and which has not yet provided practical antiprotozoal drugs. Cordycepin (3'-deoxy-adenosine) is generally trypanocidal; it cures *T. congolense* infections in mice, with a therapeutic index LD_{10}/CD_{90} (CD_{90} = dose curing 90% of the animals), higher than quinapyramine or ethidium bromide.

It may not require much effort to standardize plating procedures for *C. fasciculata* to the point where multiple-disc tests or the like are routine. Detection of synergistic or antagonistic effects would then be simple. In chemotherapy there is more use of rational drug combinations, not the arbitrary mixtures promoted in the first flush of triumphs with antibiotics. Thus Hitchings (1961), in discussing sulfonamide-pyrimethamine synergism, especially in toxoplasmosis, envisions use of more than two drugs for sequential blocking. A combination of quinine, pyrimethamine, and dapsone (4,4'-diaminodiphenylsulfone) is recommended for chloroquine-resistant *falciparum* malaria (Bartelloni *et al.*, 1967; Blount, 1967). Reports, too preliminary to cite in detail, describe combinations of four or even five drugs as being better for seeking out leukemia cells lurking in crannies of the body. Older work on drug synergism in antitumor chemotherapy foreshadowed this (Venditti and Goldin, 1964; Frei and Freireich, 1965). A great value of agar-plate tests may emerge when the drugs tested will be the forms active *in vitro*, and it will be known how to tinker with the substituents of the molecule to suit them for *in vivo* use. A halfway stage in *Leishmania-T. cruzi* work may be comparisons of the inhibition of intracellular versus extracellular forms of *T. cruzi* in chick embryo cultures (Bayles *et al.*, 1966). The familiar *p*-rosaniline hydrochloride [tris(*p*-aminophenyl)carbonium chloride] and furazolidone were promising: Both pro-tected mice given inocula uniformly lethal in untreated mice.

The diversity in responses to chemotherapeutic agents among Tryp-anosomatidae may denote diversity in evolutionary history. It is probably not wise to put all one's screening eggs in the *Crithidia* basket. An attrac-tive additional flagellate is *Herpetomonas muscarum*, whose kinetoplast moves posterior to the nucleus along with the flagellar invagination (it does not have an undulating membrane) as the temperature rises toward blood heat (Wallace, 1966b). It is thus a thermometer, and grows well in ordinary media, as does also *Blastocrithidia culicis*, which has an anterior undulating membrane (Guttman and Wallace, 1964; Guttman, 1967). The *Leptomonas* parasitizing the macronucleus of *Paramecium* used as a pure culture by Stuart and Hanson (1967) for studying acriflavine-induced dyskinetoplasty would be interesting to try since it might be the most primitive extant member of the Trypanosomatidae—if one has faith in the family tree of the Trypanosomatidae as diagrammed by J. R. Baker (1965). These phylogenetic schemes must not gloss over the big gaps in knowledge of parasite and vector (Mattingly, 1965). It would appear advantageous to learn which easily handled insect trypanosomatids most resemble each pathogen at least for *in vitro* responses to drugs. Whether phylogenetic surmises are substantiated by drugs obtained via screening or screening is improved by shrewd phylogenetic guesswork—hardly matters; the actual work is much the same.

If antiprotozoal agents follow the pattern set by antibacterial antibiotics and insecticides, resistance will be a problem for a long time, along with increased incidence of drug hypersensitivity. If this forecast should prove pessimistic—that new drugs will not have to be developed continually—prudence dictates that methods for drug development should be under constant, vigilant study. This responsibility can not be borne wholly by the pharmaceutical industry or by the laboratories of any one government; it must be shared by academic biologists and by international agencies, e.g., the World Health Organization. A sadly amusing commentary on the overdetachment of biologists from these urgencies is that the most detailed published account of commercial screening methods for antibiotics seems to be a novel (Savage, 1964) of no particular literary merit but rich in laboratory lore. Academic complacency must take some blame for physi-cians having to fall back on quinine, dangerously ototoxic, largely controlled by a cartel (Anonymous, 1967), with the burden of high prices falling on the elderly who make up a large percentage of the users of quinine alkaloids (mainly quinidine as a cardiac stimulant), and levying a heavy toll on the poor in the tropics. In Sinclair Lewis' 1924 novel "Arrowsmith," the hero, foreswearing worldly distractions, and disillusioned by life as an organization man, went forth to discover how quinine works. Quinine is still a mystery.

Sinclair Lewis once was to speak at a university on "How to Be a Writer." He addressed the overflowing crowd: "If you want to be a writer—raise your hand!" There was a forest of hands. Upon that he roared, "Then why aren't you home writing!" We have felt guilty at preaching, not working, but the obduracy of the problem makes us cry, "Help!" before returning to the bench.

ACKNOWLEDGMENT

Work from our laboratory cited in this chapter was aided mainly by Grant (to S. H. Hutner) AI-07895 from the National Institutes of Health, and U.S. Public Health Service Grant FR-05596 to Dr. F. S. Cooper. H. Fromentin was enabled to work in New York as a WHO fellow. Gathering material on screening was stimulated by U.S. Army Research Office Contract DA 49-092-ARO-153 to S. H. Hutner.

REFERENCES

Aaronson, S., and Hutner, S. H. (1966). *Quart. Rev. Biol.* **41**, 13–46.
Albert, A. (1966). "The Acridines," 2nd ed. Arnold, London.
Ames, B. N., and Whitfield, H. J., Jr. (1967). *Cold Spring Harbor Symp. Quant. Biol.* **31**, 221–225.
Andreasen, A. A., and Stier, T. J. B. (1956). *J. Cellular Comp. Physiol.* **48**, 317–328.
Anonymous. (1967). *New Republic* Apr. 8, p. 9.
Arai, S., Yoshida, K., Izawa, A., Kumagai, K., and Isheda, N. (1966). *J. Antibiotics (Tokyo)* **A19**, 118–120.
Baernstein, H. D. (1963). *J. Parasitol.* **49**, 12–21.
Baker, H., Frank, O., Hutner, S. H., Aaronson, S., Ziffer, H., and Sobotka, H. (1962). *Experientia* **18**, 224–226.
Baker, H., Frank, O., Ning, M., Gellene, R. A., Hutner, S. H., and Leevy, C. M. (1966). *Am. J. Clin. Nutr.* **18**, 123–133.
Baker, H., Frank, O., Feingold, S., and Leevy, C. M. (1967). *Nature* **215**, 84–85.
Baker, J. R. (1965). *In* "Evolution of Parasites" (A. E. R. Taylor, ed.), pp. 1–27. Blackwell, Oxford.
Ball, C., and Roper, J. A. (1966). *Genet. Res.* **7**, 207–221.
Bang, F. B. (1966). *In* "Cells and Tissues in Culture" (E. N. Willmer, ed.), Vol. 3, pp. 151–261. Academic Press, New York.
Barich, L. L., Schwarz, J., and Barich, D. (1962). *Cancer Res.* **22**, 53–55.
Bartelloni, P. J., Sheehy, T. W., and Tigertt, W. D. (1967). *J. Am. Med. Assoc.* **199**, 173–177.
Bayles, A., Waitz, J. A., and Thompson, P. E. (1966). *J. Protozool.* **13**, 110–114.
Berberian, D. A., Freele, H., Rosi, D., Dennis, E. W., and Archer, S. (1967). *J. Parasitol.* **53**, 306–311.
Bergel, F. (1967). *Hosp. Med.* pp. 881–887.
Black, H. S., and Jirgensons, B. (1967). *Plant Physiol.* **42**, 731–735.
Blount, R. E. (1967). *Arch. Internal Med.* **119**, 557–560.
Boné, G., Parent, G., and Steinert, M. (1966). *Proc. 1st Intern. Congr. Parasitol., Rome, 1964,* Vol. 1, pp. 297–298. Pergamon Press, Oxford.

Borst, P., Kroon, A. M., and Ruttenberg, G. J. C. M. (1967). In "Genetic Elements. Properties and Function" (D. Shugar, ed.), pp. 81–116. Academic Press, New York.

Boyland, E. (1967). In "Potential Carcinogenic Hazards from Drugs" (R. Truhaut, ed.), pp. 204–208. Springer, Berlin.

Bray, R. S., and Munford, F. (1967). J. Trop. Med. Hyg. 70, 23–24.

Bulder, C. J. E. A. (1964). Antonie van Leeuwenhoek, J. Microbiol. Serol. 53, 189–194.

Burgoyne, L. A., and Symons, R. H. (1966). Biochim. Biophys. Acta 129, 502–510.

Busey, J. F. (1966). Antimicrobial Agents Chemotherapy p. 1125.

Buthala, D. A. (1965). Ann. N. Y. Acad. Sci. 130, 17–23.

Buyske, D. A., and Dvornik, D. (1966). In "Annual Reports in Medicinal Chemistry, 1965" (C. K. Cain, ed.), pp. 247–266. Academic Press, New York.

Calam, C. T. (1964). Progr. Ind. Microbiol. 5, 1–53.

Carter, H. E., Gaver, R. C., and Yu, R. K. (1966). Biochem. Biophys. Res. Commun. 22, 316–320.

Castellani, O., Ribeiro, L. V., and Fernandes, J. F. (1967). J. Protozool. 14, 447–451.

Chain, E. B. (1964). In "New Perspectives in Biology" (M. Sela, ed.), pp. 205–214. Elsevier, Amsterdam.

Chain, E. B. (1967). In "Reflections on Research and the Future of Medicine" (C. E. Lyght, ed.), pp. 129–169. McGraw-Hill, New York.

Chin, W., Contacos, P. G., Coatney, G. R., and King, H. K. (1966). Am. J. Trop. Med. Hyg. 15, 823–829.

Clarke, A. J., and Shepherd, A. M. (1966). Nature 216, 546.

Cleveland, L. R. (1957). J. Protozool. 4, 168–175.

Clifford, J. I., Rees, K. R., and Stevens, M. E. M. (1967). Biochem. J. 103, 258–261.

Cosgrove, W. B. (1966). Acta Protozool. 4, 155–160.

Cosgrove, W. B., and Kessel, R. G. (1958). J. Protozool. 5, 296–298.

Creasey, W. A. (1966). Biochem. Pharmacol. 15, 367–375.

Crick, F. H. C., Barnett, L., Brenner, S., and Watts-Tobin, R. J. (1961). Nature 192, 1227–1232.

Danforth, W. F. (1967). In "Research in Protozoology" (T. T. Chen, ed.), Vol. 1, pp. 201–306. Pergamon Press, Oxford.

Davenport, J. L. (1962). Giorn. Ital. Chemioterap. 6–9, 2412–49.

Dawson, M., and Dryden, W. F. (1967). J. Pharm. Sci. 56, 545–561.

Deane, M. P., and Deane, L. M. (1961). Rev. Inst. Med. Trop. Sao Paulo 3, 149–160.

De Matteis, F., Donnelly, A. J., and Runge, W. J. (1966). Cancer Res. 26, 721–726.

Dewey, V. C., and Kidder, G. W. (1966). Arch. Biochem. Biophys. 115, 401–406.

Dickens, F. (1967). In "Potential Carcinogenic Hazards from Drugs" (R. Truhaut, ed.), pp. 144–151. Springer, Berlin.

DiPalma, J. R., ed. (1965). In "Drill's Pharmacology in Medicine," 3rd ed. McGraw-Hill (Blakiston), New York.

Dodin, A., and Fromentin, H. (1962). Bull. Soc. Pathol. Exotique 55, 797–804.

Duggan, A. J., and Hutchinson, M. P. (1966). J. Trop. Med. Hyg. 69, 124–131.

Durbin, C. G., and Robens, J. F. (1964). Ann. N. Y. Acad. Sci. 111, 696–711.

Ehrlich, J., Sloan, B. J., Miller, F. A., and Machamer, H. E. (1965). Ann. N. Y. Acad. Sci. 130, 5–16.

Elslager, E. F. (1966). In "Annual Reports in Medicinal Chemistry, 1965" (C. K. Cain, ed.), pp. 136–149. Academic Press, New York.

Emerson, R., and Fuller, M. S. (1966). Quart. Rev. Biol. 41, 303–304.

Emmelot, P. (1964). In "Molecular Pharmacology " (E. J. Ariëns, ed.), Vol. 2, Part III. pp. 53–198. Academic Press, New York.

Endo, H., Ishizawa, M., and Kamiya, T. (1963). *Nature* **196**, 195–196.
Epstein, S. S., Joshi, S., Andrea, J., Mantel, N., Sawicki, E., Stanley, T., and Tabor, E. C. (1966). *Nature* **212**, 1305–1307.
Epstein, S. S., Saporoschetz, I. B., and Hutner, S. H. (1967). *J. Protozool.* **14**, 238–240.
Eron, L. J., and McAuslan, B. R. (1966). *Biochim. Biophys. Acta* **114**, 633–636.
Evans, V. J., and Bryant, J. C. (1965). In "Tissue Culture" (C. V. Ramakrishnan, ed.), pp. 145–167. Junk, The Hague.
Fell, H. B. (1965). In "Tissue Culture" (C. V. Ramakrishnan, ed.), pp. 3–7, 9–16, and 17–26. Junk, the Hague.
Fernandes, J. F., Halsman, M., and Castellani, O. (1966). *Exptl. Parasitol.* **18**, 203–210·
Fisher, F. M., and Sanborn, R. C. (1962). *Nature* **194**, 1193.
Florini, J. R., Bird, H. H., and Bell, P. H. (1966). *J. Biol. Chem.* **241**, 1091–1098.
Foley, G. E., and Epstein, S. S. (1964). *Advan. Chemotherapy* **1**, 175–353.
Foster, J. W. (1964). In "Global Impacts of Applied Microbiology" (M. P. Starr, ed.), pp. 61–73. New York.
Foster, J. W., and Pittillo, R. F. (1953). *J. Bacteriol.* **65**, 361–367.
Fox, J. J., Watanabe, K. A., and Bloch, A. (1966). *Progr. Nucleic Acid Res. Mol. Biol.* **5**, 251–313.
Freese, E., and Freese, E. B. (1966). *Radiation Res.* Suppl. 6, 97–140.
Frei, E., III, and Freireich, E. J. (1965). *Advan. Chemotherapy* **2**, 269–298.
Friend, C., Patuleia, M. C., and Nelson, J. B. (1966). *Proc. Soc. Exptl. Biol. Med.* **121**, 1009–1010.
Fromentin, H. (1964). *Compt. Rend.* **259**, 1599–1601.
Fromentin, H. (1966). *Bull. Soc. Pathol. Exotique* **57**, 219–224.
Fromentin, H. (1967). *J. Protozool.* **14**, Suppl., 20.
Fromentin, H., and Vaucel, M. A., (1965). Unpublished experiments.
Fuller, W. (1967). In "Genetic Elements, Properties and Function" (D. Shugar, ed.), pp. 17–39. Academic Press, New York.
Gabe, M., Karlson, P., and Roche, J. (1964). *Comp. Biochem.* **6**, 245–298.
Gale, E. F. (1966). *Symp. Soc. Gen. Microbiol.* **16**, 1–21.
Galloway, C. B. (1967). House Doc. No. 10, 90th Congr., 1st Session, 34 pp. U.S. Govt. Printing Office, Washington, D. C.
Garnjobst, L., and Tatum, E. L. (1956). *Am. J. Botany* **43**, 149–157.
Gause, G. F. (1966). *Chem. & Ind. (London).* pp. 1506–1513.
Gilbert, L. I., and Schneiderman, H. A. (1961). *Am. Zoologist* **1**, 11–51.
Giles, C. L., Jacobs, L., and Melton, M. L. (1964). *Arch. Ophthalmol.* **72**, 82–85.
Gillette, J. R. (1966). *Advan. Pharmacol.* **4**, 219–261.
Goble, F. C. (1967). *Develop. Ind. Microbiol.* **8**, 132–139.
Gold, K., and Baren, C. F. (1966). *J. Protozool.* **13**, 255–257.
Goldin, A., and Hawking, F. (1964). *Advan. Chemotherapy* **1**, 1–579.
Goldin, A., and Hawking, F. (1965) *Advan. Chemotherapy* **2**, 1–330.
Goodman, L. S., and Gilman, A., (eds.) (1965). "The Pharmacological Basis of Therapeutics," 3rd ed. Macmillan, New York.
Goodwin, L. G. (1964). In "Biochemistry and Physiology of Protozoa" (S. H. Hutner, ed.), Vol. 3, pp. 495–524. Academic Press, New York.
Goodwin, L. G., and Rollo, I. M. (1955). In "Biochemistry and Physiology of Protozoa" (S. H. Hutner and A. Lwoff, eds.), Vol. 2, pp. 225–276. Academic Press, New York.
Granick, S., and Gibor, A. (1967). *Progr. Nucleic Acid Res. Mol. Biol.* **6**, 143–186.
Grant, P. T. (1966). *Symp. Soc. Gen. Microbiol.* **16**, 281–293.
Greenblatt, C. L., and Glaser, P. (1966). *Exptl. Parasitol.* **16**, 36–52.

Greenblatt, C. L., and Lincicome, D. R. (1966). *Exptl. Parasitol.* **19**, 139–150.
Griffiths, M. H. (1966). *Biochem. J.* **99**, 12–21.
Griswold, D. P., Jr., Casey, A. E., Weisburger, E. K., Weisburger, J. H., and Schabel, F. M., Jr. (1966). *Cancer Res.* **26**, Part 1, 619–625.
Gutteridge, W. E. (1966). *J. Gen. Microbiol.* **44**, Proc., 4.
Guttman, H. N. (1963). *Exptl. Parasitol.* **14**, 129–142.
Guttman, H. N. (1966). *J. Protozool.* **13**, Suppl., 18.
Guttman, H. N. (1967). *J. Protozool.* **14**, 267–271.
Guttman, N. H., and Wallace, F. G. (1964). *In* "Biochemistry and Physiology of Protozoa" (S. H. Hutner, ed.), Vol. 3, pp. 459–494. Academic Press, New York.
Hakala, M. T. (1965). *Biochim. Biophys. Acta* **102**, 210–225.
Halevy, S. (1966). *Exptl. Parasitol.* **18**, 296–300.
Hamamura, Y., Kuwata, K., and Masuda, H. (1966). *Nature* **212**, 1386–1387.
Happold, F. C., and Stephenson, D. (1935). *Parasitology* **27**, 383–393.
Hartwell, L. H. (1967). *J. Bacteriol.* **93**, 1662–1670.
Hawking, F. (1963). *Exptl. Chemotherapy* **1**, pp. 1–24.
Heckly, R. J., Dimmick, R. L., and Windle, J. J. (1963). *J. Bacteriol.* **85**, 961–966.
Heinemann, B., and Howard, A. J. (1965). *Antimicrobial Agents Chemotherapy*, pp. 126–130.
Heller, C. S., and Sevag, M. G. (1966). *Appl. Microbiol.* **14**, 879–885.
Herrmann, E. C., Jr. (1965). *Ann. N. Y. Acad. Sci.* **130**, 318–320.
Hitchings, G. H. (1961). *Trans. N. Y. Acad. Sci.* [2] **23**, 700–708.
Hitchings, G. H. (1962). *In* "Drugs, Parasites and Hosts" (L. G. Goodwin and R. H. Nimmo-Smith eds.), pp. 196–210. Churchill, London.
Hoare, C. A. (1954). *J. Protozool.* **1**, 28–33.
Hobby, G. L., ed. (1967) *Antimicrobial Agents Chemotherapy*, pp. 1–1138.
Hueper, W. C. (1967). *In* "Potential Carcinogenic Hazards from Drugs" (R. Truhaut, ed.), pp. 79–104. Springer, Berlin.
Hurst, E. W., and Paget, G. E. (1963). *Brit. J. Dermatol.* **75**, 105–112.
Hutchison, D. (1963). *Advan. Cancer Res.* **7**, 235–250.
Hutner, S. H., and Provasoli, L. (1965). *Ann. Rev. Physiol.* **27**, 19–50.
Hutner, S. H., Baker, H., Aaronson, S., Nathan, H. A., Rodriguez, E., Lockwood, S., Sanders, M., and Petersen, R. A. (1957). *J. Protozool.* **4**, 259–269.
Hutner, S. H., Aaronson, S., Nathan, H. A., Baker, H., Scher, S., and Cury, A. (1958). *In* "Trace Elements" (C. A. Lamb, O. G. Bentley, and J. M. Beattie, eds.), pp. 47–65. Academic Press, New York.
Hutner, S. H., Nathan, H. A., and Baker, H. (1959) *Vitamins Hormones* **17**, 1–52.
Hutner, S. H., Provasoli, L., and Baker, H. (1960). *Microchem. J., Symp. Ser.* **1**, 95–113.
Hutner, S. H., Cox, D., and Zahalsky, A. C. (1965). *Progr. Protozool., 2nd Intern. Conf. Protozool., London, 1965* Intern. Congr. Ser. No. 91, pp. 100–101.
Hutner, S. H., Zahalsky, A. C., Aaronson, S., Baker, H., and Frank, O. (1966). *Methods Cell Physiol.* Vol. 2, 217–228.
Hutner, S. H., Zahalsky, A. C., Aaronson, S., and Smillie, R. M. (1967). *In* "Biochemistry of Chloroplasts" (T. W. Goodwin, ed.), Vol. 2, pp. 703–720. Academic Press, New York.
Hutner, S. H., Zahalsky, A. C., and Aaronson, S. (1968). *In* "The Biology of *Euglena*" (D. E. Buetow, ed.), Vol. II. Academic Press, New York (in press).
Iwasaki, H. (1967). *J. Phycol.* **3**, 30–34.
Jaffe, J. J. (1965). *Biochem. Pharmacol.* **14**, 1867–1881.
Jaffe, J. J. (1967). *Mol. Pharmacol.* **3**, 359–369.

Janakidevi, K., Dewey, V. C., and Kidder, G. W. (1966a). *Arch. Biochem. Biophys.* **113**, 758–759.

Janakidevi, K., Dewey, V. C., and Kidder, G. W. (1966b). *J. Biol. Chem.* **241**, 2576–2578.

Jao, R. L. (1967). *Am. J. Med. Sci.* **253**, 639–649.

Jones, B. M. (1966). *In* "Cells and Tissues in Culture" (E. N. Willmer, ed.), Vol. 3, pp. 397–457. Academic Press, New York.

Jukes, T. H., and Broquist, H. P. (1963). *Metab. Inhibitors: Comprehensive Treatise* **2**, 481–534. Academic Press, New York.

Kahan, D., Hutner, S. H., Zahalsky, A. C., and Bacchi, C. J. (1967). *J. Protozool.* **14**, Suppl., 18.

Kahan, D., Zahalsky, A. C., and Hutner, S. H. (1967a). Submitted for publication.

Karrer, P. W., and Greenberg, J. (1964). *J. Bacteriol.* **87**, 536–542.

Kasten, F. H. (1967). *Intern. Rev. Cytol.* **21**, 141–202.

Kato, M., and Yamada, H. (1966). *Life Sci.* **5**, 717–722.

Kaufman, S. (1967). *Ann. Rev. Biochem.* **36**, 171–184.

Kavanagh, F. (1963). *Anal. Microbiol.* pp. 249–259.

Kidder, G. W., and Dutta, B. N. (1958). *J. Gen. Microbiol.* **18**, 621–638.

Kisliuk, R. L., Friedkin, M., Schmidt, L. H., and Rosson, R. N. (1967). *Science* **156**, 1616–1617.

Kitagawa, M., Tanigaki, N., Yagi, Y., Planinsek, J., and Pressman, D. (1966). *Cancer Res.* **26**, Part 1, 752–756.

Krassner, S. M. (1965). *J. Protozool.* **12**, 73–78.

Krassner, S. M. (1966). *J. Protozool.* **13**, 286–290.

Krassner, S. M. (1967). *J. Protozool.* **14**, Supp., 20.

Kuemmerle, H. P., and Preziosi, P. (1964). *Proc. 3rd Intern. Congr. Chermotherapy, Stuttgart, W. Germany, 1963*, Vol. II. Thieme, Stuttgart.

Kusel, J. P., and Weber, M. M. (1965). *Biochim. Biophys. Acta* **98**, 632–639.

Kusel, J. P., Moore, K. E., and Weber, M. M. (1967). *J. Protozool.* **14**, 283–296.

Layton, H. W., Kemp, G. A., and Gale, G. O. (1966). *Proc. Soc. Ind. Microbiol.* p. 13.

Legator, M. (1966). *Bacteriol. Rev.* **30**, 471–477.

LePecq, J.-B., and Paoletti, C. (1967). *J. Mol. Biol.* **27**, 87–106.

Lindegren, C. C. (1959). *Nature* **184**, 397–400.

Lindegren, C. C. (1961). *Antimicrobial Agents Ann.* pp. 520–525.

Linnane, A. W. (1965). *In* "Oxidases and Related Redox Systems" (T. E. King, H. S. Mason, and M. Morrison, eds.), Vol. II, pp. 1102–1128. Wiley, New York.

McCalla, D. R. (1965). *Can. J. Microbiol.* **11**, 185–191.

McGhee, R. B., and Hanson, W. L. (1964). *J. Protozool.* **11**, 555–562.

McGhee, R. B., Hanson, W. L., and DeBoe, J. (1967). *J. Protozool.* **14**, Suppl., 9–10.

McLoughlin, D. K. (1967). *Develop. Ind. Microbiol.* **8**, 151–156.

Malkin, M. F., and Zahalsky, A. C. (1966). *Science* **154**, 1665–1667.

Martin, G. J., and Fisher, C. V. (1944). *J. Lab. Clin. Med.* **29**, 383–389.

Mattingly, P. F. (1965). *In* "Evolution of Parasites" (A. E. R. Taylor, ed.), pp. 29–45. Blackwell, Oxford.

Meyer, H., and Holz, G. G., Jr. (1966). *J. Biol. Chem.* **241**, 5000–5007.

Michelson, A. M., Massoulié, J., and Guschlbauer, W. (1967). *Progr. Nucleic Acid Res. Mol. Biol.* **6**, 83–141.

Milder, R., and Deane, M. P. (1967). *J. Protozool.* **14**, 65–72.

Monro, R. E., Maden, B. E. H., and Traut, R. R. (1967). *In* "Genetic Elements,

Properties and Function" (D. Shugar, ed.), pp. 179–203. Academic Press, New York.

Moser, R. H. (1964). "Diseases of Medical Progress." Thomas, Springfield, Illinois.

Nagai, S., Yanagishima, N., and Nagai, H. (1961). *Bacteriol. Rev.* **25**, 404–426.

Nagata, C., Kodama, H., Tagashira, Y., and Imamura, A. (1966). *Biopolymers* **4**, 409–427.

Nakamura, H. (1965). *J. Bacteriol.* **90**, 8–14.

Nathan, H. A., and Cowperthwaite, J. (1954). *Proc. Soc. Exptl. Biol. Med.* **85**, 117–119.

Nation, J. L., and Robinson, F. A. (1966). *Science* **152**, 1765–1766.

Neipp, L. (1964). *Exptl. Chemotherapy* **2**, 169–248.

Newton, B. A., and Gutteridge, W. E. (1967). *J. Protozool.* **14**, Suppl., 41.

O'Brien, R. D., Cheung, L., and Kimmel, E. C. (1965). *J. Insect Physiol.* **11**, 1241–1246.

O'Brien, R. L., Allison, J. L., and Hahn, F. E. (1966a). *Biochim. Biophys. Acta* **129**, 622–624.

O'Brien, R. L., Olenick, J. G., and Hahn, F. E. (1966b). *Proc. Natl. Acad. Sci. U. S.* **55**, 1511–1517.

O'Connell, K. M., Hutner, S. H., Frank, O., and Baker, H. (1967). *J. Protozool.* **14**, Suppl. 10.

Ormerod, W. E. (1967). *J. Invert. Pathol.* **9**, 247–255.

Parks, L. W., and Starr, P. R. (1963). *J. Cellular Comp. Physiol.* **61**, 61–65.

Paul, J. (1967). *In* "Problems of *in vitro* Culture" (A. E. R. Taylor, ed.), pp. 71–80. Blackwell, Oxford.

Paul, J. S., Reynolds, R. C., and Montgomery, P. O. (1967). *Abstr. 12th Ann. Meeting Biophys. Soc.* p. 84.

Pittam, M. D. (1966). *Proc. 1st Int. Congr. Parasitol., Rome, 1964* Vol. 1, pp. 330–331. Pergamon Press, Oxford.

Porterfield, J. S. (1967). *In* "Methods in Virology," (K. Maramorosch and H. Koprowski, eds.), Vol. 1, pp. 525–536. Academic Press, New York.

Price, K. E., Buck, R. E., Schlein, A., and Siminoff, P. (1962). *Cancer Res.* **22**, 885–591.

Price, K. E., Buck, R. E., and Lein, J. (1965). *Antimicrobial Agents Chemotherapy* pp. 505–517.

Racker, E. (1965). "Mechanisms of Bioenergetics." Academic Press, New York.

Rahman, S. B., Semar, J. B., and Perlman, D. (1967). *Appl. Microbiol.* **15**, 970.

Reich, E. (1966). *Symp. Soc. Gen. Microbiol.* **16**, 266–280.

Rembold, H. (1964). *Z. Physiol. Chem.* **339**, 258–259.

Reynolds, E. H. (1967). *Lancet* **I**, 1086–1088.

Richmond, M. H. (1966). *Symp. Soc. Gen. Microbiol.* **16**, 301–335.

Robinson, F. A. (1966). "The Vitamin Co-Factors of Enzyme Systems," pp. 226–233. Pergamon Press, Oxford.

Roe, F. J. C., and Ambrose, E. J. (1966). *In* "The Biology of Cancer" (E. J. Ambrose and F. J. C. Roe, eds.), pp. 223–229. Van Nostrand, Princeton, New Jersey.

Rollo, I. M. (1964). *In* "Biochemistry and Physiology of Protozoa" (S. H. Hutner, ed.), Vol. 3, pp. 525–561. Academic Press, New York.

Rosenoer, V. M. (1966). *Exptl. Chemotherapy* **4**, 9–77.

Rosenoer, V. M., and Jacobson, W. (1966). *In* "Cells and Tissues in Culture" (E. N. Willmer, ed.), Vol. 3, pp. 351–396. Academic Press, New York.

Ryley, F. (1966). *Proc. 1st Intern. Congr. Parasitol., Rome, 1964* Vol. 1, pp. 41–42. Pergamon Press, Oxford.

Sacktor, B., and Childress, C. C. (1967). *Arch. Biochem. Biophys.* **120**, 583–588.

Sacktor, B., and Dick, A. R. (1965). *Biochim. Biophys. Acta* **99**, 546–549.

Savage, M. (1964). "In Vivo." Simon & Schuster, New York.

Schabel, F. M., Jr., and Pittillo, R. F. (1961). *Advan. Appl. Microbiol.* **3**, 223–256.

Schatz, G. (1965). *Biochim Biophys. Acta* **96**, 342–345.

Schnitzer, R. J. (1966). *Trans. N. Y. Acad. Sci.* [2] **28**, 923–934.

Schnitzer, R. J., and Hawking, F. (1963–1966). *Exptl. Chemotherapy* **1–4**.

Schoental, R. (1967). *In* "Potential Carcinogenic Hazards from Drugs" (R. Truhaut, ed.) pp. 152–161. Springer, Berlin.

Schwartz, J. B. (1961). *J. Protozool.* **8**, 9–12.

Shive, W. (1950). *Ann. N. Y. Acad. Sci.* **52**, 1212–2234.

Shute, P. G. (1965). *Lancet* **II**, 1232–1234.

Silver, S. D. (1966). *Exptl. Chermotherapy* **4**, 505–511.

Simard, R. (1966). *Cancer Res.* **26**, 2316–1328.

Simard, R., and Bernhard, W. (1966). *Intern. J. Cancer* **1**, 463–479.

Simpson, G. G. (1967). *Am. Scholar* 363–377.

Slonimsky, P. P., Acher, R., Péré, G., Sels, A., and Somlo, M. (1965). *Colloq. Intern. Centre Natl. Reche. Sci.* (*Paris*) **124**, 435–479.

Slotnick, I. J., Dougherty, M., and James D. H., Jr. (1966). *Cancer Res.* **26**, Part 1, 673–675.

Smith, C. G. (1966). *In* "Annual Reports in Medicinal Chemistry, 1965" (C. K. Cain, ed.), pp. 267–276. Academic Press, New York.

Smith, C. G., Grady, J. E., and Northam, J. I. (1963). *Cancer Chemotherapy Rept.* **30**, 9–12.

Steinert, M. (1958). *Exptl. Cell Res.* **15**, 431–433.

Steinert, M. (1965). *Progr. Protozool., 2nd Intern. Conf. Protozool., London, 1965* Intern. Congr. Ser. No. 91, pp. 40–41. Excerpta Med. Found. Amsterdam.

Stock, J. A. (1966a). *Exptl. Chemotherapy* **4**, 80–237.

Stock, J. A. (1966b). *In* "The Biology of Cancer" (E. J. Ambrose and F. J. C. Roe, eds.), pp. 176–222. Van Nostrand, Princeton, New Jersey.

Streisinger, G., Okada, Y., Emrich, J., Newton, J., Tsugita, A., Terzaghi, E., and Inouye, M. (1967). *Cold Spring Harbor Symp. Quant. Biol.* **31**, 77–84.

Stuart, K. D., and Hanson, E. D. (1967). *J. Protozool.* **14**, 39–43.

Swim, H. E. (1967). *In* "Lipid Metabolism in Tissue Culture Cells" (G. H. Rothblat and D. Kritchevsky, eds.), Wistar Inst. Symp. Monograph No. 6, pp. 1–14, Discussion 15–16, Wistar Inst. Press, Philadelphia, Pennsylvania.

Symons, R. H., and Ellery, B. W. (1967). *Biochim. Biophys. Acta* **145**, 368–377.

Thimann, K. V., and Radner, B. S. (1958). *Arch. Biochem. Biophys.* **74**, 209–223.

Thompson, P. E. (1967). *Ann. Rev. Pharmacol.* **7**, 77–100.

Thompson, P. E., Olszewski, B. J., Elslager, E. F., and Worth, D. F. (1963). *Am. J. Trop. Med. Hyg.* **12**, 481–503.

Trager, W., and Rudzinska, M. (1964). *J. Protozool.* **11**, 133–145.

Truhaut, R., ed. (1967). "Potential Carcinogenic Hazards from Drugs," pp. 7–27. Springer, Berlin.

Truhaut, R., Festy, B., Le Talaer, J-Y., and Jeanteur, P. (1966). *Compt. Rend.* **262**, 1905–1908.

Venditti, J. M., and Goldin, A. (1964). *Advan. Chemotherapy* **1**, 397–498.

Wagner, R. P., and Mitchell, H. K. (1965). "Genetics and Metabolism," 2nd ed., pp. 290–291. Wiley, New York.

Wallace, F. G. (1966a). *Exptl. Parasitol.* **18**, 124–193.

Wallace, F. G. (1966b). *J. Protozool.* **13**, Suppl. 23.

Waring, M. J. (1966). *Symp. Soc. Gen. Microbiol.* **16,** 235–265.

Warr, J. R., and Roper, J. A. (1965). *J. Gen. Microbiol.* **40,** 273–281.

Warwick, G. B., and Roberts, J. J. (1967). *Nature* **213,** 1206–1207.

Waymouth, C. (1965). *In* "Tissue Culture," (C. V. Ramakrishnan, ed.), pp. 168–179. Junk, The Hague.

Weinman, D. (1953). *Ann. N. Y. Acad. Sci.* **56,** 995.

Weisburger, J. H., and Weisburger, E. K. (1967). *In* "Methods in Cancer Research" (H. Busch, ed.), Vol. 1, pp. 307–398. Academic Press, New York.

West, R. A., Jr., Barbera, P. W., Kolar, J. R., and Murrell, C. B. (1962). *J. Protozool.* **9,** 65–73.

Williams, A. E., Keppie, J., and Smith, H. (1964). *J. Gen. Microbiol.* **37,** 285–292.

Williamson, J. (1966). *Trans. Roy Soc. Trop. Med. Hyg.* **61,** 8–9.

Wilson, A. C., and Pardee, A. B. (1962). *J. Gen. Microbiol.* **28,** 283–363.

Woodruff, H. B. (1966). *Symp. Soc. Gen. Microbiol.* **16,** 22–46.

Woodruff, H. B., and McDaniel, L. E. (1958). *Symp. Soc. Gen. Microbiol.* **8,** 29–48.

Zahalsky, A. C., and Marcus, S. L. (1965). *Nature* **208,** 296.

Zahalsky, A. C., Keane, M. M., Hutner, S. H., Lubart, K. J., Kittrell, M., and Amsterdam, D. (1963). *J. Protozool.* **10,** 421–428.

Zeleznick, L. D., and Sweeney, C. M. (1967). *Arch. Biochem. Biophys.* **120,** 292–295.

10

*The Fine Structure**

MARIA A. RUDZINSKA† AND KEITH VICKERMAN

I. INTRODUCTION

A. SIMILARITIES IN FINE STRUCTURE OF PROTOZOA AND CELLS OF HIGHER ORGANISMS

Electron microscope studies have revealed that Protozoa possess all major organelles found in cells of higher organisms (Beams and Anderson, 1961; Fauré-Fremiet, 1958; Grimstone, 1961; Pitelka, 1963). A brief review of the structure and function of a few most common organelles will supply evidence for this statement.

Cilia and flagella were among the first organelles found to have the same basic fine structure in all plants and animals starting with Protozoa and Protophyta up to organisms highest in evolutionary development (Fawcett and Porter, 1954; Fawcett, 1961). The characteristic structural pattern is the presence of eleven longitudinal fibers, two located in the center and nine at the periphery. The central fibers are about 250 Å in diameter and in

* This chapter covers only blood Protozoa. The fine structure of other blood Protista will be discussed in appropriate sections concerned with diseases caused by the latter parasites.

† Supported by a grant (E-1407) from the National Institute of Allergy and Infectious Diseases, United States Public Health Service.

cross-sections are circular. The peripheral fibers are double and in cross-section appear as figures of eight. Each fiber is a microtubule composed of a dense wall surrounding a less dense core and measuring about 250 Å in diameter.

The basal bodies* from which cilia and flagella arise have a similar fine structure. They are short cylinders surrounded by nine double or triple fibers continuous with the nine fibers of the cilium or flagellum. Basal bodies do not have the two central fibers, but may have an assortment of different more or less defined structures. Diversities in some details found in different organisms do not detract from the basic uniformity in structure of this organelle.

It is significant that centrioles have a fine structure very similar to basal bodies. It has been long supposed that these two organelles are homologous. Centrioles are found in most cells, including many Protozoa; they are absent from cells of higher plants. They divide before the nuclear division and are places of attachment for the spindle during mitosis.

Metazoan cells are surrounded by a thin limiting membrane, the plasma membrane. In high-resolution electron micrographs, the plasma membrane appears to be three-layered in the form of two electron-dense lines separated by a light one, each about 25 Å thick. The three-layered structure of the plasma membrane is in agreement with the theoretical lipoprotein molecular model of membranes. According to this model, the protein molecules are on the outside and adhere to the polar groups of the bimolecular lipid layer, forming the two electron dense lines. The nonpolar groups of the bimolecular lipid layer are adjacent to each other and most probably represent the electron-light middle line. This three-layered structure is named the "unit membrane" and is characteristic not only for the plasma membrane but for all cytoplasmic membranes (J. D. Robertson, 1959).

Protozoa are covered by a plasma membrane which has all the characteristics of a unit membrane. Some Protozoa possess one or more additional membranes or other protective layers outside the plasma membrane. Such extraneous coats, although quite common among Protozoa, may be found in cells of higher organisms also, as for instance, mucins on the surface of cells in the gastrointestinal tract, or cellulose forming the cell wall in plant cells. What should be emphasized is that the plasma membrane, which is an essential component of cells in higher animals and plants, is present also in Protozoa and has the structure of a unit membrane.

The ground substance of the cytoplasm contains small, dense particles

---

* Different names have been given to the structure at the base of cilia and flagella, such as basal granule in ciliated epithelium; kinetosome, or basal body in ciliates; blepharoplast in certain flagellates. Since they all have the same structure and are homologous, it has been rightly proposed (Fawcett, 1961) to use for all of them the descriptive term *basal body*.

about 100–200 Å in diameter. They are either loosely scattered or form rosettes and other configurations or are attached to membranes of the endoplasmic reticulum. They were first identified by Palade and Siekevitz in a combined electron microscope and biochemical study of liver and pancreatic cells (Palade and Siekevitz, 1956a,b) as composed of ribonucleo-protein and are termed RNP (ribonucleoprotein) particles, ribosomes, or Palade's particles. All three terms are synonymous and will be used inter-changeably in this chapter. The particles attached to membranes are engaged in protein synthesis. Recent studies have shown that among ribosomes not attached to membranes, those aggregated in groups play an important role in protein synthesis. They are named polysomes (Warner et al., 1962) and differ from other RNP particles in that they are intercon-nected by a thin strand representing most probably messenger ribonucleic acid (mRNA). Thus, the RNP particles are a very important component of cells. All types of ribosomes just described are present also in Protozoa. As in cells of higher organisms their number reflects the synthetic activity of the cell.

The endoplasmic reticulum is considered as an interconnected vacuolar system permeating the entire cytoplasm (Palade, 1956; Porter, 1957, 1961) from the plasma membrane to the nucleus, as first found by Porter (Porter and Blum, 1953) in electron micrographs of unsectioned cells in tissue culture. Its membranes are either rough or smooth surfaced; in the former, Palade's particles are attached to the membranes. The endoplasmic reticulum appears as vesicles, canaliculi, or flat, elongate cisternae and the form, distribution and amount vary, depending on the type of cell. It serves most probably as a circulatory system for transport and diffusion. It is engaged in protein synthesis, in lipid metabolism, in biosynthesis of steroid hormones, and probably in a number of other functions. The above described functions and structural features of the endoplasmic reticulum were first described in metazoan cells, and all its varieties were found also in Protozoa (Pitelka, 1963). There is no doubt that their function is similar.

The Golgi apparatus is composed of smooth-surfaced piles of flat sacs, vesicles, and small vacuoles. When the organelle is composed exclusively of flattened sacs with small vesicles budding off from the edges, forming a disc, it is known as a dictyosome, and as a parabasal body when connected with a centriole. All the forms were found in Protozoa and only the first two in other organisms. The Golgi apparatus is involved in secretion, transport, absorption, and most probably in a number of other functions.

Mitochondria vary in size and shape. They measure from 0.5–1 μ in shorter diameter and are in the form of ovoid or spherical bodies, short rods, or long filaments up to 7 μ long. They are easy to identify in electron micro-graphs due to their characteristic fine structure. They are surrounded by two unit membranes about 100 Å apart and their intramitochondrial structure is composed of cristae mitochondriales or of microvilli. Both structures derive from the inner membrane by infoldings or tubular projections. The former result in the formation of flat plates, cristae mitochondriales, the latter form fingerlike extensions, microvilli. In cross-sections, cristae look like narrow plates, microvilli have the appearance of circles or ovals. The matrix of the mitochondria varies in density and may contain fine fibrils and small dense particles.

The cristae-type mitochondria are found in metazoan and plant cells. They are also present in some Protozoa (flagellates). Mitochondria with microvilli prevail in Protozoa but were found also in cells of higher organisms (Sedar and Rudzinska, 1956).

From biochemical work on isolated mammalian mitochondria (Lehninger, 1964; Green et al., 1963) it is known that they are the site of oxidative phosphorylation. Enzymes of the respiratory chain constituting the main energy-transforming system of mitochondria are attached or built into the mitochondrial membranes in an orderly fashion, organized in compact assemblies. The mitochondrial matrix contains all the soluble enzymes and coenzymes involved in the Krebs cycle and fatty acid oxidation. Biochemical studies with isolated protozoan mitochondria (Klein and Neff, 1960; Hogg and Kornberg, 1963), although only just beginning, indicate that their function is similar to mammalian mitochondria. Both metazoan and protozoan mitochondria stain with Janus green.

The nucleus in cells of higher organisms and in Protozoa is encased in a nuclear envelope composed of two membranes, each 70–80 Å thick. The outer membrane is covered by RNP particles, is in places continuous with the endoplasmic reticulum, and is actually regarded as part of the vacuolar system (Watson, 1955). At intervals the outer and inner nuclear membranes join, forming pores. More recent studies indicate that the pores are plugged and thus free communication between nucleus and cytoplasm is rather questionable (Maggio et al., 1963). The matrix of the nucleus contains small granules and fine fibrils. The fine structure of chromosomes is rather unsatisfactory in cells of higher organisms and in Protozoa, and it is possible to detect only dense masses composed of tightly packed granules and fibrils representing the chromatin material. Nucleolei are usually in the form of separate spheroidal bodies exceeding in density the chromatin material and composed also of granules and fibrils.

The fine structure of the mitotic spindle has only recently been identified by electron microscopy in metazoan, plant, and protozoan cells (de Harven and Bernhard, 1956; Roth and Daniels, 1962; Ledbetter and Porter, 1963; Porter et al., 1964; Porter, 1966; Aikawa, 1966; Hepler et al., 1966). It is composed of microtubules which are very similar in structure to those in cilia and flagella. It was found during the last few years that microtubules have a wide distribution (particularly in Protozoa) and actually can be considered as a common component of the cell either as a permanent or a transient structure. (For more information on the structure, function, and distribution of microtubules read the excellent article by Porter, 1966.)

The few illustrative examples presented above show convincingly that Protozoa possess all the major organelles found in cells of higher organisms and that the basic structure of these organelles is very similar or almost the same in both groups. Thus from the ultrastructural point of view, Protozoa are cells. This is an important conclusion. It permits not only to bring to an end the long-lasting controversy as to the cellularity versus acellularity of Protozoa (Hyman, 1940, 1959) but leads also to a further conclusion that Protozoa are one-celled organisms. This combination of terms has been regarded as conflicting, and was finally overcome by Fauré-Fremiet (1957) and Grimstone (1961). The latter author approached the problem from the logical point of view and convincingly demonstrated that "cells" and "organisms" represent two different classes which are not "mutually exclusive."

The fact that Protozoa are one-celled organisms has far-reaching consequences. Functions which in higher organisms are distributed among specialized tissues have to be performed in and by a single cell in a protozoan. This explains the surprising complexity and richness of submicroscopic structures in Protozoa which not infrequently far surpasses those in cells of higher organisms. This applies first of all to fibrillar components which can be described as unique in their multitude of combinations and arrangements (for examples, see Pitelka, 1963).

B. ORGANELLES IN FREE-LIVING AND PARASITIC PROTOZOA

To find differences in the fine structure between free-living and parasitic forms is rather a difficult task, since the number, distribution, and level of organization of organelles vary from species to species and depend also on the physiological state of the organism and on environmental conditions. The organelle most sensitive to changes appears to be the mitochondrion. In the free-living protozoan *Pelomyxa illinoisensis* drastic changes occur in the fine structure of mitochondria during mitosis (Daniels and Breyer, 1965). Most interphase mitochondria have a poor internal structure composed of short peripheral microvilli; a large central area remains structureless. However, during prophase and during part of metaphase the intra-mitochondrial space becomes packed with microvilli which acquire a complicated zigzag pattern. In *Tokophrya infusionum*, a free-living suctorian, the number of mitochondria and microvilli decreases in old as well as in overfed organisms (Rudzinska, 1961, 1962). Changes in the mitochondrial system due to age were reported also in *Tetrahymena pyriformis* (Elliot and Bak, 1964). Experiments with yeast cells have shown that this organism can be deprived of true mitochondria when grown in anaerobic conditions (fermentation), but develops this organelle when transferred to aerobiosis (Linnane *et al.*, 1962; Yotsuyanagi, 1962). In yeast this is a reversible process.

Parasitic anaerobic Protozoa such as *Entamoeba histolytica*, (Miller *et al.*, 1961), *Entamoeba invadens* (Siddiqui and Rudzinska, 1965), *Giardia muris* (Friend, 1966), and most of the intestinal flagellates of termites (Grimstone, 1961; Grassé, 1956) do not possess mitochondria. This probably applies to all Protozoa living in an anaerobic environment since they presumably do not carry out oxidative phosphorylation. Thus, the presence and absence of mitochondria is closely related to aerobiosis and anaerobiosis, respectively.

In free-living ciliates such as *Tetrahymena pyriformis* (Sedar and Rudzinska, 1956), *Paramecium multimicronucleatum* (Sedar and Porter, 1955), *Tokophrya infusionum* (Rudzinska, 1961), *Colpoda maupasi* (Rudzinska *et al.*, 1966a), mitochondria are numerous, near the periphery of the body and also in other parts of the cell. The intramitochondrial space is rich in microvilli.

Mitochondria of blood protozoa are of special interest since they give the opportunity to study this organelle in parasites with a highly complicated life cycle including a vertebrate and invertebrate host as well as forms parasitic inside cells. Interesting differences were reported in the mitochondrial system between mammalian and bird intraerythrocytic parasites (Rudzinska *et al.*, 1965) and will be discussed in Section II,A,3 of this chapter. Excellent examples of the relationship between the structure of mitochondria and the environmental conditions created by the host are provided by the comparative studies of bloodstream and midgut forms of trypanosomes (Vickerman, 1962; see Section III,A,3,b).

All types of the endoplasmic reticulum are present in free-living and in most parasitic forms. In *Amoeba invadens* only vesicles were found (Siddiqui and Rudzinska, 1965). It seems that in most parasitic protozoa so far studied the endoplasmic reticulum is rather poorly represented, although extensive development of this organelle may take place during certain stages of the life cycle (see Sections II,A,4 and II,B,1).

The Golgi apparatus is well represented in certain intestinal flagellates (Grimstone, 1959) and in trypanosomes (see Section III,A,4,c). However, among the majority of parasitic protozoa one finds a very primitive type of Golgi apparatus composed of small vesicles aligned in rows, but this is true also of most free-living forms. The anaerobic *Amoeba invadens* (Siddiqui and Rudzinska, 1965) and, what is of great interest, most ciliates, do not possess a Golgi apparatus. Thus this organelle does not supply any clear difference between free-living and parasitic protozoa.

The same could be said about cilia, flagella, and all fibrillar structures. They are richly represented in free-living and parasitic forms of ciliates, often achieving a high degree of organization and complexity in both groups.

II. PROTOZOA INVADING BLOOD CELLS

Intracellular blood Protozoa belong to the subphylum Sporozoa except *Babesia* (Piroplasma) and *Theileria*, which in accordance with the new nomenclature are localized in the subphylum Sarcomastigophora (The Committee on Taxonomy, 1964). These parasites have a complicated life cycle including a vertebrate and invertebrate host. The alteration of the host is connected with the asexual (in the vertebrate host) and the sexual (in an invertebrate host) reproduction. *Lankesterella* is an exception, for the sexual and asexual reproduction take place in the vertebrate host (Stebbens, 1966a).

In laboratory conditions the invertebrate host can be omitted by transferring the parasite from one vertebrate host to another by blood passage; thus the sexual phase of reproduction is left out. After a number of blood passages, some major changes may occur as infectivity to the invertebrate host is lost.

Sporozoites produced in the invertebrate vector upon entering the vertebrate host do not invade erythrocytes but cells of different organs and tissues, depending on the species; e.g., mammalian malaria parasites invade hepatic cells in the liver; bird plasmodia, cells of the reticuloendothelial system. There they grow and multiply asexually by schizogony. This is the preerythrocytic or exoerythrocytic phase of the infection. After several reproductive cycles merozoites enter the peripheral bloodstream where they infect erythrocytes, starting the intraerythrocytic stage with numerous reproductive cycles and the formation of gametocytes.

The malaria parasite, *Plasmodium* of the family Plasmodidae, the cause of the malarial disease in mammals, birds, and reptiles, has been more extensively studied by means of electron microscopy than any other intraerythrocytic blood protozoan. This is due first of all to the methods of culturing and achieving high degrees of infection in laboratory animals and in tissue culture (Meyer and Xavier de Oliveira, 1947; Xavier de Oliveira and Meyer, 1955; Huff *et al.*, 1960; Davis *et al.*, 1966). It is significant that the first electron microscopy work was done on *Plasmodium knowlesi* (Fulton and Flewett, 1956) and *Plasmodium lophurae* (Rudzinska and Trager, 1957), two species which were used extensively for studies on growth requirements *in vitro* within erythrocytes (Trager, 1943, 1947; Ball *et al.*, 1945; Anfinsen *et al.*, 1946), followed by most difficult studies on the nutritional requirements of *P. lophurae* outside the host cell by Trager (1950, 1955, 1957, 1960, 1964b). A number of species in mammalian and avian hosts as well as in the mosquito vector have been investigated.

Babesia (Piroplasma) and *Theileria*, the cause of redwater fever and East Coast African fever, respectively, parasitize mammals (cattle, sheep, goats, pigs, horses) and birds. Not all developmental stages in the vertebrate host are well known. The disease is transmitted by ticks. Some of the intraerythrocytic stages have been studied by electron microscopy.

Leucocytozoon has a life cycle similar to *Plasmodium*. Asexual reproduction takes place in birds; gametocytes form in red blood cells or lymphocytes. Sexual reproduction occurs in insects. The only forms herein studied by electron microscopy are microgametes.

Hepatocystis, a parasite of lower monkeys, squirrels, bats, and hippopotami, produces gametocytes in erythrocytes. The arthropod host is one or more species of *Culicoides* (Garnham, 1966a). As in *Leucocytozoon*, the fine structure of microgametes only was studied.

Lankesterella, a parasite of birds and frogs, undergoes asexual and sexual reproduction in the vertebrate host. In *Lankesterella hylae* schizogony and sporogony take place in endothelial cells or macrophages. Sporozoites invade erythrocytes or leucocytes (Mackerras and Mackerras, 1961). It is transmitted to another vertebrate host by a blood-sucking invertebrate. Young trophozoites and newly invading merozoites and sporozoites were studied by electron microscopy.

A. Trophozoites in the Intra- and Exoerythrocytic Stage

The fine structure of most trophozoites was studied in the intraerythrocytic stage because, under laboratory conditions, the infection was maintained by blood passage in vertebrate hosts. Under these circumstances the infection is limited to the blood and therefore to the erythrocytes of the host. This is to a certain degree a convenience, since the infection can be easily tested. In addition, the parasites are not difficult to handle for electron microscopy; a few drops of blood from a heavy infection supplies enough material for embedding. The blood can be fixed either in suspension or clotted and handled as a piece of tissue (Rudzinska, 1955).

A successful method for studying exoerythrocytic forms of malaria parasites by electron microscopy has been developed by Meyer and Musacchio (1960). These investigators have been growing exoerythrocytic forms of *Plasmodium gallinaceum* in tissue culture for a number of years. They were able to identify *in vitro* all exoerythrocytic stages as known from *in vivo* observations in infected chickens by Huff and Coulston (1944). The convenience of this method lies in the availability of the material whenever needed and in the possibility of selecting the desired areas for electron microscopy. Recently the tissue culture method has been further developed by Davis *et al.* (1966).

1. Limiting Membrane

All malaria parasites in the intra- and exoerythrocytic stage are surrounded by two unit membranes, closely applied to the cytoplasm of the

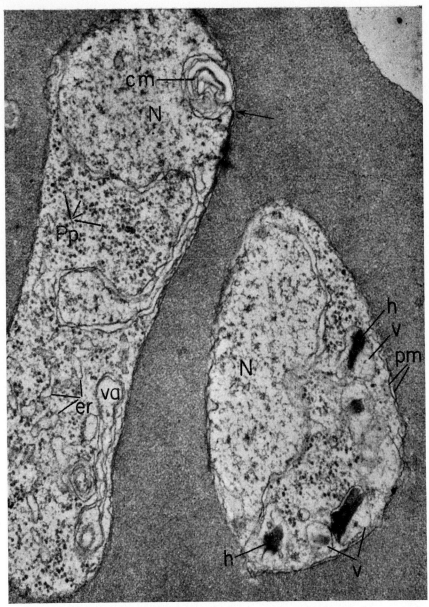

Fig. 1. Electron micrograph of two organisms of *Plasmodium coatneyi* in an erythrocyte of the rhesus monkey. The parasite is surrounded by two plasma membranes (pm). The nucleus (N) is large and of lower density than the cytoplasm. The latter is filled with Palade's particles (Pp). The endoplasmic reticulum (er) is composed of rough-surfaced vesicles. A vacuole surrounded by two membranes (va) lies near the periphery. The membranes of the double concentric structure (cm) show continuity with the double limiting membrane at arrow. Several vesicles (v) contain the pigment hemozoin (h). Magnification: × 47,800. (From Rudzinska *et al.*, 1965.)

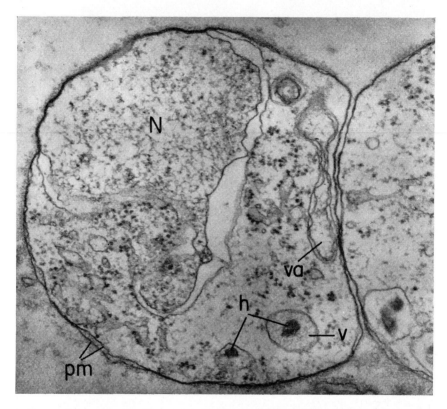

FIG. 2. *Plasmodium coatneyi* outside the host cell. The double membrane surrounding the parasite is clearly seen at pm. For key to symbols see Fig. 1. Magnification: × 48,000. (From Rudzinska and Trager, 1968.)

host (Figs. 1, 4, 7A, 18). Evidence that the two membranes belong to the parasite is supplied by electron micrographs of trophozoites of *Plasmodium coatneyi* (Rudzinska and Trager, 1968) which happen to be outside the host cell (Fig. 2) and of *P. berghei* (Ladda *et al.*, 1965) representing "migrating" trophozoites with a part of the parasite outside the erythrocyte. The electron micrographs show clearly two plasma membranes in trophozoites outside the host cell as well as around the extruded portion of the trophozoite. Some investigators believe that the outer membrane of the parasite is formed by the host cell (Aikawa, 1966; Hepler *et al.*, 1966).

The thinness of the membranes and the intimate contact with the host's cytoplasm favor a mutual exchange of materials between host and parasite. In the intraerythrocytic stage the spacing between the membranes varies from place to place, ranging from 60 to 170 Å. Usually the membranes appear as wavy lines, touching each other in places or forming loops (Fig. 1).

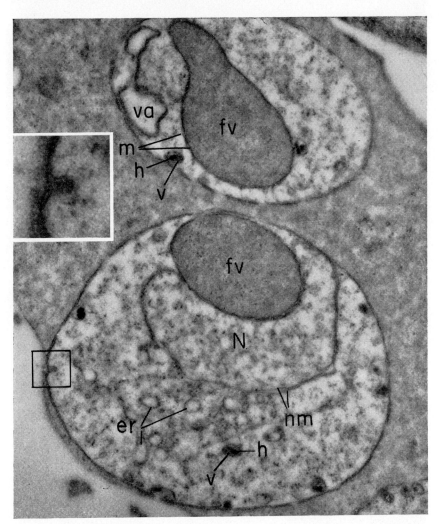

Fig. 3. Two parasites of *Plasmodium berghei* in a reticulocyte of the rat containing large food vacuoles (fv) surrounded by two membranes (m). In the lower parasite the nucleus (N), encased in a double membrane (nm) lies around the food vacuole and has a cuplike shape. Several tiny vesicles (v) contain the pigment hemozoin (h). The endoplasmic reticulum (er) is represented by rough-surfaced vesicles. In the upper parasite a vacuole (va) surrounded by two membranes is present. Magnification: × 36,000. An invagination of the double limiting membrane in the outlined area is shown at higher magnification (100,000) in the insert. (From Rudzinska and Trager, 1959.)

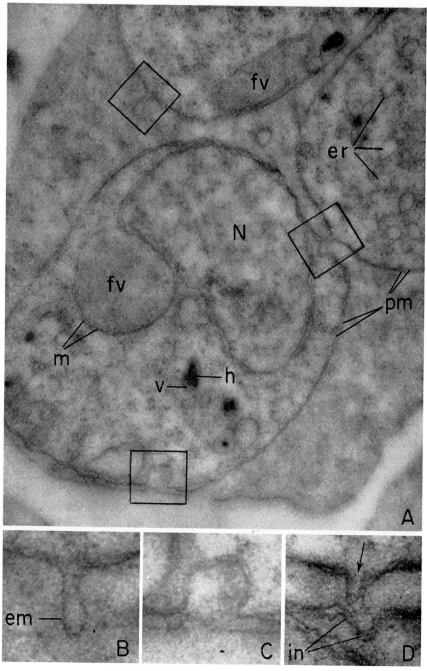

Fɪɢ. 4. Part of rat reticulocyte infected with three parasites of *P. berghei*. A: Each of the parasites is surrounded by two membranes (pm) and two membranes (m) are around food vacuoles (fv). For key to additional symbols see Fig. 3 legend. Magnification: × 40,000. The three outlined areas are shown at higher magnification (× 100,000) in B, C, and D. B: The external membrane (em) forms an evagination. C: Both membranes invaginate. D: Invagination (in) and evagination (arrow) of both membranes can be seen. (From Rudzinska and Trager, 1959.)

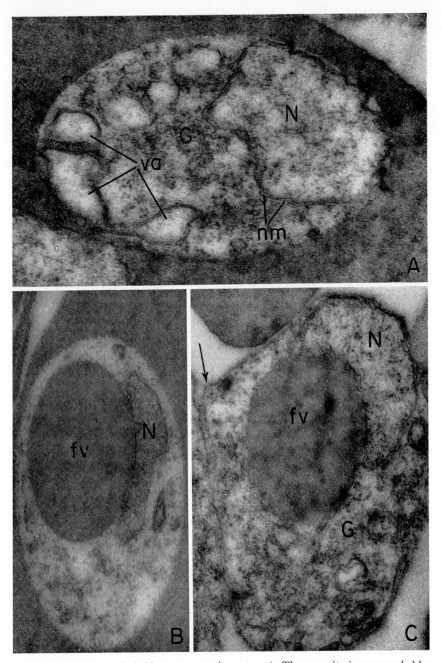

FIG. 5. *Babesia rodhaini* in mouse erythrocytes. A: The parasite is surrounded by a single membrane seen at arrow. The nucleus (N) is of low density and encased in a double membrane (nm). Nearby the Golgi apparatus (G) can be seen and at the periphery of the organism vacuoles (va) surrounded by two membranes. Magnification: × 50,000. B: *Babesia* containing a large food vacuole (fv). Magnification: × 30,000. C: *Babesia* outside the host cell. A large food vacuole (fv) deriving from the host cell cytoplasm indicates that the parasite was previously inside a red blood cell. The single membrane surrounding the parasite is clearly seen at arrow. Magnification: × 40,000. (From Rudzinska and Trager, 1962.)

In *Plasmodium berghei* in rat reticulocytes invaginations of both membranes into the cytoplasm of the parasite can be seen (Figs. 3, 4A,C,D). Occasionally papillalike evaginations of both (Fig. 4A and D) or only of the external membrane (Fig. 4A and B) can be found. Similar evaginations of the membrane were found in *Lankesterella hylae* (Figs. 15A, 16A). These observations suggest that the plasma membrane is an active and flexible structure capable of changes. This feature of the plasma membrane is of particular importance and will be stressed later.

Lankesterella garnhami (Garnham *et al.*, 1962b) and *Lankesterella hylae* (Stebbens, 1966b) are covered by two membranes in the intra- and exo-erythrocytic stage (Figs. 15A, 17). Between the outer membrane of the parasite and the host cell is a large space, a vacuole surrounding the parasite and lined by a single membrane (Figs. 15A, 17). In this respect *Lankesterella* resembles *Toxoplasma* and *Sarcocystis* (Ludvik, 1963) which are also separated from the host cell by a vacuole. In all these genera the periparasitic vacuole is lined by a single membrane.

Babesia rodhaini (Rudzinska and Trager, 1962) is covered by a single plasma membrane (Fig. 5), and there are some suggestions that *B. canis* and *B. caballi* are surrounded by two unit membranes (C. F. Simpson *et al.*, 1963). Previous electron microscope studies with *B. canis* (Bayer and Dennig, 1961) found a single unit membrane around the parasite. It would be worthwhile to reinvestigate the number of limiting membranes in *Babesia*. *Theileria parva* (Jarrett and Brocklesby, 1966) also seems to be covered by a single membrane, however *Theileria* sp. (Kreier *et al*, 1962) by two membranes.

2. Mechanism of Food Intake: Pinocytosis, Cytostome, Food Vacuoles

The feeding mechanism is of particular interest with intracellular parasites and especially with those where the entire body surface lies in intimate contact with the cytoplasm of the host cell. Even in this close relationship a separation of host and parasites remains—as shown above—through the two (or less often one) unit membranes surrounding the parasite. Although these membranes are very thin (60–100 Å each) and closely applied to the cytoplasm of the host, they are, nevertheless, a barrier which restricts the passage of intact host cytoplasm. It was generally believed that intracellular parasites prepare their food for diffusion (through the membrane) by hydrolysis or liquefaction of host cytoplasm. The electron microscopy study of intraerythrocytic parasites revealed that diffusion is not the only way an intracellular parasite gets its food supply from its host. It was found that these parasites are able to engulf large droplets of host cytoplasm by pinocytosis or phagotrophy, leading to the formation of food vacuoles.

The organism which first supplied this information was *Plasmodium lophurae* (Rudzinska and Trager, 1957). Electron microscopy studies of intraerythrocytic stages of *P. berghei* (Rudzinska and Trager, 1959; Fulton and Flewett, 1956; Peters *et al.*, 1965; Jerusalem and Heinen, 1965; Ladda and Arnold, 1965; Ladda *et al.*, 1966; Warhurst, 1965), *P. gonderi, P. jalciparum, P. ovale* (Rudzinska *et al.*, 1960, 1965; Trager *et al.*, 1967); *P. knowlesi* (Fulton and Flewett, 1956; Fletcher and Maegraith, 1962; Aikawa *et al.*, 1966b), *P. coatneyi* (Rudzinska *et al.*, 1966b; Rudzinska and Trager, 1968), *P. vinckei* (F. E. G. Cox and Vickerman, 1966), *P. gallinaceum* (Ristic and Kreier, 1964; Aikawa *et al.*, 1966b), *P. fallax, P. cathemerium* (Duncan *et al.*, 1959; Aikawa *et al.*, 1966a), *P. cynomolgi* (Aikawa *et al.*, 1966b), *Babesia rodhaini* (Rudzinska and Trager, 1962), and *Theileria mutans* (Büttner, 1966) gave full support to this finding and permitted a general conclusion to be drawn as to the mechanism of food intake in these parasites.

Malaria parasites, *Babesia rodhaini* (Rudzinska and Trager, 1962; Flewett and Fulton, 1959) and *Theileria parva* (Jarrett and Brocklesby, 1966) contain in the intraerythrocytic stage dense bodies of the same general appearance as the host cytoplasm (Figs. 3–5, 7–9, 11, 12, 14). These bodies in fact represent parts of engulfed host cytoplasm. The engulfment of host cytoplasm takes place through invagination of the plasma membrane apparently at any place of the surface (Figs. 8–12) and also at a specialized area, the so-called "cytostome" (Figs. 10, 11, 31). The cytostome* was found by Aikawa *et al.* (1966a) in several species of *Plasmodium* and it appears as a depression of the plasma membrane surrounded by two dense rings.

Evidence as to the origin of these dense bodies from host cytoplasm is provided by stages of their formation as illustrated in Figs. 6–8. An early stage of this process can be seen in Fig. 6 and more advanced stages in Figs. 7 and 8. In Fig. 8A, a small food vacuole is almost completed, but its connection with the cytoplasm of the host still persists in the form of a short narrow neck. In Fig. 8B one of the two food vacuoles appears to be in the last stage of pinching off from the erythrocyte cytoplasm. In all these electron micrographs continuity between the host cytoplasm and the food vacuole may be seen as well as continuity between the double plasma membrane of the parasite and the double membrane of the food vacuole in formation.

More evidence is supplied by electron micrographs showing food vacuoles formed in the vicinity of the nucleus, as seen in Figs. 3, 7A, 9. Apparently when the two bodies come in contact, the nucleus is forced to give up space

* This organelle was first described by Garnham *et al.* (1961) in sporozoites of *Plasmodium* and named "micropyle" and thereafter in *Toxoplasma* (Garnham *et al.*, 1962a) and in *Lankesterella garnhami* (Garnham *et al.*, 1962b).

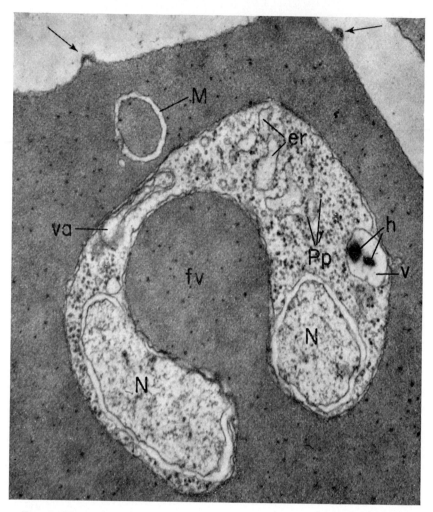

Fig. 6. *Plasmodium coatneyi* in erythrocyte of the rhesus monkey. A large drop of host cytoplasm is being engulfed and the continuity between the membranes surrounding the future food vacuole (fv) and the double plasma membrane is clearly defined. A vacuole (va) surrounded by two membranes lies at the periphery. In the cytoplasm of the host cell a Maurer's cleft (M) can be seen as well as elevations and thickening of the plasma membrane at arrows. For key to additional symbols see the legend for Fig. 1. Magnification: × 47,000. (From Rudzinska *et al.*, 1965.)

for the growing food vacuole and as a result it acquires a cuplike shape (Figs. 3 and 7A). Under the impact of the developing food vacuole the two structures become so closely approximated that fusion of their membranes takes place (Fig. 9). Instead of four membranes (two belonging to the

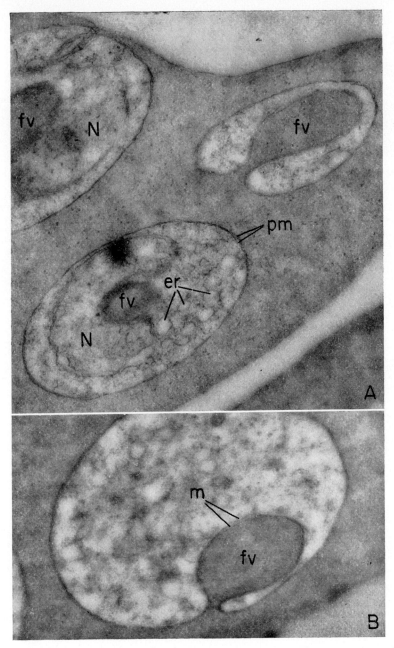

FIG. 7. Stages in food vacuole formation in *P. berghei*. A: Three young parasites in reticulocyte of the rat. In the parasite at the upper right the food vacuole being formed (fv) is in a slightly more advanced stage than in Fig. 6. The nucleus (N) in the lower left has a cuplike shape and surrounds the food vacuole (fv). Magnification: × 36,000. B: Further stage in engulfment of drop of host cytoplasm. For key to additional symbols see the legend for Fig. 1. Magnification: × 36,000. (From Rudzinska and Trager, 1959.)

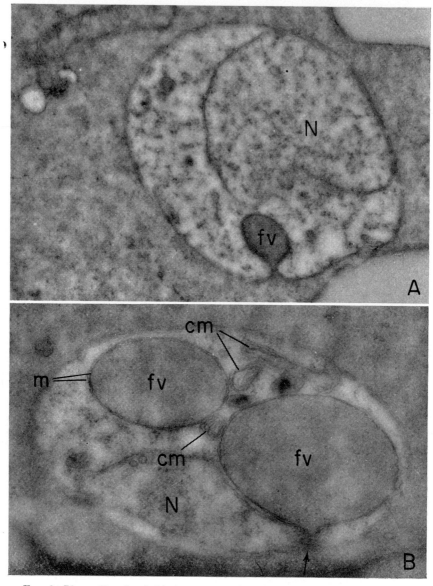

FIG. 8. *Plasmodium berghei* in last stages of food vacuole formation. A: A narrow neck connects the food vacuole being formed (fv) with the cytoplasm of the host. B: The larger food vacuole (fv) at the right is in the last stage of pinching off (arrow) from the cytoplasm of the host. Note the two vesicles containing the pigment hemozoin and two concentric double-membraned structures (cm). N, nucleus; m, membranes. Magnification: × 40,000. (From Rudzinska and Trager, 1959.)

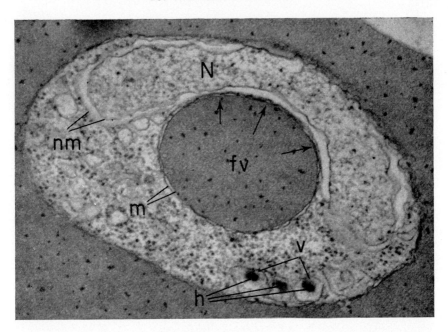

FIG. 9. *Plasmodium coatneyi* with large food vacuole (fv) surrounded by two membranes (m). The nucleus (N) lies in close proximity to the food vacuole. In places where the two bodies touch each other instead of four membranes, two belonging to the nucleus (nm) and two to the food vacuole (m), only three membranes can be seen at arrows. Pigment hemozoin, h. Magnification: × 47,800. (From Rudzinska and Trager, 1968.)

nucleus and two to the food vacuole) only three membranes may be seen, the middle being thicker and darker. Such a fusion of membranes is known as myelination.

Supporting evidence for the origin of food vacuoles from host cytoplasm is supplied by electron micrographs of *Babesia* which happen to be freed from the host cytoplasm, as seen in Fig. 5C. Around this parasite there is no trace of host cytoplasm. However, inside the parasite is a large body which undoubtedly derives from an erythrocyte, the former host. The presence of dense host cytoplasm inside the freed parasite clearly indicates its origin.

Strongest support for the origin of food vacuoles from host cytoplasm comes from the finding of the cytostome by Aikawa *et al.* (1966a). In trophozoites of the intraerythrocytic stage of *Plasmodium*, the cytoplasm of the host is being engulfed also through this organelle (Figs. 10, 11), leading to the formation of food vacuoles which look exactly like those described above. In both instances, the food vacuoles are formed through invagina-

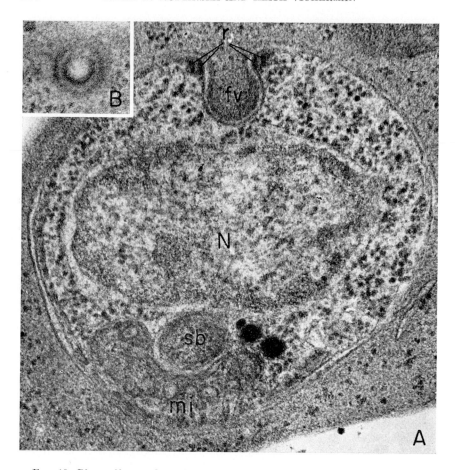

Fig. 10. *Plasmodium cathemerium* and *P. fallax* in erythrocytes. A: *Plasmodium cathemerium* in an erythrocyte of a canary engulfing a droplet of host cytoplasm through the cytostome composed of two thick rings (r), seen at longitudinal section. The food vacuole being formed (fv) is surrounded by two membranes which are continuous with the two membranes covering the parasites. The nucleus (N) is large with chromatin below the double nuclear membrane. A mitochondrion (mi) is associated with a spherical body (sb). Magnification: × 74,000. B: Tangential section through the cytostome of an exoerythrocytic merozoite of *P. fallax* to show the two concentric rings constituting the cytostome. Magnification: × 68,000. (Courtesy of Aikawa *et al.*, 1966a.)

tions of the two plasma membranes and are, therefore, encased in a double membrane which shows the same thickness, density, and spacing as the limiting membranes surrounding the parasite. The same organism is able to form food vacuoles through the cytostome and at any other part of the body, as seen in Fig. 11. The figure shows a parasite with several food vacuoles,

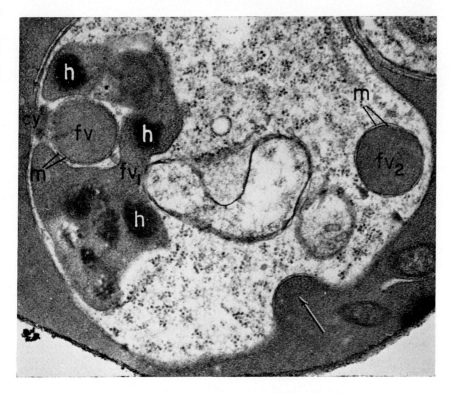

FIG. 11. *Plasmodium lophurae* in erythrocyte of the canary. Note that two newly formed food vacuoles, one (fv) at the cytostome (cy) and the other at the opposite site (fv₂), are surrounded by two membranes (m), whereas those in which digestion has started (fv₁), as seen by the presence of the pigment hemozoin (h), have only a single membrane. At arrow, most probably a food vacuole in its early stage of formation is seen. Magnification: × 36,000. (Courtesy of Aikawa *et al.*, 1966a).

two of them in the process of formation, one at the cytostome, the other on the opposite side of the body. Aikawa *et al.* (1966a,b) believe that food intake occurs only through the cytostome. The above discussion on the origin of food vacuoles (see also Trager, 1966) supplies ample evidence that pinocytosis is not limited to the cytostome but might take place at any region of the plasma membrane.

Finally, from biochemical studies it is well known that the main source of protein for malaria parasites is the hemoglobin of host cytoplasm (Moulder, 1955) and that the residue of its digestion is the pigment, hemozoin (Deegan and Maegraith, 1956; Sherman and Hull, 1960). The pigment was found inside these dense bodies (Figs. 11, 14A and B) leaving no doubt as to their origin or to their function as food vacuoles.

Although the engulfment of host cytoplasm follows the same pattern in all malaria parasites, in *Theileria* and in *Babesia*, the fate of the ingested host cytoplasm may take one of three courses. In all bird malaria parasites so far studied, *P. lophurae* (Rudzinska and Trager, 1957), *P. cathemerium*, *P. fallax* (Aikawa *et al.*, 1966a), and *P. gallinaceum* (Ristic and Kreier, 1964; Aikawa *et al.*, 1966b), digestion of hemoglobin takes place within the food vacuoles, as evidenced by the presence of hemozoin in the form of dense, coarse pigment granules inside the food vacuole (Figs. 11, 14A and B). As digestion proceeds, more pigment granules accumulate and the matrix of food vacuoles becomes less dense. Newly formed food vacuoles have the same density as the cytoplasm of the host cell.

In mammalian malaria parasites (*P. berghei*, *P. ovale*, *P. gonderi*, *P. vivax schwetzi*, *P. coatneyi*, *P. vinckei, and P. knowlesi*) digestion does not occur inside the food vacuole. In these parasites food vacuoles always have a matrix as dense as the host cytoplasm and they do not contain pigment granules (Figs. 3, 4, 8, 9, 12). The pigment, hemozoin, is, however, found in small vesicles. The membrane of some of the vesicles containing the pigment shows continuity with the membrane of the food vacuole, suggesting that the vesicles derive from food vacuoles by a pinching-off process (Fig. 12). There is reason to believe that digestion of hemoglobin takes place in the small vesicles because they are the only place in which hemozoin, the residue of hemoglobin digestion, is to be found. Support for this assumption comes from varying degrees of density of the matrix of the vesicles, most probably representing stages in digestion (Fig. 13). It could, however, be assumed that digestion takes place within the food vacuole proper and that the residue hemozoin is removed from the food vacuole into the vesicles. If this were the case, one would expect to find hemozoin granules inside the food vacuole. It would be also expected that food vacuoles surrounded by many vesicles with pigment granules should have a matrix of lower density than those not surrounded by vesicles. Finally, the matrix of all vesicles with hemozoin should be of the same density. The electron micrographs do not supply evidence for any of the enumerated possibilities.

It seems that the digestive vesicles, once they become detached from the food vacuole, are able to change their position since they are found not only in the near vicinity of the food vacuole proper (Fig. 12B and C) but also scattered throughout the cytoplasm (Figs. 12A, 13).

Pinching off of small vesicles from food vacuoles, first observed and described in *P. berghei* (Rudzinska and Trager, 1959), is by no means a unique phenomenon. It was subsequently described in a number of free-living Protozoa, *Pelomyxa* (Roth, 1960), *Tetrahymena* (Müller and Röhlich, 1961), *Paramecium* (Jurand, 1961; Müller *et al.*, 1963; Schneider, 1964), and in some peritrichous ciliates (Carasso *et al.*, 1964). Of particular interest are combined electron microscope and histochemical studies performed on

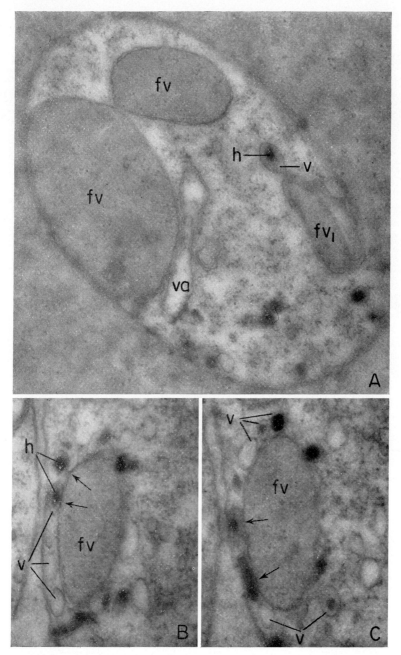

Fig. 12. *Plasmodium berghei* in rat reticulocytes. A: The smallest food vacuole (fv₁) is irregular in shape and is connected with a vesicle (v) containing a grain of hemozoin (h). Magnification: × 36,000. B and C: Two serial sections through the same organism show that the food vacuole (fv) is surrounded by numerous vesicles (v) containing hemozoin (h). Some of the vesicles (at arrows) seem to be continuous with the external membrane of the food vacuole. Magnification: × 51,750. (From Rudzinska and Trager, 1959.)

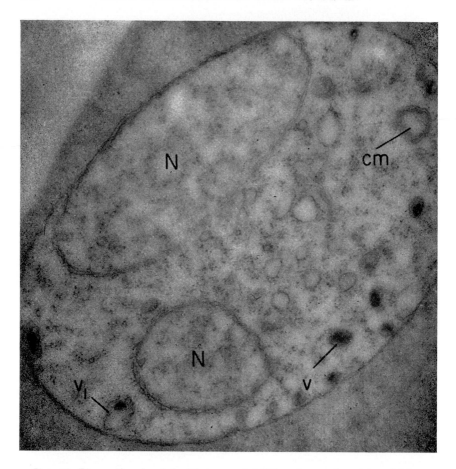

Fig. 13. *Plasmodium berghei* with two nuclei (N) and several vesicles which vary in the density of their matrix. The matrix is much denser in vesicle v than in vesicle v_1. Two concentric double-membraned structures, cm. Magnification: × 40,000. (From Rudzinska and Trager, 1959.)

Paramecium (Müller *et al.*, 1963) and on *Campanella* (Carasso *et al.*, 1964). They show that acid phosphatase activity, the localization of which is revealed by lead deposits, resides mainly in the small vesicles, whereas the food vacuole proper contains comparatively little of the enzyme.

In mammalian plasmodia so far studied, the presence of hemozoin, the residue of hemoglobin digestion, occurs in the small vesicles exclusively, and shows clearly that digestion takes place in these small vesicles only. Hemozoin, the electron-dense pigment, serves here conveniently as a natural marker of the site of hemoglobin digestion.

It is of great interest that the two different mechanisms of digestion, one inside the food vacuole, the other outside it in small vesicles, coincide very closely with the class to which the vertebrate host belongs. In bird malaria parasites the site of digestion is the food vacuole proper; in mammalian malaria parasites digestion takes place in the small vesicles pinched off from the food vacuole proper. There is one peculiar exception involving *P. falciparum*. In this species, in parasitized splenectomized chimpanzees, digestion occurs as in bird malaria (see Fig. 14A) (Rudzinska *et al.*, 1965); however, in man it is similar to that in mammalian hosts (Trager *et al.*, 1967; Ladda *et al.*, 1966). It should be stressed that the electron microscope study of *P. falciparum* in man was done on young parasites, and in chimpanzees on older, although still uninucleated, ones. It cannot be excluded that in this species both modes of digestion are possible, depending on the age of the trophozoite or that the mechanism of digestion depends on the host. It would be worthwhile to investigate this problem more extensively.

Digestion inside the food vacuole versus digestion in vesicles outside the food vacuoles proper is linked with some interesting morphological differences of the food vacuoles. In bird malaria parasites the newly formed food vacuole is surrounded by two membranes (Figs. 10, 11). However, food vacuoles in which digestion started, as seen by the presence of the pigment, have only one membrane (Figs. 11, 14) Apparently one of the membranes breaks down. It is conceivable to assume that the presence of a single membrane is a necessary prerequisite for the entrance of digestive enzymes into the food vacuole.

In mammalian malaria parasites the two membranes covering the food vacuole are permanent structures and they are always present around each food vacuole (Figs. 3, 4, 8, 9, 12). As described above, these food vacuoles never contain hemozoin, the marker of hemoglobin digestion; thus they are not engaged in digestion and apparently cannot be permeated by digestive enzymes. However, the small vesicles which contain the pigment and are obviously the site of digestion are covered by a single membrane.

The two types of food vacuoles with intra- and extravacuolar digestion differ also as to the size of hemozoin granules. In the former, they are large and coarse and measure about $0.5 \times 0.25 \mu$ (Figs. 11, 14). In the latter, they are considerably smaller and the size varies with the species. The granules are very small in *P. berghei*, $0.19 \times 0.1 \mu$ (Figs. 3, 4, 8); in *P. gonderi* and *P. vivax*, they are longer and narrower, $0.3 \times 0.06 \mu$ and $0.4 \times 0.08 \mu$, respectively (Rudzinska *et al.*, 1965). In *P. coatneyi* (Rudzinska *et al.*, 1966b; Rudzinska and Trager, 1968) they are long and narrow ($0.3 \times 0.034 \mu$) in the form of regularly shaped crystals (Figs. 6, 9). In parasites with small hemozoin granules scattered throughout the cytoplasm (as in *P. berghei*) the pigment is often difficult to detect with light optics, and its sudden appearance at segmentation is probably due to the aggregation of all vesicles

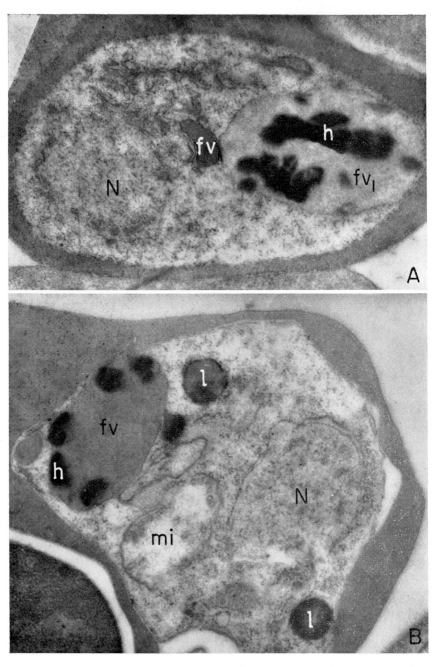

Fig. 14. Food vacuoles in *P. falciparum* and in *P. lophurae*. A: *Plasmodium falciparum* in an erythrocyte of a splenectomized chimpanzee has two food vacuoles. One (fv) is of the same density as the host cytoplasm. The other (fv₁) has a matrix of low

with hemozoin in one area to form the so-called residual body assembling all the waste products in a single vacuole surrounded by a membrane, as found in *P. lophurae* (Rudzinska and Trager, 1961); this will be discussed in more detail in Section II,B,1 of this chapter. It is only recently that by the use of reflex microscopy Westphal (1963a,b) was able to detect the tiny pigment grains in *P. berghei* and to ascertain that digestion of hemoglobin starts almost immediately after invasion.

Babesia represents the third type of digestion of hemoglobin. It feeds on the host cell in the same way as malaria parasites, engulfing large droplets of host cytoplasm through invaginations of the plasma membrane and the formation of food vacuoles (Fig. 5B and C). In *Babesia* and *Theileria* no pigment is found in the food vacuole nor outside it. Apparently the digestive pathway of host cytoplasm is different in this parasite.

In spite of differences in the mechanism of food digestion the manner of feeding is, as stressed before, the same in all malaria parasites, in *Theileria*, and in *Babesia*. In all these parasites, droplets of host cytoplasm are ingested into food vacuoles through invaginations of the parasite's plasma membrane. In this way the entire host cytoplasm becomes gradually incorporated into the body of the parasite, where it is digested.

Ingestion of droplets of fluid from the surrounding medium is a widespread phenomenon, as recently shown by electron microscopy. It became obvious from the latter studies that a great variety of cells, including those of a fixed form, are able to engulf droplets of the surrounding medium through invaginations of the plasma membrane. It appears, therefore, that osmosis is not the only way solutes enter the cell. The first observations on engulfment of fluid were made by Edwards in 1925 on amoebae. Lewis (1931), who introduced the term pinocytosis for it, had forseen its significance and universality. It might well be that the bulk of fluids finds its way to the inside of cells through pinocytosis (Palade, 1956; Trager, 1964a).

Lewis contrasted drinking, characteristic of pinocytosis (from *pinein*, to drink) with eating, characteristic of phagotrophy (from *phagein*, to eat). In the former process fluids are taken in, in the latter more dense materials. This distinction no longer holds. Recent studies show that not the fluid but the substances contained within the fluids are of primary importance for pinocytosis; the density of solutes is of no major significance. The two terms have thus become almost synonymous.

Studies on pinocytosis have shown that it can be induced by salts and certain proteins (Edwards, 1925; Mast and Doyle, 1933; Chapman-

density and contains large masses of hemozoin (h); the latter food vacuole apparently approaches the end of digestion. Magnification: × 34,000. B: The large food vacuole (fv) of *P. lophurae* in an erythrocyte of the duck contains several grains of hemozoin (h). The lipid bodies (l) are characteristic of *P. lophurae* when parasitic in ducks. The mitochondrion (mi) has circular and oval profiles of microvilli. N, nucleus. Magnification: × 28,800. (From Rudzinska *et al.*, 1965.)

Andresen and Holter, 1955; Holter, 1959; Marshall *et al.*, 1959). It is assumed that this process starts with the binding of the inducing substance to the cell surface, the latter being negatively, the former positively charged (Brandt, 1958; Schumaker, 1958). This binding in turn may stimulate the formation of additional membranous material and its subsequent invagination. Studies on the ferritin cycle by Bessis (1958) and Bessis and Breton-Gorius (1959) support the hypothesis that inducers act directly on the plasma membrane. Electron micrographs show that ferritin enters erythroblasts by pinocytosis and that this molecule is attached to the plasma membrane of the cell at various stages of its invagination. Apparently the adherence of ferritin molecules to the plasma membrane precedes its invagination and triggers off all the events which lead to the formation of pinocytotic vacuoles. Ferritin, no doubt, acts as the inducer of pinocytosis.

It seems most probable that similar mechanisms are at work in the process of food ingestion in malaria parasites, in *Theileria*, and in *Babesia*. They live intracellularly in erythrocytes and feed on the host cell by engulfing droplets of host cytoplasm. It is reasonable to assume that some proteins and salts in the erythrocyte cytoplasm act as inducers of pinocytosis, and are responsible for the invaginations in the plasma membrane of the parasite. Droplets of host cytoplasm are trapped into these invaginations with the subsequent formation of food vacuoles. This assumption is strengthened by two facts. Malaria parasites in their exoerythrocytic stage, and therefore in a different type of host cell, do not feed by pinocytosis (Meyer and Musacchio, 1960, 1965; Hepler *et al.*, 1966). There is no evidence of food vacuoles or of their formation. The cytostome, which is present at this stage also (Hepler *et al.*, 1966), does not form food vacuoles. Apparently the absence of the proper inducer is responsible for the lack of pinocytosis at this stage. The parasites most probably feed by osmosis in the exoerythrocytic stage, and it might be that the cytostome, having a single membrane, plays an important role in diffusion of nutrients from the host cell. In favor of this assumption is the presence of a branched cytostome in *Sarcocystis* (Senaud, 1966) which could be explained as a means of enlarging the single-membraned surface and thus make available a larger area for diffusion. This completely different way of feeding through the same organelle in different types of cells would strongly support the hypothesis that erythrocytes contain substances inducing pinocytosis.

The second fact in favor of the assumption that some inducers in the cytoplasm of erythrocytes are responsible for pinocytosis is that *Babesia* and *Theileria*, unrelated intraerythrocytic parasites, belonging to different subphyla (The Committee on Taxonomy, 1964), feed on the host cell in exactly the same way as malaria parasites in their intraerythrocytic stage. Both facts indicate that there is a close relationship between the surrounding medium and the manner of feeding in these parasites.

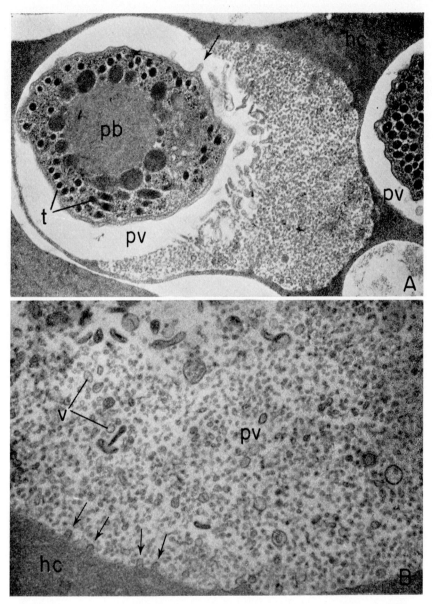

Fɪɢ. 15. *Lankesterella hylae* within erythrocytes. A: Section through two parasites, each in a separate periparasitic vacuole (pv). Numerous toxonemes (t) and a perinuclear body (pb) can be seen. The periparasitic vacuole contains granules, tiny vesicles, and elongate structures, usually found closer to the parasite. The latter might represent ultrapseudopods produced by the parasite, as seen at arrow. Magnification: × 35,000. B: Portion of periparasitic vacuole (pv) at higher magnification showing the formation of tiny vesicles at arrows from invaginations of the membrane surrounding the periparasitic vacuole. The host cytoplasm (hc) as in all erythrocytes appears to be very dense. Magnification: × 60,000. (Courtesy of Stebbens, 1966b.)

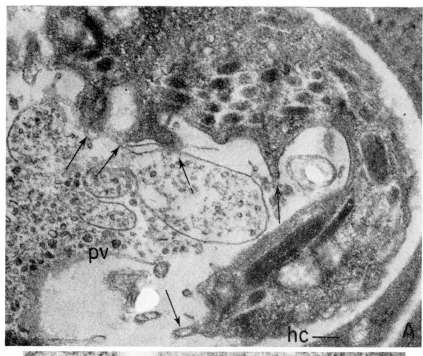

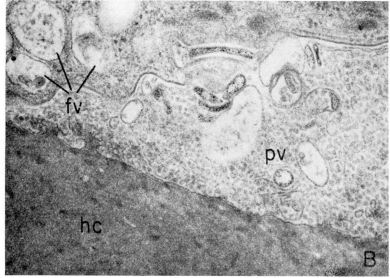

Fig. 16. *Lankesterella hylae* in erythrocytes. A: Transverse section through the parasite in the periparasitic vacuole (pv). Numerous pseudopods (at arrows) are extending from the parasite, most probably engaged in pinocytosis. B: Small vacuoles (fv) in the parasite contain granules very similar to those in the periparasitic vacuole, suggesting that they were formed by pinocytosis. Host cytoplasm, hc. Magnification: × 60,000. (Courtesy of Stebbens, 1966a.)

Of great interest in this connection is the feeding mechanism in *Lankesterella hylae* in the intraerythrocytic stage (Stebbens, 1966b). In the erythrocyte as in other host cells the parasite lies in a vacuole which separates it from the host cytoplasm (Figs. 15A, 17). At one side of the periparasitic vacuole there is a dense accumulation of fine granules, tiny vesicles, and membranes, while the rest of the vacuole appears to be deprived of this material (Fig. 15A). There is good evidence that this material derives from the host cell. Tiny invaginations of the membrane surrounding the periparasitic vacuole are apparently being pinched off to form minute vesicles, as seen in Figure 15B. These vesicles no doubt contain host cytoplasm. At places where this material is close to the parasite, numerous long, fine pseudopodlike projections extend from the parasite (Fig. 16A) and some seem to be in the process of engulfing parts of the periparasitic vacuolar content. Small vacuoles in the parasite near its periphery (Fig. 16B)

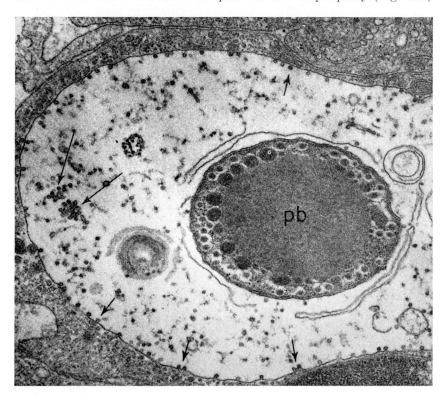

FIG. 17. *Lankesterella hylae* in an intestinal cell. Tiny vesicles are either in the process of budding off from the membrane of the periparasitic vacuole (at short arrows) or floating free in the periparasitic vacuole. Some of the vesicles are grouped together (long arrows). Perinuclear body, pb. Magnification: × 30,000. (Courtesy of Stebbens, 1966b.)

contain granules similar to those in the periparasitic vacuole, suggesting that phagotrophy or pinocytosis is actually taking place.

In *Lankesterella* no pigment is present in the intraerythrocytic stages. It might be that the host cytoplasm in the small vesicles has been partially digested and does not contain hemozoin, or that the pathway of digestion of hemoglobin is similar to that in *Babesia*.

Lankesterella hylae when parasitic inside macrophages of the liver, spleen, or gut lies in a vacuole also containing some granules, membranes, or vacuoles most probably deriving from the host cell, since the mode of their formation is the same (Fig. 17). The content of the periparasitic vacuole is evenly scattered and much less dense. In these host cells the parasite does not form pseudopods nor invaginations of the plasma membrane and most probably does not feed by pinocytosis.

3. Mitochondria

The chondriome in trophozoites of intracellular blood protozoa is of two types: one is represented by typical protozoan mitochondria, the second in which mitochondria of the usual type are lacking. All bird and amphibian blood Protozoa so far studied, (*Plasmodium lophurae, P. cathemerium, P. gallinaceum, P. fallax, Lankesterella garnhami,* and *L. hylae*) belong to the first group. Their mitochondria are surrounded by two membranes and the intramitochondrial structure is composed of microvilli which in sections appear as circular or oval profiles (Figs. 14B, 18, 21, 24, 25, 30). The shape of the mitochondria is ovoid, but short or longer rods may be encountered also, as well as branching forms (Fig. 25), although less often. The size ranges from 0.3–0.5 μ in shorter diameter. The number of mitochondria varies also with the age of the parasite. There is a single mitochondrion in newly invaded trophozoites and several in a section of older organisms.

All mammalian malaria parasites studied and *Babesia* and *Theileria* are deprived of typical mitochondria. They have, however, a structure composed of double concentric membranes (Figs. 1, 8B, 19) similar to a certain type of mitochondrion found in *Helix pomatia* during some stages of spermatogenesis (Beams and Tahmisian, 1954; Grassé *et al.*, 1956). Since mitochondria of the usual type are lacking in these parasites, it is assumed that the concentric structure might represent mitochondria. Supporting this hypothesis are recent findings in bacteria. These organisms possess a structure (Tokuyasa and Yamada, 1959; Glauert and Hopwood, 1959) similar to that in *Plasmodium* called mesosome (Fitz-James, 1960). In isolated mesosomes the electron transport system has been demonstrated (Salton and Chapman, 1962). In addition combined histochemical and electron microscope studies have disclosed in mesosomes dehydrogenase activity revealed by the accumulation of reduced tellurite in the form of an electron-dense deposit (van Iterson and Leene, 1964; van Iterson, 1965). Similar

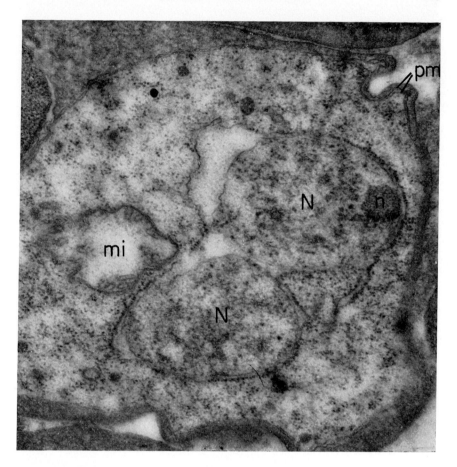

FIG. 18. *Plasmodium lophurae* in duck erythrocyte. The two membranes surrounding the parasite are clearly seen at pm. Two nuclei (N) are closely adjacent, probably in the last stage of division. In one of them a nucleolus (n) is present. The mitochondrion (mi) has oval and round profiles which represent section through microvilli. Plasma membrane, pm. Magnification: × 38,400. (From Rudzinska *et al.*, 1965.)

studies have to be performed with the concentric double-membraned structures.

Both organelles, mesosomes and the concentric double-membraned structure, derive from the plasma membrane (Figs. 1, 19B), giving support for the homology and analogy of both structures and strengthening the assumption that the concentric double-membraned structure in mammalian *Plasmodium, Babesia,* and *Theileria* also performs mitochondrial functions. The origin of mitochondria from the plasma membrane in cells of higher organisms has been suggested by other investigators (J. D. Robertson, 1961). In bacteria not possessing mesosomes the succinic dehydrogenase

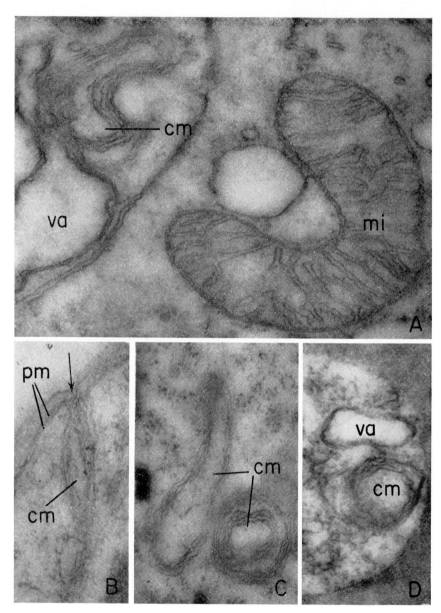

Fig. 19. Concentric double-membraned structure in *P. berghei* and in *Babesia rodhaini*. A: Small portion of *P. berghei* to show the concentric double-membraned structure (cm), vacuole surrounded by two membranes (va) and the mitochondrion (mi) of the host reticulocyte. Magnification: × 60,000. B: Formation of concentric double-membraned structure (cm) through invagination (at arrow) of the two plasma membranes (pm) in *P. berghei*. Magnification: × 80,000. C: Two concentric double-membraned structures in *P. berghei* (cm). Magnification: × 60,000. (From Rudzinska and Trager, 1959.) D: Concentric double-membraned structure (cm) and vacuole (va) surrounded by two membranes in *Babesia rodhaini*. Magnification: × 60,000. (From Rudzinska and Trager, 1962.)

activity and the total cytochrome system reside in the plasma membrane. It could be assumed that, during the evolution of bacteria, a specialized organelle evolved through proliferation and invaginations of the plasma membrane, giving rise to mesosomes, the prototype of mitochondria. The reappearance of this primitive type of mitochondrion during spermatogenesis in snails would fit well into this evolutionary hypothesis and would apply also to plasmodia where the parasitic way of life might be connected with a loss and then reappearance of certain organelles. It is difficult, however, to explain the presence of typical mitochondria in bird malaria parasites and of a primitive type in the closely related mammalian species, as both groups belong to the same genus, *Plasmodium*, and both are intracellular parasites inhabiting erythrocytes. The difference is closely related to the class to which the host belongs and this is evident in other intraerythrocytic parasites. *Babesia* and *Theileria* in mammalian erythrocytes are also deprived of typical mitochondria. It is significant that the difference is closely related to the class to which the host belongs, as other mammalian intraerythrocytic parasites are deprived of typical mitochondria, whereas *Lankesterella*, parasitic in erythrocytes of birds and frogs, possesses typical mitochondria. There is no explanation available so far for this relationship and there is no sufficient basis for speculation.

4. Ribosomes, Endoplasmic Reticulum, and the Golgi Apparatus

The ground substance of trophozoites contains small dense granules about 150–200 Å in diameter, most probably Palade's particles containing ribonucleoprotein (Figs. 1, 6, 10, 18, 24, 25, 31). Just after invasion, the particles are more densely accumulated than in the growing organism. The dense accumulations of ribosomes stems from the merozoite which always has a very compact cytoplasm (see Section II,B,1). The newly invaded merozoite appears to be as dense as the cytoplasm of the erythrocyte; it can therefore be easily overlooked in electron microscopy. However, shortly after invasion the cytoplasm becomes less condensed, most probably due to uptake of water from the host cell. It becomes condensed again at schizogony.

The endoplasmic reticulum is not too well developed. In most mammalian plasmodia it is in the form of small vesicles or less often of canaliculi or cisternae covered by smooth- or rough-surfaced membranes (Figs. 1, 3, 4, 6, 13, 26). The same applies to *Lankesterella* and to *Babesia rodhaini*. In *B. canis* (Bayer and Dennig, 1961) and *B. caballi* (C. F. Simpson *et al.*, 1963) cisternae dominate over vesicles. In bird malaria parasites roughsurfaced canaliculi and cisternae prevail over vesicles and are particularly abundant in schizonts of exoerythrocytic forms (Fig. 26).

The Golgi apparatus is poorly represented. It was found in *Plasmodium*

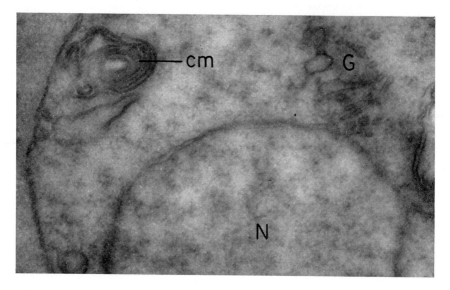

FIG. 20. Part of section of *P. berghei* to show the Golgi apparatus (G) located near the nucleus (N), and the concentric double-membraned structure (cm). Magnification: × 60,000. (From Rudzinska and Trager, 1959).

berghei (Fig. 20), and in *Babesia rodhaini* (Fig. 5A, C). It is composed of small, smooth-surfaced vesicles assembled in the vicinity of the nucleus.

5. Other Cytoplasmic Structures

In all mammalian malaria parasites a structure of low density surrounded by two membranes was found (Figs. 1–3, 6, 12, 19). This structure is elongate and often has a sausagelike shape. Its size varies. The longest so far found was 2.75 μ long with an inner diameter of 25 mμ in its narrowest part. The role and significance of this structure are difficult to determine, since no similar organelle has been found in other cells. The low density and structureless appearance of its matrix suggest that it might be a reservoir of fluid, and as such of some importance for an intracellular parasite embedded in a dense cytoplasm of the host cell. Similar structures were found in *Babesia* (Figs. 5A, 19D) and *Theileria*. It is not excluded that this structure might be a prototype of the concentric double-membraned structure and therefore function also as a mitochondrion.

Vacuoles surrounded by a single membrane with a matrix of low density are not uncommon in *B. rodhaini* (Fig. 5) and in *Lankesterella hylae* (Fig. 21).

Spherical, large bodies about 0.5 μ in diameter, of the same density as the pigment hemozoin, suspended freely in the cytoplasm, were found in *P. lophurae* (Rudzinska and Trager, 1957). They most probably represent lipid material (Figs. 14B, 23, 27). Their number and location vary, and they appear to be more numerous in mature stages. They are readily seen in stained blood films by light microscopy or in fresh preparations in phase contrast. This structure was not found in other species of *Plasmodium*.

In *Lankesterella garnhami* (Garnham *et al.*, 1962b) and *L. hylae* (Stebbens, 1966b), a number of additional structures are present. Below the double plasma membrane are microtubules (about 27) running lengthwise at intervals (Fig. 21B). On both sides of the nucleus in *L. hylae* and located bipolarly are large, dense, spherical "paranuclear bodies" up to 1.2 μ in diameter (Figs. 15A, 17, 21A). They are surrounded by round granules (up to 0.2 μ in diameter) of greater density than the paranuclear bodies. A large area of the cytoplasm in both species is occupied by long cylindrical, round or oval bodies of various diameters, some containing a central core of lesser density (Fig. 21). All these bodies, including those around the paranuclear bodies, are most probably sections through elongate structures similar to and most probably identical with so-called toxonemes first found in *Toxoplasma* by Gustafson *et al.* (1954). These structures were found later in *Sarcocystis* (Ludvik, 1960), in sporozoites (Garnham *et al.*, 1960, 1961, 1963; Ludvik, 1963), in *Eimeria* (Cheissin and Snigirevskaya, 1965), and in merozoites of *Plasmodium* (Aikawa, 1966; Hepler *et al.*, 1966).

The anterior part of the body of *L. hylae* is equipped with a set of elaborate organelles similar to those found in the above enumerated organisms. At the very tip of the parasite are two apical rings (Fig. 21B), most probably the thickened ends of the conoid, a collarlike structure, first described in *Toxoplasma* (Gustafson *et al.*, 1954). Leading to the tip of the conoid are the narrow ends of the paired organelle in *L. garnhami*, a structure first found by Garnham *et al.* (1960) in sporozoites of *P. gallinaceum*. In *L. hylae* one finds toxonemes in the same area (Fig. 21A). It seems most probable that toxonemes and paired organelles are homologous and might be identical structures. The organelles of the anterior part of the body, including the toxonemes, are characteristic for merozoites (Aikawa, 1966; Hepler *et al.*, 1966) and sporozoites of *Plasmodium*. Their possible role and function will be discussed in Section II,B,2 of this chapter.

It is worthwhile to stress here that the organelles just described (apical rings, conoid, paired organelle, toxonemes, pellicular microtubules) are present only in sporozoites and merozoites of *Plasmodium* and that they disappear in trophozoites. They seem, however, to remain in *Lankesterella* throughout all stages including trophozoites. There could be the possibility that electron micrographs of *Lankesterella* represent young merozoites and sporozoites. This, however, does not seem to be the case since the periparasitic vacuole contains structures originating from the host cell and resulting from the interaction between trophozoite and host.

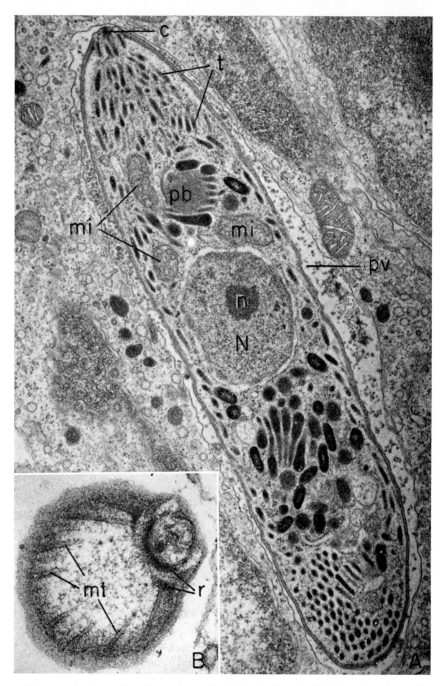

Fig. 21. Sections through *Lankesterella hylae*. A: Longitudinal section of parasite in a macrophage. Note the anterior end with the conoid (c), the numerous toxonemes (t), mitochondria (mi), the nucleus (N) with a nucleolus (n) and one of the paranuclear bodies (pb). Magnification: × 24,000. B: Oblique section through the anterior end of the parasite to show the apical rings (r) and the microtubules (mt). Magnification: × 60,000. (Courtesy of Stebbens, 1966b.)

6. Nucleus

The nucleus is surrounded by two membranes, and the outer one sometimes shows continuity with the endoplasmic reticulum (Fig. 18). The distance between the two membranes varies greatly (depending on the species) from about 17 mµ in *P. berghei* to about 60 mµ in *P. coatneyi*. It is not known whether the large spacing of the two membranes is not an artifact produced by fixation and dehydration. The matrix of the nucleus in mammalian malaria parasites and in *Babesia* is composed of loosely scattered fine fibrils and granules (Figs. 1–3, 5, 6, 8, 9). In bird malaria parasites and in *Lankesterella*, the nucleoplasm is differentiated into a nucleolus which appears as a dense, more or less spheroid body (Figs. 18, 21) and into chromatin material which is assembled at the periphery below the nuclear membrane in the form of dense granular masses particularly well defined in young trophozoites (Fig. 24) and even more so in merozoites (Figs. 28, 30). Ladda *et al.* (1965) and Ladda (1966), who stress the absence of a nucleolus in mammalian malaria, noticed its appearance in *P. berghei* treated with chloroquine. The significance of this change is not known.

During division the nuclear membrane remains intact as in most Protozoa except for the region associated with the spindle fibers as found by Aikawa (1966) and Hepler *et al.* (1966) in their recent electron microscope study of avian malaria parasites. There are also suggestions of tiny chromosomes located midway between the two poles (Aikawa, 1966). The spindle appears in the form of microtubules 200–220 Å in diameter. Microtubular spindle fibers were reported in other Protozoa (Roth and Daniels, 1962) and in a great variety of cells (see Porter, 1966).

7. Changes in Host Cells

The host cell in the intraerythrocytic stage is represented by mature erythrocytes in most cases. However, some malaria parasites, as, for instance, *P. vivax* and *P. berghei*, clearly show preference for reticulocytes. The latter are easy to identify in electron microscopy by the presence of mitochondria (Fig. 19). It is of great interest that the host cell retains at least some of its structures almost intact through a considerable length of time, and is able to survive throughout the whole cycle of the parasite, starting from the invading merozoite which grows and develops into a schizont, which in turn gives rise to a new crop of merozoites.

Some parasites induce characteristic changes in the host cell. In erythrocytes of *Plasmodium* the so-called Maurer's clefts and Schüffner's dots were described by light microscopy. The latter appear in light microscopy as even-sized granules and are characteristic for *P. vivax* infections. The former are not uniform as to their shape and size and were found first in *P. falciparum*. Recently Maurer-like clefts were described in *Babesia bigemina* and *B. argentina* (Saal, 1964).

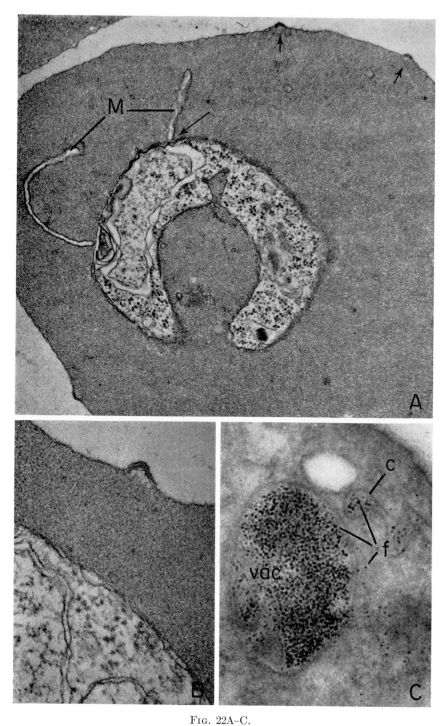

Maurer's clefts were identified in electron microscopy of *P. coatneyi* (Rudzinska *et al.*, 1966b; Rudzinska and Trager, 1968), *P. falciparum* (Trager *et al.*, 1967), *P. gonderi* (Rudzinska *et al.*, 1960), and *P. lophurae* (Rudzinska, 1966, unpublished). They appear in the form of narrow slits of low density of different shapes and sizes. They are either straight or irregular, semicircular or circular, short or long (up to 1.7 μ), and small or large (over 1 μ in diameter) (Figs. 6, 22). The great diversity in size might be due to the plane of sectioning. The straight forms prevail in *P. falciparum* (Trager *et al.*, 1967), the round in *P. coatneyi* (Fig. 6). Regardless of size and shape, their width is more or less the same and is about 40 mμ; they are lined by two membranes of the same thickness and appearance as the membranes covering the body of the parasite. In some electron micrographs they lie very close to the parasite as if touching it (Fig. 22), or as if their membranes would be continuous with the limiting membranes of the parasite (Rudzinska and Trager, 1968). The similarity in structure and close association of the membranes suggest that they might originate from the membranes of the parasite. The work of Tobie and Coatney (1961) is pertinent in this connection. These investigators, using the fluorescent antibody-staining technique, were able to demonstrate that Schüffner's dots in erythrocytes parasitized with *Plasmodium vivax* and *P. cynomolgi* fluoresce as strongly as the cytoplasm of the parasites. Nothing is known as to the role or significance of the Schüffner's dots or Maurer's clefts. They might be a means of enlarging the surface of the parasite if the connection with the parasite's body persists, or they might be a way of disposing of waste products; in such a case one would expect them to become detached.

In *P. coatneyi* the plasma membrane of the host cell shows at intervals small protrusions covered by an additional membrane, as seen in Fig. 22A and B. It is assumed that these places might be the site of agglutination of erythrocytes infected with parasites approaching schizogony. Clumped together they apparently become arrested in capillaries from which they cannot get out. This would account for the absence of schizonts and segmenters in the peripheral blood.

In a great number of rat reticulocytes infected with *P. berghei*, ferritin granules were found (Fig. 22C). Normally ferritin is present in erythro-

FIG. 22. Changes in host cells. A: Erythrocyte of the rhesus monkey infected with *P. coatneyi* shows two structures which most probably represent Maurer's clefts (M); one of the clefts (long arrow) is touching the plasma membrane of the parasite. At the periphery small protrusions (at short arrows) of the host plasma membrane can be seen. Magnification: × 35,000. B: Higher magnification (80,000) of a protrusion to show that it is composed of two membranes. (From Rudzinska and Trager, 1968.) C: Ferritin granules (f) in a rat reticulocyte parasitized by *P. berghei* are either loosely scattered or in a vacuole bound by two membranes (vac), or in a winding canaliculus (c). Magnification: × 120,000. (From Rudzinska and Trager, 1959.)

blasts, where it participates in hemoglobin synthesis. Red blood cells in which this process is completed are free of ferritin or may have only a few granules. However, large accumulations of ferritin may be found in the cytoplasm and inside mitochondria of erythroblasts and reticulocytes of certain hypochromic anemias (Bessis, 1958). In these diseases the amount of iron is normal but the amount of hemoglobin is low. Probably a block disturbs the synthesis of hemoglobin and, as a result, iron accumulates. Interestingly, in rats with malaria, large amounts of ferritin granules are assembled inside mitochondria or in vacuoles of reticulocytes. It is known that a high degree of anemia accompanies this disease (Baldi, 1950; Singer, 1953; H. W. Cox *et al.*, 1966). It is possible that this is due not only to the progressive destruction of reticulocytes invaded by the parasite but also by the process similar to hypochromic anemia described above.

B. Reproductive Forms in the Vertebrate Host

In all malaria parasites the intra- and exoerythrocytic trophozoites develop into multinucleated schizonts which develop further into segmenters composed of many merozoites arranged in the form of a rosette at the periphery of the parent cell. Some of the intraerythrocytic trophozoites give rise to macro- and microgametocytes, which, for further development, need an insect vector.

Schizonts in *Theileria parva* are found in lymphoid cells. Two types of schizonts are distinguished: macro- and microschizonts (Jarrett and Brocklesby, 1966). The latter give rise to micromerozoites. It is believed that micromerozoites, which appear late in the infection, invade erythrocytes while macroschizonts propagate in lymphoid cells.

Up to now electron microscopy studies of *Lankesterella* have not supplied information as to the structure of schizonts or segmenters.

Reproduction in *Babesia* does not involve typical schizogony. From light microscopy studies it is known that reproduction in *Babesia* starts with the formation of two and occasionally four small buds. The two or four buds grow by incorporating more and more of the nuclear and cytoplasmic matter of the parent until its whole content is exhausted. This results in the formation of two (or four) equal daughter cells. Thus reproduction would start with budding and end up with binary fission.

In electron micrographs of *Babesia rodhaini* forms representing binary fission and budding were found (Rudzinska and Trager, 1962). It would be, however, difficult to know whether the sequence of events is the same as described by the light microscopist for other species of *Babesia*, since observations in electron microscopy are based on fixed and therefore static material.

1. *Schizonts and Segmenters*

In schizogony repeated nuclear fission is the first step and leads eventually to segmentation (see Fig. 29). These stages have been studied by electron microscopy only in bird malaria parasites. It has been the general belief that cytoplasmic division starts after completion of all nuclear fissions. Electron microscopy of *P. lophurae* has shown that the last nuclear division takes place during advanced stages of cytoplasmic segmentation. In schizonts, in advanced stages of reproduction, elongated, uninucleated bodies may be found in which the nucleus and cytoplasm are in the process of division (Fig. 23), or two nuclei and the surrounding cytoplasm divide in such a way as to produce four merozoites (Rudzinska and Trager, 1961). This results in deviations in synchrony during reproduction.

The fine structure of the nucleus undergoes some characteristic changes during nuclear schizogony. Young trophozoites, just after invasion, have a nucleus differentiated into a nucleolus and into dense masses assembled below the nuclear membrane and most probably representing chromatin

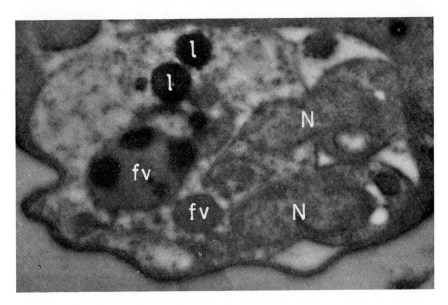

FIG. 23. Segmenter of *P. lophurae*. Two merozoites at the right side are in the process of formation. The nuclei (N) which have to be incorporated into the merozoites are in an early stage of division. All cytoplasmic components are in the vicinity of the nuclei. This accounts for the dense appearance of this part of the segmenter. The rest of the cytoplasm is of low density and contains food vacuoles (fv) and lipids (l) which will be incorporated into the residual body. Magnification: × 30,000. (From Rudzinska and Trager, 1961.)

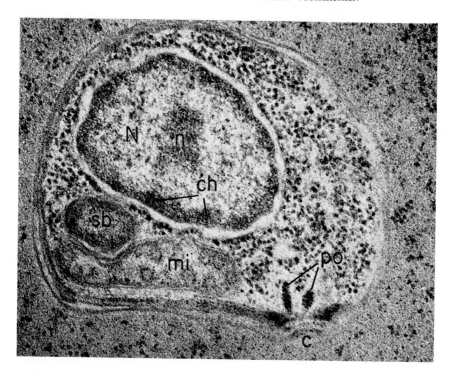

Fig. 24. *P. cathemerium* just after invasion of an erythrocyte still possesses the conoid (c), the paired organelle (po) and the mitochondrion (mi) closely associated with the spherical body (sb). The nucleus (N) is differentiated into chromatin (ch) and nucleolus (n). Magnification 67,000. (Courtesy of Aikawa, 1966.)

(Fig. 24). Nuclei of multinucleated trophozoites appear to be less differentiated. Usually the chromatin material is dispersed throughout the matrix and is not as dense (Fig. 25). It accumulates at the periphery of the nucleus again at the last nuclear division during the formation of merozoites (Figs. 28, 29). Such behavior of the chromatin might be connected with the function of the nucleus. It has been shown recently (Frenster *et al.*, 1963; Littau *et al.*, 1964) that condensed masses of chromatin are inactive and that the most active chromatin appears in the form of dispersed fibrils. This would explain very well the changes in the nucleus. In the very young organisms, the nucleus is in its resting stage with condensed chromatin; as soon as repeated divisions of the nucleus start, the chromatin material loses its compactness. It acquires it again after the last nuclear division and retains this structure through the merozoite stage and in the young invader. At these latter stages the parasite has a single, nondividing, thus resting, interphase nucleus.

Electron microscopy has been particularly instructive in revealing the role of the cytoplasm during schizogony. The active participation of the cytoplasm during this process is expressed in a number of changes which could be regarded as preparatory steps for merozoite formation. The changes start in the larger trophozoite and continue till the end of reproduction. They apply to all major organelles. The endoplasmic reticulum, which is poorly represented in the young trophozoite by a few vesicles and tubules, increases in volume in the larger trophozoite and a reversal in the form of elements occurs; long tubules prevail over vesicles (Fig. 26). The membranes are more densely covered with ribonucleoprotein particles and the matrix of the tubules shows a greater density. Both features indicate an increase in protein synthesis.

Changes can be detected also in the mitochondrial system. The young trophozoite has one or two mitochondria (Fig. 10); in the larger multinucleated parasite the number of mitochondria increases to several, indicating an increased activity of the cell (Fig. 25). The growing number of mitochondria could be regarded also as a means of preparing mitochondria for the offspring.

During schizogony entirely new structures characteristic for merozoites appear, as shown by the excellent electron microscope work of Aikawa (1966) and Hepler et al. (1966) in intra- and exoerythrocytic stages, respectively. In both forms the events during schizogony and segmentation are very similar. At many places the inner plasma membrane becomes much thicker. These places are elevated and mark the anterior ends of the budding merozoites as indicated by the appearance of the paired organelle in these areas (Fig. 26). This is followed by the formation of a conoid and the shifting of a nucleus to this area. At this stage a selection and segregation of cytoplasmic components takes place. One mitochondrion, elements of the endoplasmic reticulum, and Palade's particles assemble around each nucleus. At the same time watery materials are withheld from this part of the cytoplasm, leaving it compact and condensed. As a result, a considerable volume of watery cytoplasm accumulates, and is kept aside as an almost separate unit. The two distinctly different parts of the cytoplasm, as well as stages in segregation of materials, can be seen clearly in electron micrographs showing the formation of merozoites (Figs. 23, 27). The compact cytoplasm containing mitochondria, endoplasmic reticulum, and a condensation of ribonucleoprotein particles is incorporated into the body of the offspring. This accounts for the unusually dense appearance of merozoites.

The segregated watery part of the cytoplasm does not participate in the formation of merozoites. In the course of reproduction other elements, such as food vacuoles with the pigment hemozoin and lipid granules in *P.*

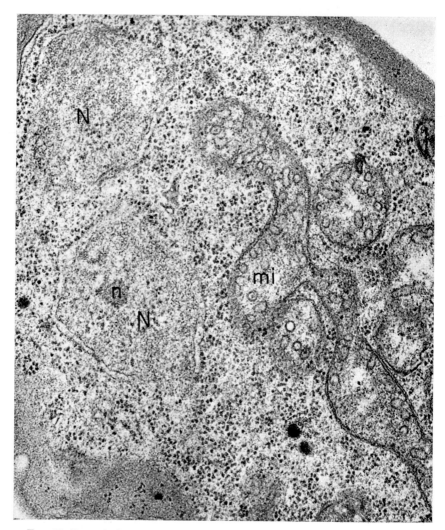

FIG. 25. Part of a large mature trophozoite of *P. cathemerium* with well developed mitochondrial system (mi). N, nucleus; n, nucleolus. Magnification × 36,000. (Courtesy of Aikawa, 1966.)

lophurae, are shifted to this part of the cell. In this way all the cytoplasmic components left behind the offspring are brought together (Fig. 27). At the end of reproduction they form a separate large vacuole surrounded by a membrane. This vacuole is the so-called residual body. It actually represents a waste basket containing watery and solid remnants of the parent cell.

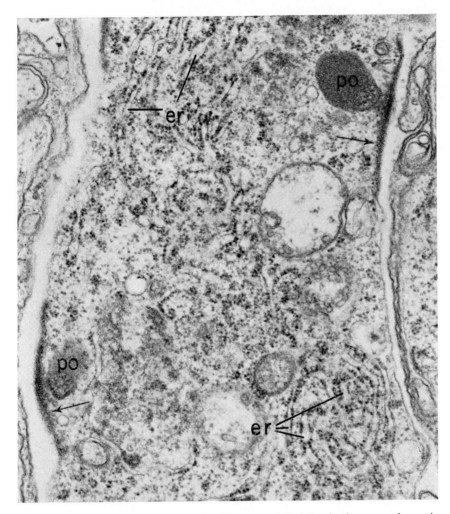

FIG. 26. Early stage in cytoplasmic schizogony of *P. fallax* in the exoerythrocytic stage. At certain places the inner membrane is much thicker (arrows), and nearby the paired organelle (po) is present. Rough-surfaced canaliculi and cisternae of the endoplasmic reticulum (er) occupy large areas of the cytoplasm. Magnification: × 36,000. (Courtesy of Hepler *et al.*, 1966.)

Theileria parva forms macro- and microschizonts. The former appear in early stages of the infection, the latter in later stages; both are found in leukocytes. The few electron microscope studies of *Theileria* available show macroschizonts as multinucleated bodies (Jarrett and Brocklesby, 1966). No macrosegmenters were found, only microsegmenters. The latter

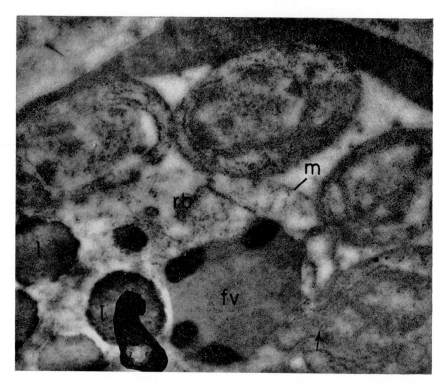

FIG. 27. Part of segmenter of *P. lophurae* with four merozoites and residual body nearing its completion. The residual body (rb) is surrounded by a membrane (m) and contains a food vacuole (fv), lipid bodies (l), and Palade's particles. At the lower right (at arrow) a mitochondrion is partially in the residual body (rb) and partially in the merozoite. Note the difference in density between the residual body and the merozoites. Magnification: × 40,000. (From Rudzinska and Trager, 1961.)

are composed of numerous merozoites which seem to arise from the parent by a process of budding, very similar to that in *Plasmodium*.

2. Merozoites

In bird malaria parasites intra- and exoerythrocytic merozoites differ in number, size, and shape. The former are less numerous (10–15 in a segmenter), smaller (1.7 × 0.85 μ), and ovoid in shape (Figs. 27, 28). The latter are much more numerous (200 in a segmenter), larger (4 × 1 μ), and oblong (Fig. 30). They have, however, a very similar fine structure. In both, the body is covered by a double plasma membrane, which forms a depression, the cytostome, midway between the anterior and posterior end (Figs. 28, 30B). This organelle has the same structure as in trophozoites. In exoerythrocytic merozoites it measures 80–100 mμ in inner diameter;

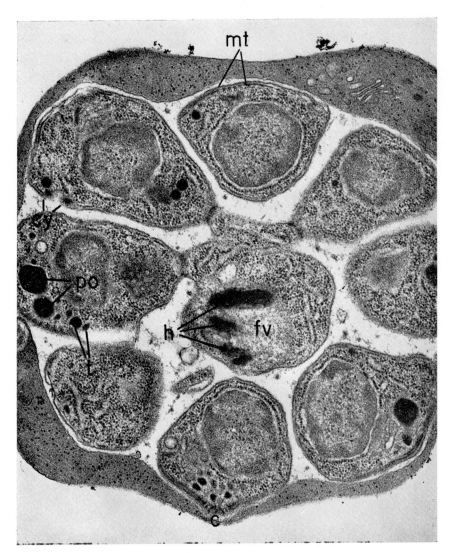

Fig. 28. Segmenter of *P. cathemerium* approaching completion. A food vacuole (fv) with hemozoin (h) is left outside the merozoites. In some of the merozoites the paired organelle (po) toxonemes (t), the cytostome (cy), the conoid (c), and the microtubules (mt) can be seen. Magnification: × 30,000. (Courtesy of Aikawa, 1966.)

in intraerythrocytic merozoites it is twice as wide. Below the plasma membrane are microtubules, about 250 Å in diameter (Figs. 28, 29) running lengthwise from one end of the parasite to the other

The cytoplasm contains numerous Palade's particles. The endoplasmic reticulum is represented by rough-surfaced tubules (Fig. 28). One mitochondrion is present in each merozoite (Figs. 28, 30). Aikawa (1966) and Hepler *et al.* (1966) found that the mitochondrion is associated with a spherical body (Fig. 30A) of unknown function. The nucleus is large and centrally located (Figs. 28, 30). It is differentiated into dense chromatin located at the periphery and into a nucleolus. The outer membrane is densely covered with ribosomes.

The apical part of the merozoite shows a high degree of differentiation. It contains a number of specialized organelles which were found previously in several parasites which are not closely related. Although these structures differ in detail, the general plan of organization seems to be the same.

At the very tip of the body is the conoid (Figs. 24, 28, 29), a cone-shaped protrusion surrounded at the rim by concentric, apical or polar rings (Aikawa, 1966). The conoid was first described by Gustafson *et al.* in *Toxoplasma* (1954) and was thereafter found in *Sarcocystis* by Ludvik (1956, 1960, 1963), who ascribed to it the function of a sucker and of a perforator at the time the parasite adheres to and penetrates the host cell (Ludvik,

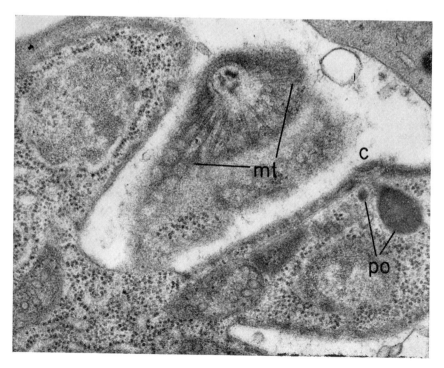

FIG. 29. Portion of a far advanced schizont of *P. cathemerium* showing the paired organelle (po), the conoid (c), and the microtubules (mt) radiating from the anterior end of the budding-off merozoites. Magnification: × 45,000. (Courtesy of Aikawa, 1966.)

1963). In the vicinity of the conoid and leading to its tip are numerous elongate cylindrical bodies named "toxonemes" in *Toxoplasma* (Gustafson *et al.*, 1954), "sarconemes" in *Sarcocystis* (Ludvik, 1956), and "convoluted tubules" in sporozoites of malaria parasites (Garnham *et al.*, 1963). Since these names are synonymous, we shall use the term "toxoneme" for all of them. In sporozoites Garnham *et al.* (1960) found in addition two pear-shaped structures of the same density as the convoluted tubules with long narrow extensions ending at the tip of the conoid, which they named "paired organelle." Garnham (1963) suggested that this organelle might contain proteolytic enzymes which "enable the sporozoite to penetrate the various cells it encounters in its journey" from the salivary gland of the insect vector to the vertebrate host. There is every reason to believe that a similar function could be ascribed to toxonemes since most probably the paired organelle represents differentiated toxonemes. In favor of this supposition is the fact that in organisms not having the paired organelle, like *Toxoplasma* and *Sarcocystis*, toxonemes are found in this area and they lead to the tip of the conoid, as does the paired organelle. In *Lankesterella garnhami*, both the toxonemes and paired organelle extend to the conoid. *Lankesterella hylae* seems to have several paired organelles and a great number of toxonemes.

Merozoites of *Plasmodium* (Aikawa, 1966; Hepler *et al.*, 1966) possess also a paired organelle (Figs. 28–30) and only a few toxonemes (Figs. 28,

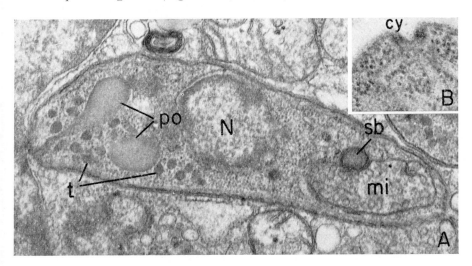

Fig. 30. Merozoite of exoerythrocytic stage of *P. fallax*. A. Note the paired organelle (po) at the apical part of the merozoite, toxonemes (t), the mitochondrion (mi) with the nearby spherical body (sb). N, nucleus. Magnification: × 28,000. B. Small portion of merozoite to show the cytostome (cy). Magnification: × 43,000. (Courtesy of Hepler *et al.*, 1966.)

30) which occasionally are seen connected with the paired organelle. The small number of toxonemes in merozoites as contrasted to their abundance in sporozoites could be related to the single penetration of the former and the multiple penetration of the latter. Both facts would favor the secretory function of the toxonemes. The disappearance of the paired organelle and toxonemes in merozoites of *Plasmodium* (Aikawa, 1966) after penetration into the host cell (Aikawa, 1966; Hepler *et al.*, 1966) favors this assumption.

The conoid, toxonemes, paired organelle, and subpellicular tubular fibrils, structures characteristic for merozoites and sporozoites of *Plasmodium*, disappear during the trophozoic stage. They are formed *de novo* during reproduction. On the other hand, it seems that in *Lankesterella* all these structures are retained during the entire life cycle. It is difficult to draw any sound conclusions from these findings. Possibly the disappearance and reappearance of these structures in *Plasmodium* might have some evolutionary significance.

The finding of these structures is considered a clue to the relationship and taxonomy of organisms equipped with these organelles by a number of investigators (Garnham, 1963; Ludvik, 1963). This applies particularly to *Toxoplasma*, *Sarcocystis*, and *Lankesterella*—organisms of uncertain taxonomy. It would appear from these studies that they all belong to the subphylum Sporozoa.

It is believed that *Theileria parva* possesses two types of merozoites, macro- and micromerozoites. Jarrett and Brocklesby (1966) question the presence of macromerozoites, and their electron microscope study led them to a tentative hypothesis, namely, that the macromerozoites are actually macroschizonts which are able to invade new host cells. The macroschizonts, and not macromerozoites, which so far have not been detected, are responsible for the maintenance of the infection in lymphocytes where they finally give rise to microschizonts. The latter appear at later stages of the infection and form micromerozoites which invade erythrocytes. Micromerozoites are similar in their fine structure to merozoites in malaria parasites. They are surrounded by two membranes and have a large, dense nucleus encased in two membranes. The cytoplasm contains numerous Palade's particles, vesicles of the endoplasmic reticulum, small vacuoles, and dense round bodies, which most probably represent the paired organelle. The preservation of structures is not good enough in the electron micrographs so far available to resolve other more delicate structures.

3. Gametocytes

After a few generations of schizonts some of the malarial merozoites take another course of development. They grow less rapidly but steadily without nuclear division, forming micro- and macrogametocytes, large uninucleated organisms. For further development they have to be transferred to a mosquito.

Not too many electron micrographs of gametocytes are available. The fine structure of the following species was studied: *P. cathemerium* (Duncan *et al.*, 1959), *P. lophurae* (Aikawa *et al.*, 1966), and *P. coatneyi* (Rudzinska and Trager, 1968). In addition Garnham *et al.* (1967) studied the fine structure of exflagellating microgametocytes of *P. berghei*, *P. cynomolgi*, *Hepatocystis kochi*, and *Leucocytozoon marchouxi*.

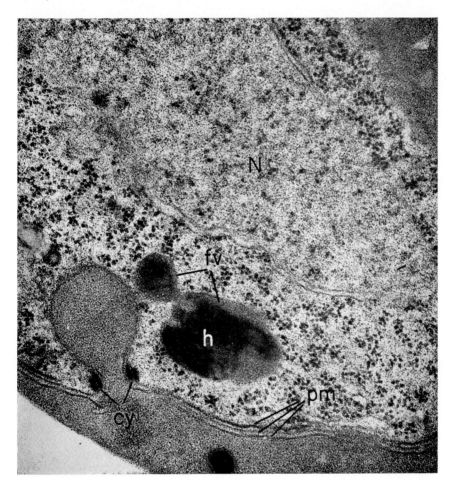

Fig. 31. Part of gametocyte of *P. cathemerium* engulfing a droplet of erythrocyte cytoplasm through the cytostome (cy). The food vacuole (fv) contains a large mass of hemozoin (h). The three membranes of the gametocyte (pm) are seen clearly. N, nucleus. Magnification: × 54,000. (Courtesy of Aikawa *et al.*, 1966a.)

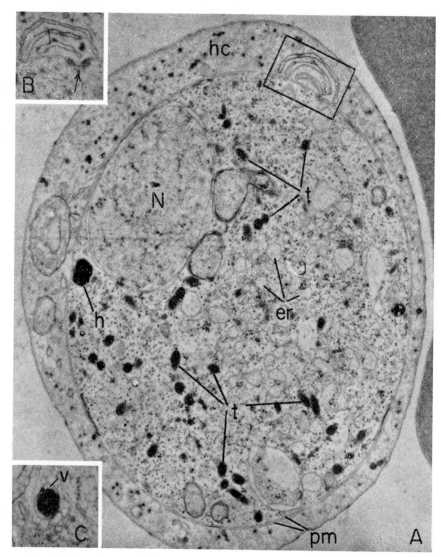

FIG. 32. Gametocyte of *P. coatneyi*. A: The parasite is surrounded by two membranes (pm). In the cytoplasm smooth- and rough-surfaced vesicles of the endoplasmic reticulum (er) are scattered throughout the cytoplasm. The ground substance of the cytoplasm contains Palade's particles. There is one hemozoin grain (h) in the section and many very dense, round, oval, or elongate bodies which most probably represent toxonemes (t). The cytoplasm of the host cell (hc) is of lower density than the cytoplasm of the parasite. N, nucleus. Magnification: × 24,000. B: A serial section of the outlined area in A shows the cytostome (at arrow). Magnification: × 30,000. C. Serial section of the hemozoin grain (h) in A shows that it lies in a vesicle (v). Magnification: × 30,000. (From Rudzinska and Trager, 1968.)

Gametocytes of *P. cathemerium* and *P. lophurae* (Fig. 31) are covered by three membranes; those of *P. coatneyi* (Fig. 32) and of the exflagellating microgametocytes are covered by two membranes only. In *P. cathemerium* (Aikawa *et al.*, 1966a) and in *P. coatneyi* a cytostome was found (Figs. 31, 32), in the former in the process of engulfing a droplet of host cytoplasm. The ground substance contains numerous ribosomes, some of them grouped together in the form of polysomes. The endoplasmic reticulum is composed of smooth- or rough-surfaced vesicles. Organelles characteristic for trophozoites of mammalian and bird plasmodia are present in the respective gametocytes. Thus, gametocytes of *P. coatneyi* have the double-membraned and the concentric double-membraned structures. The pigment hemozoin in *P. cathemerium* is inside the food vacuole, and in vesicles in *P. coatneyi*. In the latter, the pigment differs in shape and size from that in the trophozoites. It is coarse and measures about $0.23 \times 0.2 \mu$ in the gametocyte, and appears in the form of long and narrow crystals measuring about $0.3 \times 0.03 \mu$ in trophozoites. The nucleus is large, oval, and of low density. In addition, the gametocyte of *P. coatneyi* possesses very dense, round, oval, or elongate bodies measuring 60 to 116 mμ in shorter diameter. They are most probably cross- and tangential sections of toxonemes, a structure characteristic of sporozoites, merozoites, and ookinetes. Since they were found in ookinetes (Garnham *et al.*, 1962c), there is every reason to believe that these structures are toxonemes and as such characteristic for macrogametocytes. In addition, Garnham *et al.* (1967) did not find flagellar structures in the cytoplasm in more mature forms. Thus toxonemes and intracellular flagella could serve well as characteristic features of macro- and microgametocytes, respectively.

The gametocyte occupies almost the whole area of the host cell. In *P. coatneyi* the host cytoplasm is of low density as if deprived of hemoglobin.

C. STAGES IN THE INVERTEBRATE VECTOR

1. Exflagellation of the Microgametocyte

Microgametogenesis in malaria parasites presents a technically trying subject for the electron microscopist as exflagellating microgametocytes must be sought and sectioned individually. Garnham *et al.* (1967) have done this for *Plasmodium berghei*, *P. cynomolgi*, *Hepatocystis kochi*, and *Leucocytozoon marchouxi*. The remains of the degenerating erythrocyte may surround the exflagellating microgametocyte. Sections of the mature microgamete show an internal "axial filament complex" (see p. 278) of "9 + 2" fibers with an elongate nucleus lying alongside, about halfway down the long slender body; a mitochondrion may also be present in the cytoplasm. Raffaele (1939) described a thin fibril around the microgamete as the "flagellum." The electron microscope study confirmed this observation, but only partially. It has been shown that this is not a free flagellum as in *Eimeria* microgametes (Cheissin, 1964; Scholtyseck, 1965). Shadow-cast replicas of the microgamete confirm that the single flagellum is intracellular. A most surprising finding is that the axial filament complexes of the future microgametes lie coiled around the microgametocyte nucleus before exflagellation. Garnham *et al.* (1967) believe that a single centriole in the gametocyte replicates or divides until eight of them are formed. These take over the function of basal bodies giving rise to eight flagella. The details of nuclear behavior attending microgamete formation are uncertain. The preliminary observations of Garnham and co-workers indicate "endomitosis" with separation of the microgamete nuclei from the parent microgametocyte nucleus, rather than a series of complete mitotic divisions providing the microgamete nuclei.

2. Ookinetes

Garnham *et al.* (1962c) examined the fine structure of the motile zygote or ookinete of *Plasmodium gallinaceum* and *P. cynomolgi* in the mosquito gut. When burrowing into the mosquito wall, the ookinete has a two-layered pellicle differentiated anteriorly into a slit with thick lips. Beneath the pellicle lie 55–65 microtubules running the length of the body: these hollow fibrils (see p. 273) may represent the contractile structures responsible for movement in this and other motile stages of the life cycle.

The cytoplasm contains large crystalloids of uncertain significance and chemical composition, in addition to mitochondria and other membrane-bound particles, but no other distinctive structures. The intracellular location of the penetrating ookinete has been verified.

3. Oocysts

Duncan and co-workers (1960) examined sectioned oocysts of *Plasmodium cathemerium* 3–10 days after ingestion of infected blood by mosquitoes. The oocyst is encased in a "structureless capsule" which becomes thinner with age. The electron microscope work of Terzakis *et al.* (1966) shows that the segregation of the oocyst cytoplasm into sporoblasts is preceded by the appearance of numerous vacuoles at the periphery just below the capsule, with the subsequent fusion, enlargement, and penetration of these vacuoles into the cytoplasm, leading to cytoplasmic segregation into sporoblasts. Local thickenings and elevations of the sporoblast cell membrane clearly indicate the beginning of budding of sporozoites, the process being very similar to merozoite formation.

4. Sporozoites

The microanatomy of sporozoites has been described by Garnham and co-workers (1960, 1961, 1963) in several species of malaria parasites and recently by Vanderberg *et al.* (1967) in *P. berghei.* As mentioned earlier (see Section II,B,2) the fine structure of sporozoites is very similar to that of merozoites. Almost all the specialized organelles described in merozoites were found first in sporozoites by Garnham *et al.* The body is covered by two membranes and beneath them are longitudinally arranged microtubules, their number (11–15) depending on the species. Garnham (1966b) considers the microtubules as a possible structural basis of locomotion in sporozoites. At the anterior end of the sporozoite is the apical cup (Garnham *et al.*, 1960) analogous and homologous to the conoid. The paired organelle, which was first found and described in sporozoites by Garnham *et al.* (1960), is prominent and well developed but does not dilate posteriorly in all species. The cytoplasm is filled with toxonemes (named "convoluted tubules" by Garnham *et al.*, 1963). In both bird and mammalian sporozoites, a mitochondrion was found. The presence of a typical protozoan mitochondrion in sporozoites of *P. berghei* (Vanderberg *et al.*, 1967) is of particular interest, since it is known that this parasite, as well as other mammalian plasmodia, is deprived of such mitochondria in the intraerythrocytic stage (see Section II,A,3). The nucleus is ovoid, more or less centrally located, and in structure similar to the nucleus of merozoites. Just anterior to the nucleus is the micropyle, structurally identical with the cytostome (see Section II,A,2), which most probably does not function in pinocytosis at this stage. Garnham *et al.* (1963) suggested that it might be the route of emergence of a "sporoplasm" when the sporozoite has penetrated the liver cell, but this idea remains unsupported by experimental evidence.

III. PROTOZOA NOT INVADING BLOOD CELLS (TRYPANOSOMES)

A. STRUCTURE AND FUNCTION OF CYTOPLASMIC COMPONENTS AT DIFFERENT STAGES IN THE LIFE CYCLE

In their cytoplasmic organization the trypanosomes are probably simpler than any other flagellates. Their distinctive morphological features are to be found in the mastigont system, that is the entire, more or less integrated group of organelles associated with the basal body of the flagellum (see Section III,A,3,a). They share these features with other kinetoplastid flagellates, and in reviewing the cytology of trypanosomes some reference to the bodonid flagellates, the leishmanias, and the monogenetic trypanosomatids of insects is essential. The life cycles of trypanosomes do not involve the gross morphological reorganization encountered in intraerythrocytic parasites, and cyclical transformation centers around changes in the mastigont system. For this reason the fine structure of stages in the vertebrate and invertebrate host will be considered together in the present section.

The single locomotory flagellum of trypanosomes is attached to the pellicle along part or the whole of its length to form an "undulating membrane." The basal body appears to be associated on the one hand with the microtubular system of the pellicle, and on the other by virtue of some cryptic attachment with the kinetoplast. The kinetoplast is a specialized region of the single mitochondrion which contains deoxyribonucleic acid (DNA). The kinetoplastid flagellates are unique in that, although DNA is a widespread component of mitochondria (Nass, et al., 1965), only among these flagellates is it so abundant and compact that it can be readily visualized with the light microscope after appropriate staining. The mitochondrion extends from one end of the flagellate to the other and, together with the pellicular microtubules, may be mechanically responsible for maintaining the slender fusiform shape of the organism. The pocket ("reservoir") around the base of the flagellum may be the site at which materials enter and leave the flagellate. Other cellular structures such as the nucleus, endoplasmic reticulum, and Golgi apparatus are ultrastructurally similar to those found in other cells, though there are numerous cytoplasmic vacuoles, vesicles, and granules which are as yet difficult to define ultrastructurally and whose function is controversial.

1. Surface Membrane

The limiting membrane of trypanosomes (Fig. 33) has the usual "unit membrane" (J. D. Robertson, 1959), structure, but may be overlain by an osmiophilic (electron dense) layer of varying thickness and uncertain significance. Schulz and MacClure (1961) give the overall thickness of the whole pellicle as about 130 Å.

2. Microtubules and Shape

Lying beneath the surface membrane of a trypanosome and extending from one end of the flagellate to the other are microtubules (Fig. 33), first recorded as having the appearance of fibers in metal-shadowed dried ghosts and replicas (Kleinschmidt and Kinder, 1950a,b; Kraneveld et al., 1951; Meyer and Porter, 1954). In sections the microtubules are seen to be 150–250 Å in diameter, the walls of the tubules being 50–60 Å thick (Schulz and MacClure, 1961) and probably composed of subunits (W. A. Anderson and Ellis, 1965) as in other microtubules (Ledbetter and Porter, 1963). They are spaced at 200–300 Å and parallel the surface membrane at a depth of 80–100 Å. The course of these tubules has not been investigated in detail and their origin and morphogenesis is not clear. In the related bodonids (Pitelka, 1961) and in amoeboflagellates (Balamuth, 1965) pellicular microtubules appear to emanate from the bases of the flagella. In trypanosomes a row of microtubules, usually three or four, pass from the basal body along

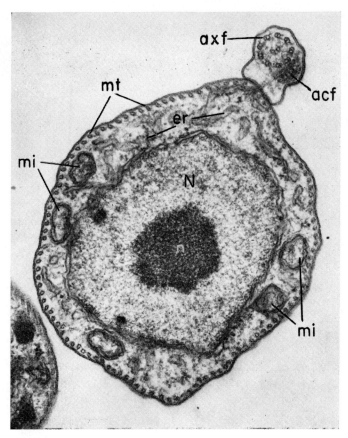

Fig. 33. Transverse section of culture form of *Trypanosoma rajae* in region of nucleus (N). Note the limiting membrane of the body and flagellum, pellicular microtubules (mt), and the axial (axf) and accessory (acf) filament complexes of the flagellum. Several profiles of the mitochondrial network (mi) lie beneath the pellicle; er, endoplasmic reticulum. Nucleolus, n, stained with lead citrate. Magnification: 45,000. (Courtesy of Preston, 1966.) All electron micrographs of trypanosomes are taken from material fixed in veronal acetate-buffered osmium tetroxide and stained with uranyl acetate unless otherwise stated.

the wall of the reservoir to the body surface (Fig. 34C). Whether all the pellicular microtubules take their origin in the dense material which lines the canal leading to the reservoir is not known, but microtubules are commonly associated with such zones of cytoplasmic density in other cells (see, for example, Tilney and Porter, 1965).

In transverse sections of trypanosomes the number of visible microtubule profiles decreases toward the extremities of the flagellate: the tubules pre-

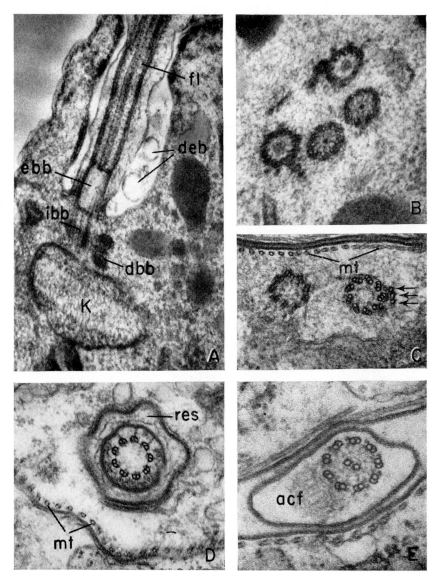

FIG. 34. Features of the flagellum and basal body complex in trypanosomatid flagellates as seen in electron micrographs. A: Section of region around base of flagellum in a lower trypansomatid (*Leptomonas* sp.). Note the intracellular (ibb) and extracellular (ebb) parts of the basal body, and the daughter basal body (dbb) forming at right angles to the parent structure. The kinetoplast (K) shows the fibrous band embedded in a denser matrix. The reservoir contains debris (deb) possibly extruded from the cytoplasm. Magnification: × 33,000. (From Vickerman, 1966b). B: Section of similar flagellate showing that 4 basal bodies (seen here in transverse section) may

sumably terminate at different levels to produce this effect; there is no evidence that they fuse together.

The function of the pellicular microtubules is debatable. Microtubules are thought to function in cell movement or contractility (de Thé, 1964; Tilney and Porter, 1965), in conduction (Slautterback, 1963), or as skeletal supports (Tilney and Porter, 1965). In trypanosomes there is no evidence that they function in conduction, though they may play a part in contractility. It seems most likely, however, that they provide a cytoskeleton which is both deformable (as in movement of the undulating membrane) and extendable (as in transformation from one stage in the life cycle to the next). That the microtubules are not the only elements responsible for shape, however, is shown by the fact that digitonin dissolves all cytoplasmic components except the microtubules (Vickerman, 1966c), but these do not prevent the trypanosome from collapsing along its longitudinal axis.

3. The Mastigont System

a. The Basal Bodies and Flagellum. The basal body of the flagellum in trypanosomes is similar in structure to the basal bodies of other flagella and cilia (Gibbons and Grimstone, 1960; Fawcett, 1961) that is, it has a cylindrical structure whose wall is formed by nine triplets of longitudinal hollow fibers, with each triplet arranged at an acute angle (about 40°) to a tangent drawn to the circumference of the cylinder (Fig. 34B and C). In trypanosomes the basal body is approximately 600–700 mμ long and 200–250 mμ in diameter. It is partly intracellular, but the distal part giving rise to the flagellum projects into the reservoir (Fig. 34A). The proximal intracellular portion has its longitudinal fibers arranged in nine triplets (Fig. 34B and C) and in transverse section shows the cartwheel structure described for higher flagellates by Gibbons and Grimstone (1960). The proximal end of the basal body is not closed by a plate and may lie closely applied to the en-

---

be present in one organism prior to division. The "cartwheel" formation is evident in these intracellular basal bodies. Magnification: × 40,000. A and B stained with phosphotungstic acid. C: Transverse section of intracellular basal bodies in *Trypanosoma vivax* to show nine triplets of subfibers in one of them. Pellicular microtubules (mt) are also evident, and, close to the right basal body, three of the microtubules (arrowed) which pass from the basal body along the wall of the reservoir. Magnification: × 50,000. D: Transverse section of "extracellular" part of basal body in *T. vivax*. The nine triplets of fibrils have given way to doublets which are connected to the surface membrane by fine fibrils or partitions; res, reservoir. Magnification: × 46,000. E: Transverse section of flagellum in *T. vivax* to show axial and accessory filament complexes. The structure of the accessory filament complex (acf) is not well shown in this picture. The section is viewed from below so that the arms on the fibers of the peripheral doublets point in an anticlockwise direction. An internal spur is evident on the nonarm-bearing fiber in some cases. Magnification: × 75,000. (B–E from Vickerman, 1966c.)

velope of the kinetoplast, as in *Trypanosoma lewisi* (Clark and Wallace, 1960) and allied flagellates, or some distance from the kinetoplast region, as in *T. vivax*. Although the basal body and kinetoplast always behave as though there were some structural bond between them (e.g., in morphogenetic movements and when the trypanosome is disintegrating), no such binding can be seen with the electron microscope (Meyer *et al.*, 1958). At about the level of the floor of the reservoir, the triplets of fibrils give way to doublets in the distal (extracellular) part of the basal body (Figs. 34D, 37B), and at the same level an incomplete plate separates the dense core of the proximal region from the electron-transparent core of the distal region of the basal body (Fig. 34A). The basal body is separated from the flagellum by one or two transverse plates. In transverse section (Figs. 33, 34E) the main shaft of the flagellum is seen to have the "axial filament complex" of nine double peripheral *plus* two central tubular fibers characteristic of cilia and flagella, and the fine-structural details of this complex are superficially the same as for other flagellates. W. A. Anderson and Ellis (1965) have reported on the details for *T. lewisi*. The peripheral doublets have the usual "arms" pointing in a clockwise direction (looking toward the base of the flagellum), but an additional feature is the presence of a counterclockwise spur traversing the lumen of the tubule which does not bear the arms (Fig. 34E). Whether this spur will prove to be of wide occurrence remains to be seen, but it is interesting to note even minor structural differences between the flagella of trypanosomes and other flagellates in view of the anomalous direction of wave propagation along the flagellum in these organisms (see Chapter 14) and our ignorance as to the significance of flagellar ultrastructure. The two central tubules of the flagellum are cross-striated (periodicity 250 Å) and are said by Anderson and Ellis to arise from different basal plates, one from a flattened plate that is attached to the peripheral tubules, the other tubule from a more proximal central disc so that this tubule must pass through the first basal plate to enter the flagellum. A further feature of the trypanosome flagellum is the paraflagellar rod (Pitelka, 1963) or accessory filament complex (Fawcett, 1961; Vickerman, 1962) that lies alongside the axial filament complex parallel with the two central tubules (Fig. 33). A similar structure is found in a variety of flagellates (see Pitelka, 1963) including most trypanosomatids but not all, for it is absent in *Crithidia oncopelti*, for example, though not in *C. fasciculata* (Vickerman, 1963a). The paraflagellar rod has a latticelike ultrastructure, though high resolution micrographs showing its structure in detail have yet to be published for trypanosomes: Mignot (1964) has described the configuration of the lattice in the phytoflagellate *Entosiphon* in some detail, and the paraflagellar rod of trypanosomes appears to have a comparable ultrastructure. The origin of the paraflagellar rod in morphogenesis is a complete mystery, as it has no homolog in the basal body and is absent from the proximal part of the flagellum, being found only in that part of the flagellum which is applied to the pellicle to form the undulating membrane. This anatomical association with the undulating membrane led to the suggestion that the rod might in some way be functionally responsible for the undulating membrane, but this is hardly likely as many flagellates possessing the rod lack an undulating membrane. It may be a device for increasing the effective diameter of the flagellum, hence its propulsive thrust.

The appearance of an undulating membrane probably derives from the deformation and stretching of the pellicle by the beating flagellum, rather than deformation and stretching of the flagellar membrane. In osmium-fixed material sectioned for electron microscopy, the flagellum is seen lying apposed to the pellicle (Fig. 33), but with no visible deformation of either structure which could indicate the nature of the undulating membrane; yet in some sections a pellicular fold similar to that recorded for tricho-monads (e.g., see Joyon, 1963) is seen, and it must be presumed that in most cases activity ceases on contact with the fixative so that this fold is not preserved. The majority of observers have failed to find any structural connection between flagellum and pellicle in the region of the undulating membrane, though W. A. Anderson and Ellis (1965) report a desmosome-like binding in their glutaraldehyde-fixed material of *T. lewisi*. In osmium-fixed trypanosomes electron-dense material lines the neck of the reservoir canal in the region where the emergent flagellum is closely applied, and this may indicate a tenuous attachment of the desmosome type.

The free flagellum of trypanosomes appears to lack the paraflagellar rod, and at its tip the two central fibers terminate before the outer nine doublets.

In addition to the basal body of the locomotory flagellum, another basal body lacking continuity with a flagellum is often visible in sections of trypanosomes. This second basal body (Figs. 36B, 37A) has the structure of the proximal part only of the first basal body and is completely intra-cytoplasmic. It could be interpreted as a vestigial structure (for discussion of vestigial organelles, see Grimstone and Gibbons, 1966) inherited from a biflagellate bodonid ancestor, or as a stage in replication of the first basal body (Grassé, 1961). Centrioles and basal bodies are commonly believed to be self-replicating structures, a daughter centriole being induced to form in the cytoplasm at right angles to the parent structure (e.g., see Bern-hard and de Harven, 1960; Gall, 1961; Dippell, 1965), though this cannot always be the case as some protozoa appear to form basal bodies in the absence of a parent structure (Vickerman, 1966b). In trypanosomatids, however, it can be inferred from electron micrographs (Fig. 34A) that a new basal body is formed from a procentriole-like structure generated near the old basal body (Vickerman, 1961, 1966b). Since sections of dividing trypanosomes often show up to four basal bodies (Fig. 34B), it would appear that both basal bodies replicate before flagellate division. Whether or not all basal bodies are capable of producing functional flagella remains to be seen.

b. *The Kinetoplast and Mitochondrion.* The literature on the kinetoplast has been extensively reviewed by Mühlpfordt (1964), and his conclusion that this term is preferable to the others used in past literature* is supported here.

In stained preparations the kinetoplast is seen with the light microscope as a basophilic structure close to the base of the flagellum. In certain strains of bloodstream trypanosomes the kinetoplast is absent and appears to be secondarily lost. Such "akinetoplastic" trypanosomes seem unaffected by their loss—an observation which made it difficult to assign a function to the kinetoplast until fairly recently, when studies on fine structure shed new light on the problem.

The cytological interest of the kinetoplast lies in the fact that it was the first example of a cellular structure outside the nucleus which was shown to contain DNA and could be observed to replicate by division. Even nowadays when evidence for DNA and self-replicating structures in cytoplasm is increasing (Vickerman, 1966b), the kinetoplast provides the most striking demonstration. Staining by the Feulgen technique (Bresslau and Scremin, 1924; Jirovec, 1927; M. Robertson, 1927; Roskin and Schlischliaiewa, 1928; Lillie, 1947; Barrow, 1954; Gerzeli, 1955; Horne and Newton, 1958; Baker, 1961) and acridine orange (Baker, 1961) combined with removal of staining properties with deoxyribonuclease, as well as the ability to incorporate tritiated thymidine as shown by autoradiography (Cosgrove and Anderson, 1954; G. Steinert et al., 1958; L. Simpson, 1965), indicate that the kinetoplast owes part, if not all, of its basophilic property to DNA.

Early studies with the electron microscope of the kinetoplast in sections (E. Anderson et al., 1956; Inoki et al., 1958; Horne and Newton, 1958; Meyer et al., 1958; Pyne, 1958, 1960a,b; Meyer and Queiroga, 1960) indicated that the kinetoplast was a double membrane-bound disc or spheroid (shape varying with species, see Fig. 35) with fibrous or lamellar inclusions. Clark and Wallace (1960) found the kinetoplast of several trypanosomatids to consist of an electron-dense bundle of fibers embedded in a mitochondrial matrix. The presence of "cristae" in the space surrounding the fibrous element suggested the mitochondrial nature of the organelle, and confirmed

---

* The nonnuclear nature of the kinetoplast speaks against the term kinetonucleus (Woodcock, 1906). The terms "micronucleus," "nucleolus," "parabasal body," "blepharoplast," and "centrosome," which have been used for this structure, are now employed by cytologists in referring to structures of entirely different homology.

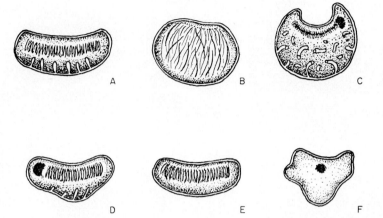

FIG. 35. Diagrams of kinetoplast in various mammalian trypanosomes in vertical section as revealed by electron microscopy. Not to scale. A: *Trypanosoma lewisi* (all stages in life cycle), *T. cruzi* (all stages except metacyclic and bloodstream trypanosomes). B: *Trypanosoma cruzi* (metacyclic and bloodstream trypanosomes). C: *Trypanosoma vivax* (bloodstream trypanosome). D: *Trypanosoma brucei* (insect midgut stage), *T. congolense* (bloodstream trypanosome). E: *Trypanosoma brucei* (bloodstream trypanosome), *T. evansi*, *T. equiperdum*. F: *Trypanosoma equinum*.

old observations of Alexeieff (1917) and others that the kinetoplast stains selectively with Janus green B.

The fibrous component of the kinetoplast (Figs. 36A and B, 37A and B) was studied at high resolution by Schulz and MacClure (1961) in *T. cruzi*, and Ris (1962) in *T. lewisi*. Schulz and MacClure described the principal element here as a fiber wound back and forth in zigzag fashion; the fiber has a thickness of 125 Å and is tubular with a 35-Å lumen. The presence of tubular structures has recently been confirmed by W. A. Anderson and Ellis (1965) in *T. lewisi*. In the same trypanosome Ris, however, found 25-Å fibers in this region, which he believed to represent the DNA of the kinetoplast. He compared the organization of the fibrous zone to that of bacterial nucleoplasm. In bacteria (prokaryotes) the nuclear apparatus (chromosome, genophore) takes the form of a single molecule of DNA in the form of a hoop which is tangled and folded in the cytoplasm (Kleinschmidt and Lang, 1962); this DNA is not bound to histone as in the nuclei of cells in plants and animals (eukaryotes). M. Steinert (1965) has demonstrated cytochemically that basic proteins of the histone type are absent from the kinetoplast, though present in the nucleus of trypanosomes. Ris has found trypanosome nuclear DNA present in 100-Å fibrils (presumably bound to histone) as in other eukaryotes. Bearing in mind the mitochondrial matrix of the kinetoplast and the recent studies of Nass *et al.* (1965), among others, who have demonstrated 20–40-Å DNA fibrils in the mitochondria of a wide

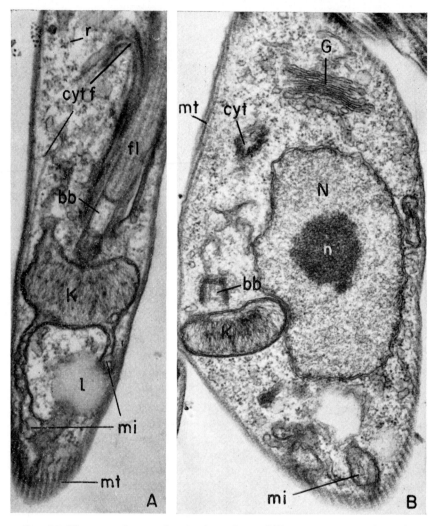

Fɪɢ. 36. Electron micrographs of culture form of *Trypanosoma rajae* from skate. A: Section through region around the base of flagellum. The fibrous band of the kineto-plast (K) fills the mitochondrial envelope at this point and tubular mitochondrial extensions (mi) are visible. The cytostomal tube supported by fibers (cytf) or, more correctly, microtubules, is shown arising at the point of emergence of the flagellum from the body of the trypanosome; this tube is deflected in a posterior direction. The dense granules associated with the fibers of the kinetoplast have been observed in other trypanosome species. These granules are too large for ribosomes (visible in the cytoplasm at r) and their nature is not known. Basal body, bb; lipid, l; pellicular microtubules, mt; nucleus, N; nucleolus, n. Magnification: × 39,000. B: Section with juxtanuclear kinetoplast, showing cytostome (cyt) in transverse section, also Golgi apparatus (G), and numerous ribosomes in cytoplasm. Some investigators have re-ported fusion of the kinetoplast and nuclear envelopes when the kinetoplast is lying apposed to the nucleus, but until such continuity has been demonstrated in well-fixed material, it is premature to attribute functional significance to these observations. Stained with lead citrate. Magnification: × 34,000. (Courtesy of Preston, 1966.)

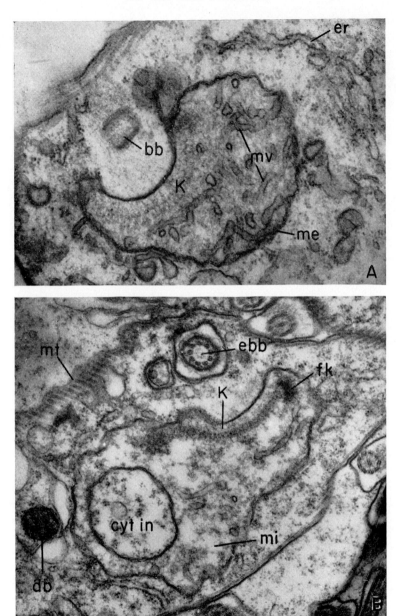

Fɪɢ. 37. A: Section of kinetoplast region of mitochondrion in *Trypanosoma vivax* (Desowitz strain). The fibrous zone of the kinetoplast (K) is narrow and occupies only a small part of the mitochondrial lumen. Tubular microvilli (mv) arise from the mitochondrial envelope (me). Note the clear cytoplasm between the kinetoplast and the second basal body (bb) (the basal body of the flagellum is not visible in this section); er, endoplasmic reticulum. Magnification: × 37,000. B: Similar section showing

variety of cells, it seems probable that Ris's identification of the kinetoplast DNA is correct, though enzymatic digestion studies are lacking. The nature of the tubular fibrils seen by other observers remains a mystery. Coiling of the DNA fibrils (as suggested for the bacterial chromosome; see Giesbrecht, 1961) might explain tubular profiles if the coils are viewed in section, but it is just as likely that the tubes have an entirely different nature. Fixation and dehydration are known to determine the appearance of mitochondrial DNA fibrils (Nass et al., 1965), uranyl acetate postfixation preventing the clumping of fibrils in osmium-fixed, ethanol-dehydrated material. Such clumping, no doubt, accounts for statements that the central zone of the kinetoplast consists of dense lamellae 190–200 Å (Sanabria, 1964) or 250–360 Å thick (Mühlpfordt, 1964). Standardization of techniques will probably result in better agreement over the interpretation of results!

The revelation of the mitochondrial nature of the kinetoplast did not provide an entirely satisfactory explanation of its function, but the discovery in electron micrographs of mitochondria appended to it, and apparently forming as buds from the kinetoplast, led M. Steinert (1960) to postulate that the kinetoplast houses the genetic information concerned in the synthesis of mitochondrial enzymes (i.e., enzymes concerned in the oxidation of pyruvic acid and reduced pyridine nucleotides produced by glycolysis). The concept of the kinetoplast as a "mother mitochondrion" (Ris, 1962) became widely accepted, and the alternative theory that the kinetoplast represents an expendable symbiont became less plausible. Meanwhile, the continuity of the kinetoplast envelope with mitochondria was demonstrated by several workers in different kinetoplastid flagellates (Meyer et al., 1958; M. Steinert, 1960; Pyne, 1960a,b; Pitelka, 1961; Schulz and MacClure, 1961; Garnham and Bird, 1962; Vickerman, 1962, 1963a; Mühlpfordt, 1963a,b; Judge and Anderson, 1964; Trager and Rudzinska, 1964; Rudzinska et al., 1964; W. A. Anderson and Ellis, 1965).

Pitelka (1961) in *Bodo*, and Mühlpfordt and Bayer (1961) in *Trypanosoma gambiense*, demonstrated from serial sections that the kinetoplast forms part of the single mitochondrion of the cell (Fig. 38), and this is probably the case in all kinetoplastid flagellates, though some authorities continue to interpret kinetoplast-associated mitochondrial structures as "buds" (W. A. Anderson and Ellis, 1965) rather than profiles of a continuous tubular system. Cytochemical staining for the enzyme nicotinamide adenine dinucleotide (NAD) diaphorase clearly demonstrates the form of the single mitochondrion (M. Steinert, 1964; Vickerman, 1965; see Fig. 39).

cytoplasmic inpocketing (cyt in). Such inpocketings are believed to be a prelude to mitochondrial division. The lateral fibrous knot (fk) visible in the kinetoplast is a constant feature of *T. vivax* and occurs also in the vector stages of *T. brucei;* mt, pellicular microtubules; ebb, extracellular part of basal body; db, dense body. Magnification: × 28,000. (Electron micrographs by Vickerman, 1966c.)

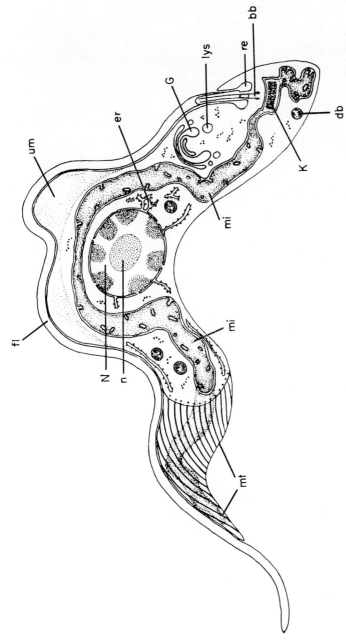

FIG. 38. *Trypanosoma brucei*, intermediate form. Diagram to show principal features of a bloodstream trypanosome: fl, flagellum; um, undulating membrane; er, endoplasmic reticulum; G, Golgi apparatus; lys, lysosome; re, reservoir; bb, basal body; db, membrane-bound dense body; K, kinetoplast; mi, mitochondrion; mt, microtubules (they are shown only at the anterior end of the flagellate); N, nucleus; n, nucleolus. (From Vickerman and Cox, 1967.)

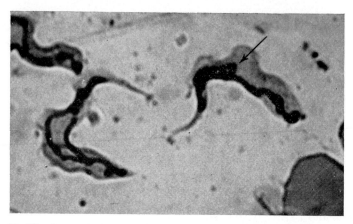

Fig. 39. The form of the mitochondrion in intermediate bloodstream stage of *Trypanosoma brucei*, demonstrated by tetrazolium (Nitro BT) staining for NAD diaphorase activity following brief glutaraldehyde fixation. Blue-black formazan deposits fill the mitochondrion, which is a single tube in the flagellate to the right, but is cleft in the one to the left. Deposit in granules close to the mitochondrial tube (arrowed) may represent granules containing enzymes of the L-α-glycerophosphate oxidase system which also give the diaphorase reaction owing to the presence of NAD-linked dehydrogenase in this system. Magnification: × 2500. (From Vickerman, 1966b.)

Since the discovery of the universal occurrence of DNA in mitochondria, the kinetoplastid flagellates are not so unique as was originally supposed: they are unusual in that the DNA is confined to a particular region of the mitochondrion close to the basal body of the flagellum (Grassé and Pyne, 1965). In the bodonid ectoparasite *Costia necatrix*, however, several fibrous bands or kinetoplasts are found within the mitochondrion, and these are not apposed to basal bodies (Joyon and Lom, 1966; Schubert, 1966).

Stages in replication of the kinetoplast by fission can be observed easily in stained preparations with the light microscope. With the electron microscope, stages in division are poorly documented, but the fibrous band appears to expand laterally, as seen in sections, and a mass of cytoplasm bulges into the mitochondrial matrix of the kinetoplast to cleave the fibrous band into two (Figs. 37B, 40). In *Trypanosoma vivax* division of the kinetoplast can be seen to be just part of the division process in the mitochondrion as a whole. Division of the mitochondrion is most clearly seen with the light microscope in *T. vivax* and *T. congolense* smears stained for NAD diaphorase (Vickerman, 1965). A series of clefts develop in the mitochondrial tube and gradually coalesce to split the tube into two halves. With the electron microscope the mitochondrial clefts are seen to take the form of cytoplasmic intrusions into the mitochondrial matrix, similar to the intrusion that divides the fibrous band of the kinetoplast. In some dividing

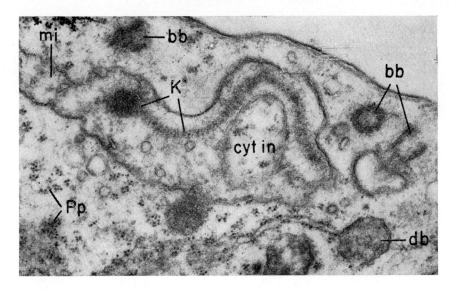

FIG. 40. Section of *Trypanosoma vivax* (Desowitz strain) to show division of the kinetoplast. The expanding fibrous zone of the kinetoplast (K) is being cleft by a cytoplasmic inpocketing (cyt in): bb, basal bodies; db, membrane-bound dense body; mi, mitochondrial tube; Pp, Palade's particles. Magnification: × 30,000. (Electron micrograph by Vickerman, 1966c.)

trypanosomes, atypical configurations of the mitochondrial cristae are occasionally observed, and these may be related to mitochondrial division; for example, the concentric arrangement of cristae figured by M. Steinert (1960) in *T. mega* and the *"Ellenbogen"* type of crista found by Schulz and MacClure (1961) in *T. cruzi*. Preston (1966) has encountered both types of configuration in the mitochondrion of *T. rajae*.

A role for the kinetoplast and its mitochondrial connections in the cyclical development of trypanosomes was suggested by Vickerman (1962, 1965) as a result of electron microscope studies on *Trypanosoma brucei*.* A comparison of the fine structure of bloodstream forms with that of fly midgut forms (as obtained in culture) showed a striking difference in the form of the mitochondrion. In slender (monomorphic) bloodstream forms, as noted by Mühlpfordt and Bayer (1961), a simple tubular mitochondrion with scarcely any cristae is found extending from the kinetoplast to the anterior end of the flagellate, and a similar but very short tube may extend to the posterior end. In midgut forms, where the kinetoplast lies further forward in the body, however, the anterior chondriome consists of a network of cristae-

---

* The trypanosomes *T. brucei*, *T. rhodesiense*, and *T. gambiense* are ultrastructurally identical at all stages in their life cycles and will be referred to as *T. brucei* throughout this section.

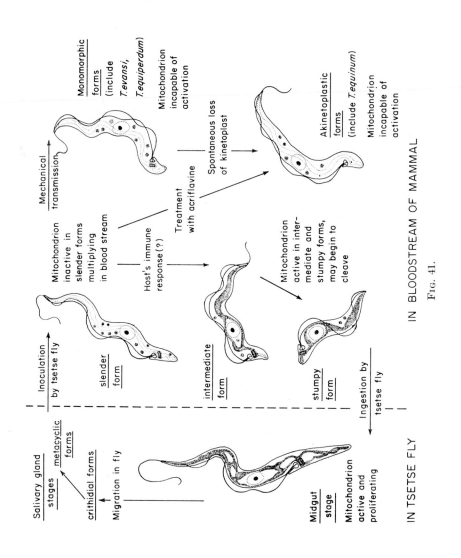

Fig. 41.

bearing tubes lying beneath the pellicle, and there is also a prominent postkinetoplastic mitochondrial tube. Comparative studies on the respiration of these two forms assisted in the interpretation of these morphological findings. In slender (monomorphic) *T. brucei*, glucose is respired only as far as pyruvate which is excreted by the flagellates; a functional Krebs cycle appears to be absent from these forms. The large amounts of oxygen consumed by the trypanosomes can be accounted for by the activity of the L-α-glycerophosphate (GP) oxidase-L-α-GP dehydrogenase cycle (see Grant, 1966). The slender bloodstream flagellates, then, derive their energy from aerobic glycolysis. The fly midgut forms, however, respire their pyruvate completely, and appear to have conventional oxidative phosphorylation linked to a Krebs cycle and cytochrome carrier system (von Brand and Johnson, 1947; Ryley, 1956, 1961; Fulton and Spooner, 1959). As Krebs cycle enzymes and cytochrome pigments are not detectable in slender bloodstream trypanosomes, it would appear that the mitochondrial remnant ("promitochondrion") observed by electron microscopy in these forms is functionally inactive, whereas in the fly midgut forms both ultrastructural and biochemical work point to marked mitochondrial activity. Von Brand *et al.* (1955) had shown that glucose and oxygen consumption by cultured midgut forms was one tenth that of bloodstream forms, and as the blood meal is digested by the fly, respirable substrate must become increasingly scarce. Vickerman (1962) suggested that on entering the fly the bloodstream forms were obliged to adopt a more economical pattern of respiration and that mitochondrial material was proliferated from the kinetoplast to bring about this switch. The outgrowth of the posterior mitochondrial tube at the time of the respiratory switch might be responsible for the shift in position of the kinetoplast observed when the trypanosomes enter the fly. The physiological and morphological changes observed in the cyclical development of trypanosomes might prove intelligible in terms of mitochondrial proliferation and regression demanded by the environment.

Further work on other stages in the life cycle of *Trypanosoma brucei* (Vickerman, 1965) has shown that the respiratory switch probably occurs, not when the trypanosomes enter the fly vector, but when slender forms transform into stumpy forms (Fig. 41). Cristae appear in the mitochondrion at this time and NAD diaphorase, indicative of mitochondrial activity, can

---

FIG. 41. Scheme showing the form and activity of the mitochondrion in various stages of the life cycle of *Trypanosoma brucei* and its derivatives. The active mitochondrion is shown stippled. The alternative terminal respiratory system (L-α-glycerophosphate oxidase-dehydrogenase cycle) of bloodstream trypanosomes is housed in extramitochondrial bodies (membrane-bound dense bodies), also indicated by stippling. Salivary gland stages in the tsetse fly are not shown. (From Vickerman, 1966b.)

be demonstrated using tetrazolium salts (Fig. 39). Although the stimulus eliciting this transformation is not clear, it is possible that the host antibody response might be responsible (Ashcroft, 1957; Wijers, 1960). If this is so, then antibodies might be expected to inhibit glucose uptake and so provoke the change in respiratory economy. It has long been maintained that stumpy trypanosomes are more likely to infect the fly than slender forms (M. Robertson, 1912; Wijers and Willett, 1960). It would appear that, by virtue of their respiratory behavior, they are preadapted to doing so.

Although nothing is known of the respiration of salivary gland forms in *T. brucei*, preliminary electron microscope studies indicate that cristae disappear from the mitochondria of crithidial forms when they transform into metacyclics (Vickerman, 1966a) and so reversion to mitochondrial inactivity might occur at this point in the life cycle. In the life cycle of *Trypanosoma brucei*, then, the mitochondrion undergoes cyclical repression and activation, as shown by studies on the fine structure and biochemistry of these flagellates. This cycle is accompanied by changes in the form of the mitochondrion and in the position of the kinetoplast—two changes that may be related to one another. The ultrastructural and biochemical evidence for repression and activation in the life cycles of other trypanosomes is more fragmentary, though a similar story has been told for the life cycle of *Leishmania donovani* (Rudzinska et al., 1964; L. Simpson, 1965). M. Steinert (1964) has recorded differences in the form of the chondriome between the two forms of *Trypanosoma mega* obtainable in culture. Wery and De-groodt-Lasseel (1967) have noted such changes for *T. cruzi* in culture, the trypanosomes corresponding to metacyclics having a poorly developed chondriome with few cristae compared to cultured crithidial forms: in this species the chondriome appears to be physiologically active in both blood-stream and culture forms (Ryley, 1956; Fulton and Spooner, 1959). It would appear that the majority of trypanosomes do not show the extremes of mitochondrial structure and metabolism witnessed in *T. brucei*. In *T. vivax* and *T. congolense* there is a single mitochondrial tube in blood-stream forms which has prominent cristae that are not platelike as in *T. brucei* but tubular (Fig. 37A and B). This mitochondrion shows diaphorase activity (Vickerman, 1965) in all trypanosomes. Yet cytochrome pigments are reputedly absent from these bloodstream trypanosomes, though glucose breakdown proceeds beyond the pyruvate stage (Ryley, 1956; Fulton and Spooner, 1959). An electron microscope study of tsetse fly proboscis stages of these trypanosomes (Vickerman, 1966c) shows that the chondriome changes from a single tube to a reticulum as in *T. brucei* so that mitochondrial development (and presumably activity) is greater in the vector.

Trypanosoma evansi and *T. equiperdum* are considered to be derivatives of *T. brucei* which have lost the ability to undergo cyclical development in the tsetse fly. They exist as monomorphic trypanosomes in the blood of

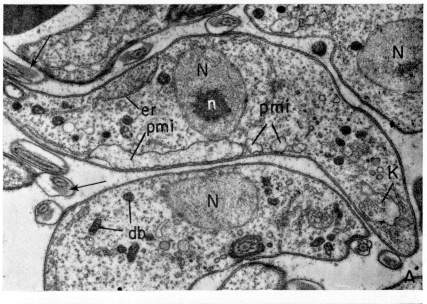

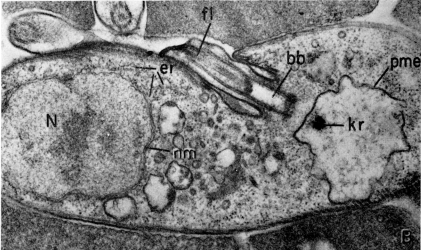

Fig. 42. A: Section of *Trypanosoma equiperdum* to show profiles of the single inactive promitochondrial tube (pmi) devoid of cristae and extending forward from the kinetoplast (K). Membrane-bound dense bodies (db) abound in the cytoplasm. Sections of several nuclei (N) are visible, one with a prominent nucleolus (n). Some sections of flagella (arrowed) show the latticelike ultrastructure of the accessory filament complex. Magnification: × 10,000. (From Vickerman, 1966c.) B: Section of *T. equinum* to show the electron-dense kinetoplast remnant (kr) within the promitochondrial envelope (pme). The nucleus (N) shows an envelope composed of two membranes (nm) with associated endoplasmic reticulum tubules (er). Free ribosomes are abundant in the cytoplasm. Magnification: × 25,000. (From Vickerman, 1966b.)

mammals and are transmitted mechanically from host to host. Ultrastructurally (Fig. 42) and biochemically they resemble slender *T. brucei* (Mühlpfordt, 1964). Continued mechanical transmission, by syringe passage, of polymorphic *T. brucei* through laboratory rodents results in strains which are all slender (monomorphic) and under the electron microscope indistinguishable from *T. evansi*: they have lost the ability to transform into stumpy trypanosomes, and presumably to activate the mitochondrion, no doubt owing to lack of selection maintaining this ability.

These monomorphic derivatives of *T. brucei* readily give rise to variants which appear, from light microscopy, to have lost the kinetoplast—the so-called "akinetoplastic" strains. This loss may occur spontaneously as a result of defective division or be induced as a result of treatment with acriflavine or other drugs. Those trypanosomes in which the mitochondrion bears cristae (i.e., is active) do not give rise to viable "akinetoplastic" strains (Mühlpfordt, 1964). *Trypanosoma brucei* which has been artificially deprived of its kinetoplast cannot activate the mitochondrion, as shown by its inability to undergo cyclical development in the tsetse fly (Reichenow, 1940). These facts indicate the indispensability of the kinetoplast for mitochondrial activity. Electron microscopy of trypanosomes and related flagellates with induced akinetoplasty shows that the mitochondrial envelope persists as a "ghost" and only the fibrous zone is missing (Mühlpfordt, 1963a,b; Vickerman, 1963b; Trager and Rudzinska, 1964). In *T. equinum*, a species which is naturally akinetoplastic, the fibrous zone is seen in electron micrographs to be missing, but its place is taken by a dense spherical body (Fig. 43) (Vickerman, 1966b). It is not known that this body represents compacted DNA.

It has been suggested (Mühlpfordt, 1964; Trager and Rudzinska, 1964) that, as the envelope of the kinetoplast persists, the term "akinetoplastic" is misleading and should be discontinued. Trager and Rudzinska offer "dyskinetoplastic" as a more suitable designation for trypanosomes with the altered form of kinetoplast.

4. Other Cytoplasmic Structures

a. *The Cytostome and Pinocytosis.* Osmotrophic nutrition has been found to involve pinocytosis in a variety of protozoa, and trypanosomes are no exception. M. Steinert and Novikoff (1960) found that in *Trypanosoma mega* the microtubules of the pellicle were deflected into a funnel-shaped cytostome near the point of emergence of the flagellum from the reservoir. They demonstrated the role of this cytostome in the ingestion of proteins from the surrounding medium by using ferritin as an electron-dense tracer.

Brown *et al.* (1965) showed pinocytotic uptake of ferritin from the base of the reservoir in *T. rhodesiense* (monomorphic bloodstream form), but no differentiated cytostomal tube. Although a cytostome identical in structure to that of *T. mega* has been observed by several workers in bodonid flagellates (Pitelka, 1961; Brooker, 1966; Joyon and Lom, 1966; Schubert, 1966), this organelle has proved to be not so widespread in trypanosomes. Preston and Vickerman (1967) found a cytostome in leishmanial, crithidial, and trypanosome stages of *T. rajae* (Fig. 36A) in culture but not in *T. brucei*, *T. vivax*, *T. congolense*, or *T. equinum*. In the African trypanosomes, however, profiles suggestive of pinocytosis are frequently encountered around the reservoir, as described by Brown and co-workers.

b. Contractile Vacuole. The contractile vacuole of trypanosomes has received comment from relatively few observers, especially at the level of the electron microscope. Clark (1959) recorded the presence of this structure in *Trypanosoma cruzi*, *T. lewisi*, and several amphibian trypanosomes. M. Steinert (1964) has noted it in *T. mega* with the light microscope. Preston (1966) has observed a contractile vacuole in electron micrographs of *T. rajae*. Like the cytostome, the contractile vacuole is a prominent feature of the free-living bodonids, and it may be that the two organelles are functionally related, the contractile vacuole serving to pump out excess water taken in during feeding through the cytostome. In systole the contractile vacuole is seen as a series of smooth-walled tubules ("spongioplasm") adjacent to the reservoir: in diastole the tubes lie around the expanding vacuole and presumably empty into it. No structures which could represent a permanent contractile vacuole pore have been recorded. Evidence for the presence of a contractile vacuole in the African trypanosomes has not been forthcoming.

c. The Golgi Apparatus; Lysosomes. The Golgi apparatus in trypanosomes (Fig. 36B) has the typical form of dictyosomes—a pile of smooth, membrane-bound, flattened sacs which appear to bud off vesicles from their margins into the cytoplasmic matrix. The sacs may be bent into cups. The Golgi apparatus usually lies close to the reservoir region, between the kinetoplast and the nucleus of the flagellate. In certain cases, however, the Golgi stack lies anterior to the nucleus in the trypanosome form. Sanabria (1963) notes such a situation in *T. cruzi*, Preston (1966) has found it in *T. rajae*, and Vickerman (1966c) in posteronucleate forms of *T. brucei*.

The function of the Golgi apparatus in trypanosomes is still a matter for conjecture, but a role in secretion seems plausible. In *T. brucei* the Golgi sacs become greatly swollen and distended as the trypanosomes transform from slender to stumpy forms (Vickerman, 1966c). As this transformation is believed to be triggered off by the host's immune response (see p. 290), and trypanosomes are known to secrete antigens ("exoantigens" of Weitz, 1960; also see Chapter 5), it is possible that the antigens have their origin in the Golgi apparatus and that hypertrophy of this structure is engendered by antibodies reacting with antigens at the site of production. The functional nature of the exoantigens is not yet known, but they appear to be proteins, and are possibly secreted enzymes (see p. 293). The ability of trypanosomes to change their antigens (see Chap. 5) in order to circumvent the host's immune response during chronic infections might imply secretion of a series of isoenzymes.

Of the enzymes which might be secreted by trypanosomes, digestive enzymes are a possibility, and digestive enzymes (acid hydrolases) are cytologically associated with membrane-bound vesicles termed lysosomes. Using acid phosphatase as a lysosome marker and the Gomori (Holt, 1959) and Burstone (1962) cytochemical reactions specific for this enzyme, discrete lysosome-like bodies have been demonstrated in the Golgi region of *T. brucei* (blood forms) under the light microscope (Brooker and Vickerman, 1964).* Enzyme activity is also evident in the reservoir itself, indicating that lysosomal enzymes are released into this cavity, possibly to initiate digestion prior to pinocytosis (Fig. 43). In *T. mega* one or two large lysosomes are prominent at the posterior end of the flagellates in culture, again showing a difference between those trypanosomes which, like *T. mega*, have a cytostome, and those which, like *T. brucei*, lack this structure. On adapting the Gomori reaction for electron microscope studies on *T. brucei*, enzyme activity is indicated in vesicles which are much smaller than those visualized with the light microscope: presumably the longer incubation times required for visualization of lysosome activity by light microscopy results in more lead sulfide being deposited not only inside but also around these bodies. There is as yet no direct evidence, however, that lysosomes are produced by the Golgi apparatus, or that the exoantigens are lysosomal enzymes.

Herbert (1965b) demonstrated several lysosomal enzymes (acid phosphatase, esterase, β-glucuronidase) in culture forms of *Trypanosoma melophagium* and *T. theileri* using cytochemical techniques, and points out that lysosomes may play an important part in the detoxification of drugs, as well as in the degradation of cellular structure during the life cycle or during degeneration of culture forms. Herbert's published electron micrographs show what look like autophagic vacuoles (cytolysomes) in the cytoplasm. This type of lysosome is believed to incorporate cytoplasmic reticulum and mitochondrial material and digest them.

d. The Reservoir or Flagellar Pocket. The pellicular pocket around the base of the flagellum is traditionally referred to as the reservoir (see Clark, 1959) but this term is of dubious functional relevance. "Flagellar pocket" (c.f. Rudzinska *et al.*, 1964; W. A. Anderson and Ellis, 1965) is perhaps more appropriate. Under the electron microscope the cavity of this pocket

---

* The structures tentatively designated lysosomes by Vickerman (1962) were misidentified, and are the "membrane-bound dense bodies" described later in this chapter.

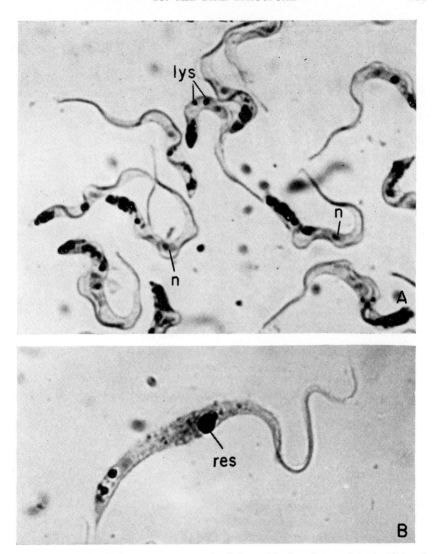

FIG. 43. Smears of trypanosomes stained for acid phosphatase by the Gomori technique. A: In *Trypanosoma brucei* the black deposits of lead sulfide indicate sites of enzyme activity. They occur in the reservoir at the base of the flagellum and in discrete bodies (lysosomes, lys) in the postnuclear region. The nucleolus (n) has taken up the hemalum counterstain. Magnification: × 1750. B: Culture forms of *T. conorrhini* show a massive deposition in the reservoir (res) and lysosome-like bodies at the posterior end of the flagellate. Magnification: × 2000. (Courtesy of Brooker, 1966.)

is seen to contain membrane-bound particles (see Fig. 34A) which appear to be extruded cytoplasmic debris—slender morphological evidence that the reservoir is a site of secretion or excretion.

e. Endoplasmic Reticulum: Ribosomes. A system of membranous tubes and vesicles, studded with ribosomes and connected to the nuclear envelope, constitutes the rough-surfaced endoplasmic reticulum of trypanosome cytoplasm (Figs. 33, 38, 42B). This reticulum is very much in evidence between the nucleus and the kinetoplast. In bloodstream trypanosomes rough-surfaced sacs are seen lying adjacent to the pellicle microtubules in the anterior (prenuclear) portion of the flagellate.

Ribosomes are abundant in the cytoplasmic matrix of trypanosomes, especially following glutaraldehyde fixation. There is evidence from biochemical studies that in trypanosomes from animals treated with Antrycide these ribosomes become aggregated to form the chemotherapy granules seen in Giemsa-stained preparations, and that in their aggregated state the ribosomes are inactive in protein synthesis (Newton, 1966).

f. Unidentified Structures. The cytoplasm of trypanosomes contains many granules, vesicles, and vacuoles whose functional identity has not been satisfactorily determined; electron microscopy has contributed little to these investigations. One problem has been that structures detected by bright field, phase, or reflex microscopy may not be preserved by the techniques employed by electron microscopists (Westphal, 1965), so that the correlation of light and electron microscope observations becomes difficult. For detailed discussion of light microscope observations on trypanosome cytoplasmic granules from several viewpoints the reader is referred to papers by Herbert (1965a,b), Michel (1964, 1966), Molloy and Ormerod (1965), Ormerod and Shaw (1963), and Westphal (1963a). Only a few notes are given here.

Basophilic granules designated "volutin" by Swellengrebel (1908) have been studied by many workers (pentose nucleic acid granules of Van den Berghe (1942), "methylene blue granules" of Michel (1964, 1966), "Type I granules" of Molloy and Ormerod (1965)). They are abundantly evident in methanol-fixed, Giemsa-stained smears of bloodstream trypanosomes over the entire course of infection. In *Trypanosoma rhodesiense* they occur throughout the cytoplasm. They are about 0.5 μ in diameter and difficult to visualize by phase microscopy. In electron micrographs their identification remains purely arbitrary (e.g., see W. A. Anderson and Ellis, 1965). The fact that they cannot be visualized after osmic acid treatment may mean that they have hitherto eluded the electron microscopist who relies solely upon this method of fixation. The relationship of volutin granules to the "chemotherapy granules" which also contain ribonucleic acid (RNA) (see above) is obscure (Ormerod and Shaw, 1963).

Phase-dense lipid vacuoles are common in both bloodstream and cultured trypanosomes. These vacuoles ("Type II" granules of Molloy and Ormerod) appear as empty spaces in Giemsa-stained trypanosomes. According to Molloy and Ormerod (1965) they are absent from slender *Trypanosoma brucei* and appear as transformation to stumpy forms takes place. Herbert (1965b) notes that these vacuoles increase in size and number in aging culture forms. Their osmophilic properties make lipid vacuoles easy to recognize in electron micrographs. Phospholipid vacuoles may also be present (Herbert, 1965b), and the "myelin forms" (concentric whorls of membranes) seen in some sections may result from hydration of phospholipid at fixation (Stoeckenius, 1957).

In the future more of the unit membrane-bound structures seen in the cytoplasm of trypanosomes with the electron microscope will undoubtedly prove to be sites of enzyme activity, but the application of cytochemical techniques to the electron microscopy of trypanosomes has scarcely started. Michel (1964, 1966) has demonstrated basophilic granules ("Azure I granules") throughout the cytoplasm of monomorphic bloodstream *T. gambiense*, and these give a positive reaction for NAD and NADP diaphorases using tetrazolium salts; reflex microscopy has been used extensively in the study of these granules. Ryley (1967) has demonstrated the presence of the L-α-glycerophosphate dehydrogenase (a component of the L-α-GP oxidase system) of monomorphic bloodstream trypanosomes in similar discrete extra-mitochondrial bodies, using L-α-GP as substrate and MTT tetrazolium salt (Hess *et al.*, 1958) as acceptor. Adaptation of this technique for electron microscopy shows that the dehydrogenase is located in membrane-bound dense bodies indicated in several electron micrographs illustrating this review (Fig. 42A and B). The dense bodies are surrounded by a unit membrane and their contents have varying electron density but always several patches of very dense material. They are prominent in the cytoplasm of bloodstream trypanosomes belonging to the *T. congolense*, *T. vivax*, and *T. brucei* groups, but are absent—or not evident—in sections of culture or vector forms. These membrane-bound dense bodies could well be the site of the particulate glycerophosphate oxidase system which is active in the terminal respiration of bloodstream forms (see Fig. 41). In culture forms elongate electron-dense bodies with the form of stout threads have been noted in electron micrographs by several workers (e.g., Herbert, 1965b) but whether these represent transformed membrane-bound dense bodies is not known. Symbionts such as have been described from the monogenetic trypanosomatid *Crithidia oncopelti* (Gill and Vogel, 1963) have not yet been demonstrated in any trypanosomes.

B. STRUCTURE AND FUNCTION OF THE NUCLEUS

The nucleus is bounded by an envelope of two apposed membranes, punctured by pores (approximately 1000 Å across). Its internal structure is not well shown by electron microscopy. The central karyosome (nucleolus), evident in the nucleus of all trypanosomatids by phase microscopy, is also prominent in electron micrographs (Figs. 33, 36B). It has a ribosome-like substructure. Chromosomal material is visible in sections stained with lead salts or uranyl acetate as dense patches lying beneath the nuclear envelope (Fig. 38). Ris (1962) has discerned 100 Å fibrils here and these may represent a nucleic acid-histone complex.

Nuclear division in trypanosomes is believed to take the form of a "promitosis", similar to that observed in euglenid flagellates: the nuclear envelope remains intact throughout division and the karyosome deforms into a dumbbell-shaped structure around the constriction of which the 6–8 chromosomes appear to form a circlet as seen by light microscopy. As the two daughter karyosomes separate each takes with it one set of daughter chromosomes and half the nuclear envelope to form a new daughter nucleus. The electron microscope shows that during division the karyosomal material lies within a bundle of tubular fibers which are identical in their mensural characteristics with the fibers of the mitotic spindle in other eukaryotic cells (Vickerman, 1966c). The fibers arise at the nuclear envelope, but no centrioles are visible at their point of origin. Contrary to the conclusions of light microscopists, then, a spindle is present in this "promitosis." Chromosomal material may show signs of attachment to this spindle but some electron-dense material lies plastered against the nuclear envelope, so it is conceivable that the envelope plays some part in the separation of the daughter chromosomes.

IV. CONCLUSION AND GENERAL REMARKS

The electron microscope study of blood Protozoa uncovered some unexpected structures and mechanisms which could not be otherwise revealed. The new information contributed to a better understanding of a number of intricate problems connected with the very specialized and complicated life cycles of these parasites. As usual new problems arose; they have to be solved by other means; most of them will need a biochemical approach.

One of the most surprising findings is the way intraerythrocytic parasites feed on the host cell. Electron microscope study has shown that the cytoplasm of the erythrocyte is incorporated into food vacuoles by pinocytosis. In malaria parasites, pinocytosis takes place through invaginations of the double plasma membrane at any region of the surface of the body and through a fixed organelle, a primitive cytostome. Hemozoin, the residue of hemoglobin digestion, a natural marker of the site of digestion, indicates that in all bird malaria parasites digestion occurs inside the food vacuole, while in mammalian species in small vesicles pinched off from the food vacuole proper. Only in *Plasmodium falciparum* were both types of digestion found. It is not yet known whether this depends on the host cell or on the age of the parasite, and more work is needed to elucidate this interesting problem.

Analogous differences occur in the nuclei of parasites of mammals when contrasted with those in birds and reptiles. In the former no nucleolus is present in the nucleus; the latter possess a nucleolus. Here again *Plasmodium falciparum* is an exception; it possesses a nucleolus. No reasonable explanation is so far available.

A highly significant difference exists in the mitochondrial system of intraerythrocytic species parasitic in mammals and those found in birds and reptiles. The former do not have typical mitochondria but a structure composed of concentric double membranes which presumably performs mitochondrial functions. This assumption has to be experimentally verified. Those parasitic in birds and reptiles possess typical protozoan mitochondria, in the intra- and exoerythrocytic forms as well as in sporozoites. Nothing is known as to the chondrion of exoerythrocytic forms of mammalian plasmodia, since no electron microscopy work is available so far. Sporozoites of the mammalian malaria parasite, *P. berghei*, do have regular mitochondria and this is most probably the case in all other mammalian species. From the studies so far performed it seems that the presence and absence of mitochondria is closely related to the taxonomic group to which the host belongs. It is hoped that biochemical studies will be able to elucidate this interesting relationship, which is difficult to understand.

Of general interest for cell biologists is the origin of mitochondria. The concentric double-membraned structure in *Plasmodium* derives from the plasma membrane. Should cytochemical and biochemical studies supply evidence that these structures perform mitochondrial functions, a significant step in the evolution of a cell organelle would have been found.

The electron microscope study disclosed that the reproductive forms of malaria parasites (merozoites, sporozoites, ookinetes) possess at the apical part of the body a set of highly specialized organelles such as a conoid, paired organelle, toxonemes, and pellicular microtubules. These structures disappear in the trophozoite stage and are formed again toward the end of schizogony. Thus the organism undergoes dedifferentiation and redifferentiation several times during its lifetime. It appears that in *Lankesterella* these structures are retained through all developmental stages. The fact that these structures are permanent in *Lankesterella* while they appear and disappear in *Plasmodium* during the lifespan may indicate that *Lankesterella* represents a more primitive stage in the evolutionary ladder.

Studies on the fine structure of trypanosomes have been largely concerned with the kinetoplast in relation to the function of this organelle. The synthesis of this work with parallel studies on metabolism has broadened considerably our understanding of the life cycles and evolution of these flagellates. These developments in the fundamental biology of trypanosomes form not only a basis for work on their chemotherapy, but also a contribution to academic cell biology. In the trypanosomes we have an outstanding illustration of the mitochondrion as an extranuclear genetic system, clearly showing replication, activation, and mutation, on the one hand, and a high DNA content, presumably representing genetic information, on the other. It is a pity that the absence of a sexual process in trypanosomes precludes studies on the relative roles of nuclear and mitochondrial genetic systems in controlling mitochondrial activity and inheritance. The application of cytochemical techniques at the level of the electron microscope and the more accurate identification of centrifugal fractions that the electron microscope offers will facilitate future work on the other cytoplasmic systems of trypanosomes.

REFERENCES

Aikawa, M. (1966). *Am. J. Trop. Med. Hyg.* **15**, 449.
Aikawa, M., Hepler, P., Huff, C. G., and Sprinz, H. (1966a). *J. Cell Biol.* **28**, 355.
Aikawa, M., Huff, C. G., and Sprinz, H. (1966b). *Military Medi.* **131**, Suppl., 969.
Alexeieff, A. (1917). *Compt. Rend. Soc. Biol.* **80**, 499.
Anderson, E., Saxe, L. H., and Beams, H. W. (1956). *J. Parasitol.* **42**, 11.
Anderson, W. A., and Ellis, R. A. (1965). *J. Protozool.* **12**, 483.
Anfinsen, C. B., Geiman, Q. M., McKee, R. W., Ormsbee, R. A., and Ball, E. G. (1946). *J. Exptl. Med.* **84**, 607.
Ashcroft, M. T. (1957). *Ann. Trop. Med. Parasitol.* **51**, 301.
Baker, J. R. (1961). *Trans. Roy. Soc. Trop. Med. Hyg.* **55**, 518.
Balamuth, W. (1965). *Progr. Protozool., Abstr. 2nd Intern. Conf. Protozool., 1965.* p. 40. Excerpta Med. Found., Amsterdam.
Baldi, A. (1950). *Riv. Malariol.* **29**, 349.
Ball, E. G., Anfinsen, C. B., Geiman, Q. M., McKee, R. W., and Ormsbee, R. A. (1945). *Science* **101**, 542.
Barrow, J. H. (1954). *Trans. Am. Microscop. Soc.* **73**, 242.
Bayer, M. E., and Dennig, K. H. (1961). *Z. Tropenmed. Parasitol.* **12**, 28.
Beams, H. W., and Anderson, E. (1961). *Ann. Rev. Microbiol.* **15**, 47.
Beams, H. W., and Tahmisian, T. N. (1954). *Exptl. Cell Res.* **6**, 87.
Bernhard, W., and de Harven, E. (1960). *Proc. 4th Conf. Electron Microscopy, Berlin, 1958*, Vol. II, p. 217. Springer, Berlin.
Bessis, M. (1958). *Bull. Acad. Natl. Med. (Paris)* **142**, 629.
Bessis, M.C., and Breton-Gorius, J. (1959). *Rev. Hematol.* **14**, 165.
Brandt, P. W. (1958). *Exptl. Cell Res.* **15**, 300.
Bresslau, E., and Scremin, L. (1924). *Arch. Protistenk.* **48**, 509.
Brooker, B. E. (1966). Unpublished data.
Brooker, B. E. (1966). Personal communication.
Brooker, B. E., and Vickerman, K. (1964). *Trans. Roy. Soc. Trop. Med.* **58**, 293.

Brown, K. N., Armstrong, J. A., and Valentine, R. C. (1965). *Exptl. Cell Res.* **39**, 129.
Burstone, M. S. (1962). "Enzyme Histochemistry and its Application in the Study of Neoplasms." Academic Press, New York.
Büttner, D. W. (1966). *Z. Tropenmed. Parasitol.* **17**, 397.
Carasso, N., Favard, P., and Goldfischer, S. (1964). *J. Microscopie* **3**, 297.
Chandler, A. C., and Read, C. P. (1961). "Introduction to Parasitology." Wiley, New York.
Chapman-Andresen, C., and Holter, H. (1955). *Exptl. Cell Res.* **3**, Suppl., 52.
Cheissin, E. M. (1964). *Zool. Zhu.* **43**, 647.
Cheissin, E. M., and Snigirevskaya, E. S. (1965). *Protistologica* **1**, 121.
Clark, T. B. (1959). *J. Protozool.* **6**, 227.
Clark, T. B., and Wallace, F. G. (1960). *J. Protozool.* **7**, 115.
Cosgrove, W. B., and Anderson, E. (1954). *Anat. Record* **120**, 813.
Cox, F. E. G., and Vickerman, K. (1966). *Ann. Trop. Med. Parasitol.* **60**, 293.
Cox, H. W., Schroeder, W. F. and Ristic, M. (1966). *J. Protozool.* **13**, 327.
Daniels, E. W., and Breyer, E. (1965). *J. Protozool.* **12**, 417.
Davis, A. G., Huff, C. G., and Palmer, T. T. (1966). *Exptl. Parasitol.* **19**, 1.
Deegan, T., and Maegraith, B. G. (1956). *Ann. Trop. Med. Parasitol.* **50**, 194.
de Harven, E., and Bernhard, W. (1956). *Z. Zellforsch. Mikroskop. Anat.* **45**, 378.
de Thé, G. (1964). *J. Cell Biol.* **23**, 265.
Dippell, R. V. (1965). *Progr. Protozool., Abstr. 2nd Intern. Conf. Protozool., 1965* p. 65. Excerpta Med. Found., Amsterdam.
Duncan, D., Street, J., Julian, S. R., and Micks, D. O. (1959). *Texas Rept. Biol. Med.* **17**, 314.
Duncan, D., Eades, J., Julian, S. R., and Micks, D. (1960). *J. Protozool.* **7**, 18.
Edwards, G. (1925). *Biol. Bull.* **48**, 236.
Elliott, A. M., and Bak, I. J. (1964). *J. Cell Biol.* **20**, 113.
Fauré-Fremiet, E. (1957). *Biol. Jaarboek Koninkl. Natuurw. Genoot. Dodonaea Gent.* **24**, 47.
Fauré-Fremiet, E. (1958). *Rev. Pathol. Gen. Physiol. Clin.* **58**, 265.
Fawcett, D. W. (1961). *In* "The Cell" (J. Brachet and A. E. Mirsky, eds.), Vol. 2, p. 217. Academic Press, New York.
Fawcett, D. W., and Porter, K. R. (1954). *J. Morphol.* **94**, 221.
Fitz-James, P. C. (1960). *J. Biophys. Biochem. Cytol.* **8**, 507.
Fletcher, K. A., and Maegraith, B. G. (1962). *Ann. Trop. Med. Parasitol.* **56**, 492.
Flewett, T. H., and Fulton, J. D. (1959). *Ann. Trop. Med. Parasitol.* **53**, 501.
Frenster, J. H., Allfrey, V. G., and Mirsky, A. E. (1963). *Proc. Natl. Acad. Sci. U. S.* **50**, 1026.
Friend, D. S. (1966). *J. Cell Biol.* **29**, 317.
Fulton, J. D., and Flewett, T. H. (1956). *Trans. Roy. Soc. Trop. Med. Hyg.* **50**, 150.
Fulton, J. D., and Spooner, D. F. (1959). *Exptl. Parasitol.* **8**, 137.
Gall, J. G. (1961). *J. Biophys. Biochem. Cytol.* **10**, 163.
Garnham, P. C. C. (1963). *Proc. 1st Intern. Congr. Protozool., Prague, 1961* p. 427. Academic Press, New York.
Garnham, P. C. C. (1966a). "Malaria Parasites and Other Haemosporidia." Blackwell, Oxford.
Garnham, P. C. C. (1966b). *Biol. Rev.* **41**, 561.
Garnham, P. C. C., and Bird, R. G. (1962). *Sci. Rept. 1st. Super. Sanita* **2**, 83.
Garnham, P. C. C., Bird, R. G., and Baker, J. R. (1960). *Trans. Roy. Soc. Trop. Med. Hyg.* **54**, 274.

Garnham, P. C. C., Bird, R. G., Baker, J. R., and Bray, R. S. (1961). *Trans. Roy. Soc. Trop. Med. Hyg.* **55**, 98.

Garnham, P. C. C., Baker, J. R., and Bird, R. G. (1962a). *Brit. Med. J.* **I**, 83.

Garnham, P. C. C., Baker, J. R., and Bird, R. G. (1962b). *J. Protozool.* **9**, 107.

Garnham, P. C. C., Bird, R. G., and Baker, J. R. (1962c). *Trans. Roy. Soc. Trop. Med. Hyg.* **56**, 116.

Garnham, P. C. C., Bird, R. G., and Baker, J. R. (1963). *Trans. Roy. Soc. Trop. Med. Hyg.* **57**, 27.

Garnham, P. C. C., Bird, R. G., and Baker, J. R. (1967). *Trans. Roy. Soc. Trop. Med. Hyg.* **61**, 58.

Gavin, M. A., Wanko, T., and Jacobs, L. (1963). *J. Protozool.* **9**, 222.

Gerzeli, G. (1955). *Riv. Parassitol.* **16**, 209.

Gibbons, I. R., and Grimstone, A. V. (1960). *J. Biophys. Biochem. Cytol.* **7**, 697.

Giesbrecht, P. (1961). *Zentr. Bakteriol., Parasitenk, Abt. I. Orig.* **183**, 1.

Gill, J. W., and Vogel, H. J. (1963). *J. Protozool.* **10**, 148.

Glauert, A. M., and Hopwood, D. A. (1959). *J. Biophys. Biochem. Cytol.* **6**, 515.

Grant, P. T. (1966). *Symp. Soc. Gen. Microbiol.* **16**, 281.

Grassé, P. P. (1956). *Arch. Biol. (Paris)* **67**, 595.

Grassé, P. P. (1961). *Compt. Rend.* **252**, 3917.

Grassé, P. P., and Pyne, C. K. (1965). *Progr. Protozool., Abstr. 2nd Intern. Conf. Protozool., 1965*, p. 131. Excerpta Med. Found., Amsterdam.

Grassé, P. P., Carasso, N., and Favard, P. (1956). *Ann. Sci. Nat.: Zool. Biol. Animale* [11] **18**, 339.

Green, D. E., Beyer, R. F., Hansen, M., Smith, A. L., and Webster, G. (1963). *Federation Proc.* **22**, 1460.

Grimstone, A. V. (1959). *J. Biophys. Biochem. Cytol.* **6**, 369.

Grimstone, A. V. (1961). *Biol. Rev.* **36**, 97.

Grimstone, A. V., and Gibbons, I. R. (1966). *Phil. Trans. Roy. Soc. London* **B250**, 215.

Gustafson, P. V., Agar, H. D., and Cramer, D. I. (1954). *Am. J. Trop. Med. Hyg.* **3**, 1008.

Hepler, P. K., Huff, C. G., and Sprinz, H. (1966). *J. Cell Biol.* **30**, 333.

Herbert, I. V. (1965a). *Exptl. Parasitol.* **16**, 348.

Herbert, I. V. (1965b). *Exptl. Parasitol.* **17**, 24.

Hess, R., Scarpelli, D. G., and Pearse, A. G. E. (1958). *Nature* **181**, 1531.

Hogg, J. F., and Kornberg, H. L. (1963). *Biochem. J.* **86**, 462.

Holt, S. J. (1959). *Exptl. Cell Res. Suppl.* **7**, 1.

Holter, H. (1959). *Ann. N. Y. Acad. Sci.* **78**, 524.

Horne, R. W., and Newton, B. A. (1958). *Exptl. Cell Res.* **15**, 103.

Huff, C. G., and Coulston, F. (1944). *J. Infect. Diseases* **75**, 231.

Huff, C. G., Pipkin, A. C., Weathersby, A. B., and Jensen, D. V. (1960). *J. Biophys. Biochem. Cytol.* **7**, 93.

Hyman, L. H. (1940). "The Invertebrates: Protozoa Through Ctenophora." McGraw-Hill, New York.

Hyman, L. H. (1959). "The Invertebrates: Smaller Coelomate Groups." McGraw-Hill, New York.

Inoki, S., Nakanishi, K., and Nakabayashi, T. (1958). *Biken's J.* **1**, 194.

Jarrett, W. F. H., and Brocklesby, D. W. (1966). *J. Protozool.* **13**, 301.

Jerusalem, C., and Heinen, U. (1965). *Z. Tropenmed. Parasitol.* **16**, 377.

Jirovec, O. (1927). *Arch. Protistenk.* **59**, 550.

Joyon, L. (1963). *Ann. Fac. Sci., Univ. Clermont* **22**, 1.

Joyon, L., and Lom, J. (1966). *Compt. Rend.* **262D**, 660.

Judge, D. M., and Anderson, M. S. (1964). *J. Parasitol.* **50,** 757.

Jurand, A. (1961). *J. Protozool.* **8,** 125.

Klein, R. L., and Neff, R. J. (1960). *Exptl. Cell Res.* **19,** 133.

Kleinschmidt, A., and Kinder, E. (1950a). *Zentr. Bakteriol., Parasitenk., Abt. I. Orig.* **156,** 219.

Kleinschmidt, A., and Kinder, E. (1950b). *Optik* **7,** 322.

Kleinschmidt, A. K., and Lang, D. (1962). *Proc. 5th Intern. Congr. Electron Microscopy, Philadelphia, 1962* Vol. **2,** 1–8. Academic Press, New York.

Kraneveld, F. C., Houwink, A. L., and Keidel, H. (1951). *Koninkl. Ned. Akad. Wetenschap., Proc.***C54,** 393.

Kreier, J. P., Ristic, M., and Watrach, A. M. (1962). *Am. J. Vet. Res.* **23,** 657.

Ladda, R. L. (1966). *Military Med.* **131,** Suppl., 993.

Ladda, R. L., and Arnold, J. (1965). *Compt. Rend.* **260,** 6991.

Ladda, R. L., Arnold, J., Martin, D., and Luehrs, F. (1965). *Trans. Roy. Soc. Trop. Med. Hyg.* **59,** 420.

Ladda, R., Arnold, J., and Martin, D. (1966). *Trans. Roy. Soc. Trop. Med. Hyg.* **60,** 369.

Ledbetter, M. C., and Porter, K. R. (1963). *J. Cell Biol.* **19,** 239.

Lehninger, A. L. (1964). "The Mitochondrion." Benhamin, New York.

Lewis, W. H. (1931). *Bull. Johns Hopkins Hosp.* **49,** 17.

Lillie, R. D. (1947). *J. Lab. Clin. Med.* **32,** 76.

Linnane, A. W., Vitols, E., and Nowland, P. G. (1962). *J. Cell Biol.* **13,** 345.

Littau, V. C., Allfrey, V. G., Frenster, J. H., and Mirsky, A. E. (1964). *Proc. Natl. Acad. Sci. U. S.* **52,** 93.

Ludvik, J. (1956). *Zentr. Bakteriol., Parasitenk., Abt I. Orig.* **166,** 60.

Ludvik, J. (1958). *Zentr. Bakteriol., Parasitenk., Abt I. Orig.* **172,** 330.

Ludvik, J. (1960). *J. Protozool.* **7,** 128.

Ludvik, J. (1963). *Proc. 1st Intern. Congr. Protozool., Prague, 1961* p. 387. Academic Press, New York.

Mackerras, M. J., and Mackerras, I. M. (1961). *Australian J. Zool.* **9,** 123.

McQuillen, K. (1956). *In* "Bacterial Anatomy" (E. T. C. Spooner and B. A. D. Stocker, eds.), pp. 127–149. Cambridge Univ. Press, London and New York.

Maggio, R., Siekevitz, P., and Palade, G. E. (1963). *J. Cell Biol.* **18,** 267.

Marshall, J. M., Jr., Schumaker, V. V., and Brandt, P. W. (1959). *Ann. N. Y. Acad. Sci.* **78,** 515.

Mast, S. O., and Doyle, W. L. (1933). *Protoplasma* **20,** 555.

Meyer, H., and Musacchio, M. de O. (1960). *J. Protozool.* **7,** 222.

Meyer, H., and Musacchio, M. de O. (1965). *J. Protozool.* **12,** 193.

Meyer, H., and Porter, K. R. (1954). *Parasitology* **44,** 16.

Meyer, H., and Queiroga, L. T. (1960). *J. Protozool.* **7,** 124.

Meyer, H., and Xavier de Oliveira, M. (1947). *Rev. Brasil Biol.* **7,** 327.

Meyer, H., Musacchio, M. de O., and Mendonca de Andrade, I. (1958). *Parasitology* **48,** 1.

Michel, R. (1964). *Z. Tropenmed. Parasitol.* **15,** 400.

Michel, R. (1966). *Z. Tropenmed. Parasitol.* **17,** 68.

Mignot, J. P. (1964). *Compt. Rend.* **258,** 3360.

Miller, J. H., Swartzwelder, J. C., and Deas, J. E. (1961). *J. Parasitol.* **47,** 577.

Molloy, J. O., and Ormerod, W. E. (1965). *Exptl. Parasitol.* **17,** 57.

Moulder, J. W. (1955). *In* "Some Physiological Aspects and Consequences of Parasitism" (W. H. Cole, ed.), pp. 15–26. Rutgers Univ. Press, New Brunswick, New Jersey.

Mühlpfordt, H. (1963a). *Z. Tropenmed. Parasitol.* **14**, 357.
Mühlpfordt, H. (1963b). *Z. Tropenmed. Parasitol.* **14**, 475.
Mühlpfordt, H. (1964). *Z. Tropenmed. Parasitol.* **15**, 289.
Mühlpfordt, H., and Bayer, M. (1961). *Z. Tropenmed. Parasitol.* **12**, 335.
Müller, M., and Röhlich, P. (1961). *Acta Morphol. Acad. Sci. Hung.* **10**, 297.
Müller, M., Röhlich, P., Toth, J., and Toro, I. (1963). *In* "Lysosomes" (A. V. S. deReuck and M. P. Camerson, eds.), pp. 201–216. Little, Brown, Boston, Massachusetts.
Nass, M. M. K., Nass, S., and Afzelius, B. A. (1965). *Exptl. Cell Res.* **37**, 516.
Newton, B. A. (1966). *Symp. Soc. Gen. Microbiol.* **16**, 281.
Ormerod, W. E., and Shaw, J. J. (1963). *Brit. J. Pharmacol.* **21**, 259.
Palade, G. E. (1956). *J. Biophys. Biochem. Cytol.* **2**, Suppl., 85.
Palade, G. E., aud Siekevitz, P. (1956a). *J. Biophys. Biochem. Cytol* **2**, 171.
Palade, G. E., and Siekevitz, P. (1956b). *J. Biophys. Biochem. Cytol.* **2**, 671.
Peters, W., Fletcher, K. A., and Staubli, W. (1965). *Ann. Trop. Med. Parasitol.* **59**, 126.
Pitelka, D. R. (1961). *Exptl. Cell Res.* **25**, 87.
Pitelka, D. R. (1963). "Electron Microscopic Structure of Protozoa." Pergamon Press, Oxford.
Porter, K. R. (1957). *Harvey Lectures* **51**, 175.
Porter, K. R. (1961). *In* "The Cell" (J. Brachet and A. E. Mirsky, eds.), Vol. 2, pp. 621–676. Academic Press, New York.
Porter, K. R. (1966). *Ciba Found. Symp., Principles Biomol. Organ.* p. 308.
Porter, K. R., and Blum, J. (1953). *Anat. Record* **117**, 685.
Porter, K. R., Ledbetter, M. C., and Badenhauser, S. (1964). *Proc. 3rd Reg. Conf. (Eur.) Electron Microscopy, Prague, 1964* p. 119. Czech. Acad. Sci., Prague.
Preston, T. M. (1966). Personal communication.
Preston, T. M. (1900). Unpublished data.
Preston, T. M., and Vickerman, K. (1967). In preparation.
Pyne, C. K. (1958). *Exptl. Cell Res.* **14**, 388.
Pyne, C. K. (1960a). *Compt. Rend.* **250**, 1912.
Pyne, C. K. (1960b). *Compt. Rend.* **251**, 2776.
Raffaele, G. (1939). *Riv. Malariol.* **18**, 141.
Reichenow, E. (1940). *Arch. Inst. Biol.* (*Sao Paulo*) **11**, 433.
Ris, H. (1962). *Proc. 5th Intern. Conf. Electron Microscopy, Philadelphia, 1962* Vol. II, Art. XX-1. Academic Press, New York.
Ristic, M., and Kreier, J. P. (1964). *Am. J. Trop. Med. Hyg.* **13**, 509.
Robertson, J. D. (1959). *Biochem. Soc. Symp.* (*Cambridge, Engl.*) **16**, 3.
Robertson, J. D. (1961). *In* "Regional Neurochemistry" (S. S. Kety and J. Eikes, eds.) pp. 497–530. Pergamon Press, Oxford.
Robertson, M. (1912). *Proc. Roy. Soc.* **B85**, 527.
Robertson, M. (1927). *Parasitology* **19**, 375.
Roskin, G., and Schlischliaiewa, S. (1928). *Arch. Protistenk.* **60**, 460.
Roth, L. E. (1960). *J. Protozool.* **7**, 176.
Roth, L. E., and Daniels, E. W. (1962). *J. Cell Biol.* **12**, 57.
Rudzinska, M. A. (1955). *J. Protozool.* **2**, 188.
Rudzinska, M. A. (1961). *J. Gerontol.* **16**, 213.
Rudzinska, M. A. (1962). *Gerontologia* **6**, 206.
Rudzinska, M. A., and Trager, W. (1957). *J. Protozool.* **4**, 190.
Rudzinska, M. A., and Trager, W. (1959). *J. Biophys. Biochem. Cytol.* **6**, 103.
Rudzinska, M. A., and Trager, W. (1961). *J. Protozool.* **8**, 307.

Rudzinska, M. A., and Trager, W. (1962). *J. Protozool.* **9**, 279.

Rudzinska, M. A., and Trager, W. (1968). *J. Protozool.* **15**. (In press.)

Rudzinska, M. A., Bray, R. S., and Trager, W. (1960). *J. Protozool.* **7**, Suppl., 24.

Rudzinska, M. A., D'Alesandro, P. A., and Trager, W. (1964). *J. Protozool.* **11**, 166.

Rudzinska, M. A., Trager, W., and Bray, R. S. (1965). *J. Protozool.* **12**, 563.

Rudzinska, M. A., Jackson, G. J., and Tuffrau, M. (1966a). *J. Protozool.* **13**, 440.

Rudzinska, M. A., Trager, W., and Bradbury, P. C. (1966b). *Program Abstr. Am. Soc. Parasitol., San Juan, 1966* p. 36.

Ryley, J. F. (1956). *Biochem. J.* **62**, 215.

Ryley, J. F. (1961). *Ann. Trop. Med. Parasitol.* **55**, 149.

Ryley, J. F. (1967). *Proc. 1st Intern. Congr. Parasitol., Rome, 1964* (in press).

Saal, J. R. (1964). *J. Protozool.* **11**, 582.

Salton, M. B. J., and Chapman, J. A. (1962). *J. Ultrastruct. Res.* **6**, 489.

Sanabria, A. (1963). *Exptl. Parasitol.* **14**, 81.

Sanabria, A. (1964). *Exptl. Parasitol.* **15**, 125.

Schneider, L. (1964). *Z. Zellforsch. Mikroskop. Anat.* **62**, 225.

Scholtyseck, E. (1965). *Z. Zellforsch. Mikroskop. Anat.* **66**, 625.

Schubert, G. (1966). *Z. Parasitenk.* **27**, 271.

Schulz, H., and MacClure, E. (1961). *Z. Zellforsch. Mikroskop. Anat.* **55**, 389.

Schumaker. V. V. (1958). *Exptl. Cell Res.* **15**, 314.

Sedar, A. W., and Porter, K. R. (1955). *J. Biophys. Biochem. Cytol.* **1**, 583.

Sedar, A. W., and Rudzinska, M. A. (1956). *J. Biophys. Biochem. Cytol.* **2**, Suppl., 331.

Senaud, J. (1966). *Compt. Rend.* **262**, 119.

Sherman, I. W., and Hull, R. W. (1960). *J. Protozool.* **7**, 409.

Siddiqui, W. A., and Rudzinska, M. A. (1965). *J. Protozool.* **12**, 448.

Simpson, C. F., Bild, C. E., and Stoliker, H. E. (1963). *Am. J. Vet. Res.* **24**, 408.

Simpson, L. (1965). *Progr. Protozool., Abstr. 2nd Intern. Conf. Protozool., 1965*, p. 41. Excerpta Med. Found., Amsterdam.

Singer, I. (1953). *J. Infect. Diseases* **92**, 97.

Slautterback, D. B. (1963). *J. Cell Biol.* **18**, 367.

Stebbens, W. E. (1966a). *J. Protozool.* **13**, 59.

Stebbens, W. E. (1966b). *J. Protozool.* **13**, 63.

Steinert, G., Firket, H., and Steinert, M. (1958). *Exptl. Cell. Res.* **15**, 632.

Steinert, M. (1960). *J. Biophys. Biochem. Cytol.* **8**, 542.

Steinert, M. (1964). *J. Cell Biol.* **20**, 192.

Steinert, M. (1965). *Exptl. Cell Res.* **39**, 69.

Steinert, M., and Novikoff, A. B. (1960). *J. Biophys. Biochem. Cytol.* **8**, 543.

Stoeckenius, W. (1957). *Exptl. Cell Res.* **13**, 410.

Swellengrebel, N. H. (1908). *Compt. Rend. Soc. Biol.* **64**, 38.

Terzakis, J. A., Sprinz, H., and Ward, R. A. (1966). *Military Med.* **131**, Suppl. 984.

The Committee on Taxonomy and Taxonomic Problems of the Society of Protozoologists. (1964). *J. Protozool.* **2**, 7.

Tilney, L. A., and Porter, K. R. (1965). *Protoplasma* **60**, 318.

Tobie, J. E., and Coatney, G. R. (1961). *Exptl. Parasitol.* **11**, 128.

Tokuyasu, K., and Yamada, E. (1959). *J. Biophys. Biochem. Cytol.* **5**, 123.

Trager, W. (1943). *J. Exptl. Med.* **77**, 411.

Trager, W. (1947). *J. Parasitol.* **33**, 345.

Trager, W. (1950). *J. Exptl. Med.* **92**, 349.

Trager, W. (1955). *11th Conf. Protein Metab., New Brunswick, 1955* p. 3.

Trager, W. (1957). *Acta Trop.* **14**, 21.

Trager, W. (1960). *In* "The Cell" (J. Brachet and A. E. Mirsky, eds.), Vol. 4, pp. 151–213. Academic Press, New York.

Trager, W. (1964a). *In* "The Cell" (J. Brachet and A. E. Mirsky, eds.), Vol. 6, pp. 81–137. Academic Press, New York.

Trager, W. (1964b). *Am. J. Trop. Med. Hyg.* **13**, 162.

Trager, W. (1966). *Military Med.* **131**, 1009.

Trager, W., and Rudzinska, M. A. (1964). *J. Protozool.* **11**, 133.

Trager, W., Rudzinska, M. A., and Bradbury, P. (1966). *Bull. World Health Organ.* **35**, 883.

Van den Berghe, L. (1942). *Acta Biol. Belg.* **2**, 464.

Vanderberg, J., Rhodin, J., and Yoeli, M. (1967). *J. Protozool.* **14**, 82.

van Iterson, W., and Leene, W. (1964). *J. Cell Biol.* **20**, 361.

van Iterson, W. (1965). *Bacteriol. Rev.* **29**, 299.

Vickerman, K. (1961). *In* "The Biology of Cilia and Flagella" (M. A. Sleigh, ed.), p. 60. Macmillan, New York.

Vickerman, K. (1962). *Trans. Roy. Soc. Trop. Med. Hyg.* **56**, 487.

Vickerman, K. (1963a). *Proc. 1st Intern. Congr. Protozool., Prague, 1961* p. 398. Academic Press, New York.

Vickerman, K. (1963b). *J. Protozool.* **10**, Suppl., 15.

Vickerman, K. (1965). *Nature* **208**, 762.

Vickerman, K. (1966a). *Trans. Roy. Soc. Trop. Med. Hyg.* **60**, 8.

Vickerman, K. (1966b). *Sci. Progr. (London)* **54**, 13.

Vickerman, K. (1966c). Unpublished observations.

Vickerman, K. (1900). Unpublished data.

Vickerman, K., and Cox, F. E. G. (1967). "The Protozoa: An Introduction." John Murray, London.

von Brand, T., and Johnson, E. M. (1947). *J. Cellular Comp. Physiol.* **29**, 33.

von Brand, T., Weinbach, E. C., and Tobie, E. J. (1955). *J. Cellular Comp. Physiol.* **45**, 421.

Warhurst, D. (1965). *Progr. Protozool. Abstr. 2nd Intern. Conf. Protozool. 1965*, p. 183. Excerpta Med. Found., Amsterdam.

Warner, J. R., Rich, A., and Hall, C. E. (1962). *Science* **138**, 1399.

Watson, M. L. (1955). *J. Biophys. Biochem. Cytol.* **1**, 257.

Weibull, C. (1956). *In* "Bacterial Anatomy" (E. T. C. Spooner and B. A. D. Stocker, eds.), pp. 111–126. Cambridge, Univ. Press, London and New York.

Weitz, B. (1960). *J. Gen. Microbiol.* **23**, 589.

Wery, M., and Degroodt-Lasseel, M. (1967). *Ann. Soc. Belge Med. Trop.* (in press).

Westphal, A. (1963a). *Z. Tropenmed. Parasitol.* **14**, 49.

Westphal, A. (1963b). "Einfuhrung in die Reflexmikroskopie und die physikalischen Grundlagen mikroskopischer Bildentstehung." Thieme, Stuttgart.

Westphal, A. (1965) *Progr. Protozool., Abstr. 2nd Intern. Conf. Protozool. 1965*, p. 96. Excerpta Med. Found., Amsterdam.

Wijers, D. J. B. (1960). Thesis, University of Amsterdam.

Wijers, D. J. B., and Willett, K. C. (1960). *Ann. Trop. Med. Parasitol.* **54**, 341.

Woodcock, H. M. (1906). *Quart. J. Microscop. Sci.* **50**, 151.

Xavier de Oliveira, M., and Meyer, H. (1955). *Parasitology* **45**, 2.

Yotsuyanagi, Y. (1962). *J. Ultrastruc. Res.* **7**, 121.

11

Development and Reproduction
(*Vertebrate and Arthropod Host*)

R. BARCLAY McGHEE

I. INTRODUCTION

Despite the rather voluminous amount of time and effort directed toward elucidating the development and life cycles of the various members of the Haemosporina and Trypanosomatidae much confusion still reigns. Many of the reports which apparently conflict with accepted concepts of development do so in large measure because of the work of inadequately trained investigators. Interpretations of the evolutionary and thus the taxonomic affinities of the various groups are often based on reports from a single individual performing a single study of a parasite in a species of animal which was not known to be parasite-free at the time of study. A review of the literature points up that now, when investigations are at the molecular level, one could certainly serve the field of parasitology well by reinvestigation of many of the apparently illogical life cycles reported in the literature.

Realizing some of the inadequacies of our knowledge of the two groups, we present this rather generalized account of what is considered the present concept of development and reproduction in the Trypanosomatidae and Haemosporidia.

II. THE TRYPANOSOMATIDAE

A common failing of parasitologists, other scientists, and indeed human beings in general, is to view their own special groups of organisms as separate—spatially, temporally, and mentally—from all others. Especially is this true of those whose studies have to do with human pathogens. The importance, speaking anthropocentrically, of human pathogens is well established but no group stands alone. Knowledge of related organisms is an indispensable adjunct to understanding any group, and those who are not so equipped miss much. This is true of the trypanosomes, a large assemblage of organisms existing in all the classes of vertebrates and among many invertebrates. It is a remarkably homogeneous mélange considering its wide dispersal and, in fact, much of the confusion relative to taxonomy may be laid to this general similarity. Thus, while it is quite easy to delimit the family, the generic designation, not to mention species assignment, presents many difficulties. One may not state with certainty, for example, whether the trypanosomatid seen in the rectum of the phytophagous hemipteron, *Oncopeltus fasciatus*, is of the genus *Phytomonas*, *Leptomonas*, *Crithidia*, *Blastocrithidia*, *Rhynochoidomonas*, *Trypanosoma*, *Leishmania*, *Herpetomonas*, or even possibly *Proleptomonas*. If belonging to any of these genera, one is unable to say with any certainty whether there is one or perhaps innumerable species of the particular genus present. Animal inoculation with the various isolates would not necessarily answer the question. We are not entirely certain that some so-called "insect trypanosomids" might not have the capability of infecting vertebrates (Laveran and Franchini, 1914; Fantham and Porter, 1915a,b,c; McGhee, 1959b).

Within the wide array of hosts, one may find the various forms of trypanosomes in the alimentary tract, the salivary glands, hemocoel, or in the ovaries (O'Farrell, 1913) of the invertebrates; these forms may exist in latex or in nonlactiferous sap of various plants; they are found in macrophages, various tissue cells, erythrocytes, and in the body fluids of vertebrates. One genus, *Proleptomonas*, has been found to exist as a free-living soil protozoan (Woodcock, 1915).

The possibility of presenting a complete survey of the development of the many species of the family in a single chapter is unrealistic. An attempt will be made, therefore, to present as general yet as comprehensive a view as possible of the family Trypanosomatidae with the hope of acquainting the reader with the various genera, subgenera, and species which present confusing, frustrating, and at times, delightful facets of the intricacies of parasitic organisms.

A. DISTRIBUTION

The family Trypanosomatidae falls under the suborder, Trypanosominae; order, Kinetoplastida; and class, Zoomastigophorea (Honigberg *et al.*, 1964). The family is described as being slender and leaflike with a single nucleus and flagellum (Kudo, 1958); it contains seven genera and innumerable species ranging from *Trypanosoma gambiense* of man to *Proleptomonas faecicola* of the soil.

Among the orders of the mammalia one finds representatives of the genus *Trypanosoma* and *Leishmania*. Within the mammalian trypanosomes are two major groups and eight subgenera (Hoare, 1964). The mammalian group contains numerous species—eighty or ninety among the *lewisi* group alone (Davis, 1952). One species, *Trypanosoma (Endotrypanum) schaudinni*, is found within the erythrocytes of the sloth, *Choloepus didactylus*.

Although many trypanosomes have been reported from birds the entire life cycle of none has been elucidated.

Trypanosoma has been found in numerous species of fish, amphibia, and, along with *Leishmania* (in lizards), such reptiles as lizards, turtles, snakes, and crocodiles.

Invertebrates serve as intermediate hosts of *Trypanosoma*, *Leishmania*, and *Phytomonas*. In addition invertebrates are themselves infected with five genera of so-called "insect trypanosomidae": *Herpetomonas* (Kent, 1880), *Blastocrithidia* (Laird, 1959), *Crithidia* (Léger, 1902), *Leptomonas* (Kent, 1880), and *Rhynchoidomonas* (Patton, 1910). Plants belonging to the families Euphorbiaceae, Asclepiadaceae, Sapotaceae, Urticaceae, and Compositae are the hosts of *Phytomonas*. Pathogenic effects are produced by *Phytomonas* in the coffee plant and the organisms may be of considerable economic importance (Stahel, 1931; Vermeulen, 1963).

One genus and one species, *Proleptomonas faecicola* (Woodcock, 1915) has been found in excreta, in rotten cabbage stalks (Fantham, 1922), and in numerous soil samples (Sandon, 1926). Whether this organism is truly free-living or represents a stage of the life cycle of some parasitic form is as yet unknown.

B. PROBLEMS OF GENERA

For years the genera of the family Trypanosomatidae seemed to be quite securely and neatly packaged into a scheme familiar to all students of parasitology (Wenyon, 1926). However, owing to a misinterpretation of Léger's work (1902) by Patton (1908), the inclusion of the error into the works of various authors, and later owing to the reinvestigation of the works of Léger by Wallace (1943), the scheme no longer holds. A major mistake was in misinterpreting the description by Léger of a short, barley-corn-shaped organism from *Anopheles maculipennis* which he named

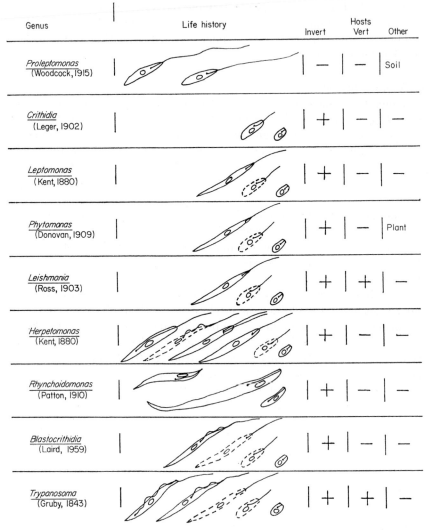

FIG. 1. Characteristics of the genera of the family Trypanosomatidae. The forms figured in dashed lines in this and the two following figures are forms which have never been demonstrated. Diagrammatic.

Crithidia fasciculata. He also described, in the same publication, another organism which seemed to have a rudimentary undulating membrane. From his description it seems logical to assume that he was aware of the existence of a doubly infected insect. Léger's work was confirmed by Novy *et al.* (1907), who extended his work by growing crithidia on culture medium.

In 1908, Patton reworked this group and decided that Léger was dealing with one genus alone and that the barleycorn-shaped organisms were in reality young forms of those with undulating membranes. He called the combined group *"Crithidia"*. Woodcock (1914) propagated the error.

These errors were incorporated into Wenyon's "Protozoology" (1926) and persist in most textbooks today.

Wallace (1943) redefined the genus *Crithidia* as conforming to that of Léger and followed this work with several papers concerning the generic designation. Still, many modern textbooks continue Patton's error. The form with the short undulating membrane has now (Laird, 1959) been placed in the genus *Blastocrithidia*. Thus, we have the following genera of insect trypanosomids, based on the predominant form seen in these organisms (Fig. 1).

C. TRYPANOSOMATIDAE OF INVERTEBRATES

1. The Genus Leptomonas

In general we adhere to the description given by Kent (1880) and by Wenyon (1926) as including flagellates which within their life cycles exhibit both leishmanial and leptomonal forms and which are confined to invertebrate hosts. In addition, as seen by McGhee and Hanson (1963) as well as by Noguchi (but not mentioned in the text) (1926), there is in some the formation of leishmaniform organisms by unequal fission.

Wallace (1966) lists over 60 species of *Leptomonas*. Of these, the life cycle has been described for but a few, and those few mostly confined to the development of *L. ctenocephali* of the flea (Wenyon, 1926). In this host the infection is limited to the intestinal tract and the Malpighian tubes. Generally, infections are in the posterior portion of the gut and it is here that the presumably resistant leishmaniform stage of *Leptomonas* is found. The forms in the dejecta of the flea are ingested by larval fleas and the parasites survive the development through the pupal stage to adulthood.

In 1963, McGhee and Hanson found that certain of laboratory-bred *Oncopeltus fasciatus* (the milkweed bug) harbored a leptomonad which was later identified as *Leptomonas oncopelti* (Noguchi and Tilden, 1926). Reproduction of this organism occurred in two ways; binary fission and budding. The former type of division produced daughter organisms approximately equal to the parent while the latter led to the formation of leishmaniform stages.

In the usual type of reproduction, nuclear division proceeds in the lateral plane. In budding, however, the nucleus either divides longitudinally or migrates forward following a lateral division. The migration anteriorly of one of the nuclei brings it in close proximity to the daughter kinetoplast. From this point division may proceed in one of three ways, all of which involve unequal cytoplasmic cleavage.

In some, the cytoplasm divides, separating the much smaller daughter from its parent. The new axoneme grows free of the existing one and the newly formed individual becomes free-swimming with, however, a change of morphology from the parent form, and more closely resembles a crithidian.

The second type is similar to the first save that the new axoneme seems to grow within the cytoplasmic sheath of the parent in a manner similar to that suggested by Wenyon (1926). Following cytokinesis the smaller organism remains attached to the parent flagellum. This smaller form may in turn divide into two smaller individuals still attached to the flagellum.

In the third type of division there is, following the first nuclear division and migration, a second division of the nucleus and kinetoplast. In this stage, two unequal divisions of the organism and a regrowth of two axonemes apparently into the cytoplasmic sheath of the parent flagellum result in the formation of two leishmaniform organisms and, upon a second division, four of these forms result.

2. The Genus Crithidia (Fig. 1)

We now adhere to the description as advanced by Léger (1902) of a truncate organism, barleycorn-shaped, which reproduces by binary fission and never, so far as we have observed, by budding or unequal fission. *Crithidia* is best known of all the insect trypanosonomids as regards physiology but its life cycle is practically unknown. It exists, so far as has been observed, either in a crithidiform or leishmaniform stage. Although we have successfully infected a variety of insects with this genus of protozoan (McGhee *et al.*, 1965) including *Rhodnius* (Hanson and McGhee, 1966), either by interrectal injection or *per os*, we have never been able to demonstrate transmission with these experimental animals. Fleming (1966) was unable to secure transmission of *C. luciliae* from housefly to housefly but was able to observe transmission of a yet unidentified crithidian from *Musca domestica* to other flies of the same species. Hanson and McGhee (1966) isolated a species of crithidia (quite fastidious as regards its growth in culture) from *Oncopeltus fasciatus* and twice observed a minimum transmission from parents to offspring (1 bug infected out of 15 exposed). Clark *et al.* (1964) noted that the flagellates of *C. fasciculata* could exist, at least temporarily, in the water inhabited by larval mosquitoes. Reproduction seems to be by binary fission alone. Cysts seem to be but slightly modified crithidiform organisms. Annear (1961) noted that *C. oncopelti* could survive drying for as long as 6 months.

3. The Genus Herpetomonas (Kent, 1880)

The description, as given by Wenyon (1926) of *Herpetomonas*, has been amended by Wallace (1966) to include those organisms which have a leptomonad stage, quite often with two flagella, and a form in which the kinetoplast is posterior to the nucleus. In the latter stage a reservoir extends almost the entire length of the body. There is no undulating membrane, a condition which removes it from the so-called "trypanosome" stage (Wallace, 1966). Only one species, *Herpetomonas muscarum*, has been thoroughly studied but even its life cycle is not fully known. No cyst forms have been observed in cultures of flies which are infected with *Herpetomonas* alone. Glaser (1926) and later Fleming (1966) demonstrated that its host (*Musca domestica*) is for all practical purposes free of flagellates during some of the winter months and that the organism infecting larval flies will not survive pupation. The latter author also showed that gnotobiotic flies were susceptible to infection; that transmission of the flagellate was from adult to adult fly; and that the presence of fresh garbage favored transmission. Corwin (1962) noted living herpetomonads in banana mash in the presence of infected *Drosophila*. It seems possible, therefore, that transmission of *Herpetomonas* may be realized by the motile flagellate.

4. The Genus Blastocrithidia (Laird, 1959)

This organism, called *Crithidia* by Patton (1908), is rounded posteriorly and tapered anteriorly. The kinetoplast lies close to and anterior to the nucleus; there is a short undulating membrane. So far as we have been able to determine there is no leptomonad form in its development. Moreover, as seen by Gibbs (1951) and in this laboratory (Hanson and McGhee, 1963), there are leishmaniform stages produced by unequal fission in its natural and experimental hosts. Crithidiform parasites apparently are not a part of its life cycle. Hanson and McGhee (1963) succeeded in infecting *Oncopeltus fasciatus* with a blastocrithidian from *Melanolestus picipes* (*Reduviidae*) and noted that the flagellate passed from parent to offspring bugs. O'Farrell (1913) described an ovarian-transmitted blastocrithidian (*B. hyalommae*) in the tick. Multiple fission and internal budding have been described from this species as well as for a blastocrithidian seen in the bug, *Euryophthalmus convivins* (McCulloch, 1917).

5. The Genus Rhynchoidomonas (Patton, 1910)

Wallace (1966) recognizes this genus as valid although other authors (Wenyon, 1926) place it in *Herpetomonas*. It is found in Diptera. The kinetoplast is posterior to the nucleus but there is no undulating membrane. The posterior end may be forked. Eight species are listed. Two of these were obtained from *Drosophila* (Chatton, 1913).

D. TRYPANOSOMATIDAE OF PLANTS

1. The Genus Phytomonas (Donovan, 1909)

Phytomonas is essentially leptomonad in form and infects plants and insects. Various schemes have been forwarded for the development of this protozoan, but with the finding of *Leptomonas oncopelti* in the milkweed bug, *Oncopeltus fasciatus*, it seemed reasonable to suspect that many of the life cycles formerly described had included parts of the cycles of two parasites. We (McGhee and Hanson, 1964) grew oncopeltids free of *Leptomonas* and allowed them to feed on infected milkweed plants. Thereafter, by daily dissections, we were able to work out the cycle of this species of *Phytomonas* in the bug. *Phytomonas* enters the alimentary tract as an actively swimming leptomonad and passes directly to the pylorus without any demonstrable change. Once in the pylorus, there is to our knowledge no division, simply growth. After a period of approximately 10 days during which time the flagellates had reached relatively enormous sizes, there was an invasion of the hemocoel. No dividing forms have been seen in this body cavity. Shortly thereafter, similar giant forms were observable in the salivary glands. Unequal cytokinesis ensues, and eventually smaller forms are produced. Bizarre forms are also found but their significance is unknown. The glands become heavily populated with active flagellates. When such bugs are fed on clean seedlings of the milkweed, the plant becomes infected (McGhee and Hanson, 1964).

E. TRYPANOSOMATIDAE OF THE SOIL

1. The Genus Proleptomonas (Woodcock, 1914)

Woodcock (1914) first reported this genus from a culture of goat feces. The description given leaves some question as to whether or not it has a kinetoplast. In 1922, Fantham saw a similar flagellate in decaying cabbage and, inasmuch as it possessed a definite kinetoplast, he called it *Herpetomonas brassicoe*. Another soil form was given the name *H. terricoloe*. Sandon (1926) describes this organism as recovered from soil samples from Greenland, Canada, and the United States. As mentioned before, some of the parasitic organisms are capable of living, for a short period at least, a free-living life, and these may be but forms of those normally found in insect guts. Until more information is available all the above free-living forms should be placed in Woodcock's original description of *Proleptomonas faecicola*.

F. TRYPANOSOMATIDAE OF VERTEBRATES

1. The Genus Leishmania (Donovan, 1909)

In man and in reservoir hosts *Leishmania* exists as a leishmaniform parasite within the cytoplasm of phagocytic cells. Within the host cell reproduction is by binary fission and continues until the escape of the organisms from the host cytoplasm. Transmission is by the sandfly *Phlebotomus*, where, after ingestion, the leishmaniform organisms change into the leptomonad forms which then undergo muliplication—again by binary fission. Eventually leptomonads reach the hypopharyngeal tube and are ready for transmission to a vertebrate host. *Leishmania tropica, L. brasiliensis,* and *L. mexicanum* attack the more superficial tissues of the body, whereas *L. donovani* is a more visceral infection and produces the disease known as kala azar.

2. The Genus Trypanosoma (Gruby, 1880)

The true trypanosomes are heteroxenous parasites with both invertebrate and vertebrate hosts. In the trypanosome stage the kinetoplast is located posterior to the nucleus and a definite undulating membrane is present. There are eight subgenera in the group infecting mammals.

a. Trypanosoma of Fish. Trypanosomes have been described from both fresh and saltwater fish. Brumpt (1906) investigated the development of trypanosomes in leeches and Robertson (1912) determined the complete cycle of those occurring in freshwater fish, in the leech, *Hemiclepsis*. The same author (Robertson, 1907) had earlier described the cycle of an elasmobranch trypanosome, *T. ragae,* in *Pontobdella*. The trypanosomes of freshwater fish displayed some slight differences as to the area of development in the leech. From 1912 to the present, with the exception of the work of Laird (1952), no concentrated attention has been given this group of organisms. The author failed to find trypanosomes in over 200 sea catfish taken from the Atlantic Ocean off the Georgia coast.

Robertson's account of the development of *T. ragae* of the carp (1907) is as follows (Fig. 2): After blood that contains trypanosomes is ingested by a susceptible leech (*P. murciata*), the trypanosomes round up and lose the flagellum. Division into two smaller forms followed by a second division into four still smaller forms occurs. After 36 hours axonemes begin to develop in the daughter organisms and blastocrithidial forms are produced. Brumpt (1906) described the presence of a leptomonad form in a trypanosome of the eel, but such forms were not observed by Robertson nor were they seen in cultured trypanosomes of the goldfish (Thomson, 1908). Eventually, in the leech, long slender trypanosomes develop. These presumably migrate forward to the proboscis sheath and are regurgitated into the next fish fed upon.

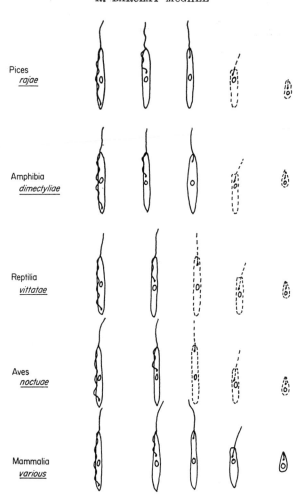

Pices
rajae

Amphibia
dimectyliae

Reptilia
vittatae

Aves
noctuae

Mammalia
various

Fig. 2. Characteristic stages of *Trypanosoma* found in various classes of verte
brates. The underlined name represents the species used to illustrate the various
morphological forms. Diagrammatic.

b. Trypanosoma of Amphibia. The bulk of research dealing with trypanosomes of amphibians, as in the case of the fish trypanosomes, was done
prior to 1930. They were first observed by Gluge in 1842.

Trypanosoma rotatorium is a parasite of frogs and is passed from tadpole
to tadpole by the leech, *Hemiclepsis marginata* (Nöller, 1913). Flagellates
are found in adult anurans but many seem to be abnormal. All stages
occur in the blood; no forms have been observed in the other tissues. In the
adult there are several forms present; the long narrow ones, the flat leaflike individuals, and the large compact organisms. The exact taxonomic

position of this trypanosome has never been truly determined. It is quite possible that its species may parasitize several different amphibian hosts. Development in the leech begins with the development of leptomonad forms, followed on the third day by blastocrithidial forms. After 7 days, narrow trypanosomes, which are apparently the infective forms, develop.

A careful study of the trypanosomes of the salamander, *Diemyctylus viridescens*, was carried out by Barrow (1954). In the vertebrate host trypanosomes occur only in the blood, where single and dividing forms may be found. In the leech there is a rounding up of the organism after 6 hours, followed by unequal division stages. By day 6 blastocrithidians and leptomonads are present and eventually trypanosomes develop from the blastocrithidian forms. After 16 days infective trypanosomes are found. Trypanosomes are regurgitated into the salamander.

c. Trypanosoma of the Reptilia. Laveran and Mesnil (1902) first described reptilian trypanosomes in the blood of a tortoise, *Damonia reevesii*. Very few attempts have been made to follow the life history of any of the forms found in lizards, snakes, chelonians, or in Crocodilia. Koch (1906) suggested that the trypanosomes of crocodiles (*T. kochi*) might be transmitted by a tsetse fly, *Glossina palpalis*. The actual cycle has never been conclusively shown, due partly to the fact that *G. palpalis*, in addition to *T. kochi*, apparently harbors a flagellate of its own, *Herpetomonas grayi*.

In aquatic reptiles, transmission is thought to be by the leech *Poecilobdella* and is marked, as are other developmental stages of trypanosomes, by the formation of blastocrithidial forms followed in turn by infective trypanosomes. No leptomonad, crithidial, or leishmaniform stages have been seen (Robertson, 1907).

d. Trypanosoma of Birds. Many avian trypanosomes, named specifically because of their occurrence in different species of birds, have been described but, as is true of so many of the trypanosomes, too little is known of their life histories to determine their exact taxonomic status. Little literature concerning experimental work is available beyond the fact that division takes place in the bloodstream in a manner similar to that of *T. lewisi* in the rat (Taliaferro, 1924). The species have little to separate themselves. The possibility exists of transmission by mosquitoes (Woodcock, 1914), but no definite proof exists. In the mosquito blastocrithidial and, later, infective trypanosomal forms occur.

e. Trypanosoma of Mammals. The mammalian trypanosomes have been subject to more classification changes than any other members of the family Trypanosomatidae. The most logical grouping of the mammalian trypanosomes is that of Hoare (1964, 1966). He divides the trypanosomes into two major sectors, the Stercoraria and the Salivaria. The latter section includes those trypanosomes which undergo their development in the anterior portion of the alimentary tract of the invertebrate. Under these two sections he lists eight subgenera, four under the Stercoraria and four under the Salivaria. In many instances the organisms listed as belonging to the various subgenera are incompletely known and they may, when and if their life cycles are elucidated, fall into another subgenus.

The four subgenera under Stercoraria, as now constituted, are grouped relative to their invertebrate vectors: hippoboscid flies, fleas, and reduviid bugs. One subgenus, *Endotrypanum*, is too little understood to define the invertebrate vector in which the organism develops, although Shaw (1964) suspects that development takes place in *Phlebotomus*. In 1964 Hoare classified *Endotrypanum* as a subgenus but in a later publication (1966) he raised it to generic rank. He does this on the basis of work done by Shaw (1964) who studied its development in the sloth, the presumed vector (*Phlebotomus*), and in culture. Its actual position in the family Trypanosomatidae remains, to quote Hoare, "enigmatic." Until more definite description is forthcoming for this parasite, it is felt that it should continue in its subgeneric rank. Inasmuch as the general scheme of development is similar where known within the subgenera, the cycles of the organisms will be considered collectively under the subgeneric designation (see Fig. 3).

i. Subgenus Megatrypanum. As implied in the name, this group includes large trypanosomes, found in Artiodactyla (*T. theileri, T. melophagium, T. mazamarum, T. tragelaphi, T. ingens, T. cephalophi*) and in Chiroptera (*T. megaderma, T. morinarum, and T. pessoai*).

Two of the better known representatives of the subgenus are *T. theileri* and *T. melophagi. Trypanosoma melophagi* may reach a size of 60 μ in its vertebrate host, the sheep. The organism occurs in the bloodstream of sheep without overt harmful effects to the host. Development in the hippoboscid fly, *Melophagus ovinus*, follows the characteristic pattern of blastocrithidial and trypanosomal forms and, in this case, leishmaniform organisms. Lambs may be infected by feeding to them the posterior gut contents of infected keds (Hoare, 1923).

Trypanosoma theileri (= *americanum*) is found quite commonly in cattle throughout the world, although its presence is only detected in most instances through cultivation. Crawley (1909) found 60% of the cattle in the Washington, D. C., area infected. Glaser (1926) found 70% of the cows in New Jersey infected. More recently Wells *et al.* (1965) isolated the organisms from cattle in Scotland, and Fleming (1966), in a survey of dairy cattle, found 4 of 54 cattle infected in Georgia, United States. The mode of transmission for this organism is not completely known but it is through mechanical transmission via *Tabanus, Haematopota,* and hippoboscid flies.

ii. Subgenus Herpetosoma. Numerous species falling within this group

Subgenus	Invertebrate host	Forms
Megatrypanum *melophagium*	Hippoboscidae	
Herpetosoma *lewisi*	Siphonaptera	
Schizotrypanum *cruzi*	Reduviidae	
Endotrypanum *schaudinni*	Phleboto – ? minae	
Duttonella *vivax*	Muscidae	
Nannomonas *congolense*	Muscidae	
Pycnomonas *suis*	Muscidae	
Trypanozoon *brucei*	Muscidae	

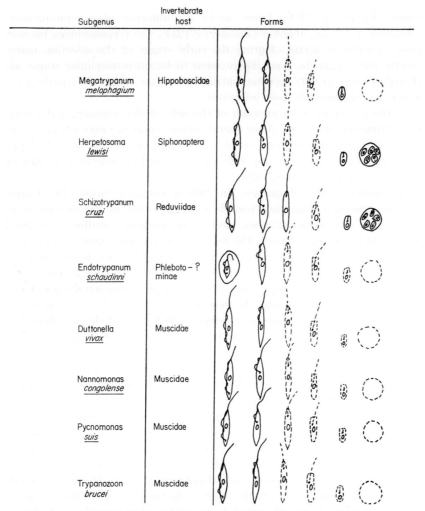

FIG. 3. Schematic representation of the various subgenera of mammalian *Trypanosoma*. The underlined names under the bold type subgenus represent the species selected to illustrate the characteristics of the subgenus. The leishmaniform parasites within circles exist in part intracellularly.

are found in Rodentia. Insofar as is known the transmitting agent is the flea Siphonaptera.

Trypanosoma lewisi is the best known member of the subgenus and has been the subject of intense investigation by numerous investigators. In the vertebrate host, *Rattus*, the trypanosomes first undergo a multiplicative phase during which there exists considerable variation in size of the organisms (Taliaferro, 1924). Later, under the influence of a reproduction-inhibiting antibody, ablastin (Taliaferro, 1924), the trypanosomes become more uniform in length. During the early stages of the infection many bizarre forms may be seen. There seem to be no intracellular stages although Carini (1910) reported intraerythrocytic stages. No pathogenic effects are apparent in most infections.

In the flea there is an invasion of the cells of the stomach, and a type of multiple fission ensues. After liberation the organisms may either invade more cells or may transform to blastocrithidial stages and later to trypanosomal stages in the rectum. The rat is infected by ingesting the flea or its infected feces (Wenyon, 1926).

iii. Subgenus Schizotrypanum. In 1909 a young scientist, Dr. Carlos Chagas, was sent from the Oswaldo Cruz Institute in Rio de Janeiro to investigate a malaria outbreak among the workers installing a railroad line in Minas Gerais, Brazil. He remained in the area more than a year before he became acquainted with the hemophagous hemipteron, or "barbeiro," *Panstrongylus megista.* Noting its habits, Chagas considered the possibility of its being a vector of some as yet unknown disease. Upon dissecting some of the bugs he reported ". . . I found in the posterior intestine of each one numerous flagellates which showed the morphological characteristics of crithidias." Aware of the existence of trypanosomes in local monkeys, he sent some of his bugs to the institute director, Dr. Oswaldo Cruz, who injected the gut contents into laboratory monkeys. Infection followed, and our knowledge of *Trypanosoma cruzi* began (Chagas, 1909; Bacellar, 1963).

Our understanding of the development of *T. cruzi* in the invertebrate host has been clouded by the probable dual infections of many of the hemipteran hosts with a second trypanosome. Chagas (1909) found nymphal bugs to be susceptible to infection and reported finding forms of the protozoan in the hemocoel and salivary glands of *Triatoma sordida* and *T. infestans.* Later Tejera (1920) reported the presence of a different species of trypanosome, *T. rangeli,* which is indeed found principally in the hemocoel and salivary glands. Brumpt (1912a) apparently possessd a homogeneous population of *T. cruzi* and, contrary to reports by Chagas (1909), secured infection in experimental animals through parenteral introduction of feces from infected bugs. Thus conflicting reports concerning the various stages existing in the insect have appeared and continue to appear. Chagas himself (1909) thought he observed schizogony in the parasites in the bug and set up a new genus, *Schizotrypanum,* and, although he later (1912) changed the genus back to its original *Trypanosoma,* one still finds the term *Schizotrypanum* in the literature.

Despite the possibility of Chagas' disease occurring naturally in other areas of the world, it is probably confined to the Western Hemisphere. It exists in numerous blood-sucking bugs and, in addition to man, in many animal reservoirs. The distribution of the organism reaches from central Chile to Maryland in the United States. Human infections have been known to exist as far north as Mexico and southward to Chile. In all probability, the organism is primarily confined to animals other than man, and perhaps man should be considered the reservoir host. Six genera and 26 species of Reduviidae have been found infected. The genera include *Panstrongylus*, *Triatoma*, *Eutriatoma*, *Rhodnius*, *Meprais*, and *Psammolestes* (Angel Torrico, 1946). Experimental infections have been produced in bedbugs, ixodid ticks, one species of hippoboscid, in the larvae of saturnid moths (Goldman, 1964), and in *Oncopeltus fasciatus* (Hanson and McGhee, 1966).

The list of vertebrates susceptible to *T. cruzi* is equally long and includes dogs, cats, monkeys, armadillos, bats, ferrets, opossums, foxes, wood rats, house mice (Packchanian, 1942) in Texas, raccoons (Walton, 1958) in Maryland, four species of mammals in Georgia (McKeever *et al.*, 1958), and opossums and raccoons in Alabama (Olson *et al.*, 1964).

No matter which invertebrate host, the development seems to be roughly the same (Hoare, 1934). Insects become infected by ingesting blood from an infected vertebrate host. In the gut, the parasite loses the trypanosomal form and develops into a rather short blastocrithidial stage. The organism becomes quite numerous and from it are produced leptomonads. After about 3 weeks, a small "metacyclic" trypanosome develops. These metacyclic forms are voided in the feces, are scratched into an area of skin adjacent to the insect bite, and infection of the vertebrate commences.

There is more or less general agreement concerning the nature of development in the invertebrate host, but many variations on the theme have been proposed. As already mentioned, Chagas (1909) thought he detected schizogony and also sex. Chagas (1927) described zygotes in the insect host. Muniz (1927) thought he detected sexual stages in cultures. Elkeles (1940) found forms which suggested sexual stages. These various claims have been discussed and rejected by Noble (1955). He concludes that the direct insect development is from trypanosome to leishmanial to trypanosome stage and that the blastocrithidial and leptomonad forms are incidental. There seems to be general agreement that the various stages of *T. cruzi* occur in the alimentary tract alone in the hemipteran host.

In the vertebrate host, *T. cruzi* makes its way into muscle tissue, including cardiac muscle, and nerve cells. It loses its flagellum, rounds up, and is found in nests as leishmanial stages. Division and continued invasion of tissue takes place. Eventually some transform to blastocrithidial forms or directly to trypanosomal forms (Elkeles, 1942). These trypanosomal forms are ingested by the susceptible insect host and the cycle is repeated.

Trypanosoma rangeli was first described by Tejera (1920) from the intestinal contents of *Rhodnius prolixus*. On the basis of forms in the insect, Tejera gave it the name of *T. rangeli*. In 1942 the same organism was isolated by allowing an uninfected insect to feed on infected patients (xenodiagnosis). Pifano and Mayer (1949) found a high incidence of *T. rangeli* in *R. prolixus* in a river valley in Venezuela. Using xenodiagnosis they reported 41 human cases, some also infected with *T. cruzi*. They also found trypanosome forms in salivary glands of infected bugs. Development in the insect has been described most recently by Tobie (1961). Following an infective meal, development begins in the bug; the flagellates soon migrate to the hemocoel and afterward to the salivary glands. Intrarectal injection of *Rhodnius prolixus* shows the same pattern of development (Hanson and McGhee, 1966). Long blastocrithidial forms are found in the hemocoel in 5 days and are followed by short recurved trypanosomal forms. The salivary glands are infected within 15 days; parasites here are principally trypanosomes. Infection may also be produced by injecting culture forms into the hemocoel of *Oncopeltus fasciatus*. Stages similar to those in *Rhodnius* are observed in the hemocoel but no forms have been observed in the salivary glands of this bug and injections of the organisms into the rectum have failed to produce infections.

Many investigators have held the belief that *T. rangeli* is transmitted to vertebrate hosts in the same manner as is *T. cruzi* (D'Alessandro, 1963). Tobie's (1964) unsuccessful attempts to infect suckling rats except through bites of infected bugs is rather convincing proof that, at least in the conditions obtaining in her experimental work, the route of infection is not the feces.

Infections seem to be confined mostly to infants and children. There are no leishmanial forms observed and no pathology has been reported.

Trypanosoma ariarii was discovered by Groot *et al.* (1951) in the valley of the Ariari River in Colombia. It differs somewhat in morphology from *T. rangeli* but is transmitted by the same species of hemipteron, *Rhodnius prolixus*, and follows the same route of development in the vertebrate and invertebrate. No pathogenic condition was produced in two human volunteers and no leishmanial stages have been demonstrated in human beings or in experimental animals. Trypanosomes are found in the salivary glands and in the proboscis of *Rhodnius*, and it is presumed that infection of the vertebrate host is through bites. Until more definitive criteria for differentiation are found, this species will have to be synonymous with *Trypanosoma rangeli* (Zeledon, 1954).

iv. *Subgenus Endotrypanum.* The single species of this subgenus, *T. schaudinni*, occurs in the edentate, the two-toed sloth (*Choloepus didactylus*). Its occurrence in red blood cells has occasioned its reception into a separate genus, although other trypanosomes have also been reported within erythrocytes (Carini, 1910). The vector is still unknown although Shaw (1964) suspects the sandfly (Phlebotominae). Within the vertebrate it exists as blastocrithidia and trypanosome forms.

v. Subgenus Duttonella. The invertebrate vectors, where such are known or required, of this subgenus are muscid flies. Chief among these is *Glossina*, the tsetse fly. Transmission is via bite. The two species, *T. vivax* and *T. uniforme*, comprise the subgenus *Duttonella*. Both species occur in Arteriodactylata and are transmitted by species of *Glossina*, in which host development takes place in the proboscis. Blastocrithidial and trypanosomal forms occur in the fly, whereas trypanosomal forms alone are found in the blood of the vertebrate. The infection is severe in cattle, sheep, and goats.

vi. Subgenus Nannomonas. Members of this subgenus undergo their development in the midgut and proboscis of *Glossina* but are not found in the salivary glands (Robertson, 1913). Cattle, horses, and sheep are the hosts for *T. congolense* and *T. dimorphon* while *T. simiae* occurs in monkeys. Forms are found in the blood of the vertebrate, and a wasting disease is produced. Blastocrithidial and trypanosomal forms alone are found in the fly.

vii. Subgenus Pycnomonas. The single species of this subgenus, *T. suis*, is found in pigs. In the invertebrate host, *Glossina*, reproduction takes place in the midgut and salivary glands. The organism is found in the blood of pigs and produces severe pathogenic effects.

viii. Subgenus Trypanozoon. The subgenus contains *T. evansi*, *T. equinum*, and *T. equiperdum*. One of these, *T. equiperdum*, requires no intermediate host. Infections are acquired venereally, although mechanical transmission may be possible (Sergent and Sergent, 1906a). It produces a fatal disease in horses known as *dourine*.

Trypanosoma evansi is worldwide in its distribution. It produces a fatal disease in horses. Transmission is mechanical and involves *Tabanus* (Musgrave and Clegg, 1903), and possibly ticks (Cross, 1923). *Trypanosoma equinum* is distinguished by the apparent absence of the kinetoplast. It produces a disease in cattle and horses known as *mal de Caderas*. Although listed as a pathogen of horses of South and Central America, there are practically no references to its existence, and in a recent publication of tropical diseases (Tropical Health, 1965) it is not listed as significant in the Southern Hemisphere. Transmission is generally by mechanical means through tabanid flies, although Hoare (1965) reports infection of vampire bats (*Demodus rotundus*) and subsequent transfer to horses and cattle on which the bats fed.

Trypanosoma gambiense and *T. rhodesiense* are the causative agents of human sleeping sickness in Africa. These organisms as well as *T. brucei* are

transmitted by *Glossina*. In the invertebrate host they multiply as trypanosomes in the midgut but eventually reach the salivary glands where
blastocrithidial and trypanosomal forms are produced (Yorke and MacFie,
1924). The differentiation of these three species remains confused. Ashcroft
(1959) reaches the conclusion that the three are morphologically identical
and may be separated only on physiological grounds.

Infections in vertebrates are severe and quite often result in fatal diseases.
A more detailed account of this subgenus as well as the others belonging to
the mammalian organism will be dealt with in other chapters.

G. PROBLEMS OF HOST SPECIFICITY

In reviewing the literature one is impressed with the plethora of conflicting reports by numerous authors relative to host specificity. Much confusion
reflects a general lack of knowledge of the basic fundamentals of biology
in general, and protozoology in particular. As a result we are faced with
the fact that we are not even completely sure of the development of most
of the species of the family Trypanosomatidae.

One particular pitfall is the tendency to describe as a trypanosome any
blastocrithidian observed in any hematophagous insect. That this may be
erroneous has been shown in our laboratory where we have successfully infected *Rhodnius prolixus* (a vector of *T. cruzi*) with parasites belonging to
the genera *Crithidia*, *Blastocrithidia*, and *Leptomonas*. Conversely, we have
infected *Oncopeltus fasciatus*, a milkweed bug, with *Trypanosoma cruzi* and
T. rangeli.

In one series of experiments Hanson and McGhee (1963) succeeded in
infecting the same milkweed bug with six species of *Crithidia*, two of *Blastocrithidia*, and two species of *Leptomonas*. The same bug is also susceptible
to infection by *Phytomonas* (McGhee *et al.*, 1965). McGhee (1959b) infected
avian embryos with *Crithidia* from phytophagous hemiptera and observed
intracellular forms in macrophages. Hanson and McGhee (1963) were
successful in infecting chick embryos with *Blastocrithidia* and *Leptomonas*.
Later Schmittner (1963) infected chick embryos with six species of *Crithidia*
and also secured infections in *Drosophila*, *Tenebrio*, and *Acheta*. *Phytomonas* infects three species of Lygeidae bugs: *Oncopeltus fasciatus*, *O.
sandarchatus*, and *Lygeus kalmii*, as well as various genera of the milkweed
plants (Family: Asclepiadaceae) (McGhee *et al.*, 1966).

Species are many times described on the sole basis of existence in a hitherto unreported host. Because of the lack of host specificity, as noted above,
this is an erroneous practice. This author has found unreported trypa-
nosomid flagellates in over ten species of hemipterons examined. Multiply
this by the possible numbers still unexamined and the number of species one
could describe is prodigious.

An additional consideration must be attached to examining any infected individual. The animal may be infected with as many as five genera of flagellates. The milkweed bug, for example, may be infected with *Trypanosoma, Crithidia, Leptomonas, Blastocrithidia,* and *Phytomonas.* Any attempt to elucidate the life cycle of a given organism in such a mélange of genera would be disastrous.

Even if the animal be infected with a single genus, one may not be certain whether there are one or innumerable species of *Trypanosoma* or *Blastocrithidia.* Ideally only the life cycle of cloned organisms should be described. Often this has proved impractical, however, and workers have resorted to serology (Noguchi, 1926; McGhee and Hanson, 1963), growth curves, biological characteristics, and physiology (Cosgrove, 1963) as aids in determining species.

Anyone engaged in a study of the trypanosomes should be familiar with the entire family Trypanosomatidae and be able to distinguish with some certainty the strictly insect forms from the *Trypanosoma;* otherwise, we may expect vast amounts of confusing information and quantities of misinformation.

III. THE PLASMODIIDAE

To understand fully malaria of higher organisms, one must be cognizant of the range of the genus in mammals, birds, reptiles, and even, perhaps, in the amphibia. In all hosts, the general principles as to development, transmission, and stages of the life cycle and physiology, so far as we now know, are similar. While the general scheme is one of agreement there are some differences; as, for example, the development of preerythrocytic stages of the human, simian, and rodent malarias in the liver cells as opposed to development of comparable avian stages in the macrophages and, later endothelium, of the host.

Individuals may find it necessary to work with malaria other than human, simian, or even rodent in order to fit the given research situation or to elucidate a specific point. Any discussion of malaria must, therefore, include some description of malaria and its allies in other hosts. Those who deal with any group of the Haemosporidia should be familiar with all the other organisms falling within the category.

A. INCIDENCE AND DISTRIBUTION OF PLASMODIA THROUGH THE ANIMAL KINGDOM

Beyond an unconfirmed report by Fantham *et al.* (1942), no references exist in the literature relative to plasmodia in amphibia. Fantham *et al.* claim to have found *Plasmodium bufonis* in the American toad and *P. catasbiana* in the American bullfrog. McGhee has examined hundreds of blood films of *Bufo* and *Rana* from various areas of the United States without finding forms similar to those described above. Frogs, however, have such a plethora of intracellular organisms that one should not be surprised if someday the two species of plasmodia may be rediscovered. Schmittner and McGhee (1961) discovered an intracellular sporozoan, *Babesiosoma stableri*, in the erythrocytes of *Rana pipiens*. This organism has asexual and, we suspect, sexual forms in the bloodstream and is transferable to Fowler's toad (*Bufo fowleri*). In this host there are pathogenic effects and, in some toads, death. There are in others definite indications of acquired immunity on the part of the host, and the organism may well be one which has promise of use as a study of immunity in poikilothermic animals.

Garnham (1966) lists 23 species of reptilian malaria. All, interestingly, are found in saurians or ophidians; none in chelonians or crocodilians. By far the most extensive and best work on the general subject of lizard malaria was done by Thompson and Huff (1944). Despite their work and that of Jordon (1964), the vector involved and the method of transmission are unknown.

Many species of malaria have been described from birds, although there are only 15 to 20 species generally accepted as valid. Morphology, which is the criterion of identification of a given avian malaria species, changes markedly with the change of host. Species much used in the laboratory are *Plasmodium lophurae* (Coggeshall, 1938), *P. relictum* (Grassi and Feletti, 1891), *P. fallax* (Schwetz, 1930) and *P. gallinaceum* (Brumpt, 1935).

The discovery by Vincke and Lips (1948) of a rodent malaria which may be transferred to laboratory rats, mice, and hamsters has resulted in a marked increase of research on the organism. Only recently has its entire life cycle been elucidated (Yoeli, 1965). Another parasite, *P. vinckei*, has somewhat similar characteristics.

Some years after Taliaferro *et al.* (1934) succeeded in infecting monkeys with human malaria, accidental laboratory infections of human beings with simian malaria occurred (Coatney *et al.*, 1961), resulting in reinvestigation of many of the monkey malarias and eventual discovery of several new species (Eyles, 1963).

In so far as we know, the various malaria parasites of Amphibia, Reptilia, Aves, and Mammalia are somewhat alike in that they are heteroxenous, developing in mosquitoes and vertebrates; they exist in at least a part of their development in the erythrocyte where they digest a portion of the hemoglobin, and they have a preerythrocytic cycle which occurs in cells other than erythrocytes. In both the preerythrocytic and erythrocytic stages they undergo schizogony, an asexual multiple fission with the production of merozoites and, in the erythrocytic stages finally produce gametocytes, the sexual forms which are to undergo, under suitable conditions, sporogony in the mosquito. In the main, differences that exist are minor and will be discussed in the ensuing section.

This similarity has proved most fortunate for investigators working with malaria. Many of the new drugs that at one time, at least, offered so much promise for malaria control were developed utilizing avian or rodent malaria as the test organism.

B. The Exoerythrocytic Cycle

1. In Allied Groups

a. Eimeriina. Among presumed relatives of malaria classified as Coccidia are the genera *Schellackia* and *Lankesterella,* alike in that they exist for some period of time in the vertebrate host erythrocyte. Although Baker (1965) states that coccidian ancestry of the Haemosporidia is generally accepted, such acceptance rests almost entirely on possible fortuitous similarities of development, i.e., life in the erythrocyte. Nevertheless, their life cycle does perhaps bear some relationship to *Plasmodium* and should be considered in any general discussion.

Schellackia bolivari (Reichenow, 1919) develops in the gut epithelium of various species of lizards. It undergoes a cycle of development common to the Eimeriina: one or two asexual cycles followed by the fertilization of macrogametes by microgametes and the formation of oocysts. The sporozoites formed from the oocysts, in contrast to those of most coccidia, enter the circulating blood, and penetrate erythrocytes or lymphocytes. They remain unchanged in these cells until ingested by a mite, *Liponyssus.* No development takes place in this latter host; reinfection of the lizards, or infection of others depends on ingestion of the mite containing the sporozoites.

The sporozoites of *Lankesterella* are also found in the erythrocytes of the frog host (Nöller, 1913). Schizogony, however, takes place in vascular endothelium, and after schizogony and formation of gametocytes and oocysts, the sporozoites leave the endothelium and penetrate erythrocytes. Here they remain until ingested by a leech in which no change ensues, other than the digestion of the host cell and perhaps a slight enlargement. Some wander into the proboscis sheath and thence into another frog.

b. Adeleina. The true Haemogregararina, *Karyolysus* and *Hepatozoon,* belong to the suborder Adeleina in which there are fewer and more isogenous gametes. They differ also from *Schellackia* and *Lankesterella* in that there is a development with reproduction in the invertebrate host, the leech. The most complete account of the life cycle of *Haemogregarina* is that of Reichenow (1910), who traced its development in the leech and water tortoise.

Sporozoites are introduced by the leech into the turtle, *Emys orbicularis*, and invade the red blood cells where schizogony takes place. Reinvasion of erythrocytes occurs until the development of gametocytes. In the leech, microgametes fuse with macrogametes and a zygote is formed. Sporozoites are liberated from the oocysts and make their way into the proboscis sheath. Infection of turtles is accomplished through the bite of the leech.

Karyolysus is considered to be a parasite of reptiles. Schizogony occurs in endothelial cells of a lizard; some of the merozoites produced penetrate erythrocytes and develop into gametocytes. Following the ingestion of lizard blood by a mite there is the formation of oocysts in the gut epithelium. The sporokinetes liberated from the resultant sporoblasts enter the ova of the host and are passed to the offspring. Spores now present in the newly hatched mites are discharged in the feces and are, in turn, ingested by the lizard where the cycle continues. *Hepatozoon* occurs in reptiles, birds, and mammals and, as the name implies, undergoes schizogony in the parenchymal cells of the liver. Transmission is through a mite, but Ball (1966) has demonstrated that the culicine mosquito, *Culex tarsalis*, is susceptible to infection by the hepatozoon found in a boa.

These accounts of development provide material for speculation relative to evolution but the parasites seem to be rather carelessly and casually transmitted, particularily *Schellackia* and *Lankesterella*. Generally nature provides a much more efficient method. The above development seems to have little survival value. The penetration of the erythrocyte by the sporozoite of hemogregarines should be reinvestigated in view of the newer knowledge of the preerythrocytic stages of malaria. The results might yield much valuable information as to the evolution of parasites. It may be that the cycle of hemogregarines as seen by Robertson (1910) in the lung capillaries is more representative.

In addition, none of the investigators has used known parasite-free animals. For this, host animals must be reared free of the potential vector. Too many reports have resulted from associating schizogony with gametocytes, oocysts, or sporozoites without any assurance they were not a part of the life cycle of another parasite.

c. *Leucocytozoon*. The genus *Leucocytozoon* is a pathogen of ducks, and falls, as does the malaria organism, in the family Plasmodiidae. Two types of schizonts are found (Huff, 1942). One, the *hepatic schizont*, occurs in the cells of the liver; the other, *megaloschizont*, is quite large and occurs in the blood vessels or extravascularly in the heart, spleen, liver and intestine. Eventually certain of the merozoites develop into gametocytes which infect and badly distort lymphocytic cells of the circulating bloodstream. The invertebrate hosts are various diptera belonging to the families Simuliidae and Chironomidae (Fallis *et al.*, 1951). In the invertebrate host, microgametes are produced by exflagellation; these fertilize the macro-gametes and oocysts develop. The sporozoites of these forms migrate to the salivary glands and are injected into other ducks after which the cycle commences.

d. Haemoproteus. *Haemoproteus* is still another parasite, occurring prin-cipally in birds, in which sexual stages, alone, are found in the circulating blood: this time, however, in erythrocytes. Schizogony takes place in lung endothelium. After several schizogonic cycles, merozoites rupture the sur-rounding membranes and penetrate red blood cells. Here they develop into macro- and microgametocytes. Pigment is present, indicating ingestion of hemoglobin and its subsequent degradation. Sporogony takes place in the hippoboscid fly *Lynchia* (Sergent and Sergent, 1906a). Exflagellation oc-curs, and the resultant microgametes penetrate macrogametes, producing zygotes. The oocyst develops on the hemocoelic side of the midgut. Sporo-zoites are formed, some of which lodge in the salivary glands ready for trans-mission to the vertebrate (Adie, 1924).

2. Exoerythrocytic Stages of Malaria

a. In Reptiles. The most complete description of exoerythrocytic develop-ment of malaria in lizards was that given by Thompson and Huff (1944). They isolated and described *Plasmodium mexicanum* from a Mexican lizard, *Sceloporus ferrariperezi.* In experimentally infected lizards it was discovered that there were two types of development: the *elongatum* type in which the exoerythrocytic stages occur in a great variety of blood and blood-forming cells, and the *gallinaceum* type in which development takes place in macro-phages and endothelium. As will be seen, this development is similar to that seen in avian embryos injected with sporozoites of *P. gallinaceum,* an avian malaria parasite (McGhee, 1949). No preerythrocytic stages have been de-scribed, however, since the vectors of saurian malaria are not known as yet.

b. In Birds. One of the most significant research efforts of the twentieth century was that of Huff and his co-workers on the pre- and exoerythro-cytic stages of avian malaria. Although suspected as far back as the work of Ben Harel (1923), it remained for Huff and his colleagues to elucidate with care and elegance the various preerythrocytic stages of the avian ma-laria parasite, *P. gallinaceum* (Huff and Coulston, 1944). The results of their findings led to newer studies of immunity, physiology, pathogenicity, and treatment of avian and mammalian malarias, and cannot be overesti-mated. The method used was simple: Massive doses of sporozoites were injected into the wing skin of domestic chickens and biopsies were made of the areas at various intervals. In such a way the first *cryptozoite* was de-scribed. After growth and schizogony of this initial stage in macrophages, the newly liberated merozoites infected other macrophages or endothelium where the *metacryptozoic* generation ensued. The offspring of this schizogony were capable of infecting erythrocytes. The development was observed by McGhee (1949) using chick and duck embryos. Later Huff and co-workers described essentially the same pattern of development for other species of avian malarias (1951). His findings set the stage for the eventual discovery of the exoerythrocytic stages of various malarias of mammals.

    c. In the Lower Mammalia. An equally painstaking, persistent, and inspired research effort relative to the pre- and exoerythrocytic stages of rodent malaria was carried out by Yoeli (1965). Its culmination was the complete description of the life history of *P. berghei* (Vincke and Lips, 1948). Injection of sporozoites into the rat, *Thamnomys*, gave rise to developing forms in the liver parenchyma. Fifty-one hours after injection schizonts containing up to 1782 nuclei were observed in liver cells. These forms differed from those found in reptiles and in birds by their location within liver cells. It is from these parasites that eventual erythrocytic stages develop.

    d. In the Primata. Development of malaria in monkeys, is, in general, similar to that in the rodent or in the human being. The original work of Schaudinn (1899), who reported the entry of sporozoites into erythrocytes, delayed the complete elucidation of the tissue stages of this and other malarias. In 1948 Shortt and Garnham fed numerous mosquitoes infected with *P. vivax* on a paralytic patient. In addition, they injected this same individual with 200 salivary glands of an anopheline mosquito infected with the same parasite. Seven days later the liver was biopsied and exoerythrocytic stages in early stages of development were found. Later Bray (1957) found similar stages in the liver of a chimpanzee following the infection of this animal with sporozoites of *P. vivax*.

    Exoerythrocytic stages of *P. malariae* have never been demonstrated in the human being. Bray (1960), however, injected chimpanzees with sporozoites of *P. malariae* and demonstrated preerythrocytic stages in the liver of this primate.

    In 1951 Shortt *et al.* demonstrated for the first time the exoerythrocytic stages of *P. falciparum* in the liver of a human volunteer who allowed himself to be bitten by great numbers of infected anopheline mosquitoes. This finding was verified by Jeffery *et al.* (1952). Garnham (1948) described the cycle of *Hepatocystis kochi* (a relative of *Plasmodium*) in the monkey and in 1955 he described the exoerythrocytic cycle of *P. ovale* in man.

C. The Transition Period

    Once preerythrocytic development is completed, parasites enter the bloodstream and penetrate the red blood cells where the erythrocytic cycle begins. Despite the lack of attention devoted to this period of the parasite's development, it offers some intriguing problems. Huff (1959) states that the parasite, in following the pattern it does, is in the grip of its own evolutionary development and this is probably true, but it does not explain how the parasite is able to make the step from macrophage, endothelium, or liver parenchyma to the erythrocyte.

It may be that some sort of immunity becomes active in the cells housing the exoerythrocytic stages and the parasite is forced to a more hospitable environment. Another possibility is suggested by the situation seen in the embryos of chicks and ducks infected with preerythrocytic stages of *P. gallinaceum* (McGhee, 1949). In these animals as well as in embryos receiving grafts of tissues containing exoerythrocytic stages (Zuckerman, 1946), both an *elongatum* and *gallinaceum* type of infection developed. This sort of condition also obtains in lizard malaria (Thompson and Huff, 1944). In the embryo the spleen is the first organ infected; this is not true of the infection in the hatched animal. In the embryo the site of earliest infection is that coincidentally of the site of blood formation. It therefore becomes possible that the parasite would be in a position of infecting a cell, such as an early member of the erythrocytic series, which could be flushed out into the blood. Such infected cells were seen in the bloodstream of the infected embryo. In the hatched chick, erythropoiesis ordinarily occurs in the bone marrow. Since there is no longer production of erythrocytes in the spleen this former avenue to the bloodstream is now closed to exoerythrocytic forms. The more abrupt type of transition now becomes operative. In both the embryo and the hatched animal, *P. elongatum* develops in the bone marrow and the more gradual transition is more apt to occur.

Exoerythrocytic stages from blood forms (phanerozoites) may develop in somewhat the reverse manner. Once humoral immunity has made itself apparent in the peripheral bloodstream, those parasites which are able to survive in cells less exposed to the immune processes would have a better chance of survival than those still in erythrocytes. It is possible that penetration of endothelium continues during the parasite's erythrocytic cycle. Upon removal of most of the erythrocytic stages, those parasites in the endothelium would constitute the reservoir for future relapses.

D. THE ERYTHROCYTIC CYCLE

1. General Outline

To the casual observer the erythrocytic cycle of malaria seems to be one which is completely known. On the other hand, realization that only one individual has actually observed the penetration of the erythrocyte by a merozoite (at least in the last 60 years) suggests that the cycle may not be so completely known. The general outline of development is, however, rather clear and holds, with minor variations, for all species of malaria. The merozoites, following their liberation from the mother schizont, penetrate erythrocytes and growth ensues. Hemoglobin is ingested in the cytoplasm (Aikawa *et al.*, 1966). Several nuclear divisions take place with eventual cytokinesis and formation of merozoites and subsequent liberation from the host cell. Somewhere during the process of schizogony, which differs from the coccidia in its continuousness, some merozoites develop into gametocytes. These undergo no further development until they reach the outside of the vertebrate body.

Parasites exist as pigmented multishaped organisms whether in lizards or in man. In the lizard, as pointed out by Thompson and Huff (1944), morphological forms paralleling those seen in birds exist. Likewise, various rodent and simian forms resemble those of human malaria. Some simian forms are practically indistinguishable from those in the human being. McGhee (1951) noted that *P. lophurae*, an avian malaria parasite, when introduced into mice, resembles somewhat that of *P. malariae* in man.

2. Penetration and Cytochemistry

McGhee (1950), while observing the infection of chick embryos by plasmodia, found that certain mammalian erythrocytes became infected when introduced into the circulating blood of the infected embryos. Of 14 species of mammals the red blood cells of 3 were found to be infected after a sojourn of 4 hours in the infected embryo bloodstream. These were the erythrocytes of the swiss mouse, the rabbit, and the pig. These cells, though harvested from three separate orders of mammals, Rodentia, Lagomorpha, and Arteriodacyla, were alike in certain chemical details, such as the sodium-potassium ratio and in concentrations of organic acid-soluble phosphate. The erythrocytes which are nearest in chemical composition to those of the susceptible animals were those of the guinea pig and rat. Inasmuch as the organic acid-soluble phosphate contents of erythrocytes of younger animals generally is higher, the content of organic acid-soluble phosphate of cells of baby rats and guinea pigs was determined and found to be well within the range of the amounts found in susceptible cells. Upon injection of erythrocytes from these two animals into infected embryos it was found that those of the baby rat were indeed susceptible but that those of the baby guinea pig were not. Later McGhee (1953) found that the susceptibility of the rat erythrocytes disappears at about 40 days of host age.

Mouse erythrocytes, although showing some degree of increased resistance concomitant with age, did not exhibit the dramatic change seen in rat erythrocytes.

Examination of duck erythrocytes following 4 and 24 hours in infected chicken embryos revealed a rather unexpected aspect of cell susceptibility.

Hitherto, all cells tested had, at best, been only lightly infected, but as many as 15 % of the introduced washed duck erythrocytes were infected after only 4 hours. This, despite the fact that the foreign erythrocytes constituted only as much as 20 % of the total cells.

Thus it was found that: (1) the ability of a parasite usually found in a nucleated avian erythrocyte to penetrate, grow, and reproduce in an anucleated mammalian cell indicates that the presence of a host cell nucleus is not necessary for the life of this parasite; (2) the similar chemical makeup of infected cells indicates a certain predilection of the parasite toward these similar cells; (3) the decrease in susceptibility of rat erythrocytes coincident with aging of the animals suggests that perhaps hemoglobin might also enter into the picture of susceptibility, inasmuch as about the same time sequence is followed in the replacement of fetal by adult hemoglobin; and (4) there is a decided preferential invasion of adult duck erythrocytes over those of the host animal—the chick embryo.

On the surface it would appear that the duck erythrocyte is a cell that might be quite easily penetrated by perhaps any malaria parasite. A comparative study of four species of malaria and the susceptibility of various erythrocytes to these parasites was next carried out (McGhee, 1957). Certain cells are more susceptible to invasion by certain parasites, but penetration is not merely a question of brute force but is in all probability occasioned by the presence of an enzyme (or enzymes), which attacks specific cells.

At the time of cell penetration the parasite is exposed to three environments. The first of these is the parent erythrocyte at the time of segmentation (Fig. 4). Trager and McGhee examined microcultures of *P. lophurae* and were able to see the actual progress of segmentation of this avian

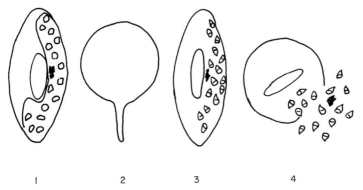

1   2   3   4

Fig. 4. Changes in shape of a duck erythrocyte infected with *Plasmodium lophurae* during schizogony and release of merozoites. Numbers 1 through 4 represent successive stages of the process. Diagrammatic.

malaria parasite. If one looked at mature parasites with several nuclei, but no cytoplasmic cleavage, the following events took place: There was suddenly a convulsive rounding up of the host cell. During this short stage the parasite, because of the concentration of host cytoplasm, became invisible. The action was so sudden that quite often a bit of cytoplasm was left behind in the process and trailed as a tail. Shortly thereafter, the process reversed and the host erythrocyte resumed its usual shape. During this obscure phase, cytokinesis had taken place and one could now see the individual merozoites dancing in the cytoplasm. Again the erythrocyte rounded up, but now did not return to its former shape. Rather, the merozoites exploded from the cell, spilling hemoglobin and infective parasites in the bloodstream. In one instance, not observed by McGhee, Trager (1956) noted the entrance of a merozoite into an adjacent erythrocyte, as did Weathersby (1966), who observed the entry of exoerythrocytic merozoites of *P. gallinaceum* into erythrocytes within brain capillaries.

Previous thinking had been that the liberation of the merozoites from the schizont was accomplished by crowding of the host erythrocyte until it burst. Even a cytochemist like Ponder (1948) says, "In these cases the destruction of the cells is almost certainly mechanical." The finding reported above does not necessarily disprove this but if one examines the literature certain interesting facts present themselves. The segmenting parasites of *P. gallinaceum* and *P. circumflexum* are large and exceed the size of the gametocytes. The segmenters, however, of *P. nucleophilum. P. rouxi, P. elongatum,* and *P. vaughni* are smaller than their gametocytes. Yet the gametocytes in the peripheral circulation do not destroy the erythrocyte while the small segmenting parasites do. The idea of strict mechanical destruction is, therefore, invalid. It is much more logical to assume that the egress of the merozoites is accomplished by some elaboration on the part of the parasites—a substance that weakens the ultrastructure of the erythrocytes.

Schaudinn's observation (1899) suggests that the merozoite enters the erythrocyte by force. Trager's and Weathersby's observations suggest an entry by means of an enzyme. The observations presented as to the comparative susceptibility of various erythrocytes to various species of parasites show that there are no "stronger" or "weaker" parasites or erythrocytes and indicate a selectivity. This selectivity is even more pronounced when the parasite preferentially invades duck to chick erythrocytes.

While there may be some reasons to accept the concept of mechanical exit and entry of merozoites, the more logical concept would be by means of an enzyme of release and penetration. Conceivably these may be the same substance.

In a preliminary study, McGhee and Hanson (1959) separated two protein components of the chick erythrocyte. One of these components, corresponding to stromatin of the human erythrocyte (Calvin *et al.*, 1946), when injected into malarious ducklings, delayed the increase in parasite numbers over that seen in control ducklings.

If the liberation of merozoites is accomplished by an enzyme some of this substance(s) would be poured into the bloodstream. If it fixed on a cell in contact with a merozoite, entry would be effected. However, other cells not penetrated by the merozoite would likewise be exposed to this enzyme. Also a foreign enzyme might act as an antigen when fixed to infected and uninfected erythrocytes and, following antibody production, render them susceptible to removal by macrophages whether infected or not. Fixation of an enzyme on uninfected cells may help to explain the apparent nonspecific removal of uninfected erythrocytes in malaria infections. A reaction simulating and perhaps being classifiable as an autoimmune response is discussed at length elsewhere.

E. Host Specificity of the Malarias

The studies of disease of the human being brought about by trypanosomatid and malaria organisms have been, quite often, hampered by the inability to secure infections of these organisms in laboratory animals. As a result, parasites of other animals resembling human parasites have been the primary experimental organisms. Yet, if one examines the literature, attempts to infect lower animals with human parasites have been surprisingly few. Indeed, studies of host specificity using any parasite has attracted, as judged by the literature, very few investigators. Beyond the value of using such experimentally infected hosts in the study of human disease, these experiments should yield valuable information regarding affinities and evolutionary trends of the various parasites.

Little work has been done with host specificity of malaria. What was done, with the exception of the infection of monkeys with *Plasmodium falciparum* (Taliaferro *et al.*, 1934), was fruitless.

The possibilities of securing infection of lower animals with human parasites and of human beings with parasites of lower animals still remain and suggest most fruitful fields of investigation.

The erythrocytes of the rabbit, pig, mouse, and baby rat are susceptible to infection by *P. lophurae* when these cells were injected into the bloodstream of avian embryos infected with this parasite. Of these animals, the mouse seemed to be most likely to be susceptible in an attempt to produce infections of the intact animal. Moreover, as it seemed that the erythrocytes of infant mice were more susceptible to infection and since younger animals would possibly contain less natural antibody in their bloodstream, the animal of choice would be baby mice.

Accordingly, less-than-one-day-old mice were given massive intravenous doses of infected embryo erythrocytes. A low-grade infection which disappeared in 8 days ensued. The parasites existing at that time were reintroduced into the chick embryo and maintained in this host until sufficient levels of parasitemia were attained to return the, hopefully, adapted parasites to the mouse. After ten such adaptive passages the levels of infection in the mouse attained heights that allowed for direct mouse-to-mouse passage. From this point on, the levels of infection gradually increased with each transfer until at times 7000 parasites per 10,000 erythrocytes could be observed. The parasite was for all practical purposes not pathogenic for the mouse.

Parasitemias of a transient nature, alone, were secured in adult mice.

Infections were maintained in baby mice for a period of 3 years at which time tests indicated that there was but slight variation from the originally introduced strain (McGhee, 1956).

Baby rats also proved susceptible to infection although no attempts were made to adapt this parasite to continuous survival in this host.

The host specificity of *P. berghei*, a parasite of mammals, was likewise found to be not so rigid as might be expected. Massive injections of parasitized mouse erythrocytes into duck and goose embryos produced a slight infection in the nucleated cells of the embryo.

Further expansion of hosts has been demonstrated by Jordon (1957), who reports a wide variety of birds to be susceptible to *P. lophurae*.

Accidental infections of human beings with *P. cynomolgi* of monkeys have been reported and experimental infections of human volunteers described (Coatney *et al.*, 1961). Recent tabulations list six species of simian malaria infective for man (Contacos and Coatney, 1963).

Human malaria species have likewise been shown to be capable of infecting "lower" mammals. Infections of *P. falciparum* have been secured in *Hylobates lar*, three species of *Mulatta*, and *Tupaia glis* (Ward and Cadigan, 1966; Hickman *et al.*, 1966; Porter and Young, 1966). The successful infection of the laboratory mouse with *P. falciparum* was reported by Weinman *et al.* (1966). Only a few mice proved to harbor the parasites for any length of time. In these, development has proceeded to a stage of nuclear division. It is possible, therefore, that after years of vain effort, the infection of laboratory animals with human malaria may be a reality.

F. The Invertebrate Cycle

1. General Considerations

The general picture of development of plasmodia in the mosquito is the same despite the mosquito species or the vertebrate host. Following the ingestion of blood containing gametocytes the cycle commences with ex-flagellation of the microgametocytes. The newly formed microgamete has the ability to swim vigorously in the surrounding medium. There are few visible changes in the macrogametocyte to macrogamete transition, other than emergence from the erythrocyte and in some instances change from an elongate to a spherical parasite. A single microgamete fertilizes the macrogamete and a zygote is formed. In some plasmodia, this changes to an ookinete which is apparently motile, although, if so, is not visible in the case of *P. gallinaceum*. A penetration either through the cell or between the cells of the midgut of the mosquito is effected and the ookinete comes to rest on the hemocoelic side of the gut. It attaches to the gut wall where the oocyst develops. Within this structure, the sporozoites develop and are eventually released. These are later found throughout the body including the salivary glands. The mosquito injects sporozoites into the vertebrate host in the act of biting.

2. Immunity

Generally speaking, the mammalian malarias develop only in Anophelinae while avian malarias are somewhat less specific in that they usually develop in Culicinae but may also undergo development to infective stages in Anophelinae.

Within the invertebrate, as in the vertebrate host, however, immune forces are at work. In several instances, beginning with the works of Huff (1927, 1930, 1931, 1934), inherited Mendelian recessive characters have been shown to be responsible for susceptibility of *Culex pipiens* to *P. cathemerium*. Unfortunately this characteristic was accompanied by a tendency to produce infertile offspring and proved difficult, therefore, to carry on to any lengths. Further work relative to the genetic basis of susceptibility and resistance to malaria has been carried out by Trager (1942). He established a strain of *Aedes aegypti* susceptible to *P. lophurae*. Micks studied the genetics of susceptibility (1949) of *P. elongatum* in *C. pipiens* and Ward (1966) of *P. gallinaceum* in *Aedes aegypti*.

In a most original work, Weathersby (1963) showed first that the attachment of the oocyst to the stomach wall is not a necessary prerequisite for development of *P. gallinaceum* in *A. aegypti*. He injected infected chicken blood into the hemocoel of the mosquito and observed that oocysts formed in any area of the mosquito body. Later he showed by parabiotic joining of susceptible mosquitoes (*Aedes aegypti*) to nonsusceptible (*Culex pipiens*) that a factor of resistance in the latter mosquito transferred to the usually susceptible one.

REFERENCES

Adie, H. (1924). *Indian J. Med. Res.* 2, 671.

Aikawa, M., Huff, C. G., and Sprinz, H. (1966). *Military Med.* 131, Suppl., 969.

Angel Torrico, M. R. (1946). *Ann. Lab. Central Bolivia* 1, 19; *Trop. Diseases Bull.* 44, 510 (*abstr.*).

Annear, D. I. (1961). *Australian J. Exptl. Biol. Med.* 39, 295.

Ashcroft, M. T. (1959). *Trop. Diseases Bull.* 56, 1073.

Bacellar, R. C. (1963). "Brazil's Contribution to Tropical Medicine and Malaria." Rio de Janeiro.

Baker, J. R. (1965). "Evolution of Parasites," p. 1. Blackwell, Oxford.

Ball, G. (1966). Personal communication.

Barrow, J. (1953). *Trans. Am. Microscop. Soc.* 72, 197.

Barrow, J. (1954). *Trans. Am. Microscop. Soc.* 73, 242.

Ben Harel, S. (1923). *Am. J. Hyg.* 3, 652.

Bray, R. S. (1957). *Am. J. Trop. Med. Hyg.* 6, 514.

Bray, R. S. (1960). *Am. J. Trop. Med. Hyg.* 9, 455.

Bray, R. S. (1964). *Bull. Inst. Franc. Afrique Noir* 26, 238.

Brumpt, E. (1906). *Compt. Rend. Soc. Biol.* 61, 77.

Brumpt, E. (1912a). *Bull. Soc. Pathol. Exotique* 5, 723.

Brumpt, E. (1912b). *Bull. Soc. Pathol. Exotique* 5, 514.

Brumpt, E. (1935). *Compt. Rend.* 200, 783.

Calvin, M., Evans, R. S., Behrendt, V., and Calvin, C. (1946). *Proc. Soc. Exptl. Biol. Med.* 61, 416.

Carini, A. (1910). *Soc. Med. Cir. Sao Paulo* (From S. von Prowazek, "Handbuch der Path. Prot." Leipzig, 1912).

Chagas, C. (1909). *Mem. Inst. Oswaldo Cruz* 1, 159.

Chagas, C. (1912). *Brasil-Med.* 26, 305.

Chagas, C. (1927). *Compt. Rend. Soc. Biol.* 97, 829.

Chatton, E. (1913). *Compt. Rend. Soc. Biol.* 74, 551.

Clark, T. B., Kellen, W. R., Lindergren, J. E., and Smith, T. A. (1964). *J. Protozool.* 11, 400.

Coatney, G. R., Elder, H. A., Contacos, P. E., Getz, M. E., Greenland, R., Rossan. R. N., and Schmidt, L. H. (1961). *Am. J. Trop. Med. Hyg.* 10, 673.

Coggeshall, L. (1938). *Am. J. Hyg.* 27, 615.

Contacos, P. G., and Coatney, G. R. (1963). *J. Parasitol.* 49, 912.

Corwin, R. M. (1962). Master of Science Thesis, University of Georgia, Athens, Georgia.

Cosgrove, W. B. (1963). *Exptl. Parasitol.* 13, 173.

Crawley, H. (1909). *U.S. Dept. Agr., Bur. Animal Ind., Bull.* 119, 5.

Cross, H. E. (1923). *Trans. Roy. Soc. Trop. Med. Hyg.* 16, 469.

D'Alessandro, A. (1963). *Proc. 7th Intern. Congr. Trop. Med. Malaria, Rio de Janeiro,* p. 146.

Davis, B. E. (1952). *Univ. Calif. (Berkeley) Publ. Zool.* 57, 145.

Donovan, C. (1909). *Lancet* II, 1495.

Elkeles, G. (1940). *Rev. Soc. Arg. Biol.* 16, 763.

Elkeles, G. (1942). *Rev. Soc. Arg. Biol.* 18, 315.

Eyles, D. E. (1963). *J. Parasitol.* 49, 866.

Fallis, M., Davies, D. M., and Vickers, M. A. (1951). *Can. J. Zool.* 29, 305.

Fantham, H. B. (1922). *S. African J. Sci.* 16, 185.

Fantham, H. B., and Porter, A. (1915a). *Proc. Cambridge Phil. Soc.* 18, 39.

Fantham, H. B., and Porter, A. (1915b). *Proc. Cambridge Phil. Soc.* 18, 137.
Fantham, H. B., and Porter, A. (1915c). *Ann. Trop. Med. Parasitol.* 9, 543.
Fantham, H. B., Porter, A., and Richardson, L. R. (1942). *Parasitology* 34, 199.
Fleming, M. E. (1966). Ph.D. Thesis, University of Georgia, Athens, Georgia.
Garnham, P. C. C. (1948). *Trans. Roy. Soc. Trop. Med. Hyg.* 41, 601.
Garnham, P. C. C. (1955). *Trans. Roy. Soc. Trop. Med. Hyg.* 49, 158.
Garnham, P. C. C. (1966). "Malaria Parasites and Other Haemosporidia." Blackwell, Oxford.
Gibbs, A. J. (1951). *J. Parasitol.* 37, 587.
Glaser, R. (1922). *J. Parasitol.* 8, 99.
Glaser, R. (1926). *Am. J. Trop. Med. Hyg.* 6, 205.
Gluge, G. (1842). *Arch. Anat. Phys. Wiss. Med.* 148 (quoted from Wenyon, 1926).
Goldman, M. (1964). *J. Parasitol.* 36, 1.
Grassi, B., and Feletti, R. (1891). *Centr. Bakteriol., Parasitenk.* 9, 402, 429, and 461.
Groot, H. F., Renjifo, S., and Uribe, C. (1951). *Am. J. Trop. Med. Hyg.* 31, 673.
Gruby, D. (1880). *Compt. Rend. Soc. Biol.* 17, 1134.
Hanson, W. L., and McGhee, R. B. (1961). *J. Protozool.* 8, 200.
Hanson, W. L., and McGhee, R. B. (1963). *J. Protozool.* 10, 233.
Hanson, W. L., and McGhee, R. B. (1966). Unpublished data.
Hartman, E. (1927). *Arch. Protistenk.* 60, 1.
Hickman, R. L., Gochenour, W. S., Jr., Marshall, J. D., Jr., and Guilloud, N. B. (1966). *Military Med.* 131, 935.
Hoare, C. A. (1923). *Parasitology* 15, 365.
Hoare, C. A. (1934). *Trop. Diseases Bull.* 31, 757.
Hoare, C. A. (1964). *J. Protozool.* 11, 200.
Hoare, C. A. (1965). *Proc. 2nd Intern. Conf. Protozool.*
Hoare, C. A. (1966). *Ergeb. Mikrobiol.* 39, 43.
Honigberg, B., Balamuth, W., Bova, E. C., Corliss, J. O., Gojdico, M., Hall, R. P., Kudo, N. C., Levine, N., Loeblich, A. C., Weiser, J., and Wenrich, D. H. (1964). *J. Protozool.* 11, 7.
Huff, C. (1927). *Am. J. Hyg.* 7, 706.
Huff, C. (1930). *Am. J. Hyg.* 12, 424.
Huff, C. (1931). *J. Prevent. Med. (Baltimore)* 5, 249.
Huff, C. (1934). *Am. J. Hyg.* 19, 123.
Huff, C. (1942). *J. Infect. Diseases* 71, 18.
Huff, C. (1951). *J. Infect. Diseases* 88, 17.
Huff, C. (1954). *J. Infect. Diseases* 94, 173.
Huff, C. (1959). *Exptl. Parasitol.* 8, 163.
Huff, C. G., and Coulston, F. (1944). *J. Infect. Diseases* 77, 224.
Jeffery, G. M., Wolcott, G. B., Young, M. D., and Williams, D., Jr. (1952). *Am. J. Trop. Med. Hyg.* 1, 917.
Jordon, H. B. (1957). *J. Parasitol.* 43, 395.
Jordon, H. B. (1964). *J. Protozool.* 11, 562.
Kent, W. S. (1880). "A Manual of the Infusoria." Bogue, London.
Koch, R. B. (1899). *Z. Hyg. Infektionskrankh.* 32, 1.
Koch, R. B. (1906). *Deut. Med. Wochschr.* 32, 51.
Kudo, R. B. (1958). "Protozoology." Thomas, Springfield, Illinois.
Laird, M. (1952). *Proc. Zool. Soc. London* 121, 285.
Laird, M. (1959). *Can. J. Zool.* 37, 749.
Laveran, A., and Franchini, F. (1914). *Bull. Soc. Pathol. Exotique* 7, 580.
Laveran, A., and Mesnil, F. (1902). *Compt. Rend.* 135, 609.

Léger, L. (1902). *Compt. Rend. Soc. Biol.* **54**, 398.

Levine, N. D. (1965). *J. Protozool.* **12**, 225.

McCulloch, I. (1917). *Univ. Calif. (Berkeley) Publ. Zool.* **18**, 75.

McGhee, R. B. (1949). *J. Infect. Diseases* **84**, 105.

McGhee, R. B. (1950). *Am. J. Hyg.* **52**, 42.

McGhee, R. B. (1951). *J. Infect. Diseases* **88**, 86.

McGhee, R. B. (1953). *J. Infect. Diseases* **93**, 4.

McGhee, R. B. (1956). *J. Protozool.* **3**, 122.

McGhee, R. B. (1957). *J. Infect. Diseases* **100**, 92.

McGhee, R. B. (1959a). *J. Parasitol.* **45**, 48.

McGhee, R. B. (1959b). *J. Infect. Diseases* **105**, 18.

McGhee, R. B. Unpublished data.

McGhee, R. B., and Hanson, W. L. (1959). *J. Parasitol.* **45**, 103.

McGhee, R. B., and Hanson, W. L. (1962). *J. Protozool.* **9**, 488.

McGhee, R. B., and Hanson, W. L. (1963). *J. Protozool.* **10**, 239.

McGhee, R. B., and Hanson, W. L. (1964). *J. Protozool.* **11**, 555.

McGhee, R. B., Schmittner, S. M., and Hanson, W. L. (1965). *Proc. 2nd Intern. Conf. Protozool.*, p. 135.

McGhee, R. B., Hanson, W. L., and Blake, J. D (1966). *J. Protozool.* **13**, 324.

McKeever, S., Gorman, G. W., and Norman, L. (1958). *J. Parasitol.* **44**, 583.

Mayer, H., and da Rocha-Lima, H. (1912). *Arch. Schiffs- Tropenhyg.* **16**, 376.

Micks, D. W. (1949). *J. Natl. Malaria Soc.* **8**, 206.

Muniz, J. (1927). *Compt. Rend. Soc. Biol.* **97**, 821.

Musgrave, W. E., and Clegg, M. T. (1903). *Bur. Govt. Lab. Manila Bull.* **5**.

Noble, E. R. (1955). *Quart. Rev. Biol.* **30**, 1.

Noguchi, H. (1926). *J. Exptl. Med.* **44**, 327.

Noguchi, H., and Tilden, E. B. (1926). *J. Exptl. Med.* **44**. 307.

Nöller, W. (1913). *Arch. Protistenk.* **31**, 169.

Novy, F. G., MacNeal, W. J., and Torrey, H. N. (1907). *J. Infect. Diseases* **4**, 223.

O'Farrell, W. R. (1913). *Ann. Trop. Med. Parasitol.* **7**, 545.

Olson, P. F., Shoemaker, J. P., Turner, H. F., and Hays, K. L. (1964). *J. Parasitol.* **50**, 599.

Packchanian, A. (1942). *Am. J. Trop. Med.* **22**, 623.

Patton, W. S. (1908). *Arch. Protistenk.* **12**, 131.

Patton, W. S. (1910). *Bull. Soc. Pathol. Exotique* **3**, 300.

Pifano, F., and Mayer, M. (1949). *Arch. Venezolanos Patol. Trop. Parasitol. Med.* **1**, 153.

Ponder, E. (1948). "Hemolysis and Related Phenomena." Grune & Stratton, New York.

Porter, J. A., and Young, M. (1966). *Military Med.* **131**, Suppl., 952.

Reichenow, E. (1910). *Arch. Protistenk.* **20**, 251.

Reichenow, E. (1919). *Sitzber. Ges. Naturforsch. Freunde Berlin*, 440.

Robertson, M. (1907). *Proc. Roy. Soc. Edinburgh* **B17**, 83.

Robertson, M. (1910). *Quart. J. Microscop. Sci.* **53**, 741.

Robertson, M. (1912). *Phil. Trans. Roy. Soc. London* **B202**, 29.

Robertson, M. (1913). *Phil. Trans. Roy. Soc. London* **B203**, 161.

Rudzinska, M. A., and Trager, W. (1957). *J. Protozool.* **4**, 190.

Russell, P. L., West, L. S., and Manwell, R. D. (1946). "Practical Malariology." Saunders, Philadelphia, Pennsylvania.

Sandon, H. (1926). "The Composition and Distribution of the Protozoan Fauna of the Soil." Oliver & Boyd, Edinburgh and London.

Schaudinn, F. (1899). *Sitzber. Ges. Naturforsch. Freunde Berlin* **7**, 159.
Schmittner, S. M. (1963). Thesis.
Schmittner, S. M., and McGhee, R. B. (1961). *J. Protozool.* **8**, 381.
Schwetz, J. (1930). *Arch. Inst. Pastur Algerie* **8**, 289.
Sergent, Ed., and Sergent, Et. (1906a). *Compt. Rend. Soc. Biol.* **61**, 494.
Sergent, Ed., and Sergent, Et. (1906b). *Ann. Inst. Pasteur* **20**, 665.
Shaw, J. J. (1964). *Nature* **201**, 417.
Sherman, I. W. (1966). *J. Parasitol.* **52**, 17.
Shortt, H. E., and Garnham, P. C. C. (1948). *Trans. Roy. Soc. Trop. Med. Hyg.* **41**, 785.
Shortt, H. E., Fairley, N., Hamilton, C. G., Shute, P. G., and Garnham, P. C. C. (1951). *Trans. Roy. Soc. Trop. Med. Hyg.* **44**, 405.
Stahel, G. (1931). *Phytopathol. Z.* **4**, 65.
Taliaferro, W. H. (1924). *J. Exptl. Med.* **39**, 171.
Taliaferro, W. H., Taliaferro, L. C., and Cannon, P. R. (1934). *Am. J. Hyg.* **19**, 318.
Tejera, E. (1920). *Bull. Soc. Pathol. Exotique* **13**, 527.
Thompson, P. E., and Huff, C. G. (1944). *J. Infect. Diseases* **74**, 48.
Thomson, J. D. (1908). *J. Hyg.* **8**, 75.
Tobie, E. J. (1961). *Exptl. Parasitol.* **11**, 1.
Tobie, E. J. (1964). *J. Parasitol.* **50**, 593.
Trager, W. (1942). *J. Parasitol.* **28**, 457.
Trager, W. (1956). *Trans. Roy. Soc. Trop. Med. Hyg.* **50**, 419.
Trager, W. (1964). *J. Parasitol.* **50**, 593.
"Tropical Health—A Report on a Study of Needs and Resources." *Natl. Acad. Sci.*, Natl. Res. Council, Washington, D. C., 1965.
Vermeulen, H. (1963). *J. Protozool.* **10**, 216.
Vincke, I. H., and Lips, M. (1948). *Ann. Soc. Belge Med. Trop.* **28**, 97.
Wallace, F. G. (1943). *J. Parasitol.* **29**, 196.
Wallace, F. G. (1966). *Exptl. Parasitol.* **18**, 124.
Walton, B. C. (1958). *Am. J. Trop. Med. Hyg.* **7**, 603.
Ward, R. A. (1966). *Military Med.* **131**, Suppl., 923.
Ward, R. A., and Cadigan, F. C. (1966). *Military Med.* **131**, Suppl. 944.
Weathersby, A. B. (1963). *Proc. 7th Intern. Congr. Trop. Med. Malaria, Rio de Janeiro*, Abstr. p. 432.
Weathersby, A. B. (1966). Personal communication.
Weinman, D., Cavanaugh, D. C., and Desowitz, R. (1966). *Trans. Roy. Soc. Trop. Med. Hyg.* **60**, 562.
Wells, E. A., Lumsden, W. R., Hardy, G. J. C., and James, D. (1965). *Proc. 2nd Intern. Congr. Protozool.*, p. 158.
Wenyon, C. M. (1926). "Protozoology. A Manual for Medical Men, Veterinarians and Zoologists," 2 vols. Baillière, London.
Woodcock, H. M. (1914). *Zool. Anz.* **44**, 26.
Woodcock, H. M. (1915). *Brit. Med. J.* **11**, 704.
Yoeli, M. (1965). *Trans. Roy. Soc. Trop. Med. Hyg.* **59**, 255.
Yorke, W. (1920). *Trans. Roy. Soc. Trop. Med. Hyg.* **14**, 31.
Yorke, W., and MacFie, J. W. S. (1924). *Trans. Roy. Soc. Trop. Med. Hyg.* **18**, 125.
Zeledon, R. (1954). *Rev. Biol. Trop., Univ. Costa Rica* **2**, 231.
Zuckerman, A. (1946). *J. Infect. Diseases* **79**, 1.

12

Preservation and Storage in Vitro

I. Preservation by Freezing

H. T. MERYMAN

II. Preservation by Freeze-Drying

H. T. MERYMAN

III. Results Obtained

DAVID WEINMAN

I. PRESERVATION BY FREEZING

A. THEORY

When the body temperature of a living organism is reduced, alterations in metabolic processes result. In general, this is without ill effect unless excessively prolonged. There does not appear to be any lethal "direct action of cold" and if freezing did not occur, all organisms could probably be reversibly cooled to very low temperature and indefinitely suspended animation easily achieved. However, save for those few plants and microorganisms containing no freezable water, freezing represents a massive alteration of the cell both anatomically and chemically.

An ice crystal begins with a crystal nucleus. Nuclei which initiate ice at moderate subfreezing temperatures are thought to be particles which have a surface configuration similar to the crystal structure of ice to which water is absorbed in crystalline array. The surface of an ice crystal is constantly both losing and gaining water molecules. At low temperatures the tendency to gain and keep molecules is enhanced, while at higher temperatures, the molecules are more energetic and tend to escape from the crystal surface more readily. The radius of curvature of the crystal will also influence its growth. If the radius is small, as in a very small particle, there is a greater tendency for molecules to leave the crystal. When a crystal is of such a size and temperature that the rate of loss of molecules exactly equals the rate of gain, it is said to be "critical" in size. If a momentary fluctuation reduces it in size, i.e., reduces its radius of curvature, the rate of loss increases and the crystal will rapidly vanish, whereas with a momentary increase in size the rate of water loss would be reduced and the crystal would grow. Since the rate of loss of water molecules is reduced with a lowering in temperature, critical size is therefore also reduced and a nucleus too small to nucleate ice at one temperature may do so when the temperature is further lowered.

The expectation of finding a nucleus large enough to initiate crystallization at or near the melting point is very poor and usually a solution must be supercooled well below its melting point before crystallization is initiated. For any given solution the distribution of various sizes of nuclei is presumably random, so that reducing the volume of the aliquot frozen increases the expectation of supercooling. The probability of substantial supercooling in a volume as small as a single cell is very good. Individual mammalian erythrocytes, for example, are reported to supercool to between $-15°$ and $-20°C$ (Rinfret, 1964), although a 1 ml sample of cells will rarely supercool below $-6°C$.

When a cell suspension is cooled below the melting point it will thus almost invariably supercool several degrees before the largest nucleus present becomes of critical size. When this occurs, the crystal will then grow rapidly, branching through the solution, generally giving the impression of multiple crystals. In fact, however, most freezing which occurs at moderate rates of cooling results from one or very few initial nuclei. Whether they originate inside or outside a cell is of little consequence since a crystal originating within a cell will concentrate the intracellular solution leading to osmotic rupture of the cell so that the crystal promptly becomes extracellular at the cost of a single cell.

As ice grows in the extracellular solution, solutes are concentrated. If the external membranes of the cells are permeable to water, osmotic transfer will take place, dehydrating the cells which become concentrated between

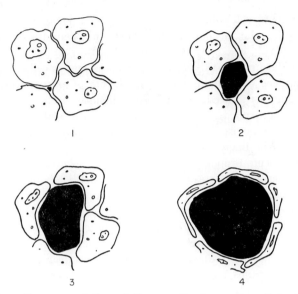

FIG. 1. Schematic representation of the growth of an extracellular ice nucleus into a large crystal. This has plasmolyzed the surrounding cells. (Reprinted from Meryman, 1963, by permission.)

the ice crystals (Fig. 1). If the cell membrane is impermeable to water, there will be no initial effect on the cell from the presence of extracellular ice. However, with further reduction in temperature crystal nuclei within the cells become of critical size, leading to intracellular freezing. Intracellular ice has almost universally been found lethal.

As the first ice crystal grows, it liberates its latent heat of fusion. If no heat were being withdrawn from the sample, the temperature would rise to the melting point and further crystallization would cease. However, if heat is continually withdrawn from the specimen (for example, by immersing it in a low temperature bath), crystal growth will continue. If the rate of heat removal (freezing rate) is very high, heat will be removed from the specimen at a faster rate than it can be supplied by the initial growing crystal, the temperature will continue to fall, and additional nuclei will become critical. Carrying this to its extreme one can see that, at ultrarapid rates of cooling, the number of nuclei becoming critical could be very large and lead to crystals uniformly distributed both inside and outside the cells (Fig. 2).

The injurious effects of extracellular ice do not appear to be primarily mechanical in nature. Ice crystals grow in the direction of minimum solute concentration and do not pierce cells, as was once commonly believed. The injury is probably entirely on the basis of dehydration and the resulting concentration of cell solutes (Meryman, 1967). There are three principal hypotheses for the mechanism by which this dehydration leads to cell death.

a. It has been proposed that, within pores and micelles of the cell and on the surfaces of proteins, there are layers of water which have been stabilized and structured through the influence of the adjacent surfaces; that this structuring is, in turn, essential for the stability of protein and that freezing removes this water, resulting in protein denaturation (Sinanogulu and Abdulner, 1965).

b. It has been proposed that it is the concentration of electrolytes, primarily sodium and potassium chloride, which leads to the denaturation of protein, particularly lipoprotein (Lovelock, 1953a,b).

c. It has been proposed that a liquid phase separates structural components of the cell. Dehydration permits the contact of surfaces normally separated by liquid. The formation of disulfide bonds by the oxidation of sulfhydryl groups or by disulfide interchange between proteins or portions of protein molecules normally separated by liquid phase has been specifically proposed (Levitt, 1962).

Resistance to freezing injury is found in nature only in those organisms which either have so low a water content that freezing does not occur or somehow bind sufficient water to limit dehydration to a tolerable degree. Certain of the intertidal mollusks are exposed to freezing temperatures at low tide and have been shown to possess a resistance to freezing injury at temperatures as low as $-10°C$. Mollusks which are susceptible to freezing injury have nearly two thirds of their body water frozen at temperatures around $-6°C$, whereas in the resistant variety, a temperature of $-10°$ is reached before the same proportion of water has been frozen (Williams and

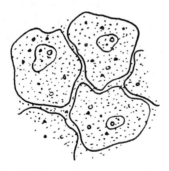 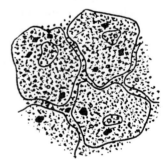

FIG. 2. Schematic representation of the development and growth of intracellular crystals through the imposition of a rate of heat withdrawal (cooling rate) in excess of that which can be supplied by the first nuclei to become critical, thus leading to further cooling and the appearance of more nuclei. (Reprinted from Meryman, 1963, by permission.)

Meryman, 1965). Some organisms also avoid freezing injury by extensive supercooling. Insects, for example, have been reported to supercool to as low as $-35°C$ (Salt, 1964).

Most organisms, however, are lethally dehydrated by extracellular ice formation and the preservation by freezing of the majority of virtually all cells of higher animals was not possible until the discovery in 1949 of the protective properties of glycerol (Polge *et al.*, 1949). Glycerol, and the more recently reported cryoprotective agent, dimethylsulfoxide (Lovelock and Bishop, 1959), probably prevent freezing injury solely through a re-duction of the amount of water frozen. In order to be in vapor pressure equilibrium with ice at a given temperature, a solution must possess a certain water activity (ratio of free water molecules to effective total molecules). Increasing the concentration of solute molecules (by adding glycerol) means that, to maintain the necessary activity, proportionately more free water must be unfrozen. This water is fully available as a solvent for electrolytes and other solutes. In this fashion, the addition of a solute such as glycerol can serve to decrease the amount of water frozen at any given temperature and decrease the concentration of other solutes.

For cryoprotective agents to function, they must penetrate the cell; otherwise the concentration of other solutes within the cell will not be reduced. It is in this regard that dimethylsulfoxide (DMSO) is superior to glycerol since it penetrates cell membranes more rapidly and more uni-versally. Webb (1965) has proposed that the effectiveness of a protective agent may be also related to its "water likeness" and ability to substitute for water in the ordered layers on protein. Whether cryoprotective agents have such a specific role or whether they merely maintain a liquid phase which limits salt concentration and prevents contact between cell com-ponents has not yet been determined, but from the practical point of view this is purely of academic concern. The important point is that the intro-duction of a compound which will penetrate cell membranes, which is miscible with water in all proportions, and which has no toxic effects when concentrated to 6 to 8 M at low temperature, can function as a protective agent against freezing injury at low rates of cooling.

Under rare special circumstances rapid freezing also provides a method for preserving life. The rapid freezing either permits some intracellular water to remain unfrozen in the cell, a goal facilitated by extracellular cryoprotective agents such as polyvinylpyrrolidone (PVP), or simply provides insufficient time for the effects of dehydration to take place. Optimum rapid cooling rates are of the order of $50°–100°C$ per second. Among mammalian tissues, only erythrocytes and epithelium can be successfully treated in this manner. Other cells of higher animals are invariably destroyed by rapid freezing, presumably due to the formation of intracellular ice. Microorganisms are far more flexible in their behavior during freezing. Many possess such low water content that they are unaffected by freezing temperatures. Others will survive rapid freezing only, while still others must be frozen in the presence of a penetrating cryoprotective agent in order to achieve any reasonable degree of survival (Mazur, 1966). Protozoa do not appear amenable to rapid freezing procedures. Some can be preserved by slow freezing with cryoprotective agents, many cannot, apparently because their external membranes are impermeable to the necessary protective compounds.

B. TECHNIQUES

The techniques for the preservation of living cells by freezing can best be considered in five subdivisions: Preparation of the sample, freezing, storage, thawing and resuspension. In general, rapid freezing techniques are not suitable for cell preservation save for certain hardy strains of bacteria and viruses. This means that a conventional procedure for slow freezing in the presence of a penetrating cryoprotective agent will be the rule.

1. Preparation of Sample

In practice, glycerol and dimethylsulfoxide are the only penetrating cryoprotective agents in general use, although Webb (1965) has reported good results with inositol, particularly as a protective agent against aerosol dehydration injury in microorganisms. Both glycerol and DMSO have excellent cryoprotective properties. The principal difference between them lies in their ability to penetrate the cell membrane, an essential requirement for protection. Glycerol is considerably more variable in its penetrating characteristics for different cells than is dimethylsulfoxide, and even where glycerol penetrates satisfactorily, DMSO appears to pass the membrane more rapidly. During the preparation of the sample this may not be a particularly important factor since the actual times required for penetration are still very brief, being of the order of seconds or minutes. As water passes the cell membrane more rapidly than the protective agent, the latter will exert a momentary osmotic effect causing a temporary shrinkage of the cell, the effect being more marked with glycerol than DMSO. As both of these compounds, like water, ultimately achieve concentration equilibrium on both sides of the cell membrane, it is necessary that the suspending medium contain, in addition to the cryoprotective agent, all those elements necessary to maintain osmotic stability of the cell.

The concentration of cryoprotective agent necessary for good preservation will vary over a range of about 3 M to as low as 0.5 M. Unfortunately it is not possible to give hard and fast rules for the choice of optimum concentration. Like most practices in applied cryobiology, the procedures are empirical and new applications must be explored on that basis.

2. Freezing

There are three methods of freezing which have been found empirically to permit good recovery of certain cells: rapid freezing, slow freezing, and two-step freezing.

With the exception of mammalian erythrocytes and epidermis, cells of higher organisms do not appear to withstand rapid freezing, presumably because their high degree of hydration favors the development of intracellular ice at high cooling rates. Many bacteria and viruses can be effectively fast frozen but their selection is a matter of trial and error. Rapid freezing of small quantities of material can be achieved in glass capillaries or between glass cover slips by immersion in liquid nitrogen or isopentane at temperatures between $-100°$ and $-150°C$, producing freezing rates of from fifty to several hundred degrees per second. Where larger quantities are to be handled, it becomes increasingly difficult to maintain high cooling rates within closed containers. Meryman and Kafig (1955) reported a technique useful for the rapid freezing of blood in which the cell suspension is sprayed through an oscillating 27-gage needle down into the surface of liquid nitrogen in an open-top tray. The small droplets, ranging in size from 0.1 to 1.0 mm in diameter, float on the surface for about 2 seconds before reaching bath temperature and sinking to the bottom. Following freezing, the droplets can be collected from the bottom of the tray with a fine mesh screen. This is a simple and effective method for handling large volumes but the maintenance of sterility is difficult.

Another method of rapid freezing originally designed for whole blood consists in shaking a partially filled metal container under the surface of liquid nitrogen in order to shell freeze the contents (Doebbler *et al.*, 1966). The rates of cooling are not as rapid as those attainable with thin films or droplets.

The standard method for freezing sensitive cells is slow freezing in the presence of a penetrating cryoprotective agent. Although many studies of rates of cooling have been reported, most of these have been concerned with animal cells. It has been found empirically that a cooling rate of approximately 1°C per minute gives good results in the majority of situations. Commercial apparatus is available for controlled rate freezing at 1°C per minute as well as other rates. However, it should be appreciated that the freezing rate of 1°C per minute, which was originally selected as best for bovine spermatozoa, is by no means a magic number to be worshipped inflexibly. Furthermore, it is a recommendation which refers to the temperature of the bath in which specimens are frozen and not to the specimen temperature itself. Attempts to force the specimen temperature through the freezing process at a smooth 1°C per minute are futile and without justification (Meryman, 1963).

On the theory that the slow cooling rate is necessary to permit the maintenance of osmotic equilibrium in the cell as the external medium is concentrated by freezing, it follows that rate restrictions should no longer be essential once the majority of cell water has been frozen. It is therefore customary to freeze at a low rate to between −20° and −25°C, following which the cooling rate can be substantially increased to minimize further exposure to intermediate temperature. Carried to an extreme, this procedure, known as two-step freezing, consists in slow freezing to −20° or −25°C followed by very rapid cooling to very low temperature. For some cells, this is the only technique which will produce good recovery (Asahina, 1966), the probable explanation being that, at −20°C, there is still sufficient unfrozen water remaining in the cell to permit survival. Further slow freezing will result in further dehydration whereas ultrarapid freezing at this stage will preserve the status quo. At a temperature higher than −20°C, the cell will contain more water, perhaps sufficient to permit intracellular crystallization, so that very rapid freezing without the preliminary partial dehydration would be lethal. This technique has not received the attention it might and it is possible that some cells unable to tolerate cryoprotective agents in sufficient concentration to protect against slow freezing might tolerate two-step freezing at a low additive concentration.

3. Storage

The selection of a stabilizing storage temperature will be based on two considerations. First, the extent of injury and the rate at which it proceeds will depend in part on the degree of dehydration of the cell. Second, the rate at which the injury progresses during storage will also be proportional to temperature. This leads to two alternative approaches toward storage. Either the degree of dehydration can be minimized, reducing the potential for injury, or storage can be at a very low temperature where deleterious reactions will not take place regardless of the degree of dehydration. Storage temperature is generally considered to be stabilizing below −100°C.

Care should always be taken not to confuse freezing injury and storage injury. Freezing injury can be defined as that resulting from freezing followed by immediate thawing. Any additional progressive decay that occurs with storage time is storage injury and can always be prevented by storing at a lower temperature.

In cells which will tolerate high concentrations of cryoprotective agents, the dehydration of freezing can be sufficiently reduced to provide good stability at almost all temperatures. However, it is not possible to predict stability at intermediate storage temperature over very long periods of time since, in many cases, the cryoprotective agents themselves possess some degree of toxicity.

One approach to storage which has been almost totally ignored is to store at relatively high temperature with concentrations of cryoprotective agent which would be inadequate to confer protection at lower temperature following slow freezing. Temperatures from $-10°$ to $-20°C$ appear to be particularly satisfactory, being high enough to prevent the excessive dehydration of the specimen in the presence of low additive concentration, yet low enough to confer a substantial reduction in the rate of potentially injurious processes. Such compromise storage temperatures should not be considered adequate for indefinite preservation but may provide an attractive alternative or perhaps the only solution for storage of moderate duration of special material. Temperatures between $-30°$ and $-50°C$ are probably the worst for storage since they produce substantial dehydration but are insufficiently low to be stabilizing.

One other alternative for moderate prolongation of storage time is supercooling. The temperatures to which solutions may be supercooled below their melting point can be further reduced by the addition of solute, just as the melting point itself is lowered. Furthermore, since spontaneous crystallization depends on the presence of a nucleus of critical size, reductions of specimen volume will enhance the probability of supercooling. Storage temperature of $-10°C$ supercooled can be achieved by the addition of 1 to 1.5 M of most solutes.

Three choices of equipment present themselves for low temperature storage: The mechanical refrigerator, the dry ice chest, and the liquid nitrogen refrigerator. Mechanical refrigeration is most effective at temperatures down to about $-40°C$. Below this temperature the apparatus becomes more specialized, double and triple cascade compressors become necessary, and capital expenditures increase rapidly. Possibly a more vital consideration is that of reliability. Material stable at $-40°C$, or above, presumably can tolerate breaks in storage temperature provided they are not of excessive duration. Mechanical refrigeration is not indefinitely reliable and failures of the apparatus or the power supply will ultimately occur. When lower temperatures of storage are used and, particularly, when the storage material cannot tolerate even brief breaks in storage temperature, other means of refrigeration should be considered.

Dry Ice refrigerators are satisfactory for many materials which are relatively stable. However, Dry Ice at $-79°C$ cannot be considered to produce a completely stabilizing temperature and is inadequate for indefinite preservation. Carbon dioxide gas is a poor thermal conductor and the coldest gas will tend to layer in the bottom of the refrigerator. The temperature just above the level of the Dry Ice can run as high as $-30°$ or $-40°C$. Where Dry Ice is used for refrigeration, the actual temperature in the vicinity of the specimen should be measured.

For indefinite preservation liquid nitrogen is the refrigerant of choice. Liquid nitrogen refrigerators are large metal Dewar flasks and those designed for storage range in size from 25 to over 1000 liters. The evaporation loss of liquid nitrogen during storage is from about 1 liter per day for the 25-liter container to about 10 liters per day for the 1000-liter tanks. The cost of liquid nitrogen in laboratory quantities is approximately $.50 a liter. There are certain hazards associated with the storage of material immersed in liquid nitrogen. Sealed ampules which leak will admit liquid nitrogen and on thawing the ampules become dangerously explosive. Safety glasses should always be worn during the removal of ampules from liquid nitrogen immersion. A convenient method for checking the seal on an ampule is to immerse it prior to freezing in any cold, colored liquid such as ink and water. The negative pressure produced by cooling the ampule will draw in liquid if a leak exists. After 5 or 10 minutes the ampules are rinsed and those containing color are discarded. One inconvenience of storage in liquid is the tendency for ampules to float. In the larger containers it is practical to maintain a tank only partially filled so that samples may be stored in the gas phase, the temperature of which will run around $-160°C$. Automatic level control devices are available to maintain the level from a supply tank.

An alternate method of refrigeration by liquid nitrogen is to admit the liquid to a vapor-filled refrigerated chamber on demand of a thermostat. Liquid nitrogen may be obtained in pressurized containers so that a solenoid valve can control the supply. This approach enables the user to operate the refrigerator at any temperature from ambient to nearly the temperature of the liquid nitrogen. Control circuits can be made relatively fail-safe and driven by a battery supply. Such a refrigeration apparatus can be operated at between $-100°$ and $-130°C$, temperatures much less demanding in terms of chamber insulation and refrigerant supply than when the chamber contains the liquid nitrogen itself.

It is important to know if lapses in storage temperature have occurred. As an alternative to a continuous recording of chamber temperature, one can freeze selected liquids in small flasks or test tubes which are distributed inverted throughout the storage chamber. If the melting point of the liquid is just above that of the temperature of storage, any break in storage temperature will permit the liquid to thaw and run from its container.

4. Thawing

It has generally been stated that, where the rate of thawing influences the recovery of cells following freezing, rapid thawing is superior. This is obviously true where rapid freezing has been necessary to preserve viability. Where cells have been suspended in a fully protecting concentration of cryoprotective agent, the rate of thawing is generally immaterial. Recent observations on erythrocytes (Meryman, 1967) have indicated that, when borderline concentrations of cryoprotective agents are used, recovery can be substantially improved by slow freezing coupled with very slow thawing of the order of 1°C per minute or less.

5. Resuspension

The hazards of resuspension following freezing and thawing with cryoprotective agents are generally overlooked. Since these agents, particularly glycerol, pass through cell membranes less rapidly than water, they will have a momentary osmotic effect when concentrations are changed suddenly. When cells which have been equilibrated in glycerol solution are suddenly transferred to an isotonic glycerol-free solution, this will have the momentary effect of a transfer to a hypotonic medium. When recovery following freezing and thawing is poor, this is one aspect of the procedure which should be considered and the benefits of a stepwise reduction in additive concentration investigated. It is during resuspension that the virtues of the rapid penetration of dimethylsulfoxide are most apparent.

II. PRESERVATION BY FREEZE-DRYING

A. THEORY AND TECHNIQUES

The freeze-drying process consists essentially in the sublimation of ice from a previously frozen specimen. The principle virtues of this form of dehydration are twofold. First, the absence of a liquid phase during drying prevents shrinkage and results in a porous product which rehydrates readily. Second, the dehydration can be conducted at a temperature low enough to minimize chemical reactions which might otherwise occur during the removal of water. The technique of freeze-drying consists simply in exposing the frozen specimen to an atmosphere with a water vapor pressure lower than that of the specimen. The process is usually conducted in the presence of a partial vacuum in order to facilitate the flow of water vapor from the drying boundary of the specimen to the dehydrating agent, whether it be pump, desiccant, or refrigerated condenser. By definition, the specimen must remain frozen during the process of freeze-drying. The transition from ice to water vapor requires about 670 calories per gram of latent heat. If the drying progresses rapidly, the loss of latent heat may be sufficient to keep the specimen frozen even though its environment may be substantially above freezing. This is the situation which prevails in many applications of freeze-drying. In fact, in most industrial applications heat must be supplied to the specimen in substantial quantities in order to support continued sublimation, otherwise the specimen temperature may fall too low to permit a continued rapid rate of dehydration.

Heat cannot be appreciably stored in the specimen, and energy introduced in the form of heat from the environment can only leave the specimen in the form of the heat of sublimation, i.e., as water vapor (see Fig 3). The rate at which water vapor escapes through the dried shell of the specimen will depend on the vapor pressure gradient between the surface of the specimen and the drying boundary within, and on the resistance to vapor flow of the dried shell between these two locations. The temperature of the ice at the drying boundary will determine its vapor pressure while the efficiency of the vacuum system and the vapor trap will determine the vapor pressure at the specimen surface. If the external vapor pressure and the resistance of the dried shell are such that there is insufficient flow of vapor to utilize all the heat energy being introduced, then the specimen temperature must rise until the vapor pressure at the drying boundary is sufficiently high to produce the necessary increase in vapor flow. Thus, the temperature of the specimen at the drying boundary will be completely dependent on the rate at which heat is introduced to the specimen and on those factors that influence the removal of water vapor.

Fig. 3. Electrical analog of the freeze-drying system. This figure illustrates the manner in which the freeze-drying system must be visualized as a continuous flow of energy from heat input to vapor removal, analogous to a simple series resistance. The heat necessary to supply the latent heat of sublimation is comparable to the electrical energy from a battery, the temperature of the heater being represented by the voltage at A. This heat must flow to the specimen across either a radiative or conductive resistance, comparable to R_1 in the figure. On reaching the specimen the heat must then flow to the drying boundary within the specimen, traversing either the frozen portion or the dried shell depending on the circumstances of drying. At the drying boundary, heat will be consumed to supply latent heat of sublimation, so that the quantity of water vapor released is equivalent to the amount of heat energy available. Thus, although there is a change of state at the drying boundary, in thermodynamic terms there is no interruption in the energy flow. Leaving the drying boundary, the vapor encounters the resistance to flow of the dried shell and the space between specimen and condenser, completing the energy flow which began at the heater. Looking at this system in terms of the simple electrical analog, it can be seen that only the values of A and E, heater and condenser temperature, are fixed by the operator. All others, particularly the important value of C, drying boundary temperature, will vary depending on the magnitude of R_1 through R_4, the resistances to heat and vapor flow. This illustrates that control of specimen temperature can only be achieved on the basis of direct measurement or by limiting A to a very low value. (From Meryman, 1966.)

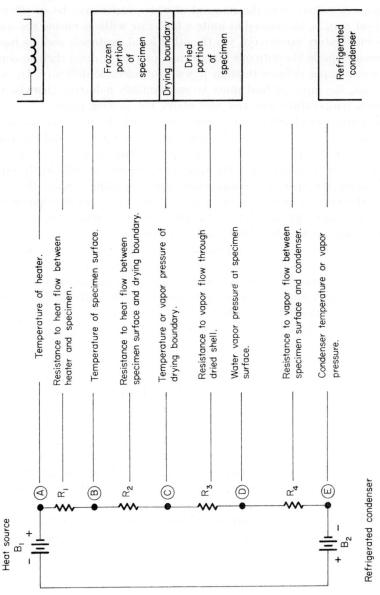

FIG 3.

In a specimen which is thin, and produces a porous dried shell of low resistance to vapor flow, the removal of water vapor may be so efficient that heat may be introduced at quite a high rate without raising the specimen temperature excessively. On the other hand, materials such as tissue or viscous liquids like vitreous humor may produce a dried shell so dense that water vapor diffuses through it with considerable difficulty. In such situations, the input of heat must be substantially reduced; otherwise the specimen temperature may rise well above the melting point.

The conventional laboratory freeze-drier, in which specimens are dried in ampules attached to a manifold, relies on heat supplied by radiation and conduction from the room to provide the necessary heat of sublimation. This technique presumes that the rate of drying will be sufficiently rapid to maintain the specimen temperature below freezing despite the heat input. However, if heat is introduced at a constant rate throughout the drying procedure, drying will proceed over a range of temperatures, starting at a low temperature when the dried shell is thin and vapor removal efficient, but gradually increasing in temperature as drying progresses and the dried shell introduces an increasing resistance to vapor flow. This is clearly not a logical procedure. If there is a maximum safe temperature for drying, then drying at a lower temperature is inefficient while drying at a higher temperature is deleterious to the specimen. It should be evident that for every situation there is an ideal temperature of drying and, furthermore, that this temperature should be maintained throughout the entire primary drying procedure and that the temperature should be permitted to rise only after all the ice has been sublimed.

Optimum temperatures for drying have been established for many materials. For some, such as foodstuffs, the temperature utilized is a compromise between a drying temperature low enough to produce a superior product and one high enough to permit economical operation. It has been shown for some specimens that recovery is improved by drying at quite low temperatures. Greaves (1956) and Muggleton (1962) have reported a substantial increase in the survival of several microorganisms when the drying was carried out at $-30°$ as opposed to $-20°C$ (see Meryman, 1966). It should, therefore, be evident that control of the temperature of the specimen during freeze-drying is essential to obtain the optimum recovery of the sample in the minimum time. Most small-scale laboratory apparatus available on the commercial market does not make such control possible. Where feedback control of heat input based on specimen temperature is not available, the best compromise for the laboratory investigator is to surround the specimen with the environmental temperature above which the specimen should not rise during drying. Microorganisms or tissues for histological study, for example, should be dried below $-30°C$

for best results unless experimental evidence indicates otherwise. The ampules in which drying is conducted can be immersed in a bath at $-30°C$ during drying to assure that no higher temperature is possible. Drying may then be at a lower temperature and take longer than necessary, but the quality of the product will not be endangered.

The completion of drying is most conveniently determined by measuring the vapor pressure near the specimen with a vacuum gage. During drying, the pressure will be high but will fall to approach the vapor pressure of the vapor trap on completion of drying.

Greaves (1949) has demonstrated the importance of secondary drying for many applications where maximum storage stability is desired. Secondary drying consists in subjecting specimens, after initial drying, to a prolonged exposure to an environment of negligible water vapor pressure, usually over phosphorus pentoxide in vacuum. Greaves (1949) has reported an increase in the stability of typing sera from a half-life of 30 minutes at 100°C following 24 hours of secondary drying to a 5-hour half-life after 1 week over P_2O_5. Similar results were reported with guinea pig complement. No commercial apparatus of United States manufacture provides facilities for secondary drying, but the apparatus is easily assembled.

B. APPLICATIONS

The application of freeze-drying to living cells other than bacteria and viruses is negligible. Most of these latter organisms, however, will not withstand complete desiccation, but require the presence of water-binding solutes which retain water at a level of about 1 % of dry weight (Fry, 1966).

Algae have been reported to withstand drying in some cases (Daily and McGuire, 1954; Holm-Hansen, 1963). Generally, it is the smaller organisms of 10 μ or less in diameter which survive.

No reports of recovery of protozoa following freeze-drying are known to the author, although Annear (1956) vacuum-dried *Strigamonas oncopelti* using a technique later shown by Greaves (1962) to have been rapid drying without freezing. Burns (1964) has reported unsuccessful attempts to freeze-dry protozoa.

Burns (1964) has also reported attempts to freeze-dry nematodes, obtaining success only with organisms in the larval stage and then only when they were apparently sufficiently dehydrated to prevent freezing. It would appear that, other than suitable bacteria and viruses, one should not generally turn to freeze-drying for the preservation of living organisms. A review of the theory and practice of freeze-drying has been published by Meryman (1966), of the freeze-drying of bacteria by Fry (1966), and of viruses by Greiff and Rightsel (1966).

REFERENCES

Annear, D. I. (1956). *Nature* **178**, 413.

Asahina, E. (1966). *In* "Cryobiology" (H. T. Meryman, ed.), pp. 479–483. Academic Press, New York.

Burns, M. E. (1964). *Cryobiology* **1**, 18.

Daily, W. A., and McGuire, J. M. (1954). *Butler Univ. Botan. Studies* **11**, 139.

Doebbler, G. E., Rowe, A. W., and Rinfret, A. P. (1966). *In* "Cryobiology" (H. T. Meryman, ed.), pp. 407–450. Academic Press, New York.

Fry, R. M. (1966). *In* "Cryobiology" (H. T. Meryman, ed.), pp. 665–696. Academic Press, New York.

Greaves, R. I. N. (1949). *Brit. Sci. News* **2**, 175.

Greaves, R. I. N. (1956). *Can. J. Microbiol.* **2**, 365.

Greaves, R. I. N. (1962). *In* "Progrès Récent en Lyophilisation" (L. Rey, ed.), p. 167. Hermann, Paris.

Greiff, D., and Rightsel, W. (1966). *In* "Cryobiology" (H. T. Meryman, ed), pp. 697–728. Academic Press, New York.

Holm-Hansen, O. (1963). *Nature* **198**, 1014.

Levitt, J. (1962). *J. Theoret. Biol.* **3**, 355.

Lovelock, J. E. (1953a). *Biochim. Biophys. Acta* **10**, 414.

Lovelock, J. E. (1953b). *Biochim. Biophys. Acta* **11**, 28.

Lovelock, J. E., and Bishop, M. W. H. (1959). *Nature* **183**, 1394.

Mazur, P. (1966). *In* "Cryobiology" (H. T. Meryman, ed.), pp. 213–315. Academic Press, New York.

Meryman, H. T. (1963). *Federation Proc.* **22**, 81.

Meryman, H. T. (1966). *In* "Cryobiology" (H. T. Meryman, ed.), pp. 609–663. Academic Press, New York.

Meryman, H. T. (1967). In press.

Meryman, H. T., and Kafig, E. (1955). *Proc. Soc. Exptl. Biol. Med.* **90**, 587.

Muggleton, P. W. (1962). *Prog. Ind. Microbiolo.* **4**, 191.

Polge, C., Smith, A. U., and Parkes, A. S. (1949). *Nature* **164**, 666.

Rinfret, A. P. (1964). Personal communication.

Salt, R. W. (1964). *In* "Handbook of Physiology" Sect. 4, Chap. 21, Am. Phys. Soc., Washington, D. C.

Sinanogulu, O., and Abdulner, S. (1965). *Federation Proc.* **24**, Suppl. 15, S-12.

Webb, S. J. (1965). "Bound Water in Biological Integrity." Thomas, Springfield, Illinois.

Williams, R. J., and Meryman, H. T. (1965). *Cryobiology* **1**, 317.

III. RESULTS OBTAINED

A. GENERAL CONSIDERATIONS

Low temperature methods are preferred to other preservation procedures because they are thought to offer unique stability in the material frozen. Numerical loss in freezing, storage, and thawing is conceded, but otherwise the material frozen is assumed to be unchanged when retrieved on thawing. This may or may not be true, and variables to be considered are (1) pre-selection, (2) correlated death in freezing and thawing, and (3) changes induced by the preservation procedure.

1. Preselection

Experiments on microorganisms deal with one or a few of their many characteristics, i.e., enzyme reaction, drug resistance, infectivity. If this characteristic is carried by all members of the population, then selection of the aliquot to be frozen may not be important, for recovery of even one viable microorganism signifies that the identifying characteristic will also be recovered. But when populations are heterogeneous in regard to the characteristic, all the microorganisms do not carry it and it may be absent from lots haphazardly chosen. If testing is to be made immediately after thawing, the marked organisms should be present at the start in sufficient numbers to withstand freezing-storage-thawing losses, and still have sufficient survivors to manifest the characteristic sought for. If the final test requires 10,000 marked organisms for a positive reaction, loss is 50%, and the initial representation of the marked organism is 1 in 10,000, then a population of at least 2×10^8 organisms should be frozen if successful tests are to be obtained immediately after thawing.

2. Selection by Nonrandom Death

Freezing, storage, and thawing of microorganisms cause death of a variable proportion of the population. Survivors are selected for resistance to these procedures. It has been generally assumed that selection for resistance to freezing and thawing damage is not correlated positively or negatively with other biological characteristics under study. It seems clear that this assumption cannot always be correct, and whether the correlation exists or is of importance in any particular study may require testing. Varied recovery rates from different populations of the same kind of single cells has been noted in sperm banks. This "will mean intense selection for what we often call 'freezability': the capacity, which varies enormously between donors, for their sperm to withstand the freeze-store-thaw procedure relatively unharmed, and subsequently to show a high rate of fertility" (Ackerman and Behrman, 1967). In single populations of mixed types of *Trypanosoma rhodesiense* Ormerod (1963) found that the long thin forms of *T. rhodesiense* resisted freezing and thawing far better than did the stumpy forms; these two types are now considered to have different mitochondrial structure and presumably a different metabolism (see Chapter 10).

3. Cell Changes Detected after Thawing

These are changes in constituents or properties of the viable organisms recovered at the end of the preservation period; it is not always known which step in the process induced the change.

The new features may appear as either losses or gains. Loss in enzyme activity during storage at $-196°C$ was reported for aldolase and aspartate aminotransferase in stored erythrocytes (Lehmann, 1965). Extrapolation of reaction rates and observed storage decay rates at higher temperature, however, would indicate that there should be no storage changes possible (except radiation injury) at liquid nitrogen temperature; the explanation of these reported low temperature storage effects remains to be elucidated (Meryman 1966, 1967).

Gains result in variants which may appear as altogether new. Or the variant may represent a statistical change in distribution of preexisting types, in which case additional analysis is required to distinguish between a selective killing effect on the nonvariant type and a favorable process acting on the variants.

Markert (1963) described an example of the gain type dealing with enzyme breakdown and reassembly. Two of the five lactate dehydrogenase (LDH) isoenzymes which are distinguishable electrophoretically were separated from the others. These two, LDH_1 and LDH_5, were then frozen separately; each yielded the parent compound on thawing. However, when LDH_1 and LDH_5 were frozen together, the subunits (monomers) reassembled to give a mixture of $LDH_{1,2,3,4}$ and LDH_5, showing that freezing caused dissociation of the enzyme into subunits and that these reassembled to give compounds not originally detectable.

Peterson and Stulberg (1964), working with the same enzyme in two lines of monkey kidney cells, noted changes in proportion of the constituents after six weeks of storage. Both lines showed an increase in LDH_5 isoenzyme; this was the only change in one line, the other had changes in all five constituents. Chilson et al. (1965) analyzed factors at work and pointed out that physical and chemical conditions may change markedly during freezing, i.e., sodium phosphate solutions which are at pH 7.0 and 0.02 M at room temperature may reach a pH of 3.5 at $-9.9°C$, the eutectic point, and concentrations of 3.42 M in unfrozen water.

A most interesting postfreezing change of potential great significance was recently reported by McKee et al. (1966). This concerned chromosomal abnormalities in normal mammalian cells and is an example of change in proportion of preexisting types.

Three samples of primary explants of human foreskin tissue were examined and chromosomes counted. In addition to the normal diploid count, approximately 1% of the cells counted were tetraploids (range 0 to 2 in three sets of 100 cells of each explant). The three sets were frozen, and after thawing the tetraploid counts of the explants was increased to 5–7%. There was some suggestion that tetraploidy increased with storage age. The authors concluded that when normal chromosome counts are wanted the use of frozen tissue fragments is excluded. Whether the increase in abnormal chromosome counts was an accentuation of a preexisting process observed in unfrozen material is not known. Nor is it established which of the various factors involved in freezing and storage were active.

What is perhaps a different type of alteration is reported by Wodinsky *et al.* (1965) as a change in histocompatability or in the minimal infective doses of 7 of 17 lymphomatous tumors observed after storage for 20–24 months in liquid nitrogen. Shulman *et al.* (1967) described the production of autoantibodies which developed after freezing tissue *in vivo* for 3 minutes with liquid nitrogen; the alteration was detected by hemagglutination in this experiment; the changes were clearly not due to storage.

B. PRESERVATION

1. In the Frozen State

a. History. Short-term preservation of bacteria at $-190°C$ was demonstrated as early as 1900 (MacFadyen, 1900), but storage in the frozen state was not systematically investigated for other protista until much later. In 1939 Turner and collaborators reported on the value of frozen storage for spirochetes, and Coggeshall for plasmodium the same year. Generalization of the method to most of the protozoa was made in the decade 1940–1950, and the creation of a reference center for protozoa, analogous to the American Type Culture Collection for Bacteria, was proposed in 1947 (Weinman and McAllister, 1947). Subsequent investigations have dealt with the merits of protective substances, which may give increased survival rates or permit freezing-storage of otherwise too-fragile organisms, and with the advantages of different storage temperatures. A recent discussion of practical procedures is in Walker (1966). Preservation of protozoa after drying apparently awaits technical development; the few successes are cited in Walker's Table XX.

b. General Validity of Method. Viable organisms are recovered from such a variety of protista stored in the frozen state that the presumption is now that all microorganisms survive for long periods when frozen, and that failures, when encountered, are not attributable to frozen storage per se. This conclusion is reinforced by successful frozen storage of organisms previously recalcitrant, provided appropriate protective agents are added. As noted above, viability is no guarantee that the material is unchanged.

c. Records of Preservation. Table I provides a list of protista discussed in this volume which have been successfully stored. In general, viability was the test employed. The listing is selective, and confirming duplications have been omitted. Citations for a genus are usually of a single species; when of unusual interest, plural species are listed. This information is

TABLE I

MICROORGANISMS PRESERVED IN FROZEN STATE[a]

Microorganism	Source	Protective agent	Storage temperature (°C)	Storage time (days)	Authors
Treponemataceae					
Borrelia novyi	Blood	None	−78	1 year	Turner and Fleming (1939)
B. recurrentis (duttoni)	Blood	None	−78	1 year	Turner and Brayton (1939)
Treponema pallidum	Rabbit testes	None	−78	3 years	Turner and Fleming (1939)
	Tissue	Glycerine	−64−−78	12 years	Weinman (this publication)
Leptospira icterohemorrhagiae	Culture	None	−78	6 months	Turner and Fleming (1939)
L. icterohemorrhagiae	Liver tissue	None	−70	238	Weinman and McAllister (1947)
	Cultures	None	−70	901	Weinman and McAllister (1947)
L. biflexa	Cultures	None	−64−−78	8 years	Weinman (this publication)
Bartonellaceae					
Bartonella bacilliformis	Cultures	None	−70	Years	Weinman (1965)
Babesiidae					
Babesia canis	Blood	Glycerol	−76	Weeks	Reusse (1956)
Plasmodidae[b]					
Plasmodium inui	Blood	None	−72−−80	70	Coggeshall (1939)
P. knowlesi	Blood	None	−72−−80	70	Coggeshall (1939)
P. falciparum	Blood	None	−70	1921	Collins *et al.* (1963)
	Mosquito	None	−70	183	Jeffrey and Rendtorff (1955)
P. ovale	Mosquito	None	−70	70	Jeffrey and Rendtorff (1955)
P. vivax	Mosquito	None	−70	375	Jeffrey and Rendtorff (1955)
Toxoplasmatidae					
Toxoplasma gondii	Peritoneal exudate	Glycerol	−70	60	Weinman and Chandler (1954)
	Peritoneal exudate	Glycerol	−70	184	Chandler and Weinman (1956)
	Peritoneal exudate	Glycerol	−70	209	Eyles *et al.* (1956)
Trypanosomatidae[c]					
Crithidia fasciculata	Culture	DMSO	−170	155	Diamond (1964)
Crithidia sp.	Culture	DMSO	−170	288	Diamond (1964)
Leishmania donovani	Tissue	None	−70	276	Weinman and McAllister (1947)
L. tropica	Culture	None	−70	794	Weinman and McAllister (1947)
Trypanosoma brucei	Blood	None	−72−−80	135	Coggeshall (1940)
T. brucei	Blood	None	−64−−70	4 years	Weinman (this publication)
T. brucei	Blood	Glycerol	−79	250	Polge and Soltys (1957)
T. congolense	Blood	None	−70	100	Levaditi (1952)
T. cruzi	Blood	None	−70	234	Weinman and McAllister (1947)
T. cruzi	Culture	None	−70	653	Weinman and McAllister (1947)
T. cruzi	Culture	DMSO	−170	987	Diamond *et al.* (1963)
T. equiperdum	Blood	None	−190	21	de Jong (1922)
T. evansi	Blood	None	−72−−80	62	Coggeshall (1940)
T. gambiense	Blood	None	−70	561	Weinman and McAllister (1947)
T. gambiense	Culture	None	−70	189	Weinman and McAllister (1947)
T. gambiense	Blood	Glycerol	−79	250	Polge and Soltys (1957)
T. lewisi	Blood	None	−70	531	Weinman and McAllister (1947)
T. ranarum	Culture	DMSO	−170	987	Diamond (1964)
T. rhodesiense	Blood	None	−70	16 years	Weinman (this publication)
T. rhodesiense	Cultures	None	−70	293	Weinman and McAllister (1947)
T. rhodesiense	Blood	DMSO	−79	30	Walker and Ashwood-Smith (1961)

TABLE I—*Continued*

Microorganism	Source	Protective agent	Storage temperature (°C)	Storage time (days)	Authors
Incertae sedis *Aegyptianella* *Anaplasma* *Ehrlichia* *Eperythrozoon* *Grahamella* *Haemobartonella*, etc.					See respective chapter sections

[a] We wish to thank Dr. Louis S. Diamond for authorization to reproduce material included in this table.
[b] Twenty-two confirmations, using either other species or other techniques, are cited by Diamond.
[c] Other species and techniques given by Diamond.

derived in part from the tabulation of Diamond (1964), q.v., for more detail; also see individual chapters.

d. Additives. Protective agents are not required for *Plasmodium*, *Leishmania*, and *Trypanosoma*, although quantitative recovery rates may be improved with them. The additives are required for detectable survival of *Toxoplasma*, which survives freezing to −70°C but not storage at that temperature. Whether required or not, additives are usually now used routinely. Glycerol and dimethyl sulfoxide are the most widely employed at present; no general statement can be made as to respective merits for any given organism; consult Diamond (1964) for references to the original literature.

e. Longevity. Protozoa frozen under favorable circumstances remain viable for years. Storage temperature is important, and best success has been reported at temperatures of −70° to −80°C or below. Emphasis has been placed on the advantage of storage in the range −170° to −190°C obtained with liquid nitrogen over the −70° to −80°C range given by solid carbon dioxide, and the statement has been made that at a temperature of −79°C the limits of preservation of living cells can be reckoned in months, at most a few years. This has not been the experience of the author. No systematic exploration has been made, but in random testing, the following protista were found viable after 4 or more years of storage at temperatures ranging between −64° and −79°C: *Trypanosoma rhodesiense* blood forms, 16 years; *Treponema pallidum* tissue, 12 years; *Leptospira biflexa* cultures, 8 years; and *Trypanosoma brucei* blood forms, 4 years. *Trypanosoma rhodesiense* and *L. biflexa* were stored with no added protective agent. These results suggest that the choice of a storage temperature below −60°C is not the only factor of critical importance in obtaining prolonged survival. Some microorganisms cannot be stored successfully at −80°C but will survive indefinitely at −197°C, provided that they withstand freezing at that temperature (Meryman, 1967).

2. In the Dried State

Drying methods have not yet been generally successful with protozoa. Therefore, interest attaches to the report of Annear (1956a), who dried *Strigomonas oncopelti*, stored the protozoon for 1 year at $-4°$ and at $+20°C$ and found it viable thereafter. Greaves (1962) repeated the successful drying of *S. oncopelti* and on analysis of Annear's technique found that drying took place from the supercooled state at about $-8°C$, and that when specimens were frozen there was no recovery. With techniques used at present, a high residual moisture content (10–20%) has been necessary for survival, but there is no reason to suppose that freeze-drying of protozoa, even with low eventual moisture content, is an a priori impossibility (Greaves, 1962). Annear later showed that *Strigomonas* could be successfully dried without providing external cooling (1961). *Leptospira* has also survived drying (Annear, 1956b; Otsuka and Manako, 1961).

REFERENCES

Ackerman, D., and Behrman, S. J. (1967). *Yale Sci. Mag.* **41**, 6.

Annear, D. I. (1956a). *Nature* **178**, 413.

Annear, D. I. (1956b). *J. Pathol. Bacteriol.* **72**, 322.

Annear, D. I. (1961). *Australian J. Exptl. Biol. Med. Sci.* **39**, 295.

Chandler, A. H., and Weinman, D. (1956). *Am. J. Clin. Pathol.* **26**, 323.

Chilson, O. P., Costello, L. A., and Kaplan, N. O. (1965). *Federation Proc.* **24**, Part III, Suppl. 15, S55–S65.

Coggeshall, L. T. (1939). *Proc. Soc. Exptl. Biol. Med.* **42**, 499.

Coggeshall, L. T. (1940). *J. Bacteriol.* **40**, 559.

Collins, W. E., Jeffery, G. M., and Burgess, R. W. (1963). *Mosquito News* **23**, 102.

de Jong, D. A. (1922). *Arch. Neerland Physiol.* **7**, 588. (Cited by Diamond, 1964.)

Diamond, L. S. (1964). *Cryobiology* **1**, 95.

Diamond, L. S., Meryman, H. T., and Kafig, E. (1963). *In* "Culture Collections: Perspectives and Problems" (S. M. Martin, ed.), pp. 189–192. Univ. of Toronto Press, Toronto.

Diamond, L. S., Bartgis, I. L., and Reardon, L. V. (1965). *Cryobiology* **1**, 295.

Eyles, D. E., Coleman, N., and Cavanaugh, D. J. (1956). *J. Parasitol.* **42**, 408.

Greaves, R. I. N. (1962). *In* "Progrès Récent en Lyophilization" (L. Rey, ed.), pp. 167–179. Hermann, Paris.

Jeffery, G. M., and Rendtorff, R. C. (1955). *Exptl. Parasitol.* **4**, 445.

Lehmann, H. (1965). *Federation Proc.* **24**, Part III, Suppl. 15, S66–S69.

Levaditi, J. C. (1952). *Compt. Rend. Soc. Biol.* **146**, 179.

MacFadyen, A. (1900). *Proc. Roy. Soc.* **B66**, 180.

McKee, M. E., Harris, S. E., Kihara, H. (1966). *Proc. Soc. Exptl. Biol. Med.* **123**, 499.

Markert, C. L. (1963). *Science* **140**, 1329.

Meryman, H. T. (1966). *In* "Cryobiology" (H. T. Meryman, ed.), pp. 70–75. Academic Press, New York.

Meryman, H. T. (1967). Personal communication.

Ormerod, W. E. (1963). *Exptl. Parasitol.* **13**, 374.

Otsuka, S., and Manako, K. (1961). *Japan. J. Microbiol.* **5**, 141.

Peterson, W. D., Jr., and Stulberg, C. S. (1964). *Cryobiology* **1**, 80.

Polge, C., and Soltys, M. A. (1957). *Trans. Roy. Soc. Trop. Med. Hyg.* **51**, 519.

Reusse, U. (1956). *Z. Tropenmed. Parasitol.* **7**, 99.

Shulman, S., Yantorno, C., and Bronson, P. (1967). *Proc. Soc. Exptl. Biol. Med.* **124**, 658.

Turner, T. B., and Brayton, N. L. (1939). *J. Exptl. Med.* **70**, 639.

Turner, T. B., and Fleming, W. L. (1939). *J. Exptl. Med.* **70**, 629.

Walker, P. J. (1966). *Lab. Pract.* **15**, 423.

Walker, P. J., and Ashwood-Smith, M. J. (1961). *Ann. Trop. Med. Parasitol.* **55**, 93.

Weinman, D. (1958). *Trans. Roy. Soc. Trop. Med. Hyg.* **52**, 294.

Weinman, D. (1965). *In* "Bacterial and Mycotic Infections of Man" (R. J. Dubos and J. G. Hirsch, eds.), 4th ed., pp. 775–784. Lippincott, Philadelphia, Pennsylvania.

Weinman, D., and Chandler, A. H. (1954). *Proc. Soc. Exptl. Biol. Med.* **87**, 211.

Weinman, D., and McAllister, J. (1947). *Am. J. Hyg.* **45**, 102.

Wodinsky, I., Meaney, K. F., and Kensler, C. J. (1965). *Cryobiology* **2**, 44.

13

Investigational Problems and the Mechanisms of Inheritance in Blood Protozoa

PETER J. WALKER

I. INTRODUCTION

A. THE GENERAL SITUATION

Rapid development of molecular biology has tended to obscure the meaning of the word genetics, principally owing to the new-found breadth. Despite the situation that a broadly based subject has more facts, and therefore more scope, it is still true to say that astonishingly little is known about any aspect of the genetics of blood parasites. Clearly, for completeness, the current state of our knowledge should be summarized in a work of this kind; it is the intention of the reviewer to cover all aspects in this summary. In addition, and perhaps most timely, there is a critical discussion

of the types of experiment and approaches which seem worthwhile, and
also the relative merits of the necessary experimental techniques. Indeed
as so much of the past difficulties in the subject have been due to techniques being extended beyond their proper bounds, comment on such
instances should form an essential part of a critical review.

Only in *Plasmodium* sp. and *Trypanosoma* sp. has sufficient evidence
been accumulated to warrant a review. It should be acknowledged that
even in these genera the paucity of fact is alarming, especially as these
parasites are of such great economic importance. However, as much of this
review is a discussion of general principles it is felt that the material presented will be of use in the study of the genetics of other genera which share
the common habitat and usually are transmitted by an arthropod host.

B. Definitions and Limits

For the present purposes a broad definition is proposed for the term
"genetics." Genetics is the study of inheritance—especially the acquisition
and inheritance of "fitness." More limited views are frequently expressed.
These views range from Mendelian to molecular. While it is fashionable
to use ultracentrifuges and electron microscopes (indeed it is necessary to
do so) one should not lose sight of the general concepts of genetics which
not only deal with events occurring within the cell but also with the interaction between the cell and the habitat. A successful interaction for the
cell is indicated by the degree of fitness. The meaning of fitness has a
special meaning for blood parasites (see below).

We may look upon the various aspects of genetics as follows: (*a*) A total
of information is stored in a cell; with this information, and not one gene
more, the cell has to deal with all aspects, including vicissitudes, of the
habitat. (*b*) The information by itself is useless without a means of expression. During complex life histories phased expression is all important, i.e.,
the ability to produce the right response at the right time. The parallel
with embryological development should not be overlooked. (*c*) The inheritance of characters is the feature that occupied classic geneticists. For
blood parasites we consider that the inheritance is important from cell to
cell as well as from infection to infection. Only the latter corresponds more
closely with inheritance in Metazoa where one considers individuals of a
generation as the units into which information is transferred. This is an
important matter which will be dealt with in greater detail later. (*d*) Population genetics and reproductive dynamics are the material of ecological
studies. They are essential in our understanding of the spread of genes in
populations of parasites, especially with the rise of reported cases of drug
resistance.

It is not out of place here to add a word of caution concerning three concepts which frequently cause some difficulty. The first is that division at any phase of the life cycle should be by this or that method (e.g., orthomitosis) and according to methods which have been studied in detail in the Metazoa. To make the assumption is as much as to say that for the past 600 million years during the evolution of the Chordata these parasites have done nothing to better themselves. Clearly they have not evolved complicated morphology. Indeed, who is to say that the universal finding of mitosis and meiosis in Metazoa and Metaphyta is not a common reduction of an earlier and more complicated process found in Archean Protista. If this were true then the present-day blood parasites may well still carry some early methods of going about life which have been lost for all this time in other groups. Such an extreme possibility cannot be denied even if unlikely to our present way of thinking. Another and distinct possibility is that the special habitats of blood parasites have evoked new methods of division and genetic control. Certainly they have had a long time to do so, and it is equally certain that the habitat has been hostile enough to demand change if the parasites are to be represented in the present-day fauna. Evolution of the vertebrate host alone has provided a changing scene for parasite life, for not only have new classes arisen but the homiothermic classes (Aves and Mammalia) have powerful immune systems not possessed by Reptilia and Amphibia. Nor should one forget that the invertebrate host has evolved and possibly changed during this period (Baker, 1963). Another evolutionary problem is specificity for the invertebrate host. It would appear that such specificity could only harm transmission. This in turn may indicate that insects have evolved a different class of response for dealing with parasitic attack. This argument is not overstrong because it may well be that it is the parasites which have evolved so that they have succeeded in infecting only one insect species. This in turn would suggest that they have only recently arrived at this state of bliss. The implication would be that either the blood parasites are recent groups—which is unlikely, because of the wide distribution of species over the whole of the vertebrates—or that they have recently become able to infect one (or a few) insect species of comparable mode of life. The *reductio ad absurdum* of this second argument is that before such adaptation occurred there were parasites without insect vectors. This seems most unlikely and I feel that the mechanisms of host specificity of the parasite-insect relation should be studied much further by insect physiologists as well as by protoparasitologists. Another possibility is that optimum fitness may well be obtained by infrequent transmission, possibly to prevent "overkill."

The crux of the first caution is that the utmost openness of mind should be cultivated when trying to unravel the intricacies of so highly integrated biologies as are found in these parasites.

The second caution springs from the small size of the parasites involved. No observations can be made without microscopes and other equipment. Thus our observations can be as good as our techniques and no better. So much of the earlier work seems, rightly or wrongly, to fall into the category of "diligent but of poor technical quality." I have examined closely some relevant arguments on the subject of reproduction and heredity of trypanosomes (Walker 1964a) and concluded that much of the earlier work should be looked at again in the light of present knowledge and methods. Further I am convinced that reviewers can do great harm by interminable representation of old evidence and old controversy as well as by attempting to sort good from bad. Suffice it to say that the often-repeated phrase "There is nothing new in trypanosomiasis" is as nearly true for this parasite as it is for malaria. It is best for the individual worker to reexamine the earlier evidence according to his own light. The reviewer, however, can and should help by indicating where such evidence is to be sought and by providing general cautions and evidence about the reliability of techniques.

The third caution concerns sampling. It is too easy to infect a mouse on one day and find parasites in the blood a few days later and assume that the relation between cause and effect is simple. By all means the patent infection is due to the infective inoculum, but what is the course which has been followed? In malaria we now know of the exoerythrocytic cycle which is essential in the development of a blood infection from sporozoites. But how certain are we where infection of immature red cells takes place in *Plasmodium berghei*, Alger (1963) found an accumulation of schizonts in the bone marrow. This is suggestive that infection of new reticulocytes occurs in the bone marrow but this is not conclusive evidence, nor do we know which stage in maturation from erythroblast to erythrocyte becomes infected or even if the exact stage is critical to the parasite. This example is cited to expose the problems of sampling and trying to draw conclusions from data which are not formal evidence of anything more than a bare statement "the such and such stage was found in the nonsuch organ." Critical appraisal should be made when thinking of an infection in the blood, for although the total number of parasites is increasing in the host, their distribution may be subject to change during the division cycle and during the course of infection. It follows that while we may take blood for chemical or biological analysis of the parasites, the results we obtain are not necessarily statements about the whole infection. Therefore, in such circumstances unless the distribution is known and the timing, stages, degrees of synchronization, age of infection, effect of immune response, etc., are taken into account, the worth of such results tends to be low. Few workers, including the author, have gone to sufficient lengths to ensure a truly solid foundation for their studies. Even in a virulent laboratory-adapted infection of *Trypanosoma brucei* I am not convinced that all the division cycle is found in the circulating blood. The evidence so far suggests to the author that between 50 and 70 % of the generation time is spent by parasites *not* circulating in the blood. Thus those which circulate are in partial synchronization. Even such a basic hypothesis is remarkably difficult to verify; such evidence as exists was given in Walker (1964a).

Thus the validity of sampling in the study of blood parasites is of great importance. The difficulties surrounding our work are incomparably greater than those facing the bacterial or free-living protozoan geneticists.

C. Previous Reviews

1. Review material on trypanosome genetics is not extensive; only Walker (1964a) has attempted a critical synthesis of past and present work in the light of new ideas from other fields of cell biology. Much of the subject is contained in odd remarks included in otherwise unrelated papers on drug action, immunity, pathology, and cytology. Noble (1955) presents a valuable and extensive review on the cytological aspects of cell division and morphology. In some respects this developed from an original paper by Noble *et al.* (1953) which contains a survey of the literature on spindle formation and chromosome number. The reader is strongly recommended to consult the original papers for detailed discussion. Grassé (1952) also deals extensively with systematics and morphology.

Baker (1963) reviews the speculations regarding the evolution of the Trypanosomatidae. Although not relevant to the factual content of this review, no account of genetics is complete if it does not recognize the ultimate in long-term experiments.

Electron microscope and biochemical studies which have a strong bearing on the phased expression of characters have been made by Vickerman (1965, 1966). Both papers have review character and should be consulted in the original.

2. Access to the literature on malaria parasites is now much easier owing to the work of Garnham (1966). This work deals with most aspects of the biology and systematics of these parasites; since it is well indexed and covers most literature up to 1966 it is pointless to duplicate references here.

3. General discussions on the possibilities of sexual behavior in Protozoa may be found in Sonneborn (1957), Wenrich (1954), and Hawes (1963a).

4. Since drug resistance is one marker that is relevant for genetic analysis, the pertinent reviews are Schnitzer and Grunberg (1957), Bishop (1959), Williamson (1962), Hawking (1963), and J. Hill (1963).

II. CYTOLOGICAL ANALYSIS OF GENETIC SYSTEMS

Despite the assertion that "gene-like aggregations of chromosomes are found in the macronuclei of ciliates" (Garnham, 1966, p. 41), cytology can reveal little of the fine workings of chromosomes. Genes, when defined as a functional unit of the information store, are not visible to the electron or light microscope. Indeed it would be most unlikely for any instrument yet devised to visualize a transient chemical process at the molecular level. The place of cytology is to provide a short cut to obtaining worthwhile interpretations of breeding experiments. When once a firm cytological model of cell division, including meiosis, has been established it should be possible to infer the relations between characters observed in individual cells in an infection. If it is technically possible, it is much easier to see, measure, and count chromosomes than to do the necessary host of breeding experiments to establish the number and size of linkage groups.

A. OBSERVATIONAL METHODS

1. Microscopy

The resolution of the light microscope is not sufficient to permit unequivocal statements to be made about chromosome numbers and it will always be impossible to visualize chromosome function of the kind found in puffs of dipteran salivary glands or lampbrush chromosomes of amphibia. It is, however, still possible to see more than we do at present. Westphal (1965) recommends the use of reflex microscopy. This method makes use of metallurgical microscopes and suitable stains. The present author has confirmed Westphal's claims in every way. The resolution is twice that obtainable with transmitted light. Thus the theoretical limit of resolution for green light (the color reflected from deoxyribonucleic acid (DNA) stained with Giemsa) is 540/4 times numerical aperture. For a normal oil immersion objective the distance between two particles which can just be seen as distinct is 1040 Å, though much smaller particles can be seen providing they are well separated. In practice this means that a sharp photomicrograph can be produced at a magnification of 3000.

The ultraviolet microscope has become unfashionable with the advent of the electron microscope. The resolution in transmitted light using light at 2536 Å from a mercury resonance arc is the same as that of a reflex microscope using green light. Observations on the distribution of nucleic acid are the principal reason for the use of the ultraviolet microscope. Smiles (1933) experimented with dark ground illumination which would have a theoretical resolving power of about 500 Å. It would probably be simpler to use incident light reflex microscopy at this wavelength if the technical problems of internal reflexions could be overcome. Such high-resolution developments seem to depend on finding reflectance stains which function at such low wavelengths.

2. Staining and Other Methods of Preparation

The light microscopist can take advantage of the development in fixatives for electron microscopy (EM), providing the axiom is accepted that if a structure appears well fixed by EM it is well fixed for light microscopy (LM) work. The justification for continuing to use LM is to avoid using ultrathin sections and for cytochemistry.

The new methods of spreading chromosomes and their constituents by layering on to hypotonic solutions and capturing the exploded products on grids for examination under the electron microscope may well have applications for studying fine structure under the light microscope (Gall, 1963). Electron cytochemistry, except for osmophilic elements, is still in its infancy while light microscope cytochemistry is well established. Cytochemical methods in general tax the contrast of the system and only to a lesser extent the resolution. Thus ultrathin sections and dispersed (as seen at very high magnifications in the electron microscope) products of reaction are conducive to low contrast and hence negative results. Although this seems true at present the possibilities for development of the electron microscope in this field are large and new techniques will no doubt be available to future protozoologists.

With regard to microspectrophotometric measurements, Baker (1961) found that the quantity of DNA in nuclei of Trypanosoma evansi from blood was 0.2×10^{-12} gm. This is at the lower limit of sensitivity of Feulgen estimations. It seems that measurements have not been made of DNA content during the observable phases of division or of the content of each polar mass of DNA or of the content of nuclei at different phases of the life history. All these measurements would be of great value in the interpretation of the nuclear interrelationships during complex life histories. No evidence has been found for attempts at microspectrophotometry of the nucleus of any malaria parasites. The observations that gametocyte and trophozoite nuclei are mainly Feulgen negative parallel the indication above that this technique is not sensitive enough to be of use on such small quantities of DNA (Garnham, 1966, p. 101).

Generally there seems to be a need for more sensitive methods of demonstrating and estimating small quantities of DNA. Shortly before the death of Dr. R. S. J. Hawes in 1963 evidence appeared during collaboration with the author that the intensity of nuclear staining was greatly impaired by the use of crude stain; in this case azure A (Ed. Gurr Ltd.). This dye was fractionated and one fraction gave intense, well-differentiated pictures of granules at the polar peripheries of the nucleus. This distribution was in agreement with the distribution of electron-dense regions but contrary to the distribution of DNA by Feulgen, acridine orange, or acriflavine fluorescence, acetocarmine or Giemsa staining after acid hydrolysis. Such observations clearly need rigorous and carefully controlled testing to obtain information on the true composition of electron-dense particles in all nuclei of trypanosomes in sections photographed with the electron microscope.

B. Malaria

Cytogenetic observations are summarized in Table I. With the sole exception of the zygotic division, the chromosome number is uniformly low throughout the life history. The number of chromosomes appears to be based on a primitive haploid nucleus of two with the sole exception of *Plasmodium gonderi*. Recently Canning and Anwar (1967) have confirmed the observation of four chromosomes in the zygote of *Plasmodium gallinaceum;* they believe that reduction division takes place as the first division of the zygote. These workers examined their preparations with a view to finding chromatids as might be expected in a meiotic metaphase. No

TABLE I

Chromosome Number in Malaria and Related Parasites

Group[a]	Species	Stage[b]	Number	Condition	Shape	Reference
Sauramoeba	*Plasmodium floridense*	e	2	Haploid	Dot, equal	Wolcott (1957)
Giovannolaia	*P. lophurae*	e	2	Haploid	Dot, equal	Wolcott (1957)
Haemamoeba	*P. relictum*	e	2	Haploid	Dot, equal	Wolcott (1957)
Haemamoeba	*P. gallinaceum*	zd	4	Diploid	2 J-rod and 2 dot	Bano (1959)
Rodent species (*Vinckeia*)	*P. berghei*	e	2	Haploid	Dot, equal	Wolcott (1957)
Rodent species	*P. berghei*	zd	4	Diploid	Dot	Bano (1959)
Human species D (*Laverania*)	*P. falciparum*	e	2	Haploid	Rod and dot	Wolcott (1955)
Simian species D (*Plasmodium*)	*P. knowlesi*	zd	4	Diploid	Dot	Bano (1959)
Simian species C (*Plasmodium*)	*P. inui*	zd	8	Diploid	Dot	Bano (1959)
Human species C	*P. malariae*	e	2	Haploid	Rod and dot	Wolcott (1955)
Simian species B (*Plasmodium*)	*P. gonderi*	zd	6	Diploid	4 rod and 2 dot	Bano (1959)
Human species B	*P. ovale*	e	2	Haploid	Rod and dot	Wolcott (1955)
Simian species A (*Plasmodium*)	*P. cynomolgi*	zd	8	Diploid	4 rod and 4 dot	Bano (1959)
	P. cynomolgi	pz	4	Haploid	2 rod and 2 dot	Bano (1959)
Human species A	*P. vivax*	e	2	Haploid	Rod and dot	Wolcott (1955)
Human species A (*Plasmodium*)	*P. vivax*	zd	4	Diploid	2 Rod and 2 dot	Bano (1959)
Haemoproteidae	*Nycteria medusiformis*	ee	4 to 8	Haploid	Rod	Garnham and Heisch (1953)

[a] In most cases the group refers to a subgenus as used by Garnham (1966).

[b] Key to abbreviations: e, erythrocytic division; zd, zygotic division; pz, postzygotic division; ee, exoerythrocytic division.

evidence for chromatid formation was obtained and it was concluded that reduction was not a conventional meiosis. This observation has led to the suggestion that there may be two linkage groups which sort at random during the brief existence of the zygote. Other speculations can be offered that it is possible that the observed "chromosomes" are only associations of convenience at the time of division of a greater number of smaller DNA strands (linkage groups). On such a hypothesis a random rearrangement of genes is possible without having numerous chromosomes. The observed low and relatively constant "chromosome" number is in agreement with this idea. Moreover, such constancy covering so many apparently unrelated species is not usual by comparison with the variability of chromosome number in other groups.

Leaving this type of speculation and returning to more solid ground the more recent evidence indicates that there is some form of reduction division in the zygote. Thus the rest of the life history in the mosquito and vertebrate is presumably in the haploid condition. This is contrary to the views of some earlier workers in which one or two polar bodies were "seen" on the surface of mature gametocytes, especially in *P. falciparum*. These observations were made with Romanowsky-stained slides in which the polar bodies appeared red. Red is the typical color for stained DNA in species of *Plasmodium*. More recently, Lüdicke and Piekarski (1952) observed Feulgen-positive reduction bodies.

In the author's opinion, working by analogy with higher organisms, it is inconceivable that a classic haploid organism can overcome the vicissitudes of development in a mosquito, transfer to vertebrate host, pre-erythrocytic cycle, erythrocytic cycle, and gametogenesis. In addition it has been shown that a single asexual phase parasite can give rise to an infection producing micro- and macrogametocytes (Bishop, 1955). Although there is good evidence for a type of reduction in the zygote, the author considers it probable that additional mechanisms for genetic control and exchange exist in the rest of the life cycle.

It is interesting to note that cross-fertilization is unlikely to occur because of mixed strains in the mosquito owing to the short span of viability of the gametes and the feeding behavior of the host. Only in mixed strains in the vertebrate are gametes produced which are capable of coming together at zygote formation. Therefore only in synchronous mixed infections of two strains of the same species is crossing likely. In the absence of cross-fertilization clone strains spring up and the results are "speciation" at the level of individual organisms. Yet the characteristic morphology and clinical behavior of the several human malarias (the most studied) are almost identical the world over and deny the existence of clones or persistent inbreeding which would appear most likely. Does this indicate ultrastable

DNA replication (despite the facility to develop drug resistance) or a rapid and probably unusual method of genetic exchange?

C. Trypanosomes

Noble (1955) gives a table entitled "The Reported Chromosome Numbers of the Trypanosomidae." He comments, "A wide discrepancy in reports of chromosome numbers is due to the small size of trypanosomes, and the difficulty of obtaining a satisfactory nuclear stain during mitosis." Of the more recent reports, quoted by Noble, three chromosomes seemed to be favored. This number has never been observed by the author. Nuclei in division have been photographed after Giemsa staining using ultraviolet light (λ = 2536 Å) (Walker, 1964a, p. 73). The polar regions of the nuclei contained numerous, minute, densely stained granules. Estimates of the number of granules are between 15 and 30 in each end of the nucleus but are uncertain because of dense packing. From evidence of two kinetoplasts and two flagella one would presume that these cells were in late anaphase or telophase. These preliminary observations were confirmed by Hawes and Walker (1963) in a series of experiments using Hawes's celloidin technique to make the parasites adhere to the slides after wet fixation by injection into fixative (Hawes, 1963b). Staining with Feulgen or mixed azures showed the same cluster of small dots at the poles of the nuclei. In the visual light microscope these dots are at the limit of resolution. The author is not satisfied; crucial experiments will have to show: (*a*) that the number of dots is more or less constant, (*b*) that the numbers of large, medium, and small dots also remain constant, (*c*) that some recognizable pattern emerges in cell division with regard to the time of duplication of number of dots and the sequence of separation of dots within newly formed nuclei, and (*d*) that the number of dots remains constant, or at least explicable, during the whole life history, especially in the insect divisions and changes to metacyclic trypanosomes.

Morphological studies on culture or insect phase trypanosomes, especially of the *brucei* group, have not been made frequently. The reasons for lack of observations is that morphological preservation is usually poor with these forms and, until recently, tsetse flies have been difficult to obtain (Nash *et al.*, 1966). Further investigation of nuclear changes should be rewarding when the newer techniques are applied.

Vickerman (1967) has shown in electron micrographs a spindle which consisted of characteristic microtubules. The spindle was narrow and lay on the long axis of the nucleus, which had a persistant nuclear membrane. One dense granular area was at the midpoint of the spindle and was the endosome. Other dense granular areas were observed against the nuclear membrane and unattached to the spindle. These areas were thought to be chromatin. The observation confirms the extraordinary nature of the trypanosome nucleus and supports the theory of multiple dot chromosomes. However, if the nucleus divides with the chromatinic bodies stuck to the nuclear membrane, visualization is made more difficult as the bodies are on a cup-shaped support and the function of the spindle is as mysterious as ever.

III. GENETIC MARKERS

A. Types of Markers and Their Potential Use

1. Morphological Markers

a. Length Measurements. Classic genetics was founded on the use of morphological markers; hence it is not surprising that these have been investigated in blood parasites. Fortunately the malaria parasite is too small to encourage such use; the essentially variable shape throughout the life cycle would make claims seem absurd. Such has not been the case with trypanosomes. Several workers in the late forties and early fifties described experiments in which they had attempted to cross long and short trypanosomes or to differentiate polymorphic types. The results needed statistical analysis of the lengths of the trypanosomes which were not much different from one another. The question raised is how far can averages based on large numbers of measurements become so refined that not only can one detect small differences in character but also one can be sure that other factors were not influencing the results. It is impossible for one worker to make an accurate appraisal of the work of others without knowing all the seemingly trivial details, but it is relevant to point out two other observations.

The length of the flagellum seems to be unexpectedly variable. Taliaferro (1923) made a detailed study of *Trypanosoma lewisi* and found the length varied with the age of the infection. Beckwith and Reich (1922) measured about 100 *T. brucei* from a guinea pig and found the length to be 18.46 μ. When the strain was in rats the length appeared to be 13.26 μ. Godfrey (1960) confirmed and extended this observation using three strains of *T. congolense* in several host species. The lengths of the trypanosomes of one strain depended upon the host, and would readily return to the original length when passaged into the former host. The change in length was from 8–14 μ to 10–19 μ and covered the ranges of both species *T. congolense* and *T. dimorphon* (Hoare, 1959).

Secondly, the length of *T. brucei* (NIMR 2), a virulent laboratory-adapted strain, is dependent upon the stain used for visualization. This was discovered by accident but investigated simply by Walker (1963). Eight stains were used under identical conditions and the average lengths were calculated from about 100 trypanosomes. The mean lengths varied from 18.6 μ to 21.03 μ and represents an increase of 11.3%. Two conclusions can be drawn from this observation. Such a change between batches of Giemsa's stain, one old and one new, must surely find a parallel in the literature. It would, therefore, be unwise to use changes in length of this magnitude for any genetic or taxonomic work. The second conclusion is based on statistical reasoning on any observations made from measurement with a light microscope. The resolution under the conditions of a good immersion objective is not better than 0.2 μ. Since the essence of scientific observation is that others should be able to repeat the results and the luxury of analysis of variance for the factor "between observers" is not usually possible, it follows that the accuracy of mean lengths need not be better than the limiting accuracy of the method. Thus the standard error should be about 0.2 μ. In the case cited with a length of about 20 μ and a standard deviation of 2 μ, the measurement of 100 trypanosomes gives a standard error of 0.2 μ. Workers who measured hundreds or even thousands of trypanosomes of a unimodal infection were obtaining little but personal satisfaction.

With these severe restrictions in mind morphological characters such as length could be of use in genetic experiments. There is no conclusive work to report.

b. The Kinetoplast of Trypanosomes. The kinetoplast is now known to be a DNA-containing organelle which is closely associated with a mitochondrial structure. The size of the latter depends on the species of trypanosome and, in the case of Salivaria trypanosomes, on the stage of the life cycle (Vickerman, 1962, 1965). The kinetoplast would seem to function as some reserve of information for in some strains of *T. evansi* and in *T. equinum* the kinetoplast is lost and these strains are unable to be passaged cyclically or even to go into culture. The akinetoplastic condition has been described as a mutation by Hoare and Bennett (1937). As it is an obvious marker it has been used to study generation times of mixed infections of kinetoplastic and akinetoplastic strains. While such studies have produced the results that one or another strain reproduced faster the method has been thought of as a possible way of demonstrating sexual behavior (Mühlpfordt, 1960). All-or-nothing characters of this kind are unsuitable for genetic work because half a kinetoplast has no reality and occasional genetic exchange could not possibly be observed against a background of rapid asexual reproduction. Evidence against sexuality in trypanosomes based on this type of experiment is considered invalid.

2. Immunological Characters

The successful use of antigenic typing in species and strains of *Paramecium* (Ciliata) has prompted the use of such techniques for investigation of inheritance in blood parasites. Antigenic analysis is the most sensitive indicator of protein species presently available. So far no evidence has been obtained to support or deny the possibility of genetic exchange. The reasons are, however, worth discussing.

Antigenic variation occurs in the normal course of infection in both trypanosomiasis and malaria (Lourie and O'Connor, 1937; Brown and Brown, 1965). One effect is to prolong the life of the infection by preventing the death of the host. It seems that only surface antigens are capable of variation; those within the cell are stable and characteristic of the species or genus. Thus the antigens which are potentially available as genetic markers are not suitable because of variability which can occur naturally during the essential growth of sufficient parasites for experimental purposes. This view will probably be shown to be too superficial when further work has been done on the identification of proteins (G. C. Hill and Guttman, 1965) and on the antigenic characterization of strains. At present it is impossible to be certain that variation has not taken place in the surface antigens nor is it known under what type of control variable protein synthesis exists. Perhaps a parallel can be drawn between the case of antigenic variation and the ease with which drug resistance develops to most chemotherapeutics.

3. Drug-Resistance Characters

Such characters suggest themselves partly from their distinctness and partly from the practical value of this type of investigation. When compared with other Protozoa, e.g., *Amoeba* (Cole and Danielli, 1963), the magnitude of the change to resistance is large. In the case of *brucei* group trypanosomes, resistance to aromatic arsenical drugs may increase by 100- to 600-fold, while *Plasmodium cynomolgi* has been reported to develop 2000-fold resistance to proguanil.

Walker (1964b) drew attention to the problems of using drug-resistance character for genetic work by distinguishing between "population" and "clinical" resistance. When parasites are exposed to a toxic substance either *in vitro* or *in vivo* the number of parasites which are killed will depend on such factors as the sensitivity of the different stages of the division or life cycle, the proportions of each stage, the levels of each population of resistance (i.e., the mean sensitivities of each population), the standard deviation for sensitivity of each population and the proportions of each population as a fraction of the total infection. All this is in addition to the chosen value

of drug concentration. Experimental justification for this complicated view
is given by Hawking and Walker (1966). Therefore to go through a certain
protocol of drug treatment tells one no more than that there were one or
more parasites which could survive the drug level used. This type of information is commonly derived from clinical observations and is termed
"clinical" resistance. It should be noted that where a few resistant parasites
are present in a generally sensitive population and the sensitive trypanosomes take up the drug, a sufficient amount can be removed from the
medium to make the resistant parasites appear more resistant than they
truly are (Walker, 1966). Similarly if repeated treatment of trypanosomes
is made in a mouse then the suppressed infection appears to become resistant but is not so when it is transferred to a new host (Browning, 1908).
Thus "clinical" resistance is of little value, unless experimental design is
such as to allow for these inherent difficulties—as, for example, in the work
of Amrein and Fulton (1959).

"Population" resistance is a concept of quantitative information about
drug sensitivity which needs to be expressed as a mean and standard deviation. The concept is observational in that it takes no account of possible
difference between stages of a division or life cycle. If such differences occur
two or more "populations" of resistance would be observed, and it would
then be necessary to ascertain the relationships between the populations.

It follows that if genetics of these microorganisms is to develop along
the lines of population genetics it is necessary, in general, to evolve systems
of analysis which will give estimates of mean and standard deviation for
each population. Skeptics should remember that the great advances in
bacterial genetics have been made using markers of an all-or-nothing
character which can be applied to the organism cultured *in vitro*. Thus
enzyme markers are of great value and deficient mutants can be studied.
The blood parasites, on the other hand, make poor subjects for culture as
yet, and consequently the whole method of investigation is denied. Studies
in vivo are complicated not only by the sampling and observational problems already outlined but also by the immunological response made to
infection. The inadequacy of *in vivo* measurements is made clear by the
somewhat extreme example reported by Soltys (1959), who found that
antibody resistant *T. brucei* was three times less sensitive to Antrycide *in
vivo* and fifty times less sensitive to exposure *in vitro* than the normal strain.

Cross-resistance provides an interesting opportunity to study intracellular reactions to a variety of chemical structures. The papers of Williamson and Rollo (1959) and of Goble *et al.* (1959) summarized our knowledge
on this subject. The ability of trypanosomes to become resistant to acriflavine and aromatic arsenicals is clearly indicative of underlying structural
properties probably relating to the mode of entry of the drug. Further

work is required to quantify the response to others in the acridine series which is made by particular populations of trypanosomes. Such work as attempted by the author is handicapped by inaccurate labeling or gross impurity or unavailability of the essential related dyestuffs.

B. RESULTS

1. Malaria

Despite the numerous studies on drug resistance in these parasites, there appears to be no work which is of significance to the would-be geneticist except in the following spheres.

Drug resistance is presumed to be a mutation, that is a "heritable change" (Dobell, 1912) by virtue of the stability of certain types of resistance when no selection pressure is applied and to the occurrence of drug-resistant strains in the field and the laboratory. The stability in the case of resistance to proguanil in *Plasmodium gallinaceum* has been demonstrated also after cyclical development in *Aëdes aegypti*. The strain was passed through four groups of chickens which received no drug between five transmissions (Bishop and Birkett, 1948).

Experiments of Young and Moore (1961) in which a chloroquine-resistant strain of *P. falciparum* was passed to a human volunteer by mosquito transmission also showed the stability of this resistant character.

Further confirmation of the mutation-like character of resistance is obtained from single parasite infections. Bishop (1958) has demonstrated that clone infections develop resistance to metachloridine.

Cross-resistance studies have not thrown light on the genetic control of resistance as the drugs pyrimethamine, proguanil, and some sulfonamides all act as folic acid antagonists. It would appear that in general the mechanism for resistance involves an alternative path to be free of the inhibitor. Exceptions have been found and for a detailed discussion the reader should consult Bishop (1959).

2. Trypanosomes

As in malaria many features of drug resistance demonstrate the mutation-like nature of the phenomenon. The stability of Atoxyl resistance in a virulent laboratory infection of *T. rhodesiense* has been demonstrated by Amrein and Fulton (1959). The stability has also been found after cyclical development in *Glossina* of strains resistant to tryparsamide, suramin, and Antrycide (reviewed by Bishop, 1959). Some loss of resistance was noted in a suramin-resistant strain.

Infections derived from single trypanosomes are capable of becoming resistant (Oehler, 1913; Inoki et al., 1961) and are also capable of developing polymorphic infections (Oehler, 1913; Walker, 1960).

Confirmation of the possibility of selection *in vivo* of a few resistant trypanosomes from an infection of sensitive ones by drug treatment has been shown by Hawking and Walker (1966). An artificial mixture contained one resistant trypanosome in one million sensitive parasites. After application of drug selection either *in vivo* or *in vitro* an infection developed in mice.

Attempts at the analysis of the development of resistance by "population" methods has been successful in one aspect of resistance. The author refined the method of Hawking (1938) in which the combined effect of acriflavine and light kills trypanosomes. When, under standardized conditions, only the concentration of drug is varied, the percentage mortality after exposure to light can be used as a direct measure of sensitivity to aromatic arsenical drugs. Experiments following the development of resistance showed that any of five resistant populations may appear either singly, mixed, or mixed with the sensitive population. It has been suggested that the populations could all be accounted for if three genes had additive properties and the nuclear condition (or expression) was haploid (Hawking and Walker, 1966). Elsewhere the author speculated on the nucleus being diploid and on two genes with incomplete dominance to account for the populations and the pattern of their arrival (Walker, 1962). The disturbing features of the evidence to date are that when clone infections are made from newly resistant strains there may appear two or three populations of resistance and also that when analyses are made during the development of resistance there have been several occasions when populations have appeared which are more sensitive than could have survived the drug levels administered. It is believed that this method of approach is correct but it needs developing to other resistant characters and extending to take into account possible changes in sensitivity during the division cycle.

Amrein and Fulton (1959) failed to demonstrate genetic exchange after mixing a strain of trypanosomes resistant to Atoxyl with a strain resistant to suramin in the same mouse. Passage was made to further mice and groups of these were treated either with Atoxyl or with suramin. A second passage was made and the drug used for treatment was reversed. All mice were cured and this demonstrated no double resistance. While the results of this experiment are clear-cut, the likelihood of getting double resistance has been discussed unfavorably by the writer (Walker, 1964a).

Immunological markers have not been used on account of variable antigens, but Gray (1962, 1965) reported after that cyclical development in *Glossina* the antigens of *T. brucei* in the new blood infection were of a common first type. It appeared to be immaterial how far along the line of relapse variants his strains were before fly transmission; there was a switchback to an initial antigen. This valuable observation suggests that the control of development of the insect phase is without regard to the previous immunological experiences in blood infections.

IV. DEVELOPMENTAL GENETICS

A. MALARIA

There appears to be no available evidence on the timing of biosynthetic activities which can be correlated to nuclear changes. Only antigenic variation during blood infections may possibly come within this section but as yet there is no evidence for genetic control.

It should be noted that single trophozoite infections can give rise to both male and female gametocytes (reviewed by Bishop, 1955). Garnham (1966) points out that we are still ignorant of the type of schizont involved and of the trigger to stimulate gametogenesis. Such observations await attention for their explanation.

B. TRYPANOSOMES

A similar state would also be true for trypanosomes were it not for the work of Vickerman (1962, 1965). We now know that during cyclical development a mitochondrion develops from the kinetoplast. The functions of this structure have been demonstrated by bulk biochemical methods and by cytochemical observation. The change of structure with passage from mammals to flies is impossible in those strains which do not have kinetoplasts and this has given rise to the mechanically transmitted species, *T. equinum*. The kinetoplast contains DNA and this establishes the link between structural observations and genetics.

Trypanosoma equinum represents an extreme example in loss of fly transmissibility. The route to such a situation is known in the other members of the *evansi* subgroup and is paralleled in the laboratory during the change from transmissible to virulent laboratory infections. The first indication of these changes is the loss of polymorphism in the blood infection. There is only rarely a loss of kinetoplast. However, the virulence increases and passage must be rapid to keep the strain in rodents. It is not clear whether there is an earlier stage in which a strain is culturable, i.e., can infect the gut of the fly, but cannot develop metacyclic infection of the salivary glands.

From the above, the kinetoplast is quite clearly implicated in the ability of a trypanosome to develop an insect infection. However, because kinetoplast-containing trypanosomes do not all possess this ability it is clear that there is either or both a controlling gene (which may be in the nucleus or in the kinetoplast) and the descriptive DNA necessary for the production of mitochondrial ribonucleic acid (RNA). If either of these fails, then loss of transmissibility results.

In current ideas about the control of gene function histones are widely considered to be involved. Beck and Walker (1964) investigated the fluorescent antibody reaction by the sandwich technique in which human autoantibodies from lupus patients were used. The results showed that staining of mammalian cells persisted after deoxyribonuclease (DNase) and ribonuclease (RNase) treatment of nuclei; this suggested that some sera had antihistone antibodies. These sera failed to stain trypanosomes and it was concluded that either the histone was absent or it was antigenically different from mammalian and other histones. Dodge (1965) (see also discussion and photographs in Vickerman, 1966) described the lack of histones in a dinoflagellate. It is possible that control at the molecular level is radically different from that found in other (higher) organisms.

V. POPULATION DYNAMICS

In population dynamics it should be clear that one is intending to study the reproduction and fate of the parasite population (infection) and not those features of the host population. This section is included because where direct Mendelian crosses are not technically possible it should still be possible to follow populations of parasites and infer the genetic relationship of characters (Falconer, 1960).

A. METHODS

The essence of any study of population dynamics is that one should be able to estimate numbers in an infection and the characters of each population. Such estimates should be made simultaneously and accurately. In the foregoing sections the difficulties attached to the use of genetic markers have been discussed. Estimation of number is of comparable difficulty.

Generally, when an infection has reached the level when it can be counted in a hemocytometer or as a percentage of infected cells it is already in a late stage. For trypanosomes the lowest count that can be made with the customary accuracy of 5% (400 cells) is 4000 per μl. Estimates can be made at lower certainty to perhaps 500 per μl. Figures below 2000 per μl are seldom quoted due partly to the time-consuming nature of such counts and partly to the confusion between misshapen trypanosomes and red cell debris when white cell counting methods are used. But for argument if the lowest count is 2000 μl and if the blood volume of a mouse is 3 ml and if all the trypanosomes are circulating then the total number of trypanosomes is 6,000,000.

One can not gloss over the fact that to get 6,000,000 trypanosomes from a single trypanosome infection there have been, if none have died on the way, some 22 generations and no less than 5,999,999 divisions. During this time it has been impossible to count or seriously observe the infection. For a virulent laboratory infection with a generation time of 3.3 hours the infection has been incommunicado for over 3 days. Any departure from the ideal situation renders observation more liable to error.

In malaria the position of counting is as bad if not worse. It is assumed that there are 200 red cells in a field of a $\times$ 100 objective, and that one is prepared to spend 15 minutes on a slide. An experienced microscopist has been timed at 22 fields per minute. A total of 66,000 cells are examined. To obtain a reasonable accuracy 400 of these require to be infected. This is an infection rate of 0.6 %. Assuming the usual number of red cells and blood volume for mice the number of infected cells in *P. berghei* at this level of infection is about 8×10^7. If the asexual cycle lasts 24 hours (Thurston, 1952) and the average number of viable merozoites per trophozoite calculated from Cox *et al.* (1964) is 5.3, then the infection could not be accurately counted until some 11 days after inoculation of a single trophozoite. During this time at least 2×10^7 red cells have already been destroyed. Such calculations do not take into account the effects of immune response on the parasite, immune hemolysis, or the possibility of damage to cells by the merozoites which seem unable to establish themselves. Possibly such merozoites amount to 35 % of any one generation.

The counting methods are clearly inadequate in themselves and progress will be hampered until these have been improved, and also until more is known of host-parasite relations. It is not forgotten that the above calculations have been made considering only the blood phase of an infection under nearly ideal laboratory conditions.

Various methods have been advocated recently as improvements in counting. Walker (1964a) used a procedure modified from Wijers (1960) in which the ratio of trypanosomes to white cells was established and also the number of white cells per microliter was counted. By multiplying the two figures a minimum estimate of 200 per μl was possible but errors other than those of random sampling were introduced.

Ormerod *et al.* (1963) advocated a counting method in which trypanosomes were observed as they swam at the edge of a drop of blood imposed on a thin agar film. Trypanosomes were estimated as numbers of parasites per field. A correction factor can be determined to bring the count to parasites per microliter. It is emphasized that the method was designed to provide a method for morphological observation as well as for counting and the authors suggest the method should be restricted to the comparison of experimental infections with their controls. Less direct methods of estimating numbers of trypanosomes have been used. Cantrell (1956) estimated the infecting proportions of normal and resistant trypanosomes by observing the rate at which trypanosomes appeared to reproduce and extrapolating back to the time of infection or treatment.

Lumsden *et al.* (1963) introduced a titration technique for counting the number of viable trypanosomes. The number of parasites is estimated from the proportion of mice which become infected from each serial dilution of an original suspension. Such a method assumes that any trypanosomes can start an infection and that the procedures do not harm the parasites even at high dilutions. The results given in the paper and those obtained later support the confidence in the method. It is advisable to test the longevity of a strain of trypanosomes before a major experiment by using the same balanced salts solution to dilute the suspension as in the main experiments. The writer (Walker, 1964c) has found that for a virulent infection of *T. brucei* a "Ringer's" solution made up with nitrates instead of chlorides is satisfactory and avoids difficulty of precipitation of various drugs and ions.

Attempts to apply the titration technique to count *P. berghei* were not successful using a medium of phosphate-buffered Ringer's with 5% inactivated horse serum. Only a few per cent of the cells remained infective after dilution. Warhurst (1964) obtained satisfactory counts when the diluting medium contained 50% calf serum.

B. Malaria

Direct evidence for the stability of drug resistance or any other character has not been obtained in human malaria owing to the impossibility of performing appropriate experiments or even of getting enough parasites to develop tests. In animal malarias the details of insect transmission of *P. berghei* have only just been worked out (Wery, 1967). The transmissibility of avian malarias does not seem to have led to population dynamic studies because of the unavailability of character determination except by *in vivo* tests.

Two papers serve to emphasize the experimental problems and the difficulty of using "clinical" evidence in the interpretation of genetic behavior. The first by Clyde and Shute (1959) shows the development of resistance to pyrimethamine due to prophylactic treatment in an area with a high incidence of *P. falciparum*. Twenty-six months after prophylaxis ended drug resistance was still detectable in patients who could be presumed to be newly infected. This indicates the permanent nature of this resistant mutation and also that resistance was transmissible through the vector. The second paper, by Ramakrishnan *et al.* (1961), is concerned with the size of a population of *P. berghei* and the apparent resistance to sulfadiazine.

Resistance to sulfadiazine apparently proceeds by selection; thus the strain becomes resistant more quickly when there are plenty of potentially resistant parasites from which to select and when selection pressure is high.

C. TRYPANOSOMES

Population dynamics have occasionally been used to interpret experiments intended to examine the possibility of genetic exchange or mutation rate. Cantrell (1956) reported on the behavior of a mixed infection of normal and oxophenarsine-resistant strains of *T. equiperdum* in rats. This method of counting has been discussed above and the general conclusions have been reviewed by the writer (1964a, p. 90). Von Brand and Tobie (1960) made similar experiments in a attempt to obtain a genetic cross and hence evidence for sexuality. Both of these papers suffer from the same simple defect, namely, that if occasional genetic exchange had occurred then the experimenter would have been unable to recognize it within the protocols used. This paper has also been discussed at some length in Walker (1964a).

Attempts to determine the mutation rate to antibody resistance have been made by Cantrell (1958) and by Watkins (1964). It seems that Watkins was unaware of the earlier work which is also technically superior: both workers used modifications of the fluctuation test of Luria and Delbruck and found that about 1 division in 1,000,000 gave a mutation in the specified direction. It is difficult to use such tests which were developed for determination of bacterial mutation rates because a large number of replicate inoculations are necessary. The size of such experiments is daunting and deterioration of the trypanosome suspension may well occur during the time of inoculation of the first to the last mouse. In both cases it is unfortunate that clone strains were prepared by dilution instead of by mechanical isolation. Watkins observed that after he had inoculated 50 mice with an average of 1 trypanosome per mouse, 47 mice died. From the Poisson distribution it is apparent that Watkins' clone strain was obtained from an average of 2.8 trypanosomes per inoculation. Chance may have provided him with a single trypanosome infection; the probability is only 0.17.

An observation which has hitherto been inexplicable and is therefore all the more worth recording is the strange appearance of very resistant trypanosomes in experiments by Walker (1964a) and Hawking and Walker (1966). During processes of *in vitro* selection by trivalent tryparsamide and analysis of resistance by acriflavine-induced photosensitivity (Walker, 1966) extreme resistance was found as the total measurable population. Selection was repeated and at the next test the extreme resistance occurred in only 13 % of the population. This observation confirms the accuracy of the measurements based on a total population with extreme resistance. However, at the next test, despite continued selection, the extreme level could not be detected. The loss of this high level, or indeed any observations on levels of resistance more than five times greater than can be observed in mice, awaits further investigation.

D. Discussion

In the end our knowledge of these parasites must involve an appreciation of their total biology. Ecology is not a fashionable word for use with organisms at this size level, but it is appropriate. Of course, the methods and thoughts of molecular biologists are necessary when the organisms under study react to their habitat largely at the molecular level. The outstanding problems would seem to be the control of host specificity and whether mutation is directional.

It was mentioned in Section I,B that there was a possibility of "over-kill" of the vertebrate host. Owing to the high incidence of blood-sucking arthropods and of warm-blooded vertebrates the chance of transmission by mechanical or cyclical means is very high. If all vectors and verte-brates were equally susceptible the parasite population would overwhelm the vertebrates and with the death of the host death would also follow for the parasite. It is obvious that this has not occurred in evolutionary time; one possible mechanism is the development of host specificity. By this means the number of potential hosts or vectors in an area for any one species of parasite is vastly reduced and transmission becomes sufficiently infrequent for the host's immune response to prevent immediate death. The concept of fitness for blood parasites is very different from that in general for metazoa. A metazoan of high fitness has high utilization of available food and a high reproductive capacity. If these criteria are extended to blood parasites the host dies. Thus, a "fit" blood parasite is one with a limited possibility for transmission, with certain antigens freely exposed to the host to provoke immune response, and with the possibility of a long period in the vertebrate to carry over the infection during times without the insect vector. Population control by the parasites of them-selves is already well advanced. One wonders at the possibility of extinct vertebrates and extinct blood protozoa that occurred in past times and how far blood protozoa have shaped the course of evolution of mammals.

The second problem is whether mutations are random or directional. Both malaria and trypanosomes are capable of becoming resistant to drugs and to antibodies. The possibility that these parasites can recognize an insulting substance and make modification to overcome it should not be ignored even though the inheritance of such modifications would be Lamarckian. It is believed that the events leading to synthesis of antibody by lymphocytes is just such a process of recognition and adaptation. However, lymphocytes seem to be nonreproductive and consequently the adaptation cannot be passed to daughter cells as would be required for blood potozoa. Nevertheless if one cell type can recognize specific chemical groupings on substances then the design and interpretation of experiments should not preclude the possibility in other cell types.

REFERENCES

Alger, N. E. (1963). *J. Protozool.* **10**, 6–10.
Amrein, Y. U., and Fulton, J. D. (1959). *J. Protozool.* **6**, 120–122.
Baker, J. R. (1961). *Trans. Roy. Soc. Trop. Med. Hyg.* **55**, 518–524.
Baker, J. R. (1963). *Exptl. Parasitol.* **13**, 219–233.
Bano, L. (1959). *Parasitology* **49**, 559–584.
Beck, J. S., and Walker, P. J. (1964). *Nature* **204**, 194–195.
Beckwith, T. D., and Reich, W. W. (1922). *Trop. Diseases Bull.* **19**, 156–157.
Bishop, A. (1955). *Parasitology* **45**, 163–185.
Bishop, A. (1958). *Parasitology* **48**, 210.
Bishop, A. (1959). *Biol. Rev.* **34**, 445–500.
Bishop, A., and Birkett, B. (1948). *Parasitology* **39**, 125.
Brown, K. N., and Brown, I. N. (1965). *Nature* **208**, 1286–1288.
Browning, C. H. (1908). *J. Pathol. Bacteriol.* **12**, 166–190.
Canning, E. W. and Anwar, N. (1967). (In press.)
Cantrell, W. (1956). *Exptl. Parasitol.* **5**, 178–190.
Cantrell, W. (1958). *J. Infect. Diseases* **103**, 263–271.
Clyde, D. F., and Shute, G. T. (1959). *Trans. Roy. Soc. Trop. Med. Hyg.* **53**, 170–172.
Cole, R. J., and Danielli, J. F. (1963). *Exptl. Cel. Res.* **29**, 199–206.
Cox, F. E. G., Bilbey, D. L., and Nicol. T. (1964). *J. Protozool.* **11**, 229–230.
Dobell, C. (1912). *J. Genet.* **2**, 201–220.
Dodge, J. D. (1965). *Prog. Protozool., Abstr. 2nd Intern. Conf. Protozool., 1965*. p. 264. Excerpta Med. Found., Amsterdam.
Falconer, D. S. (1960). "Introduction to Quantitive Genetics." Oliver & Boyd, Edinburgh and London.
Gall, J. (1963). *Science* **139**, 120–121.
Garnham, P. C. C. (1966). "Malaria Parasites and other Haemosporidia." Blackwell, Oxford.
Garnham, P. C. C., and Heisch, R. B. (1953). *Trans. Roy. Soc. Trop. Med. Hyg.* **47**, 357–363.
Goble, F. C., Ferrell, B., and Stieglitz, A. R. (1959). *Ann. Trop. Med. Parasitol.* **53**, 189–201.
Godfrey, D. G. (1960). *Ann. Trop. Med. Parasitol.* **54**, 428–438.
Grassé, P.-P. (1952). "Traité de Zoologie," Vol. 1, Sect. 1. Masson, Paris.
Gray, A. R. (1962). *9th Meeting I.S.C.T.R.*
Gray, A. R. (1965). *J. Gen. Microbiol.* **41**, 195–214.
Hawes, R. S. J. (1963a). *Quart. Rev. Biol.* **38**, 234–242.
Hawes, R. S. J. (1963b). *J. Roy. Microscop. Soc.* [3] **83**, 1–3.
Hawes, R. S. J., and Walker, P. J. (1963). Unpublished observations.
Hawking, F. (1938). *Ann. Trop. Med. Parasitol.* **32**, 367–381.

Hawking, F. (1963). *In* "Experimental Chemotherapy" (R. J. Schnitzer and F. Hawking, eds.). Academic Press, New York.

Hawking, F., and Walker, P. J. (1966). *Exptl. Parasitol.* **18**, 63–86.

Hill, G. C., and Guttman, H. N. (1965). *Prog. Protozool. Abstr. 2nd Intern Conf. Protozool, 1965*, p. 132. Excerpta Med. Found., Amsterdam.

Hill, J. (1963). *In* "Experimental Chemotherapy" (R. J. Schnitzer and F. Hawking eds.), Vol. 1, p. 513. Academic Press, New York.

Hoare, C. A. (1959). *Parasitology* **49**, 210–231.

Hoare, C. A., and Bennett, S. C. J. (1937). *Parasitology* **29**, 43–56.

Inoki, S., Sakamoto, H., Ono, T., and Kubo, R. (1961). *Biken's J.* **4**, 67–73.

Lourie, E. M., and O'Connor, R. J. (1937). *Ann. Trop. Med. Parasitol.* **31**, 319–340.

Lüdicke, N., and Piekarski, G. (1952). *Zentr. Bakteriol., Parasitenk., Abt. I. Orig.* **157**, 522–538.

Lumsden, W. H. R., Cunningham, M. P., Webber, W. A. F., van Hoeve, K., and Walker, P. J. (1963). *Exptl. Parasitol.* **14**, 269–279.

Mühlpfordt, H. (1960). *Z. Tropenmed. Parasitol.* **11**, 265–287.

Nash, T. A. M., Jordan, A. M., and Boyle, J. A. (1966). *Trans. Roy. Soc. Trop. Med. Med. Hyg.* **60**, 183–191.

Noble, E. R. (1955). *Quart. Rev. Biol.* **30**, 1–28.

Noble, E. R., McRary, W. L., and Beaver, E. T. (1953). *Trans. Am. Microscop. Soc.* **72**, 236–248.

Oehler, R. (1913). *Trop. Diseases Bull.* **2**, 359.

Ormerod, W. E., Healey, P., and Armitage, P. (1963). *Exptl. Parasitol.* **13**, 386–394.

Plimmer, H. G., and Bateman, H. R, (1908). *Proc. Roy. Soc.* **B80**, 477–487.

Ramakrishnan, S. P., Prakash, S., Chowdhury, D. S., and Basu, P. C. (1961). *Indian J. Malariol.* **15**, 95–106.

Schnitzer, R. J., and Grunberg, E. (1957). "Drug Resistance of Microorganisms." Academic Press, New York.

Smiles, J. (1933). *J. Roy. Microscop. Soc.* [3] **53**, 203–212.

Soltys, M. A. (1959). *Parasitol.* **49**, 143–152.

Sonneborn, T. M. (1957). *In* "The Species Problem," Publ. No. 50, pp. 155–324. Am. Assoc. Advance. Sci., Washington, D. C.

Taliaferro, W. H. (1923). *J. Exptl. Zool.* **37**, 127–168.

Thurston, J. P. (1952). *Exptl. Parasitol.* **2**, 311–332.

Vickerman, K. (1962). *Trans. Roy. Soc. Trop. Med. Hyg.* **56**, 487–495.

Vickerman, K. (1965). *Nature* **208**, 762–766.

Vickerman, K. (1966). *Sci. Progr. (London)* **54**, 13–26.

Vickerman, K. (1967). *Trans. Roy. Soc. Trop. Med. Hyg.* **61**, Part 3, 303.

von Brand, T., and Tobie, E. J. (1960). *J. Parasitol.* **46**, 129–136.

Walker. P. J. (1960). Unpublished data.

Walker, P. J. (1962). Ph.D. Thesis, University of London.

Walker, P. J. (1963). *Trans. Roy. Soc. Trop. Med. Hyg.* **57**, 237.

Walker, P. J. (1964a). *Intern. Rev. Cytol.* **17**, 51–98.

Walker, P. J. (1964b). *Proc. 7th Intern. Congr. Trop. Med. Malaria, Rio de Janiero, 1963* Vol. 2, p. 254.

Walker, P. J. (1964c). Unpublished data.

Walker, P. J. (1966). *J. Gen. Microbiol.* **43**, 45–58.

Warhurst, D. (1964). Personal communication.

Watkins, J. F. (1964). *J. Hyg.* **62**, 69–80.

Wenrich, D. H. (1954). *In* "Sex in Microorganisms," p. 134. Am. Assoc. Adv. Sci., Washington, D. C.

Wery, M. (1967). *Ann. Soc. Belge Med. Trop.* **46,** 755–788.

Westphal, A. (1965). *Progr. Protozool. Abstr. 2nd Intern. Conf. Protozool., 1965,* p. 96. Excerpta Med. Found., Amsterdam.

Wijers, D. J. B. (1960). Thesis, Amsterdam.

Williamson, J. (1962). *Exptl. Parasitol.* **12,** 274–322 and 323–367.

Williamson, J., and Rollo, I. M. (1959). *Brit. J. Pharmacol.* **14,** 423–430.

Wolcott, G. B. (1955). *J. Heredity* **46,** 53–57.

Wolcott, G. B. (1957). *J. Protozool.* **4,** 48–51.

Young, M. D., and Moore, D. V. (1961). *Am. J. Trop. Med. Hyg.* **10,** 317–320.

14

Locomotion of Blood Protists*

THEODORE L. JAHN AND EUGENE C. BOVEE

The protists which cause blood or blood-borne diseases of man and animals are two major groups: (1) the bacteria, including the spirochetes; and (2) the protozoa—especially (a) the sporozoa, and especially among them the hemosporids and hemogregarines; (b) the toxoplasmids (here called sporozoa because certain of their fine structures are similar to those of malarial sporozoa); and (c) the trypanosomal flagellates.

Most of the bacteria which cause blood diseases are without locomotor organelles and without any known means of motility (e.g., the "strep" and "staph" coccids); they are therefore ignored in this discussion. The spirochetes, however, are highly motile.

Since blood-vascular systems of both vertebrates and invertebrates are more than adequate to translocate the blood parasites to the tissues which they affect, the motion and locomotion of the parasites probably are mainly involved with penetration into and migration within tissues.

* Based partly on experimental work supported by NIH Research Grant 6462, NSF Research Grants GB 1589 and GB 5573, and ONR Contract NONR 4756.

I. HYDRODYNAMIC AND THERMODYNAMIC PROBLEMS

In either the blood fluid, or in the tissues of the host, the parasites encounter (1) energy expenditure problems, i.e., thermodynamic ones, involved in (2) problems of translocating through a viscous medium against frictional resistance, i.e., hydrodynamic problems.

A. HYDRODYNAMIC PROBLEMS

Virtually the only hydrodynamic problem encountered by blood parasites is how to overcome the viscous drag of the medium on the cell surface of the protist. The cell does not have to overcome any significant turbulent drag at its rear, as does a boat. The Reynolds' number (ratio of the retarding forces of turbulence to those of viscosity) for even so relatively large a protist as *Paramecium* is only about 10^{-3}, and for a single cilium about 10^{-4} or less. These figures signify a virtual absence of real turbulence. Furthermore, pseudoturbulent drag, which exists at the posterior end of *Tetrahymena* and other ciliates (Jahn and Bovee, 1967) and also of some flagellates (Gittleson, 1966), apparently can be neglected for blood parasites.

Organisms of microscopic size, such as the protists with which we are here concerned, vary but little in predictable total surface drag from the computations for a sphere of similar surface dimensions and mass (Taylor, 1951, 1952; Hancock, 1953; Jenson, 1959).

For example, a plasmodial sporozoite 15 μ long, having an average diameter of 3 μ with a volume of approximately 100 μ^3, and a surface area of about 143 μ^2 could be treated mathematically as a sphere of equivalent size to determine the viscous drag acting on its surface. The viscous drag (D) is equal to 6 πmrV, where m = the viscosity of the fluid; V = the reference velocity and r = the radius of the particle. For a sphere of 100 μ^3, radius 1.5 μ, moving 10 μ/sec in water (m = 1 centipoise at 20.5°C), $D = 1.5 \times 10^{-8}$ dynes/cm², there is enough drag to stop the sphere of that volume (or protist of that volume) instantly upon *its* cessation of *active* movement. The drag is relatively so great, and the mass (1×10^{-6} gm) relatively so small, that the protist *cannot* progress *through* the water *except by actively expending energy to produce motive force well in excess of the viscous drag force*. A bacterium, for example, has such small mass (4×10^{-8} gm for *Escherichia coli*) and therefore so little effective inertia that it is incapable of progressing forward even as much as one half of its body length through the water by inertia alone (Taylor, 1951).

This viscous drag especially affects the movements of the cilium or the flagellum, since the surface thereof upon which drag forces act to retard movement is relatively so very great compared to its miniscule mass and inertia. Therefore, it cannot produce enough force solely at its attached base in its cell body to undulate along its entire length as a passively elastic cylinder (Gray, 1955). It can develop movement only by actively generating an advancing wave of physical force via energy utilization. This wave of energy progresses in a segment-to-segment sequence along short linear units within the flagellum. Each so-activated unit, as it bends, activates the adjacent one. This produces an undulatory wave which actively travels along the flagellum (or cilium) and exerts force against the water (Machin, 1958, 1963; Gray and Hancock, 1955; Brokaw, 1965; Rikmenspoel, 1965 a,b; Silvester and Holwill, 1965). Furthermore, the force generated by this wave (or that of the sum of several sequential waves) must be sufficient to overcome not only the viscous drag force upon the flagellum, but also that upon the body.

For organisms without flagella, e.g., a sporozoite of a malarial parasite, the body itself must actively expend energy sufficient to push against the water with a force in excess of the drag forces and its own inertia.

B. THERMODYNAMIC PROBLEMS

Overcoming the major hydrodynamic problem caused by the relatively great viscous drag of the medium on the body requires not only the continuous production of abundant energy, but also continuous distribution of the energy-containing source throughout the motile system. There must also be a morphological organization and distribution of the energy-using units, as some form of actively contractile elements, through the motile parts.

This is especially true of the flagellum where it is the principal or only motile organelle. Recent cytochemical studies indicate a high concentration of proteins resembling muscle actin and myosin in the peripheral fibrils of certain flagella, strongly suggesting an actomyosin adenosinetriphosphatase (ATPase), similar to that in muscle (Nelson, 1962; Nelson and Plowman, 1963; Rikmenspoel, 1965a,b). In some spermatozoa, the Mg-activated ATPase is 10 times as active in the flagellar fraction, as in the cellular ("head") fraction of the spermatozoan (Uesugi and Yamozoe, 1966), suggesting energy concentration for motile purposes. This also suggests a high concentration of motile materials, probably actomyosin, in the flagellum. In certain protozoans and in spermatozoa, the ATPases are much more concentrated in the basal bodies of the cilia and flagella than elsewhere in the cell body, and are also present within the entire length of the flagellum or cilium (Honigberg, 1955; Levine, 1960; Nagano, 1965; Rikmenspoel, 1965a; Gibbons, 1963, 1965; Gibbons and Rowe, 1965). Furthermore, live flagella or cilia broken free of the protistan cell body (and also severed from the flagellar basal apparatus, which remains in the cell body) are able to use extraneous adenosine triphosphate (ATP) autonomously and can bend, even swim, in such solutions, by autonomously generated undulatory waves, without the presence of the basal apparatus (Hoffmann-Berling, 1955; Tibbs, 1957; Brokaw, 1961).

Silvester and Holwill (1965) suggest that the wave form is related to the spacing of ATP concentrations along the flagellum, with wave amplitude due to rate of conversion of adenosine diphosphate (ADP) to ATP to relax the actively bent units, and wave frequency due to rate of restoration of the depleted ATP.

Energy computations for spermatozoan locomotion indicate that the total capacity for ATP energy production well exceeds the requirements for both locomotion and other metabolic activities (Carlson, 1962; Rikmenspoel, 1965a,b). For the parasitic flagellate, *Trypanosoma brucei*, calculations suggest that a planar wave moving along the flagellum develops a collision force against the surface of a red blood cell (Fig. 22F) varying from 3 to 60 $\times$ 10^{-8} dynes/cm^2 (Holwill, 1965b). This agrees well with other calculations for other flagellates of a bending moment for a single wave on a flagellum of 3 to 5 $\times$ 10^{10} dynes/cm at the viscosity of water, with greater force developed at higher viscosities (Brokaw, 1965). These computations suggest that the flagellum must be "stiff," with an elastic modulus of 5 $\times$ 10^{-8} dynes/cm (Brokaw, 1965) or more (Rikmenspoel, 1966), except when bending. Therefore, bending at *any* point requires energy concentration and its active use at the point(s) which produce the bending, one estimate being 6 $\times$ 10^{-6} dynes per fiber of the flagellum (Rikmenspoel, 1966).

These requirements for flagella also apply to the elongated bodies of spirochetes, and to the slender bodies of sporozoan trophozoites. They, too, require a distribution of motile morphological units along the body, and a distribution and utilization of energy-producing units and energy-using motile materials from one end of the body to the other. The body, also, must have a certain degree of both rigidity and elasticity.

II. LOCOMOTION BY ACTIVE BODY FLECTIONS

Many parasitic protists move and locomote by means of active flections which generate undulatory waves moving along the body, thereby propelling it. Spirochetes, and the trophozoites and ookinetes of sporozoa use such means, as do also the trophozoites of toxoplasmids. The flagellated trypanosomal protists are able to flex and undulate the body, as well as the flagellum.

A. GENERAL PRINCIPLES INVOLVED

The principal role of the undulations from one end to the other along the body is the propelling of the straight parts of the body between bends as inclined planes pushing against the water. These inclined planes move with sufficient speed and force to overcome the viscous drag of the water.

B. EFFICIENCY

In general, the greater that portion of the body length which acts as a moving inclined plane(s) during an undulatory wave progression, the more efficiently the organism locomotes. A swimming cylindrical body which undulates in a series of *planar* waves has, at any moment, only about half its length inclined as planes pushing against the medium; the remainder is undergoing a bending movement. A swimming cylindrical body undulating in *helical* waves has, however, almost *one entire lateral surface* inclined and pushing against the medium. Its efficiency in producing locomotory advance depends upon the relation between the amplitude and wavelength of any one turn of the helix (Jahn and Landman, 1965). At best, i.e., with a minimal rotary component of force and a maximal linear thrust, the helical undulations are at least twice as efficient as a numerically equivalent series of planar undulations (Gray, 1955), and probably more so, perhaps by a factor of 4.

Several other factors of importance are involved in the efficiency of the undulatory movement of a more-or-less cylindrical body in swimming, whether it be the entire cell, or its flagellum. One factor is the ratio between the diameter of the body and its length. Another is the number of undulatory planar waves, or of turns of the helix, moving along the body at any one moment. A third factor is the amount of torque developed in the body. A fourth factor is contact with a surface.

According to Taylor (1952) planar waves must develop at least $1\frac{1}{4}$ wavelengths along the length of the body (or the flagellum) in order to produce sufficient force which results in any forward propulsion; and at least 2 or more complete waves are necessary if any rapidity of translocation is accomplished. If the waves are either too few and too broad in amplitude, or too numerous and too narrow in amplitude, the efficiency may be reduced below that necessary for forward movement. However, nematodes can swim by means of a traveling planar wave only $\frac{1}{2}$ wavelength long. The efficiency is also a function of the force expended in the sequentially bending portions of the undulations. It is the force and rate of progress of the bending sequences which result in motion of the stiff, inclined region and supply the force the latter exerts against the medium.

For helically moving undulations several factors are involved in their efficiency: (1) the number of turns of helix; (2) the pitch of the helix in relation to amplitude; (3) the relative diameter of and length of the body; and (4) whether the helix developed is irrotational, i.e., torqueless, or rotational, producing torque (Jahn and Landman, 1965). A long slender organism, with many irrotational helical waves moving along its body, produces a maximal thrust against a minimal drag, and may approach, closely, a theoretically perfect mechanical efficiency of 100% (Jahn and Landman, 1965).

The problem is solved with greater or lesser efficiency by each organism, via movements of its body or its locomotor organelles, producing thrust sufficient to more than overcome the viscous drag of the medium upon the body surface, thereby producing translocation of the body.

C. Roles of Fibrils in Body Flections

It has been stated that at least two actively contractile, force-producing, parallel linear fibrils are necessary along the length of an undulating cylindrical body to produce a planar undulatory wave (Taylor, 1952; Gray, 1955).

To produce a helical wave it has been postulated that at least 4 fibrils are necessary, acting sequentially in pairs, to produce alternating undulatory waves each of which is 90° out of phase with the preceding wave (Gray, 1955; Doran *et al.*, 1962). Jahn and Landman (1965), however, suggest that a minimal number of 3 fibers is necessary, providing any two of them act together to produce sequential waves 90° out of phase with one another.

However, if each fibril is capable of active bending (in contrast to a wave of contraction) then an individual fibril could act in the same way as an individual flagellum and therefore could produce a wave of almost any type. In this case only a single fibril would be necessary. However, that fibril probably might have to have a structure comparable to that of a flagellum, i.e., it would have to be composed of several individual fibrils. There is evidence that each tubular fibril of the flagellum (or cilium) *is* composed of about 10 to 13 subfibrils (Pease, 1963; Gall, 1965; Acton, 1966). Subfibrils have also been described as composing the microtubules just underneath the cell surfaces of various cells, including sporozoan and trypanosome blood protists (Garnham *et al.*, 1961; Lumsden, 1965; Burton, 1966). It is also known that the fibrils of the *"axenfaden"* of spirochetes are each composed of several subfibrils (Listgarten and Socransky, 1964).

There is still some argument concerning the roles of tubular fibrils, especially as to which, if any, are really contractile (Nagai and Rebhun, 1966); or whether they are primarily supporting elements secondarily capable of contraction (K.R. Porter *et al.*, 1964); or if they may act as supporting elements which provide more or less elastic rigidity, with their deformations during undulations being due to forces generated in them and in adjacent materials. It is clear, however, that wherever protoplasmic movements occur in and by the protistan cell, the motile portions and/or organelles contain more-or-less permanent fibrils during the development of the motion (Ballowitz, 1908; Kavanau, 1964–1965; Pitelka and Child, 1964; Tilney et al., 1966; Wohlfarth-Bottermann, 1964; Grimstone, 1966; Jahn and Bovee, 1964, 1965, 1967). Furthermore, when high hydrostatic pressure is applied to the same protists, it usually causes the fibrils to disintegrate, and the cell becomes rounded and immobile (Kitching, 1956, 1957; Marsland, 1964; Landau, 1965), suggesting both structural and motile roles for the fibrils. The presence of an ATPase associated with the fibrils suggests a motile capacity (Moses, 1966).

1. Fibrils ("Flagella, Axenfaden") of Spirochetes

All spirochetes have linear fibrils. Early investigations (e.g., Zuelzer, 1912) indicated that a bundle of straight fibrils appeared to extend from one end to the other of the cylindrical spirochete body. The body, apparently longer than the fibrillar bundle, appeared to be coiled helically around the fibrils. Hence, the fibrillar bundle was long termed the "axenfaden," or axial bundle. More recent studies with the electron microscope show clearly that the separate fibrils of the bundle are of the same appearance and dimensions as the individual, freely undulating flagella of motile bacteria (Swain, 1957; Grimstone, 1963; Jahn and Bovee, 1965). When frayed out in some preparations, the spirochete fibrils have been erroneously described as free flagella (Zettnow, 1906; Morton and Anderson, 1942; Magerstedt, 1944). The most recent detailed electronmicrographs suggest that there are two bundles of flagellalike fibrils in some spirochetes, one bundle originating at each end of the spirochete's body. Each bundle winds around the body, with ends of the fibrils of the opposite bundles overlapping midway along the length of the body (Listgarten and Socransky, 1964, Ritchie and Ellinghausen, 1965) (Fig. 1). These fibrils are under the surface membrane and are adjacent to the cytoplasmic body mass (Grimstone, 1963; Listgarten and Socransky, 1964; Ritchie and Ellinghausen, 1965; Ovčinnikov and Delektarsky, 1965) (Fig. 2). There may be only 1 fibril composing the bundle, e.g., in Leptospira sp. (Swain, 1957); or 5 to 20, e.g., Borrelia spp. (Hanapp et al., 1948; Mölbert, 1956; Swain, 1955) or from 2 to 100, e.g., various Treponema spp. (Hanapp et al., 1948; Watson et al., 1951; Bradfield and Cater, 1952; Swain, 1957; Grimstone, 1963); or over 100, e.g., Cristispira spp. (Bradfield and Cater, 1952).

2. Fibrils in Trophozoites of Sporozoa

More-or-less parallel longitudinal fibrils existing adjacent to the cell surface of sporozoan trophozoites have been intermittently reported during the past half century, especially for the large gregarines (Fowell, 1936; Troissi, 1940; Kümmel, 1958; Beams et al., 1959) but also for eimerian coccidia (Tyzzer et al., 1932) and Toxoplasma (Goldman et al., 1957).

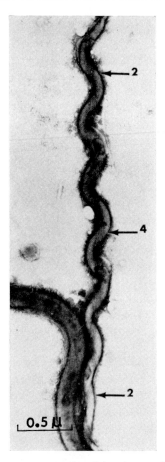

FIG. 1. The overlap of the axial fibrils originating at either end of this spirochete is clearly shown at the middle of the body where 4 fibrils are clearly seen, while near either end only 2 fibrils are present. (From Listgarten and Socransky, 1964.)

Since the advent of electronmicroscopy such fibrils have been found in a number of other sporozoans, e.g., 12 fibrils in *Plasmodium gallinaceum* (Garnham *et al.*, 1960), 15 in *P. falciparum* (Garnham *et al.*, 1961), 4 fibrils in *Eimeria acervulina* (Doran, 1966), 24 fibrils in *Eimeria perforans* (Scholty-seck and Piekarski, 1965), 22 fibrils in *Eimeria bovis* (Sheffield and Hammond, 1965), 22 in either *Sarcocystis tenella* or *Toxoplasma gondii* (Senaud, 1965), and about 30 fibrils in *Lankesterella garnhami* (Garnham *et al.*, 1962) (Fig. 3). In all cases the fibrils are attached to a ring near the blunt end of the body, and the fibrils extend about $\frac{2}{3}$ the length of the body. They may be noticeably spiralled in some species (Bringmann and Holz, 1953, 1954). The fibrils are tubular, composed of subfibrils (Garnham *et al.*, 1961) and

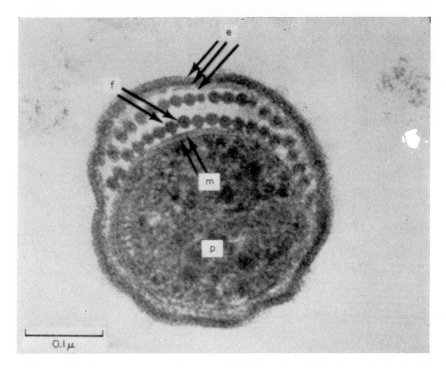

FIG. 2. In this cross-section of a large spirochete each fibril (f) of the "axial fila ment" is made up of 7 subfibrils, 6 of which are peripheral, and 1, central. Note also that the fibrils are in two layers, one group from one end apparently overlying the group originating from the other end. (e, External sheath; m, membrane; p, proto plasmic cylinder.) (From Listgarten and Socransky, 1964.)

resemble in dimensions a wide range of such fibrils found in other cells (Roth, 1964). There may also be numerous nontubular fibrils between and underlying the tubular fibrils (Gustafson *et al.*, 1954; Ludvik, 1960; Senaud, 1965; Stehbens, 1966) (Fig. 4).

Most of the electronmicroscopists who have made these observations have assumed (without proof) that these fibrils are involved in the move ments of the trophozoites, mainly on observing the arrangement of the fibrils. At any rate, there is a sufficiency of them, providing that they are motile or assist motility, to permit the sporozoan to perform a wide variety of movements.

D. LOCOMOTION OF SPIROCHETES

Spirochetes most often have been observed to follow a helical path along a surface, or through a jellylike medium (Zuelzer, 1912; and many others). They are not limited to that type of movement. They shorten and lengthen (A. Porter, 1910; Jahn and Landman, 1965); locomote with low resistance

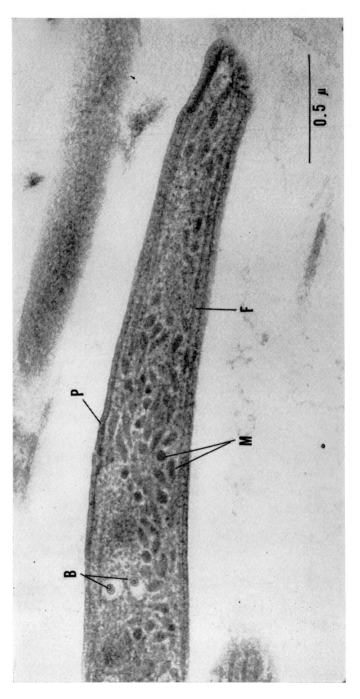

Fig. 3A.

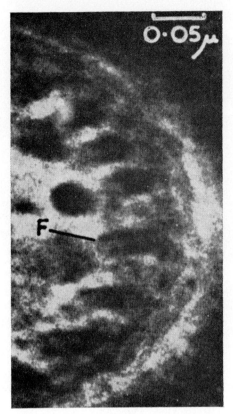

FIG. 3B.

FIG. 3. In the longitudinal section (Fig. 3A) through the body of a sporozoan trophozoite (*Plasmodium falciparum*) peripheral longitudinal tubular fibrils (F) can be easily seen. In the enlarged segment of the cross-section (Fig. 3B), other nontubular fibrils are evident in the internal cytoplasm. In the cross-section (Fig. 3C) the ends of tubular peripheral fibrils are shown at F. Fig. 3A: B, "fenestrated buttons"; M, mitochondria; P, pellicle. (From Garnham *et al.*, 1961).

to the medium (Jahn and Landman 1965); stop, and reverse along the same path, with or against a current (Perrin, 1906; Jahn and Landman, 1965); loop, bend, twist, and rotate around the long axis (Dobell, 1912); swim (Dyar, 1947), synchronously with adjacent spirochetes (Perrin, 1906; Gray, 1928); or move metachronally in rows attached to the body surface as symbionts of a trichomonad protozoan (Fauré-Fremiet, 1909; Cleveland and Grimstone, 1964; Jahn and Bovee, 1965). *Leptospira* (0.2 μ $\times$ 5 μ) can pass through the tortuous channels of Millipore MF filters with a mean pore size of 0.45 μ $\pm$0.02 μ, and which do not permit passage of bacteria.

Zuelzer (1912) noted that spirochetes move most efficiently when trav-

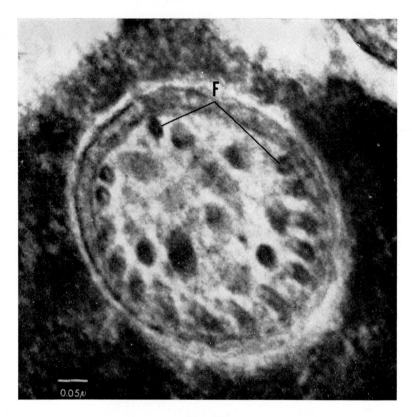

0.05μ

FIG. 3C.

eling along a helical path in contact with a surface, and assumed that surface contact is obligate in order to produce locomotion. The fact that they
swim (Dyar, 1947) suggests otherwise. However, the helical progression is
probably its prevalent mode of translocation, whether swimming, boring,
or traveling on substrate.

Those spirochetes which have only one or two fibrils in the "axial bundle",
e.g., *Leptospira pomona* (Ritchie and Ellinghausen, 1965), and other
leptospiras (Bradfield and Cater, 1952), and *Treponema carateum* (Watson
et al., 1951) generate only planar waves (Jeantet and Kermorgant, 1925).
Other spirochetes with 3 or more fibrils in the "axial bundle," e.g., *Borrelia*
spp. (Swain, 1955), other *Treponema* spp. (Swain, 1955), *Cristispira* spp.,
generate either planar or helical undulations (Perrin, 1906; Norris *et al.*,
1906; Fantham, 1908; Dimitroff, 1926; DeLamater *et al.*, 1950; Jahn and
Landman, 1965). These observations are consistent with the assumptions
of Gray (1928, 1955) and Jahn and Landman (1965) that at least 3 and
usually 4 or more fibrils are required to generate helical undulations

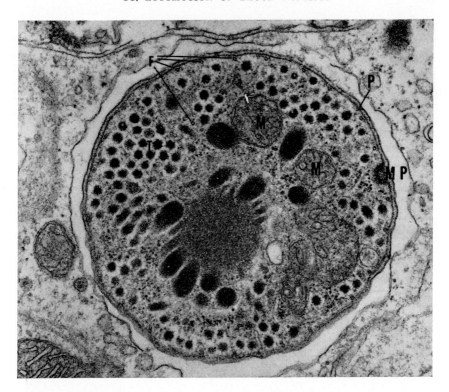

Fig. 4. A cross-section through the body of the toxoplasmid sporozoon *Lanke-sterella hylae*. Tubular subpellicular fibrils are indicated at F. Nontubular heavy fibrils ("toxonemes") are very numerous (T) throughout the internal cytoplasm; MP, micropyle; M, mitochondrion; P, pellicle. (From Stehbens, 1966.)

While it has been generally assumed for over 50 years that the bundle of "axial filaments" is responsible for the motility of the spirochetes, only generally descriptive suggestions have been offered as to how they do so, e.g., by "contractions and relaxations" of the axial bundle of filaments (Noguchi, 1928), "screwlike and serpentine" motion (Dobell, 1912), a true spiral, snakelike series of contractions (Zuelzer, 1912), boring like flexible corkscrews (Fletcher, 1927–1928), boring spirally by means of myonemes (A. Porter, 1910), or peristaltic movements of myonemes (Dimitroff, 1926).

Fantham (1908) first commented on the sine wave progression of the planar waves, and the helical nature of the "corkscrew" waves. These observations later were repeated by Jeantet and Kermorgant (1925), and by DeLamater *et al.* (1950). There appears to have been no attempt to analyze the hydrodynamic principles involved until the recent work of Jahn and Landman (1965), although A. Porter (1910) stated that the pointed end of the spirochete, *Cristispira* sp., contributed to its high mechanical efficiency of locomotion by reducing the resistance it encountered in advancing.

Jahn and Landman (1965) proposed that locomotion of *Cristispira* (and presumably of most or possibly all spirochetes) is caused by an active wave of helical bending, and that a spirochete swims in exactly the same manner as an unattached free-swimming flagellum (i.e., separated from its cell body) which is generating a helical wave.

To explain their observations that spirochetes progress with each helical turn of the body almost exactly following the trace (Fig. 5) of its predecessor, i.e., with seemingly 100% efficiency, Jahn and Landman (1965) assume that: (1) the wave of bending of the body is a true helical wave, resolvable into two planar sine waves, at right angles to each other and 90° out of phase; (2) each wave is *irrotational*, i.e., no cross-section of the body rotates around any point within the body; (3) little, if any, torque is produced because of the very low resistance of the medium to the small cross-section of the body as it advances through the medium.

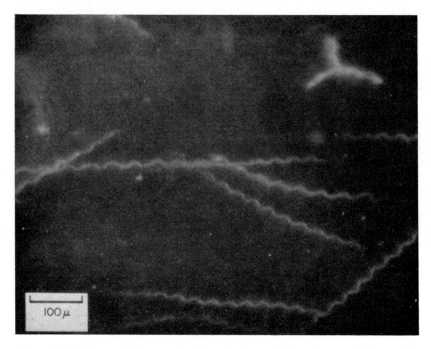

100 μ

FIG. 5. Multiexposed photograph of a spirochete, *Cristospira* sp., in locomotion. Five separate exposures were taken on the same film at 5-sec intervals. Although the images overlap more than 50%, there is scarcely any detectable longitudinal slippage. (From Jahn and Landman, 1965.)

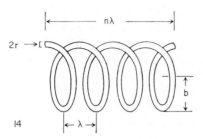

Fig. 6. A geometric sketch of a helical spirochete showing the meaning of symbols used in the text and the formula in the footnote. (From Jahn and Landman, 1965.)

If these assumptions are accurate, as appears likely, then the efficiency of the spirochete locomotion depends upon the total length of the body, which can be stated in terms of the number of turns of the helix simultaneously present along the body, the greater the number of them the greater the efficiency.* With 10 turns the efficiency may reach 96.44%; with 60 to 100 of them, over 99% (Fig. 6).

The equations of Jahn and Landman (1965) (and the resulting calculations of efficiency) are based on the ratio of the cross-section of the body to the area of one half the total lateral surfaces, and are not corrected for longitudinal viscous drag. Such a correction may become necessary as more data become available. Possibly the same equations developed for helical waves in flagella (Taylor, 1951, 1952; Hancock 1953; Machin, 1958, 1965; Holwill and Silvester, 1965) are applicable. However, the basic principle introduced by Jahn and Landman, namely, that spirochetes locomote by means of a traveling helical body wave, seems to be completely valid, regardless of what refinements of the equation may later seem desirable.

This efficiency, however, depends on the pitch of the helix and the viscosity of the medium. If the angle of pitch with the main axis of the body is too small, i.e., too small an amplitude/wavelength ratio, then the linear drag on the body becomes too great to be overcome by the bending force exerted by the body, and slippage results. If the angle is too great, i.e., too broad an amplitude/wavelength ratio, then the linear component of thrust

* The percent of mechanical efficiency (E) is calculated by Jahn and Landman (1965) thus:

$$\%E = \frac{1}{1 + \dfrac{r\sqrt{4\pi^2b^2 + \lambda_2}}{\pi nb\lambda}} \times 100$$

where r (radius of a body cross-section) = 0.1 μ; λ (wavelength of the helix) = 6.0 μ; b (wave amplitude of the helix) = 2.10 μ; n (number of helices along the body) = 10; the $\%E$ (efficiency) = 96.44. With 100 helices the $\%E$ would be 99.63.

is too low to overcome the viscous drag force, and the lateral component of force developed by each wave therefore produces mainly a rotation of the body, but little or no locomotion. It is also evident that if the path of a locomoting spirochete is blocked, it will rotate with its forward end in contact with the object, since only the lateral force components of the helical undulations can be effective. Furthermore, it will rotate in the direction *opposite* that taken by the undulations, i.e., it rotates counter-clockwise, if the undulations travel clockwise. This is also true of the sporozoa which move by means of helical waves (see below).

The many other types of spirochete movement listed above have not been explained, but probably can be if variations of undulations (i.e., planar or helical), local bendings and straightenings be taken into account. Space allotted here does not permit those considerations. The energy source for spirochete locomotion is not known. However, *Borrelia vincenti* converts ADP to ATP through an arginine to ornithine cycle, so that ATP is available (Nevin and White, 1960). Also ATPases are concentrated at the ends of the body where the *axenfaden* filaments are anchored, and along the length of the body (Hasegawa and Sasahira, 1966). These facts suggest a motile mechanism using ATP.

E. LOCOMOTION OF SPOROZOAN TROPHOZOITES

It has been known for over half a century that the various sporozoan trophozoites of certain bloodstream-invading species are motile and loco-mote when they are extracellular within their hosts, e.g., those of *Plasmo-dium vivax* (Schaudinn, 1902); other *Plasmodium* spp. (Grassi, 1901; Trager, 1956; Huff *et al.*, 1960; Yoeli, 1964; Freyvogel, 1965, 1967; Garn-ham, 1966); hemogregarines (Ball, 1957; Jahn, unpublished); *Sarcocystis* spp. (Koch, 1902; Darling, 1909; Jettmar, 1953; Westphal, 1954; Manwell and Drobeck, 1955); also there are motile forms of *Anaplasma* (España *et al.*, 1959). The variety and modes of their movements are similar to those reported for other sporozoan trophozoites, e.g., *Adelea mesnili* (Perez, 1903); *Archeobius herpobdella* (Kunze, 1907); *Eimeria necatrix* (Tyzzer *et al.*, 1932); other *Eimeria* spp. (Doran *et al.*, 1962; Bovee, 1965, 1966); various gregarines (Schultze, 1865; J. F. Porter, 1897; Mühl, 1921; Prell, 1921; Wu, 1939; Troissi, 1940; Misra, 1941; Allegre, 1948; Loubatieres, 1955; Watters, 1962; Cox, 1965; MacGregor and Thomasson, 1965; and others). Grassé (1953) considers the flexing movements of all coccidio-morphs to be due to myonemes, and their helical progression to be a type of gliding, in both cases similar to such movements in gregarines.

No critical analysis of trophozoitic movements appears to have been made until very recently (Huff *et al.*, 1960; Freyvogel, 1965, 1967; Garn-ham, 1965; Jahn, unpublished). There are, however, several older often-cited descriptions of their movements. Schaudinn (1902) observed that: (1) sporozoites of *Plasmodium vivax* swim with the pointed end forward; and (2) glide sporadically with intermittent bendings and flexings; (3) the ookinete moves with a slow gliding locomotion and also by peristaltic contractions; and (4) the microgametes demonstrate an elegant, winding, wriggling, agitated and very well-coordinated snakelike locomotion. Knowles (1928) states that: (1) sporozoites move by slow continuous (never jerky) gliding accompanied by bending movements; (2) the ookinete contracts, bends, and glides with slow wormlike movements; (3) the microgamete swims by active lashing movements. Wenyon (1926) says that sporozoites glide by means of peristaltic waves which pass along the body; and they also perform flections.

Yoeli (1964) terms the movements of sporozoites to be swift, reversible, eellike, graceful, gliding movements. Chao (1966) has observed both arcate and helical flections, and crawling by the ookinetes of *Plasmodium relictum* and by the trophozoites of *Haemoproteus* sp. and *Hemogregarina* sp. He has also noted gliding, with either end forward, repeated contraction and extension, gliding with a planar undulation, gliding intermittently in extended position, and (rarely) peristaltic movements by the trophozoites.

Freyvogel (1965, 1967) has recently analyzed time-lapse motion pictures he has made of ookinetes which show that those of several species of *Plasmodium* can glide with either end forward and are capable of reversing the direction of progress. He has also found that the ookinete glides with a snakelike or helical progression, usually with the pointed end forward, and can retrace its path exactly (Fig. 7). He clearly shows that the ookinete is motile, confirming the early contentions of Schaudinn (1902) and the general assumptions of its motility by malarialogists (e.g., Mackerras and Ercole, 1948), thus dispelling the older counter arguments of Huff (1934) and the recent ones of Howard (1962). Freyvogel (1967) also describes swiveling and revolving movements for the ookinete of *Plasmodium gallinaceum* when it is attached by one end (Fig. 8). We have cinematographed in our laboratories similar swaying movements by the ookinete of *Plasmodium relictum* (Fig. 9A).

Huff *et al.* (1960) report that their time-lapse cinematographs show that the merozoites of *Plasmodium gallinaceum* perform rotating movements when they are attached by their slender filamentous tips. They rotate either clockwise or counterclockwise, 2 to 5 rotations per second, and the direction of rotation may be changed from either direction to the other.

We have been able to confirm these movements for the sporozoites of *Plasmodium relictum*, in our laboratories, by cinematographing them at 8

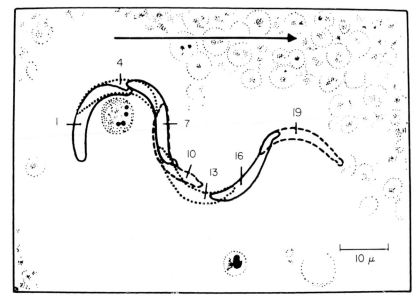

Fig. 7. Undulating gliding by the ookinete of *Plasmodium relictum*. Undulations appear to be helical. (From Freyvogel, 1967.)

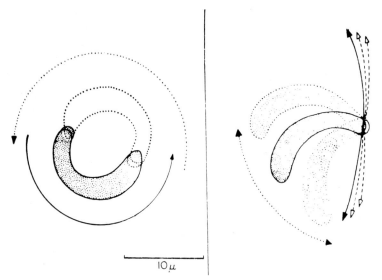

Fig. 8. Swiveling movements by the ookinete of *Plasmodium gallinaceum* fixed only at one end, and gliding in a circuitous path when unattached. (From Freyvogel, 1967.)

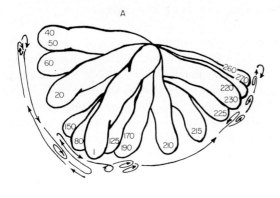

Fig. 9. The sequence of tracings at A depicts the swinging and rotating movements of the ookinete of *Haemoproteus* sp. from 250 pictures of a cinephotomicrographic film photographed at 6 pictures per sec. Each tracing is numbered according to the picture traced in its order in the sequence.

The first swinging movement, to the left (tracing of picture 40), required 6.8 sec, with a rotation in 10 pictures (1.6 sec) at the end of the swing. This was followed by a sudden return swing to the right almost to the original position (tracing of picture 80) in about 5 sec. A rotation during pictures 80 to 100 then occurred, followed by another short swing to the right during the next 4 sec (pictures 110 to 128). Another swing to the left (tracing of picture 150) took place in the next 2.8 sec. Another swing to the right followed in 3.6 sec to just beyond the original position, and included a quick rotation (2.5 sec) and was followed by two more rotations at the same rate (pictures 170 to 200). A short rapid swing to the right through an arc of about 90° then occurred in 3.3 sec (pictures 200 to 220). This was followed by a series of four rotations (pictures 225 to 270), in 7.5 sec, carrying the free tip of the ookinete farther to the right and to a position 180° away from the position taken during the farthest swing to the left.

As the tracing of picture number one indicates, the ookinete is helically coiled, counterclockwise. As the arrowed line of the overall path shows, some of the rotations were clockwise, some were counterclockwise. Whether the ookinete reversed its coil to accomplish this could not be clearly determined from the film track tracings. Changes in the peripheral contours in the tracings suggest undulations or peristaltic waves (probably the former, perhaps both), but these were not clearly determined.

Sequence B represents a reconstruction of the undulatory and rotary movements of an "exflagellating" microgamete of *Haemoproteus* sp., attached to the parent cell. It is based on tracings of six successive pictures from a cinephotographic film, photographed at 17.5 pictures per sec.

The rotation is counterclockwise, and the helical coil of the body is clockwise. Movement is produced by undulations which move from the attached end to the free tip. Each rotation takes about 0.4 sec.

As each undulation passes the center of the body (which there is stiffened and enlarged by the nucleus) and enters the slender lower end of the body, the upper part and center of the body are whipped sharply in a broad arc (tracings 3, 4).

This gives the appearance of a jerky movement which *erroneously* appears to be pulling the microgamete; whereas actually the undulations *push* the microgamete against the parent cell to which it is attached, also causing the counterclockwise rotation of the microgamete, as is hydrodynamically predictable (see text, and Fig. 11).

411

to 16 frames per second. A group of 16 sporozoites liberated from the salivary gland of a culicine mosquito will usually remain attached together by their filamentous ends for some time. Each sporozoite coils into a helical form of about $1\frac{1}{2}$ turns from the blunt free end to the attached slender end. The initial helical coil may be either clockwise or counterclockwise, more often the latter, as seen from the blunt end. Each sporozoite so coiled rotates one or more times per sec. If it is coiled clockwise, it rotates counterclockwise, and vice versa. While rotating it may reverse the direction of the helical coil of the body, and simultaneously reverses its direction of rotation (Fig. 10).

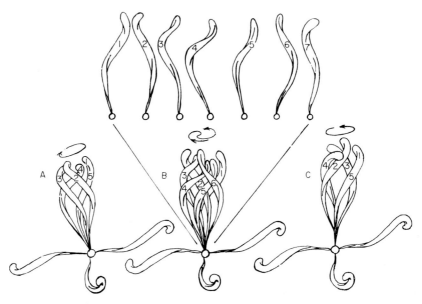

Fig. 10. Rotating sporozoites of *Plasmodium relictum*, cinephotomicrographed at 16 pictures per sec. A single rotation was completed in from 0.5 to 2 sec. During the continued rotary movements the sporozoite rotated faster, then slower, repeatedly, with rotation continuing for many minutes. Sometimes, as rotation slowed down, both the direction of the helical coil of the body and the direction of rotation were reversed. Figure 10 shows such a sequence.

At A, four tracings (each 3rd picture) from a sequence of 15 frames show one clockwise rotation, the successive positions of the single sporozoite being numbered sequentially. The body is coiled counterclockwise.

In B, reversal of both helical coil and direction of rotation is shown immediately following sequence A, every other picture from a sequence of 12 being traced and sequentially numbered. Those successive pictures are drawn separately and sequentially above to show the change in the helical coil from counterclockwise to clockwise.

Sequence C is the counterclockwise rotation immediately following sequence B, again consisting of tracings of every 3rd picture from the film track, numbered sequentially as to position. The body is coiled clockwise.

This rotation, in the direction *opposite* to that of the coil of the helix is exactly what should be expected if the body were undergoing active bending in the form of a traveling helical wave (Fig. 11C), and it would not be expected if the organism were a rigid helix rotated by flagella, as in bacterial spirilla (Metzner, 1920; Jahn and Bovee, 1965). Furthermore, as the orga-

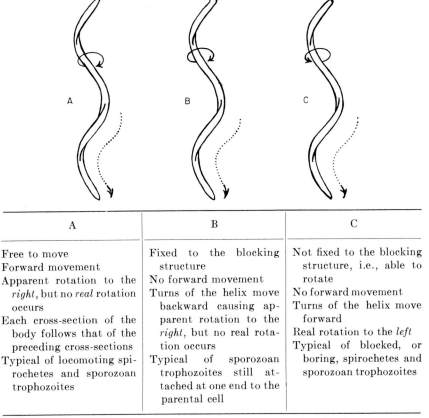

A	B	C
Free to move	Fixed to the blocking	Not fixed to the blocking
Forward movement	structure	structure, i.e., able to
Apparent rotation to the	No forward movement	rotate
right, but no *real* rotation	Turns of the helix move	No forward movement
occurs	backward causing ap-	Turns of the helix move
Each cross-section of the	parent rotation to the	forward
body follows that of the	*right*, but no real rota-	Real rotation to the *left*
preceding cross-sections	tion occurs	Typical of blocked, or
Typical of locomoting spi-	Typical of sporozoan	boring, spirochetes and
rochetes and sporozoan	trophozoites still at-	sporozoan trophozoites
trophozoites	tached at one end to the	
	parental cell	

FIG. 11. Diagram shows what happens to an organism swimming freely by means of a clockwise helical wave (A), and when its path is blocked (C). For comparison, an organism which is fixed to the blocking material and not free to rotate (B) is shown with a traveling helical wave which is identical to those of (A) and (C). Solid arrows indicate swimming path, and apparent or real rotations. Dotted arrows indicate the direction of the traveling helical wave.

nisms rotate they do *not* pull themselves backward (or stretch the helix) as would be indicated by the relative direction of the turns of the helix and of the rotation, which would occur if the helix were rigid and were being rotated by flagella or other external organelles. They continue to push against the surface that is blocking their path, and they therefore must rotate in relation to this surface. One amazing observation is that this rotation may be continued through many revolutions and many minutes. This may be one of the mechanisms for tissue penetration (see below).

A free-swimming sporozoite, which normally travels forward through the water in the same rotational direction as its helix, i.e., clockwise if it is coiled clockwise, will reverse the direction of rotation on encountering an obstacle without reversing the direction (coil) of its traveling helical waves, and it will return to its original rotation and again advance if the obstacle is removed (Fig. 12).

If the organisms were firmly attached so that they could not rotate, the turns of the helix would move directly backward along the axis of the helix, and the helix would appear to rotate (Fig. 11B). This apparent rotation (as in Fig. 11B) would be distinguishable from a real rotation (Fig. 11C); because it would be in the opposite direction in relation to the coils of the helix. However, a real rotation (as in Fig. 11C) can not be explained on the basis of a rigid helix, and the observation that the organism rotates or appears to rotate in the direction opposite that of the helix strongly supports the theory that locomotion of these organisms is by means of traveling helical waves in the body. Furthermore, this concept does not exclude the simultaneous existence of a gliding mechanism as discussed below.

Haemogregarina sp. has been cinematographed in our laboratories in the process of gliding on substrate. If one lateral surface of the body adheres tightly against the glass of the microscope slide, the body bends in an arc, and gliding tends to be in a broad arc or a circular path, with reversals sometimes occurring (Fig. 13C).

The organisms are sensitive to the shorter wavelengths of light, and to heat, so that the photographing of continued sequences of gliding is difficult. This has been solved by photographing with dark-field microscopy, and especially then with filtered orange-red light. Under those conditions the hemogregarines will glide continuously and regularly for many minutes, with the blunt end forward. The initial response to the light—apparently a shock reaction—is a slow flexing of the blunter end of the body into a helix of about $1\frac{1}{2}$ turns, followed by a very sudden elastic relaxation and straightening (Fig. 14). This is repeated one to several times. In full-spectrum, high-intensity light this is often the only movement performed. In the dark field in orange-red light the initial shock reaction is promptly followed by the coiling of the whole body into a helix of about $1\frac{1}{4}$ turns;

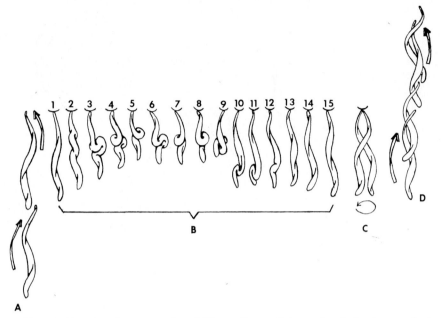

FIG. 12. A swimming sporozoite of *Plasmodium relictum*, cinephotomicrographed at 6 pictures per sec. At A, each 10th picture of the last 20 pictures of a 40-frame swimming sequence has been traced. The sporozoite is coiled in a clockwise helix, and appears to rotate clockwise as it swims. It was impeded by hitting a salivary gland cell, but continued to undulate.

The pictures bracketed at B are a sequence of each of the next 15 on the film, from the reader's left to right, depicting a flexing and straightening during a reversal of the direction of rotation, i.e., from clockwise to counterclockwise, without changing the direction of the helix of the body which continues clockwise.

At C, one counterclockwise revolution is depicted, completed in 10 pictures (1.25 sec), while advance was still impeded by the salivary gland cell.

At D the sporozoite slipped off the impeding cell and again swam forward, also again with an apparent *clockwise* rotation (tracings of each 8th picture from a sequence of 24 pictures).

The sporozoite swims in short spurts lasting 3 to 6 sec, usually. It can swim in either direction; and may swim forward in one spurt, then reverse direction in the next; or may swim in one direction for several successive spurts, with rests of 20 sec or more between, before reversing direction.

that is followed by a continuous gliding of the helical body. The body helix is in contact with substrate often at only one, and never at more than two points simultaneously. It progresses blunt end forward, and every point along the body helix passes through the point of contact originally established and only there (Fig. 13A and B). There is no apparent difference in the rate of progress whether there is only one, or if there are two points of

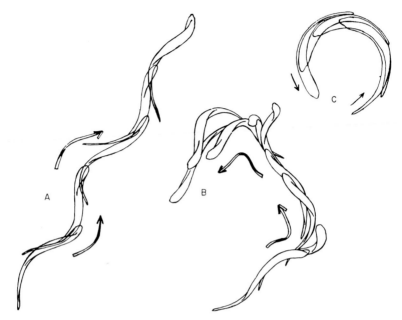

FIG. 13. Sequence A consists of tracings (each 60th picture) from 300 pictures of a cinephotomicrographic film of a hemogregarine trophozoite in steady gliding progress along a clockwise helical path, photographed at 2 pictures per sec. The organism makes contact with the substrate once with each completion of a helix; and the body glides over each contact point, without interruption, from one end to the other.

Sequence B consists of tracings from another part of the same film (each 50th picture) showing the same hemogregarine making a 180° turn by performing two 90° turns in succession. The organism bent sharply to the left following passage through each of two new contact points, but advanced in a normal helical manner between the two sharp turns. The new pathway of continuous advance is nearly 180 degrees from the original route.

Sequence C shows another hemogregarine trophozoite from another part of the same film. It is adherent to the substrate all along one side, is bent in an arc, and glides in a circuitous path. Selected tracings are from each 50th picture in a sequence of 200.

contact made by the body with the substrate. While gliding, the hemogregarine is able to reverse the helical twist of the body, without releasing contact, also therewith reversing the helical path of progress without disturbing the linear direction of gliding. If the helical coil of the body from blunt forward end to the rear is clockwise, the organism describes a clockwise helical path as it advances.

Propulsive force appears to be generated between a mucoid sheath and the body surface, with the sheath adherent to substrate at the point(s) of contact. This suggests an active shearing between the body surface and whatever the adhesive material it produces may consist. Jarosch (1964; and

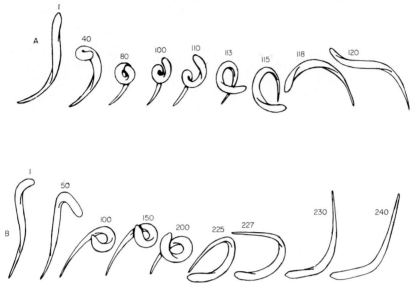

Fig. 14. Active flexing and elastic extension of a hemogregarine, a shock reaction to bright light, cinephotomicrographed at 2 pictures per sec. In sequence A, the blunt end is free to move throughout the sequence, while the slender end is attached. Active coiling requires 50 sec, while elastic extension requires only 10 sec. In sequence B, movement of the blunt end was impeded from frames 220 onward. The attached smaller end was jerked free at frame 226 and elastic extension was largely completed within 2 sec. In both sequences coiling takes nearly 4 times as long as required for elastic extension. This suggests that probably the longitudinal tubules just under the cell membrane, while elastic, also have considerable rigidity and tensile strength.

other recent papers) has refurbished in modern terms a century-old theory by Schultze (1865) which proposes that the gliding organism moves over a circulating track of cytoplasmic exudate ("mucus") which is extruded at one end, presumably the forwardly progressive end. A type of energy-using, active-shearing mechanism is proposed by Jarosch (1964) which operates between the mucus and the body surface and propels the body along the mucoid exudate if the latter is adherent to the substrate, or propels the exudate along the body surface if the body is not attached and floats free in the water. The force generated is adequate also to transport an adherent red blood cell along the helical body from one end to the other (Fig. 15).

In our dark-field cinematographs of *Haemogregarina* sp. a circulation of internal granules can be seen clearly in the thicker advancing end of the gliding trophozoite. The granules move forward in parallel streams near the cell surface to the anterior end, then turn inward and travel in rearwardly moving streams adjacent to, but internally, in relation to the for-

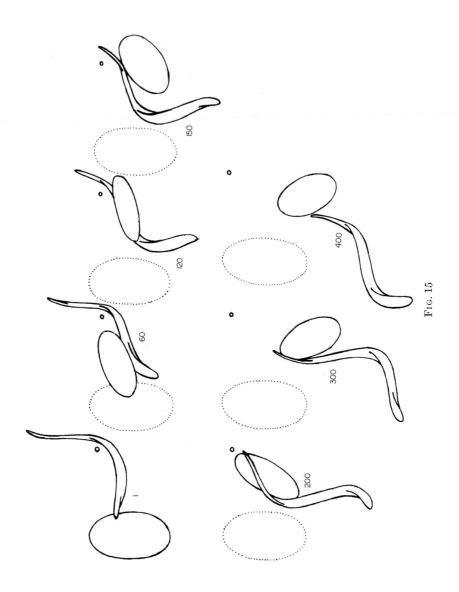

FIG. 15

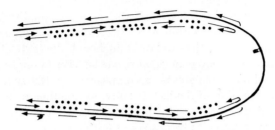

FIG. 16. Diagram of a longitudinal section through the anterior end of a gliding sporozoan trophozoite. The cytoplasmic granules stream forward internally against the periphery of the cell close to the cell membrane, until they reach the approximate position of the conoidal ring. There they turn inward and move rearward in streams inwardly adjacent to the forwardly moving streams. Externally a flow of mucoid material progresses rearward along the surface of the body.

Interspersed arrows and dots indicate granular streams. Interspersed arrows and lines indicate streaming mucoid material. These movements occur around the entire periphery of the anterior end, but the numerous other streams are omitted for the purpose of diagrammatic clarity.

ward moving streams (Fig. 16). Although the parallel tubular fibrils revealed by electronmicrographs of other investigators, e.g., Garnham *et al.* (1960), cannot be seen in the cinematographs, the spacing of the forward-moving streams is such as to suggest that they move upon or between similar fibrils. This movement of granules conceivably may be related to an active peripheral sliding of cytoplasmic material carrying the granules along the stiff peripheral fibrils to pores near the advancing blunt end (pores shown in the electronmicrographs of *Plasmodium gallinaceum*, by Garnham *et al.*, 1960) where the cytoplasmic material may be extruded as a mucoid surface coating, the granules being turned back on an inward and oppositely flowing track.

Jarosch (1964) assumes that rapid changes in pitch of coiled-coil protein fibrils in the cytoplasm (or the exudate) cause the fibrils to rotate like worm gears as metabolic energy reacts with them. These movements, he

FIG. 15. Selected tracings from a cinephotomicrographic sequence of a gliding hemogregarine photographed at 2 pictures per sec. It picks up a red blood cell at the anterior end (tracing, picture 1), and transports it along the body, the red blood cell following the turns of the body helix (tracings, frames 60, 120, 150, 200, and 300). The red blood cell is cast off at the rear (tracing, frame 400), and the hemogregarine continues its uninterrupted gliding. This indicates that some sort of a surface sheath moves to the rear of the hemogregarine body as the organism moves forward. Considerable shearing force must be generated to both propel the hemogregarine past the point of body attachment, and also lift and transport the mass of the red blood cell. The dotted ovals indicate the original position of the red blood cell; and the small circle, the position of a granule which did not move.

says, may produce undulations of greater or lesser amplitude and wave-length along the body or within it. If the body is anchored, the exudate moves along the body. If the exudate is anchored, the body glides along it. Unanchored filaments may glide against one another in opposite directions.

Penetration of tissue might be accomplished by the same mechanism which produces rotation of the helical body, as in either swimming or glid-ing, especially if the penetrating tip of the rotating cell were pointed and strengthened, as it seems to be in some cases (Garnham et al., 1960; Frey-vogel, 1967). This would permit the cell body, as it rotates, to serve as a simple mechanical drill, complete with a self-contained power supply and a chemical-mechanical transducer.

Plasmodium spp. sporozoites and merozoites enter host cells by attaching the slender filamentous end to the host cell, and glide into the cell via a "pore" (Huff et al., 1960) with or without rotating and in a few seconds (Trager, 1956). Ookinetes of *Plasmcdium berghei* take the helical form (Freyvogel, 1967) and drill their way into host tissue with the pointed end forward in the same manner as the helical gliding of other trophozoitic stages (Freyvogel, 1967) (Fig. 17). The ookinete of *Plasmodium berghei* forms a penetration tube into which it glides (Vanderberg and Yoeli, 1965). Freyvogel (1967) shows that in penetrating the peritrophic membrane of

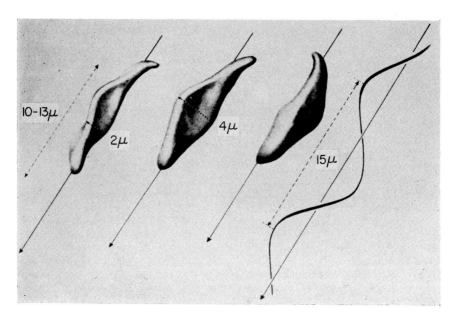

Fig. 17. A model of the ookinete of *Plasmodium cynomolgi bastianellii* in three different positions, showing it from three aspects, indicating that it is helically twisted. A diagram of the clockwise helical pathway it follows is shown overlying the arrow which indicates the direction of progress. (From Freyvogel, 1967.)

the host the ookinete is traversed by a wave of constriction which starts at the entering end and passes posteriorly. This wave is stationary in respect to the substrate, and the body passes through the constriction (Fig. 7). However, Freyvogel (1967) disclaims a role for the waves of constriction so far as locomotion otherwise is concerned. Stöhler (1957) also records peristaltic movements by the vermicule, as well as undulatory ones.

The flexing movement of trophozoites to form a helix is apparently a separate mechanism from that involved in the gliding. It may be accounted for if it be assumed that the peripheral tubules are of an actinoid protein, containing and surrounded by adjacent fibrils of a myosinoid protein, as is apparently the case in the tubular fibrils of certain cilia and flagella (Nelson and Plowman, 1963; Gibbons, 1965; Renaud et al., 1966). The actinoid tubules are assumed to serve both structural and elastic functions, as evidenced by the faster elastic intermittent straightening of the body occurring between the slower repetitive flections. The more actively motile myosinoid proteins (which can form coiled-coil network complexes in reaction with ATP; Cohen and Holmes, 1963) presumably provide tensile forces as they develop supercoils, bending the more rigid actinoid tubules (which may also be elastic tightly bonded coils; Hanson and Lowy, 1963).

Once the actinoid tubules and the body have been bent into a helix, this may be maintained until available energy is depleted and relaxation occurs; or perhaps may be maintained so long as the myosinoid supercoils actively stream or rotate against them, as Jarosch (1964) says may occur in gliding locomotion, also perhaps causing the irrotational waves which progress along the body in swimming or in rotating when forward movement is blocked.

Schaudinn's (1902) observation that trophozoites of *Plasmodium vivax* swim with the filamentous end forward has been confirmed by time-lapse cinematography for *Plasmodium gallinaceum* (Huff et al., 1960) and by direct observation of *Plasmodium lophurae* (Trager, 1956). Similar swimming has been observed for eimerian merozoites (Bovee, 1965) and has been cinematographed for the sporozoites of *Eimeria acervulina* (Doran et al., 1962). In order to drive the body forward, the undulations must push water backward. To do so the undulations must, therefore, begin at the slender advancing tip of the swimming trophozoite, and progress backward to the blunt trailing end. How they originate at the slender tip is not known.

The hydrodynamic problems and principles applicable to swimming sporozoa are exactly the same as for the swimming of spirochetes. The mechanical efficiency achieved is not as great as in the case of the spirochete partly because of the relatively greater frictional resistance of the thicker part of the body, but mostly because the number of turns of the helix along the body is fewer. Some "slippage" would be expected, and does occur.

In a very recent paper, Garnham et al. (1967) have shown by electron-micrographs that the microgamete of *Plasmodium falciparum* is a truly flag-

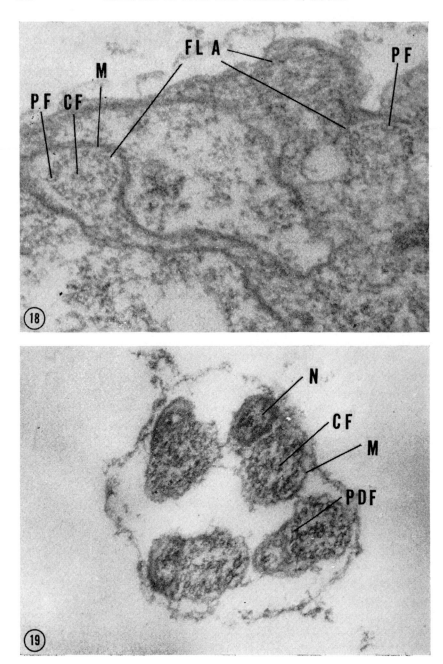

Figs. 18 and 19.

ellated organism. Each has the typical flagellar axoneme composed of the 9 peripheral doublet fibrils and 2 central fibrils covered by the cell membrane. They differ morphologically from the "typical" spermatozoan, however, by having no centriolar midpiece nor any basal body which is evident in the electronmicrographs, and the nucleus is under the flagellar membrane midway along the length, and not at one end of the flagellum (Figs. 18, 19). The flagella of the microgametes of *Eimeria intestinalis* also are said to lack basal bodies (Cheissen, 1964). Garnham *et al.* (1967) assume that the flagellum of the microgamete arises from a centriole in the parent cell, however.

There is very little information available concerning the movements of microgametes, either during "exflagellation," or swimming, or in penetrating the macrogamete. Some general descriptions cite "lashing" movements (Knowles, 1928), or an "elegant, agitated, but well-coordinated snake-like wriggling" (Schaudinn, 1902). Preliminary analysis of low-speed cinemicrographs in our laboratory of the exflagellation of the microgametes of *Haemoproteus* sp. indicate that intermittent undulations, probably helical, travel from the attached proximal end of the microgamete to the distal free end producing the "lashing" and "wriggling" movements. Higher speed cinematographs are needed, however, to determine the exact manner in which the undulations produce their effects, and these have not yet been done either by us or by others. An attempt to interpret the phenomena is shown in Fig. 9B.

Ingestive movements of an ameboid character, phagocytic and/or pinocytotic, are performed by trophozoites of malarial organisms once they are inside the host's red blood cells and as they devour the stroma (Rudzinska and Trager, 1957, 1962; Rudzinska *et al.*, 1960). Similar reports are also made for *Babesia rodhaini* (Rudzinska and Trager, 1962). These are unexplained.

F. LOCOMOTION OF TOXOPLASMIDS

Toxoplasmids are reported to be as variably motile as spirochetes. *Toxoplasma* spp. swim along helical paths at speeds of 1–100 μ/sec, somersault, rotate on the long axis, or with the blunt end against a cell, and de

FIG. 18. Axial cylinders of the flagella of microgametes in a section through the peripheral portion of a microgametocyte of *Plasmodium falciparum*. FLA, flagellar cylinders; PF, peripheral fibrils; CF, central fibrils; M, cell membrane. (Electron photomicrograph, courtesy of P. C. C. Garnham.)

FIG. 19. Group of four microgametes sectioned at the level of the nucleus, midway along the length of the microgamete. N, nucleus; CF, central fibril of flagellar cylinder; PDF, peripheral doublet fibril (note "arms" on the lower cylinder of the fibril); M, membrane of the flagellar cylinder (note that it also encloses the nucleus, and that the nucleus lies adjacent to the axial cylinder). (Electron photomicrograph, courtesy of P. C. C. Garnham.)

velop peristaltic waves (Manwell and Drobeck, 1955); or glide, dance, jerk, undulate, rotate with body twisted, and undulate the thin filamentous end like a flagellum (Jettmar, 1953); or turn over, bend and straighten, twist, or locomote as a spiral (Scott, 1930). The helical swimming movements for *Sarcocystis* were early reported by Koch (1902) and Darling (1909), and twisting, turning, gyration and swimming were cited by Scott (1930). Swimming by either *Sarcocystis* or *Toxoplasma* is with the pointed end forward (Koch, 1902; Manwell and Drobeck, 1955; Guimarães and Mayer, 1942), and the helical swimming has also been reported for *Anaplasma marginale* (España *et al.*, 1959). Motility has also been observed in *Lankesterella* (Lankester, 1882; Fantham *et al.*, 1942).

Electron microscopy demonstrates tubular, subpellicular, longitudinal fibrils in sarcoplasmids and toxoplasmids much like those in malarial sporozoa, and a network of finer fibrils farther within (Bringmann and Holz, 1953, 1954; Gustafson *et al.*, 1954; España *et al.*, 1959; Ludvik, 1960; Lainson *et al.*, 1961; Senaud, 1965).

To our knowledge no explanation of the mechanical principles involved in the above types of locomotion has been published. However, three principles seem to be involved: (1) a traveling helical body wave, exactly as described above for spirochetes and for *Haemogregarina* and *Plasmodium;* (2) a gliding mechanism, similar to that described above for *Haemogregarina*, of which the details are unknown; and (3) a peristaltic wave, probably caused by body contractions. The known similarities in fine structure also suggest that these locomotor systems are similar to those of other sporozoa.

III. LOCOMOTION OF FLAGELLATED PROTOZOA

The only flagellated protozoa found in the vertebrate bloodstream are the trypanosomes. They do not invade blood cells as do the malarial protozoa, but do invade most other tissues. However, certain stages of their life cycles are found in the bloodstream, where their locomotion is probably a factor in penetration as well as swimming.

FLAGELLAR ORGANIZATION AND MOVEMENTS IN TRYPANOSOMES

1. Fibrillar Structures

a. In the Flagellum. The trypanosomes develop only a single flagellum. As in most other eucellular flagella this is composed mainly of the typical axoneme, which is a ring of 9 peripheral bitubular fibrils and 2 central fibrils extended from a basal body composed of 9 tritubular components capped by a plate (Anderson and Ellis, 1965). Paralleling the flagellum, there is a fibrillar rod nearly as long as the body which in the bloodstream form ("trypanosome") of the organism lies between the flagellum and cell body, within the undulating membrane. It is fastened to the axial cylinder of the flagellum by extensions of the "arms" of certain of the bitubular axonemal fibrils (Anderson and Ellis, 1965); and it is also fastened to the adjacent membranes of both the flagellum and the body by cytoplasmic "rivets" (Boisson *et al.*, 1965) (Fig. 20). Thus, except for the free tip of the flagellum, the body, the accessory rod, and the flagellum must interact in the locomotor process.

b. In the Body. The body of the trypanosome has many subpellicular fibrils (about 100) running the length of the body (Kleinschmidt, 1951; Kleinschmidt and Kinder, 1950). These are hollow tubules about 250 Å in diameter, and about that same distance apart (Anderson and Ellis, 1965). The tubules are adjacent to, but not part of nor fastened to, the cell membrane

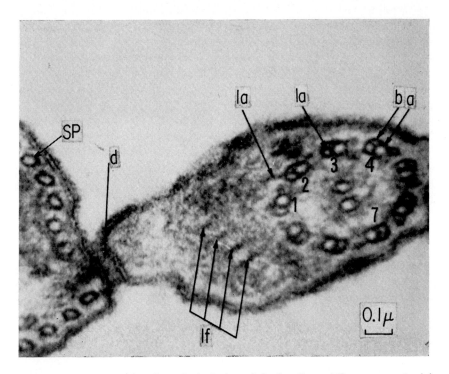

FIG. 20. A cross-section through the body and the flagellum of *Trypanosoma lewisi*. The paraxial rod of fibrillar material (lf) lies between the axial cylinder of subfibrils and the body. The flagellar membrane is attached to the body at d, and the tubular subpellicular fibrils of the body are clearly seen close to, but not attached to, the surface membrane; la, lateral arm of subfibril; a, subfibrila; b, subfibril b; d, dense intermembranous layer; sp, subpellicular microtubule. (From Anderson and Ellis, 1965.)

(Judge and Anderson, 1964). They have been considered to be possibly both myonemal and cytoskeletal, with both contractile and supporting roles, affecting the undulations of the body (Judge and Anderson, 1964; Anderson and Ellis, 1965). Their fine structure suggests they are elastic, like coiled springs (Burton, 1966). There are also endoplasmic reticular tubules and non-tubular fibrillar material in the cytoplasm which may have some roles in body movements (Boisson *et al.*, 1965). At any rate, there is a sufficiency of fibrillar potentially contractile cytoplasmic elements which could interact to produce the undulatory movements and flections that the trypanosome body performs.

2. Energy Sources for Flagellar Movements

Some evidence suggests that, as in other flagellates, the motile system of the trypanosomes is an ATP-using actomyosin complex. Ivanov and Uman-skaya (1945) suggest that a chemical transformation, probably of ATP, is the source of energy for trypanosome movements; and an ATPase was found by Chen (1948) in *Trypanosoma equiperdum*. Hoffmann-Berling (1954, 1955) showed that both body and flagellum of glycerin-extracted cell "models" of *Trypanosoma brucei* contract in response to ATP more sensitively than do sperm flagella.

Since respiration of the pathogenic trypanosomes is at least primarily fermentative, the necessary ATP probably is not generated in these organisms by a tricarboxylic acid cycle, but rather by glycolytic sequences of which the end products vary from species to species (von Brand, 1951; Hutner and Provasoli, 1955; Guttman and Wallace, 1964; Danforth, 1967).

Some of the mechanism is intrinsic in the flagellum, since acriflavine-ultraviolet (UV) spot-exposed flagella are inactivated only where the flagellum has been irradiated (Walker, 1961), and may be temporarily accelerated by acriflavine-UV treatment. The kinetoplast has been suggested as a source of enzymes and ATP (Vickerman, 1963), but kinetoplast-lacking strains are as motile as those with kinetoplasts (Clark and Wallace, 1960; Jahn, unpublished). Some sort of origin and transfer of metabolic materials must occur to and from the body cytoplasm and flagellum, but the mechanisms are unknown.

3. Flagellar Undulations

Studies on the flagellar movements of trypanosomes have been partly responsible for recent resurgence of interest in mechanisms of flagellar movements, and for recent alteration and augmentation of theories of flagellar movements.

Observations by Walker and Walker (1963) and motion pictures by

Pipkin (1962) indicated that undulations may begin at the distal tip of the flagellum of *Trypanosoma cruzi*, moving from the tip to the anchored basal apparatus in the body, pulling the body forward in the direction of the flagellar tip.

The present authors and their associates confirmed these observations with high-speed cinephotography of 350 to 600 fps for *T. cruzi*, *T. lewisi*, and *T. equiperdum* (Jahn and Fonseca 1963; Jahn *et al.*, 1964a,b) (Fig. 21). The tip-to-base undulations of the flagella in various trypanosomes have been further confirmed by Holwill (1964, 1965a,b) and by Gillies and Hanson (1963).

These studies negate the contentions of Lowndes (1941a,b, 1943a,b, 1944) that undulations move only from attached base to free tip of the flagellum. Lowndes (1943a) had urged discarding the term "tractellum" because of his contentions concerning the assumed obligatory base-to-tip progression of the undulations which can only push but do not pull the organism through the water. Since the high-speed cinematographs clearly show that tip-to-base waves *do* pull the organism forward, Jahn and Bovee (1964–1967) have reinstated the term "tractellum" as a valid one.

Trypanosomes, however, are not limited to a single type of flagellar undulation. In fact flagellates, generally, seem able to perform a variety of flagellar movements, as noted in the past century by Verworn (1890) and Bütschli (1889), and early in this century by Metzner (1920), Krijgsman (1925), and Gray (1928), and recently by Jahn and Bovee (1964–1967) and others. Although the trypanosome flagellum ordinarily generates tip-to-base oriented planar waves (Fig. 22B), it also will intermittently intersperse single base-to-tip lashing movements (Jahn and Fonseca, 1963; Gillies and Hanson, 1963) (Fig. 22D) or, occasionally, base-to-tip undulations (Pipkin, 1962; Jahn and Fonseca, 1963; Gillies and Hanson, 1963; Holwill, 1964) (Fig. 22C). Holwill (1964) describes the tip-to-base waves as symmetrical, and of increasing amplitude, and the base-to-tip waves as asymmetrical. He also (Holwill, 1965a) considers both types of waves to be nonrandom, i.e., controlled in sequential physical and chemical reactions. Holwill (1966) has recently and thoroughly reviewed these and other flagellar movements.

A trypanosome therefore may swim in either direction, but usually its body is pulled along by planar tip-to-base undulations of the flagellum. These push against the water, toward the body, exerting the force which pulls the body forward.

To what extent active undulations of the body may supplement those of the flagellum is not clear. Usually the body undulates in phase with the flagellum and presumably is bent by the flagellum. Because of the greater diameter of the body these undulations are probably more important than those of the flagellum alone, regardless of whether the body bends actively.

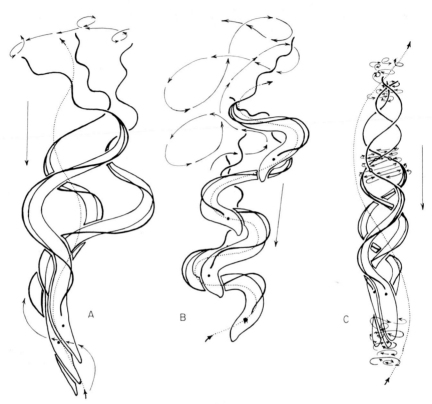

FIG. 21. Swimming trypanosomes, semidiagrammatic, based on tracings from cinephotomicrographic films photographed at 340 pictures per sec. A: *Trypanosoma lewisi*, overlay of three pictures, each 8th one in a series of 24. B: *Trypanosoma cruzi*, a like series. C: *Trypanosoma cruzi*, leptomonad form, a like series of diagrams.

In A and B amplitude of (planar) flagellar undulation increases, and two out of three waves are inhibited, and the third augmented, by undulation of the body. The body does not undulate beyond the base of the flagellum, but oscillates about a point approximating the position of the kinetoplast. The free part of the flagellum is whipped about in a complicated pathway approximating a sequence of figures of eight.

In C, increasing amplitude of tip-to-base planar undulations shows a gradual increase related to the increasing thickness of the paraflagellar protoplasm. Both the trailing end of the body and the free flagellar tip perform oscillatory movements. A granule attached to the flagellum about midway along the length oscillates back and forth in an approximate figure-eight pathway.

The straight arrow indicates the tip-to-base direction of flagellar undulation. The dotted lines indicate the general path of advance. Other solid arrows indicate whipping and oscillatory movements.

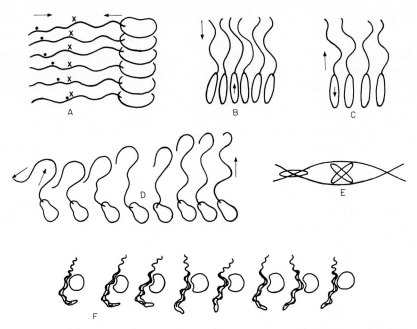

FIG. 22. Undulatory waves along the flagellum in trypanosomes. A–E: *Strigo-monas oncopelti*. (After Holwill, 1965c.) A: Flagellum generating both a tip-to-base wave (black dot), and a base-to-tip wave (X), which collide and cancel one another. B: Tip-to-base symmetrical planar waves along the flagellum, increasing in amplitude as they progress, which pull the body through the water. C: Base-to-tip symmetrical planar waves along the flagellum, increasing in amplitude as they progress, which push the body through the water. D: Asymmetrical base-to-tip waves along the flagellum. E: The envelope traced by the body as it progresses through the water. Although waves along the flagellum are planar, and the body vibrates back and forth, it also rotates about the axis of progression once about every 4 flagellar beats. At the wide point of the envelope vibration of the body is in the plane of the page; at "cross-over" points vibration is in the plane perpendicular to the paper. Arrows beside flagella indicate direction of undulatory waves. Arrows in body indicate direction body moves. F: *Trypanosoma brucei* shown deforming a red blood cell by the undula-tions of its flagellum and body. Sequential series from left to right. (After Holwill, 1965b.)

However, if the organism is attached to a substrate the body may bend out of phase with the flagellum (Jahn and Fonseca, 1963), thereby indicating that active body bending is possible.

The undulations of the free tip of the flagellum of the trypanosome stage have about half to one third the wavelength and amplitude of those of the body and the attached portion of the flagellum, and the increase in ampli-tude is immediate upon entry of the waves along the attached portion of the flagellum (Jahn, unpublished; Holwill, 1965b) and (Figs. 21A and 13; 22E).

The undulations progress along the attached flagellum and body at rates of from 5–17 cycles/sec (Holwill, 1964), or 14–23 cycles/sec (Jahn and Fonseca, 1963). These rates may vary with temperature (Holwill, 1965b) and probably other conditions. The force generated by a moving wave has been estimated as of the order of 3×10^{-8} to 60×10^{-8} dynes (Holwill, 1965b). The amplitudinal changes suggest that active bending of the body damps out two of every three waves originating at the tip of the flagellum, and augments the third, increasing both its wavelength and amplitude (Fig. 21B). Another possibility is that the body is passively bent in response to flagellar undulations, the wavelength and amplitudes along the body being due to limits of passive, elastic undulation on the part of the peripheral, subpellicular, cytoskeletal tubules.

The leptomonad stage, with the flagellum free of body attachment and extended forward, shows a *gradual* increase in wavelength and amplitude as the undulations progress from tip-to-base (Figs. 21C; 22B). This increase is probably due to the tapered accessory fibrillar rod which parallels the true axial cylinder of flagellum for about 2/3 of its length (Fig. 20).

The tip of the flagellum is jerked about in a complicated pattern of loops instead of following a sinuous pathway through the water (Fig. 21A and B).

Under some circumstances such loops may be considered as evidence of hydrodynamic inefficiency. However, it can be shown mathematically that even under conditions of 100 % efficiency such loops will occur if: (1) all wavelengths in a given flagellum are equal; and (2) the actively bending region is an odd number of quarter wavelengths long. Since the wavelength increases from tip-to-base this theory is difficult to apply to real cinematographs.

How the movements of the flagellum are originated is not clear. The old assumption that the basal apparatus must be the source of stimulus and coordination (Lowndes, 1941a,b, 1943a,b, 1944, 1947) has been eliminated by studies which show that flagella freed from the cell and lacking the basal apparatus swim in coordinated undulatory fashion if supplied with ATP as an energy source (Brokaw, 1961). Also, Holwill (1965c) has cinephotographed the undulations moving along trypanosomal flagella, and has recorded both tip-to-base and base-to-tip undulations moving simultaneously along the flagellum, cancelling one another out as they collide (Fig. 22A). He suggests that initiation of the wave is due to mechanical stimulus of the flagellar membrane, but the conduction and coordination are intrinsic, and the waves are not random waves. The first wave initiated, he finds, is that which is maintained and propagated along the length of the flagellum. The rate of energy utilization in relation to temperature, according to Holwill and Silvester (1965), and Burge and Holwill (1965), follows the Arrhenius equation for velocity of a chemical process. Although the rate of undulation varies, the character of the waves does not change as the temperature is altered within the range of viability.

Some cinephotographs made and analyzed in the authors' laboratory also suggest other employments of the trypanosome flagellum and body in translocation (Bovee *et al.*, 1966). One sequence clearly shows that *Trypanosoma equiperdum* may attach the tip of the flagellum and "somersault" the body up, over, and past the point of attachment. Whether this may be a common or even normal mode of locomotion has not yet been determined. However, leptomonad stages are known to attach to the host tissue by the flagellar tip, and they are motile under such conditions.

IV. SUMMARY

The motion of blood protists has only recently begun to receive critical study due to various technical difficulties in making the observations. However, since some recent advances indicate that the difficulties can be overcome, and since interest concerning the movements of these organisms is increasing, we can expect more illustrative and critical data in the near future.

It appears likely that mechanisms evolutionarily developed prior to the parasitic habit and used for propelling a cellular body in water by its own undulations, or by those of its flagella, have proved adequate, with minor adaptations, to the needs of penetration into, and locomotion within, the tissue of the host. The structural nature of free-living spirochetes, or of flagella of free-living flagellates, differs little from those of their parasitic relatives in form or function. Also, the gliding of sporozoan trophozoites has a seemingly identical counterpart in the gliding of free-living algae and diatoms. It appears that the hydrodynamic and thermodynamic problems to be overcome in the locomotion performed by parasitic protists differ perhaps only in magnitude from those encountered by free-living protists.

As is also true for free-living protists, there remains much to be solved in the study of the movements and locomotion of blood protists. New studies of the motile mechanism and locomotion of either free-living or parasitic protists will be useful in general studies of cellular movements, as well as providing specific data, and may yield reciprocally applicable principles.

REFERENCES

Acton, A. B. (1966). *J. Cell Biol.* **29**, 366.
Allegre, C. F. (1948). *Trans. Am. Microscop. Soc.* **67**, 211.
Anderson, W. A., and Ellis, R. A. (1965). *J. Protozool.* **12**, 483.
Ball, G. H. (1957). *J. Protozool.* **4**, Suppl., 6.
Ballowitz, E. (1908). *Arch. Ges. Physiol.* **46**, 433.
Beams, H. W., Tahmisian, T. N., Devine, R. L., and Anderson, E. (1959). *J. Protozool.* **6**, 136.

Boisson, C., Mattei, X., and Boisson, M. E. (1965). *Compt. Rend. Soc. Biol.* **159**, 228.

Bovee, E. C. (1965). *Excerpta Med. Intern. Congr. Ser.* **91**, 152.

Bovee, E. C. (1966). *Trans. Am. Microscop. Soc.* **85**, 162.

Bovee, E. C., Jahn, T. L., and Fonseca, J. R. (1966). *Am. Zoologist* **6**, 311.

Bradfield, J. R. G., and Cater, D. B. (1952). *Nature* **169**, 444.

Bringmann, G., and Holz, J. (1953). *Z. Hyg. Infektionskrankh.* **137**, 86.

Bringmann, G., and Holz, J. (1954). *Z. Tropenmed. Parasitol.* **5**, 54.

Brokaw, C. J. (1961). *Exptl. Cell Res.* **22**, 151.

Brokaw, C. J. (1965). *Abstr. 9th Ann. Meeting Biophys. Soc., San Francisco, 1965* p. 148.

Burge, R. E., and Holwill, M. E. J. (1965). *Symp. Soc. Gen. Microbiol.* **15**, 250.

Burton, P. R. (1966). *J. Cell Biol.* **33**, 18A.

Bütschli, O. (1889). *In* "Klassen und Ordnungen das Thierreichs" (H. G. Bronn, ed.), Vol. I, Part 1, pp. 1–1097. Winter, Leipzig.

Carlson, F. (1962). *In* "Spermatozoan Motility," Publ. No. 56, pp. 137–146. Am. Assoc. Advance. Sci., Washington, D. C.

Chao, J. C. (1966). Personal communication.

Cheissin, E. M. (1964). *Zool. Zh.* **43**, 647–50.

Chen, G. (1948). *J. Infect. Diseases* **82**, 226.

Clark, T. B., and Wallace, F. G. (1960). *J. Protozool.* **7**, 115.

Cleveland, L. R., and Grimstone, A. V. (1964). *Proc. Roy. Microscop. Soc.* **B159**, 668.

Cohen, C., and Holmes, K. C. (1963). *J. Mol. Biol.* **6**, 423.

Cox, V. A. (1965). *J. Protozool.* **12**, Suppl., 3.

Danforth, W. F. (1967). *In* "Research in Protozoology" (T. T. Chen, ed.), Vol. I. Pergamon Press, Oxford (in press).

Darling, S. G. (1909). *A.M.A. Arch. Internal Med.* **3**, 183.

DeLamater, E. D., Wiggall, P. H., and Haines, M. (1950). *Exptl. Med.* **92**, 239.

Dimitroff, W. T. (1926). *J. Bacteriol.* **12**, 135.

Dobell, C. (1912). *Arch. Protistenk.* **26**, 117.

Doran, D. J. (1966). Personal communication.

Doran, D. J., Jahn, T. L., and Rinaldi, R. A. (1962). *J. Parasitol.* **48**, 32.

Dyar, M. T. (1947). *J. Bacteriol.* **54**, 483.

España, C., España, E. M., and Gonzales, D. (1959). *Am. J. Vet. Res.* **20**, 795.

Fantham, H. B. (1908). *Quart. J. Microscop. Sci.* **52**, 1–73.

Fantham, H. B., Porter, A., and Richardson, L. R. (1942). *Parasitology* **34**, 199.

Fauré-Fremiet, E. (1909). *Compt. Rend. Soc. Biol.* **67**, 113.

Fletcher, W. (1927–1928). *Trans. Roy. Soc. Trop. Med. Hyg.* **21**, 265.

Fowell, R. R. (1936). *J. Roy. Microscop. Soc.* [3] **56**, 12.

Freyvogel, T. A. (1965). *Excerpta Med. Intern. Congr Ser.* **91**, p. 107.

Freyvogel, T. A. (1967). *Acta Trop.* **23**, 201.

Gall, J. G. (1965). *J. Cell Biol.* **27**, 32A.

Garnham, P. C. C. (1965). *Am. Soc. Belge Med. Trop.* **45**, 259.

Garnham, P. C. C. (1966). *Biol. Rev.* **41**, 561.

Garnham, P. C. C., Bird, R. G., Baker, J. R., Bray, R. S., and Healy, P. (1960). *Trans. Roy. Soc. Trop. Med. Hyg.* **54**, 274.

Garnham, P. C. C., Bird, R. G., Baker, J. R., and Bray, R. S. (1961). *Trans. Roy. Soc. Trop. Med. Hyg.* **55**, 98.

Garnham, P. C. C., Baker, J. R., and Bird, R. G. (1962). *J. Protozool.* **9**, 107.

Garnham, P. C. C., Bird, R. G., and Baker, J. R. (1967). *Trans. Roy. Soc. Trop. Med. Hyg.* **61**, 58.

Gibbons, I. R. (1963). *Proc. Natl. Acad. Sci. U.S.* **50**, 1002.

Gibbons, I. R. (1965). *J. Cell Biol.* **27**, 33–34A.

Gibbons, I. R., and Rowe, A. J. (1965). *Science* **149**, 424.

Gillies, C., and Hanson, E. D. (1963). *J. Protozool.* **10**, 467.

Gittleson, S. M. (1966). Ph.D. Thesis, University of California, Los Angeles, California.

Goldman, M., Carver, R. K., and Sulzer, A. J. (1957). *J. Parasitol.* **43**, 490.

Grassé, P.-P., ed. (1953). "Traité de Zoologie" (P.-P. Grassé, ed.), Vol. I, Part 2 pp. 691–797. Masson, Paris.

Grassi, B. (1901). "Die Malaria—Studien eines Zoologen."
Fischer, Jena.

Gray, J. (1928). "Ciliary Movement." Cambridge Univ. Press, London and New York.

Gray, J. (1955). *J. Exptl. Biol.* **32**, 795.

Gray, J. (1958). *J. Exptl. Biol.* **35**, 96.

Gray, J., and Hancock, G. J. (1955). *J. Exptl. Biol.* **32**, 504.

Grimstone, A. V. (1963). *Quart. J. Microscop. Sci.* **104**, 145.

Grimstone, A. V. (1966). *Ann. Rev. Microbiol.* **20**, 131.

Guimarães, F. N. and Mayer, H. (1942). *Rev. Brasil. Biol.* **2**, 123.

Gustafson, P., Agar, H. D., and Cramer, D. I. (1954). *Am. J. Trop. Med. Hyg.* **3**, 1008.

Guttman, H. N., and Wallace, F. G. (1964). *Biochem. Physiol. Protozoa* **3**, 460.

Hanapp, E. G., Scott, D. B., and Wyckoff, R. W. G. (1948). *J. Bacteriol.* **56**, 755.

Hancock, G. J. (1953). *Proc. Roy. Soc.* **A217**, 96.

Hanson, J., and Lowy, J. (1963). *J. Mol. Biol.* **6**, 43.

Hasegawa, T., and Sasahira, T. (1966). *J. Electronmicroscopy (Tokyo)* **14**, 338.

Hoffmann-Berling, H. (1954). *Biochim. Biophys. Acta* **15**, 332.

Hoffmann-Berling, H. (1955). *Biochim. Biophys. Acta* **16**, 146.

Holwill, M. E. J. (1964). *J. Protozool.* **11**, Suppl., 41.

Holwill, M. E. J. (1965a). *J. Protozool.* **12**, Suppl., 3.

Holwill, M. E. J. (1965b). *Exptl. Cell Res.* **37**, 306.

Holwill, M. E. J. (1965c). *J. Exptl. Biol.* **42**, 125.

Holwill, M. E. J. (1966). *Physiol. Rev.* **46**, 696.

Holwill, M. E. J., and Silvester, N. R. (1965). *J. Exptl. Biol.* **42**, 537.

Honigberg, B. M. (1955). *J. Protozool.* **2**, Suppl., 4.

Howard, L. M. (1962). *Am. J. Hyg.* **75**, 287.

Huff, C. G. (1934). *Am. J. Hyg.* **19**, 123.

Huff, C. G., Pipkin, A. C., Weatherby, A. B., and Jensen, D. V. (1960). *J. Biophys. Biochem. Cytol.* **7**, 93.

Hutner, S. H., and Provasoli, L. (1955). *Biochem. Physiol. Protozoa* **2**, 17.

Ivanov, I. I., and Umanskaya, M. Y. (1945). *Compt. Rend. Acad. Sci. URSS* [N.S.] **48**, 337.

Jahn, T. L., and Bovee, E. C. (1964). *Biochem. Physiol. Protozoa* **3**, 62.

Jahn, T. L., and Bovee, E. C. (1965). *Ann. Rev. Microbiol.* **19**, 21.

Jahn, T. L., and Bovee, E. C. (1967). *In* "Research in Protozoology" (T. T. Chen, ed.), Vol. I, pp. 39–198. Pergamon Press, Oxford.

Jahn, T. L., and Fonseca, J. R. (1963). *J. Protozool.* **10**, Suppl., 11.

Jahn, T. L., Bovee, E. C., Fonseca, J. R., and Landman, M. D. (1964a). *Proc. 10th Intern. Botan. Congr., Edinburgh, 1964* Abstr., Part I, pp. 218–219.

Jahn, T. L., Bovee, E. C., Fonseca, J. R., and Landman, M. D. (1964b). *J. Parasitol.* **50**, Suppl., 41.

Jahn, T. L., and Landman, M. D. (1965). *Trans. Am. Microscop. Soc.* **84**, 395.

Jarosch, R. (1964). *In* "Primitive Motile Systems in Cell Biology" (R. D. Allen and N. Kamiya, eds.), pp. 599–622. Academic Press, New York.

Jeantet, P., and Kermorgant, Y. (1925). *Compt. Rend. Soc. Biol.* **92**, 1036.

Jenson, V. G. (1959). *Proc. Roy. Soc.* **A249**, 346.

Jettmar, H. M. (1953). *Arch. Hyg. Bakteriol.* **137**, 477.

Judge, D. M., and Anderson, M. S. (1964). *J. Parasitol.* **50**, 757.

Kavanau, J. L. (1964–1965). "Structure and Function in Biological Membranes," 2 vols. Holden-Day, San Francisco, California.

Kitching, J. A. (1956). *Protoplasma* **46**, 475.

Kitching, J. A. (1957). *J. Exptl. Biol.* **34**, 511.

Kleinschmidt, A. (1951). *Zentr. Bakteriol., Parasitenk. Abt. I. Orig.* **157**, 42.

Kleinschmidt, A., and Kinder, E. (1950). *Zentr. Bakteriol., Parasitenk., Abt. I. Orig.* **156**, 219.

Knowles, R. (1928). "Introduction to Medical Protozoology," pp. 416–417. Thacker, Spink & Co., Calcutta, India.

Koch, M. (1902). *Proc. 5th Zool. Congr., Berlin, Germany, 1902*, pp. 674–684.

Krijgsman, B. J. (1925). *Arch. Protistenk.* **52**, 478.

Kümmel, G. (1958). *Arch. Protistenk.* **102**, 501.

Kunze, A. (1907). *Arch. Protistenk.* **9**, 382.

Lainson, R., Baker, J. R., Bird, R. G., Garnham, P. C. C., and Healy, P. (1961). *Trans. Roy. Soc. Trop. Med. Hyg.* **55**, 9.

Landau, J. V. (1965). *J. Cell Biol.* **24**, 332.

Lankester, E. R. (1882). *Quart. J. Microscop. Sci.* **22**, 53.

Levine, L. (1960). *Anat. Record* **138**, 364.

Listgarten, M. A., and Socransky, S. S. (1964). *J. Bacteriol.* **88**, 1087.

Loubatieres, K. (1955). *Ann. Sci. Nat. Zool.* [11] **17**, 73.

Lowndes, A. G. (1941a). *Nature* **148**, 198.

Lowndes, A. G. (1941b). *Proc. Zool. Soc. London* **A111**, 111.

Lowndes, A. G. (1943a). *Nature* **152**, 51.

Lowndes, A. G. (1943b). *Proc. Zool. Soc. London* **A113**, 99.

Lowndes, A. G. (1944). *Proc. Zool. Soc. London* **A114**, 325.

Lowndes, A. G. (1947). *Sci. Progr. (London)* **35**, 61.

Ludvik, J. (1960). *J. Protozool.* **7**, 128.

Lumsden, R. D. (1965). *J. Parasitol.* **51**, 929.

MacGregor, H. C., and Thomasson, P. A. (1965). *J. Protozool.* **12**, 438.

Machin, K. E. (1958). *J. Exptl. Biol.* **35**, 796.

Machin, K. E. (1963). *Proc. Roy. Soc.* **B158**, 88.

Machin, K. E. (1965). *Excerpta Med. Intern. Congr. Ser.* **91**, 111.

Mackerras, M. J., and Ercole, Q. N. (1948). *Australian J. Exptl. Biol.* **26**, 439.

Magerstedt, C. (1944). *Arch. Dermatol. Syphilis* **185**, 272.

Manwell, R. D., and Drobeck, H. P. (1955). *J. Parasitol.* **39**, 577.

Marsland, D. (1964). *In* "Primitive Motile Systems in Cell Biology" (R. D. Allen and N. Kamiya, eds.), pp. 173–188. Academic Press, New York.

Metzner, P. (1920). *Jarhb. Wiss. Botan.* **59**, 325.

Misra, P. L. (1941). *Trans. Rec. Ind. Mus. Calcutta* **43**, 43.

Mölbert, E. (1956). *Z. Hyg. Infectionskrankh.* **142**, 510.

Morton, H. E., and Anderson, T. F. (1942). *Am. J. Syphilis, Gonorrhea, Venereal Diseases* **26**, 563.

Moses, M. J. (1966). *Science* **154**, 424.

Mühl, D. (1921). *Arch. Protistenk.* **43**, 361.

Nagai, R., and Rebhun, L. I. (1966). *J. Ultrastruct. Res.* **14**, 571.

Nagano, T. (1965). *J. Cell Biol.* **25**, 101.

Nelson, L. (1962). *Biol. Bull.* **123**, 468.

Nelson, L., and Plowman, K. (1963). *Abstr. 7th Ann. Meeting Biophys. Soc. New York, 1963* p.MD4.

Nevin, T. A., and White, E. M. (1960). *Bacteriol. Proc.* **60**, 188–89.

Noguchi, H. (1928). *In* "The Newer Knowledge of Bacteria" (E. D. Jordan and I. S. Falk, eds.), pp. 452–497. Univ. of Chicago Press, Chicago, Illinois.

Norris, C., Pappenheimer, A. M., and Flourney, T. (1906). *J. Infect. Diseases* **3**, 266.

Ovčinnikov, N. M., and Delektarsky, V. V. (1965). "Anatomy of *Treponema pallidum*." World Health Organization/VDT/RES/79, 65 (in English).

Pease, D. C. (1963). *J. Cell Biol.* **18**, 313.

Perez, C. (1903). *Arch. Protistenk.* **2**, 1.

Perrin, W. S. (1906). *Arch. Protistenk.* **7**, 31.

Pipkin, A. C. (1962). *J. Parasitol.* **48**, 50.

Pitelka, D. R., and Child, F. M. (1964). *Biochem. Physiol. Protozoa* **3**, 131.

Porter, A. (1910). *Arch. Zool. Exptl. Gen.* **43**, 1.

Porter, J. F. (1897). *J. Morphol.* **14**, 1.

Porter, K. R., Ledbetter, M. C., and Batenhausen, S. (1964). *Excerpta Med. Intern. Congr Ser.* **77**, 36.

Prell, H. (1921). *Arch. Protistenk.* **42**, 157.

Renaud, F. L., Rowe, A. J., and Gibbons, I. R. (1966). *J. Cell Biol.* **33**, 92A–93A.

Rikmenspoel, R. (1965a). *Biophys. J.* **5**, 365.

Rikmenspoel, R. (1965b). *Exptl. Cell Res.* **37**, 312.

Rikmenspoel, R. (1966). *Abstr. 10th Ann. Meeting Biophys. Soc., Boston, 1966* p. 119.

Ritchie, A. E., and Ellinghausen, H. C. (1965). *J. Bacteriol.* **89**, 223.

Roth, L. E. (1964). *In* "Primitive Motile Systems in Cell Biology" (R. D. Allen and N. Kamiya, eds.), pp. 527–547. Academic Press, New York.

Rudzinska, M. A., and Trager, W. (1957). *J. Protozool.* **4**, 190.

Rudzinska, M. A., and Trager, W. (1962). *J. Protozool.* **9**, 279.

Rudzinska, M. A., Bray, R. S., and Trager, W. (1960). *J. Protozool.* **7**, Suppl., 24.

Schaudinn, F. (1902). *Arb. Kaiserl. Gesundh.* **19**, 169.

Scholtyseck, E., and Piekarski, G. (1965). *Z. Parasitenk.* **26**, 91.

Schultze, M. (1865). *Arch. Mikroskop. Anat. Entwicklungsmech.* **1**, 376.

Scott, J. W. (1930). *J. Parasitol.* **16**, 111.

Senaud, J. (1965). *Excerpta Med. Intern. Congr. Ser.* **91**, 189.

Sheffield, H. G., and Hammond, D. M. (1965). *Excerpta Med. Intern. Congr. Ser.* **91**, 157.

Silvester, N. R., and Holwill, M. E. J. (1965). *Nature* **205**, 665.

Stehbens, W. E. (1966). *J. Protozool.* **13**, 63.

Stöhler, H. (1957). *Acta Trop.* **14**, 302.

Swain, R. H. A. (1955). *J. Pathol. Bacteriol.* **69**, 117.

Swain, R. H. A. (1957). *J. Pathol. Bacteriol.* **73**, 155.

Taylor, G. (1951). *Proc. Roy. Soc.* **A209**, 447.

Taylor, G. (1952). *Proc. Roy. Soc.* **A211**, 225.

Tibbs, J. (1957). *Biochim. Biophys. Acta* **23**, 275.

Tilney, L. G., Hiramoto, Y., and Marsland, D. (1966). *J. Cell Biol.* **29**, 77.

Trager, W. (1956). *Trans. Roy. Soc. Trop. Med. Hyg.* **50**, 419.

Troissi, A. (1940). *J. Morphol.* **66**, 561.

Tyzzer, E. E., Theiler, H., and Jones, E. E. (1932). *Am. J. Hyg.* **15**, 319.

Uesugi, S., and Yamazoe, S. (1966). *Nature* **209,** 403.

Vanderberg, J., and Yoeli, M. (1965). *Ann. Soc. Belge Med. Trop.* **45,** 419.

Verworn, M. (1890). *Arch. Ges. Physiol.* **48,** 149.

Vickerman, K. (1963). *J. Protozool.* **10,** Suppl., 15.

von Brand, T. (1951). *Biochem. Physiol. Protozoa* **1,** 177.

Walker, P. J. (1961). *Nature* **189,** 1017.

Walker, P. J., and Walker, J. C. (1963). *J. Protozool.* **10,** Suppl., 132.

Watson, J. H. L., Angulo, J. J., Léon-Blanco, F., Varela, G., and Wedderburn, C. C. (1951). *J. Bacteriol.* **61,** 455.

Watters, C. D. (1962). *Biol. Bull.* **123,** 514.

Wenyon, C. M. (1926). "Protozoology." Baillière, London.

Westphal, A. (1954). *Z. Tropenmed. Parasitol.* **5,** 145.

Wohlfarth-Bottermann, K. E. (1964). *In* "Primitive Motile Systems in Cell Biology" (R. D. Allen and N. Kamiya, eds.), pp. 79–109. Academic Press, New York.

Wu, C. F. (1939). *Peking Nat. Hist. Bull.* **13,** 283.

Yoeli, M. (1964). *Nature* **201,** 1344.

Zettnow, E. (1906). *Deut. Med. Wochschr.* **32,** 376.

Zuelzer, M. (1912). *Arch. Protistenk.* **24,** 1.

Author Index

Numbers in italics refer to the pages on which the complete references are listed.

Hinnant, J. A., 26, *35*
Hiramoto, R. N., 42, *59*
Hiramoto, Y., 399, *435*
Hirato, K., 101, *118*
Hitchings, G. H., 185, 207, *212*
Hoare, C. A., 124, 127, 133, *135*, *146*, 152,
  161, *172*, 202, *212*, 309, 317, 318, 321,
  323, *339*, 377, 378, *390*
Hoffmann-Berling, H., 396, 426, *433*
Hogg, J. F., 220, *302*
Hollingsworth, J. W., 73, *75*
Holm-Hansen, O., 357, *358*
Holmes, K. C., 421, *432*
Holt, S. J., 294, *302*
Holter, H., 244, *301*, *302*
Holwill, M. E. J., 395, 396, 407, 427, 429,
  430, *432*, *433*, *435*
Holz, G. G., Jr., 204, *213*
Holz, J., 400, 424, 432
Honigberg, B., 309, *339*
Honigberg, B. M., 8, *21*, *146*, 395, *433*
Hood, L., 41, 42, *58*, *59*
Hopwood, D. A., 248, *302*
Horibata, K., 42, *58*
Horne, R. W., 280, *302*
Horsfall, W. R., 13, *21*
Houwink, A. L., 274, *303*
Howard, A. J., 182, *212*
Howard, L. M., 409, *433*
Howie, J. B., 68, *75*
Hraba, T., 38, *59*
Hueper, W. C., 181, *212*
Huff, C., 328, 330, 331, 337, *339*
Huff, C. G., 5, *21*, 26, 28, 31, 32, 33, *34*,
  165, *172*, 220, 223, 224, 226, 231, 235,
  236, 237, 238, 244, 253, 255, 261, 263,
  266, 267, 268, 269, 271, *300*, *301*, *302*, 326,
  329, 331, 332, *338*, *339*, *341*, 408, 409,
  420, 421, *433*
Hughes, W. L., 50, *60*
Hull, R. W., 27, *35*, 80, 81, 82, 87, *120*,
  237, *305*
Hulliger, L., 100, *118*, 164, 165, *172*
Humphrey, J. H., 50, 51, *59*
Hurst, E. W., 182, *212*
Hutchinson, M. P., 175, 190, *210*
Hutchison, D., 206, *212*
Hutner, S. H., *146*, 152, 155, 158, 160,
  171, *172*, *173*, 180, 183, 185, 187, 188,

190, 191, 192, 196, 198, 199, 201, *206*,
  209, *211*, *212*, *213*, *214*, *216*, 446, *433*
Hyman, L. H., 221, *302*

I

Ilardi, A., 80, 81, *117*
Imamura, A., 201, *214*
Ingram, D. G., 98, *118*
Ingram, R. L., 84, *118*
Ingram, V. M., 31, *35*
Inoki, S., 280, *302*, 381, *390*
Inouye, M., 202, *215*
Isheda, N., 186, *209*
Ishizawa, M., 182, *211*
Ivanov, I. I., 426, *433*
Iwasaki, H., 187, *212*
Izawa, K., 186, *209*

J

Jackson, G. J., 222, *305*
Jacobs, L., 185, *211*, *302*
Jacobs, R. L., 25, *35*
Jacobson, W., 192, *214*
Jaffe, J. J., 181, 199, *212*
Jahn, T. L., 394, 397, 398, 399, 401, 403,
  404, 405, 406, 407, 408, 413, 423, 427,
  429, 430, 431, *432*, *433*
James, D., 318, *341*
James, D. H., Jr., 196, *215*
Janakidevi, K., 190, *213*
Jandl, J. H., 71, *75*
Jao, R. L., 186, *213*
Jarosch, R., 416, 417, 419, 421, *434*
Jarrett, W. F. H., 230, 231, 258, 263, 268,
  302
Jatkar, P. R., 111, *118*
Jawetz, E., *146*
Jeantet, P., 404, 405, *434*
Jeanteur, P., 201, *215*
Jeffery, G. M., 84, 85, 89, *116*, *121*, 330,
  339, 362, *364*
Jensen, D. V., 165, *172*, *173*, 223, *302*, 408,
  409, 420, 421, *433*
Jenson, V. G., 394, *434*
Jerne, N. K., 44, 46, 57, *59*
Jerusalem, C., 32, *35*, 231, *302*
Jettmar, H. M., 408, 424, *434*
Jirgensons, B., 201, *209*
Jirovee, O., 280, *302*

Yamada, E., 248, *305*
Yamada, H., 193, *213*
Yamazoe, S., 395, *436*
Yanagishima, N., 204, *214*
Yantorno, C., 361, *365*
Yoeli, M., 26, 32, *36*, 88, *119*, 165, *174*, 273, *306*, 326, 330, *341*, 408, 409, 420, *436*
Yorke, W., 324, *341*
Yoshida, K., 186, 209
Yoshimura, M., 55, *61*
Yotsuyanagi, Y., 222, *306*
Young, M., 336, *340*
Young, M. D., 30, *35*, 133, *136*, 330, *339*, 381, *391*
Yu, R. K., 190, *210*

Z

Zahalsky, A. C., 183, 187, 188, 196, 199, 201, *212*, *213*, *216*
Zeledon, R., 96, *116*, 130, *136*, 153, 152, *174*, 322, *341*
Zeleznick, L. D., 201, *216*
Zettnow, E., 399, *436*
Ziffer, H., 185, 198, *209*
Zinkham, W. H., 72, *77*
Zuckerman, A., 26, 30, 32, *36*, 64, 65, 66, 67, 72, *77*, 80, 81, 82, 86, 89, 90, 97, 98, *121*, *122*, 331, *341*
Zuelzer, M., 399, 401, 403, 405, *436*
Zumpt, F., 126, *136*
Zuscheck, T., 110, 111, *122*

Subject Index

A

Acquired immunity, 37, 49, *see also* Immunity

Acquired tolerance, 38, *see also* Immunological tolerance, Tolerance

Acriflavine
destruction of trypanosomatid kinetoplast by, 202
effect on kinetoplast of, 202
single-nucleotide mutations from, 202

Actinoid protein, functions, 421

Actinomycin
binding of, 198
DNA model building with, 200
Trypanosoma in vitro sensitivity to, 181

Adelea mesnili, locomotion, 408

Adenosine diphosphate (ADP), in locomotion, 396

Adenosine triphosphatase, in locomotion, 395–396

Adenosine triphosphate (ATP), 204
of *Borrelia vincenti*, 408
glycolytic sequences, 426
in locomotion, 395–396
in plasmodial development, 169–170

Additives, in preservation of microorganisms, 363

Adjuvant, folinic acid as, 185

ADP, *see* Adenosine diphosphate,

Aegyptianella
general classification within Protista, 144
occurrence, 8, 144

Agglutination
capillary tube, technique, 111
tube latex, technique, 86–87, 103

Agglutinogens, genetic control of, 70

Algae, general classification, 143–144

Amastigote forms
criteria for distinguishing from promastigotes, 159–160
definition of, 159
mitochondrial development in, 160, 162–163
morphogenesis of, 160, 162

p-Aminobenzoic acid
in metabolite identification, 180, 190–191
suppressive effect, 25
target metabolite of sulfanilamide, 194
in vitamin assays, 180

Aminopterin, as lysogen, 182

Amphibia, malaria parasites of, 325–327

Amprolium, coccidiostatic antimetabolite, 184

Anaplasma
agglutinogens of, 110
antibodies in, 114
autohemagglutinins and opsonins in, 112–113
capillary tube agglutination test for, 111
characteristics of, 110
complement fixation test for, 111
effect of chemotherapy on anemia and parasitemia of, 65
erythrocytic antigen, detection of, 111–112
erythrophagocytosis in spleen and bone marrow of calves infected with, 72
fluorescent antibody techniques for, 113–114
gel precipitation test for, 111–112
general classification within Protista, 143
hemagglutinins in serum of host infected with, 70
mesoxenous, 9
motile forms, 408
occurrence, 9
relation to vertebrate host, 9
serum antigens, 112
stenoxenous, 9
transmission, 9
vectors, 12

Anaplasma marginale
anemia in infection with, 64
locomotion of, 424